STUDENT STUDY GUIDE FOR

BIOLOGY

CONCEPTS & CONNECTIONS

Fifth Edition

Campbell • Reece • Taylor • Simon

Richard M. Liebaert

Linn-Benton Community College
Albany, Oregon

PEARSON

Benjamin
Cummings

San Francisco Boston New York
Cape Town Hong Kong London Madrid Mexico City
Montreal Munich Paris Singapore Sydney Tokyo Toronto

Editor-in-Chief: Beth Wilbur
Acquisitions Editor: Chalon Bridges
Project Editor: Joan Keyes, Dovetail Publishing Services
Assistant Editor: Nora Lally-Graves
Managing Editor, Production: Erin Gregg
Production Supervisor: Jane Brundage
Marketing Manager: Jeff Hester
Manufacturing Buyer: Stacy Wong
Production Service/Compositor: TechBooks/GTS
Text Printer: Bradford & Bigelow
Cover Printer: Phoenix Color

Cover Credit: James Balog, The Image Bank

The following figures are adapted from Neil Campbell, Jane Reece, Martha Taylor, and Eric Simon, *Biology: Concepts and Connections,* 5th edition (San Francisco: Benjamin Cummings), © 2006 by Pearson Education, Inc.: 4.3, 4.6, 5.4, 6.3, 6.5, 6.7, 6.10, 6.11, 7.4, 8.3, 8.10, 10.7, 11.5, 16.8, 17.4, 17.7, 18.13, 19.2, 21.3, 23.3, 24.2, 24.5, 24.6, 24.8, 26.1, 26.3, 26.4, 27.2, 28.1, 28.10, 29.7, 30.3, 30.5, 31.2, 34.8.

The following figures are adapted from Neil Campbell, Jane Reece, Larry Mitchell, and Martha Taylor, *Biology: Concepts and Connections,* 4th edition (San Francisco: Benjamin Cummings), © 2003 by Pearson Education, Inc.: 21.2, 21.3, 22.4, 29.3, 31.4, 31.8, 34.6.

The following figures are adapted from Neil Campbell and Jane Reece, *Biology,* 7th edition (San Francisco: Benjamin Cummings), © 2005 by Pearson Education, Inc.: 3.8.

Photo 8.5 courtesy of Carolina Biological Supply/Phototake/NYC.
Photo 31.6 copyright © Ed Reschke.

ISBN 0-8053-7116-8

PEARSON

Benjamin
Cummings

Contents

A Note to Students

I once heard a teacher remark that he likes to see a textbook that is tattered and torn, because then he knows it is being used. I could think of no better end for this book than for it to be worn out—blanks filled in, notes and sketches in the margins, phrases highlighted, and corners of pages turned down. It is a book that is meant to be used.

How to Use This Book

This Study Guide is intended to help you develop a real understanding of biology. It directs you to and helps you learn about important biological concepts and connections in the textbook, lecture and lab sessions, the website and CD that accompany the text, and in the biological environment around you. Each Study Guide chapter accompanies a chapter in *BIOLOGY: Concepts & Connections*, Fifth Edition. Here is an example of how you might use the text, Study Guide, and Web/CD Activities: First, read the one-page story at the beginning of the text chapter, skim over the module headings, and read the summary at the end. Next, read the modules, alternating between the textbook and the Study Guide exercises, and work through relevant Web/CD activities. Or, you may choose to read the entire text chapter first and then complete the Study Guide exercises and Web/CD activities. Then finish up by answering the questions at the end of the text and Study Guide chapters.

Each chapter in the Study Guide has five parts:

1. *Introduction.* This short paragraph gets you started.

2. *Organizing Your Knowledge.* This section, which is the heart of the Study Guide, contains activities such as questions, diagrams, crossword puzzles, problems, and fill-in-the-blank stories that will help you master the concepts in the text. Some exercises in this section relate to single modules; others connect the information and concepts of several modules. (Module numbers are given at the beginning of each exercise.) If you get stuck on a question, refer to the text and your lecture notes. You will often find that Web/CD activities will help you work out the exercises in this Study Guide. Check your work by looking in the Answers section at the back of the book. Try to resist the desire to "peek" at the answers; the value in doing the exercises is in actually thinking about them. If you answer questions incorrectly, you may want to review relevant modules.

3. *Testing Your Knowledge.* These multiple choice and essay questions require you to recall information from the chapter. When doing the multiple choice questions, look at all the possible answers and choose the best one. Only one correct answer is given. Some answers are tempting but wrong; others are only partially correct. The essay questions are worth your effort, as exams often include essay questions. Check your answers with those at the back of the Study Guide, but again, do not simply look up the answers.

4. *Applying Your Knowledge.* The multiple choice and essay questions in this section go beyond recall of facts, and test the depth of your understanding by asking you to apply the material from the chapter to new situations, make comparisons, and put ideas together to solve biological problems. Again, there are suggested answers at the back of the Study Guide.

5. *Extending Your Knowledge.* This section suggests activities that you might pursue to supplement what you have learned. These exercises are optional and open-ended, with a range of possible outcomes, so there are no suggested answers in the back of the book.

Studying for Success in Biology

You will learn more in your biology course and enjoy it more if you have an effective study plan. First, it is important to have **a comfortable, well-lit place to study.** Try to **get enough sleep, good food, regular exercise, and time to play and relax.**

It is important to study regularly. Every day you have class you should hit the books for an hour or two. This book is intended to guide your study and help you build a more sturdy and usable understanding of biology. Each chapter, each lecture, and each concept connects with those you have already studied. If you wait until a day or two before an exam and try to "cram," you don't have time to really understand the concepts, to explore all the connections. Facts are easier to cram than concepts, but not as important and useful; if you fill up on mental junk food, there is no room for learning important stuff that is good for you (and your grade in the course!). Develop a study routine.

Studying starts with class lectures and reading, but **note that reading and studying are not the same thing.** Reading is only one of the things you do to study. Simply reading a chapter in the text does not guarantee you will learn anything, or even remember it. Don't just try to swallow whole what is stated in your textbook or lecture notes. You need to take a more active role, to get your brain engaged. A good way to do this is to **ask questions as you read.** Make the information your own by asking "What? Where? When? How? How much?" And especially **ask "Why?"** Why do we need to breathe? Why are cells so small? Why are saturated fats harmful? Why do plants bend toward the light? Scientists **look for cause and effect;** they **ask how things work.** If you want to understand biology and succeed in your biology course, this is the most productive approach. Many of the exercises in this study guide will show you how.

When reading and studying, note that **facts are different from concepts,** and **not all facts are equally valuable.** Facts are names, statistics, and simple relationships: The upper arm bone is called the humerus. A human cell contains 46 chromosomes. A sugar is a carbohydrate. Concepts are generalizations, rules, explanations: Living things naturally vary, and those variants best-adapted to their environments produce the most offspring and become more numerous. A certain hormone affects some body organs and not others because only certain organs have receptors that enable them to receive the hormone message. Simply **recalling facts is only a starting point** for learning biology. **Your goal should be understanding and explaining concepts—how things work, how they fit together.** To reach this goal, **focus on the big picture.** Don't simply make a list of everything you think you might have to know, from one end of a chapter to the other. It works better to stand back and look at all the main ideas first. Start by reading the text module headings and scanning the illustrations to **get an overview of the main concepts and how they connect.** Then dig for the key facts that support each concept. If you run out of time, at least you'll have a useful framework. Use your lecture notes to help you focus on what your professor considers the most important material. How does your teacher want you to use the text and Study Guide? Are the books to be used as your primary information source for learning on your own, as preparation for lecture, as background material, or simply as references?

Some students find biology intimidating because it seems to involve so much vocabulary. Although you have to learn some terminology in order to discuss biology intelligently and do well on exams, you don't need to name and define every part to understand how a cell, or an enzyme, or a muscle works. Rather than starting with names, it is better to **first**

figure out how things work, then learn the terms you need to describe them. If you read a description of the heart, and look at the illustrations, and perhaps see an animation on the Web or CD, you can comprehend how the heart works and then learn the names of its most important parts. In fact, the names will come easily as you need to use them. The important thing is to concentrate on the cause-and-effect relationships among the parts—how they function and interact—not just the parts themselves. Incidentally, **it is more important (and easier) to be able to use biological terms than define them.** You are much more likely to have to say "Oxygen diffuses out of the air in the lung and into the blood" than "Diffusion is the movement of molecules from an area of higher concentration to an area of lower concentration."

This brings us to another important point: **Don't memorize anything unless you have to.** Remembering and understanding—"knowing"—are two different things. For example, you don't "remember" how to drive, you just "know" how! And you can remember something without understanding it! It is more important to know diffusion when you see it and clearly understand what make it happen than to state a definition from memory. If you find yourself spending most of your study time memorizing—simply trying to remember the definition of diffusion without thinking about and understanding what diffusion really is—you are probably focusing too much on facts and vocabulary and not enough on concepts and connections. (I wonder why the authors chose that title?) An efficient way to organize facts and concepts is to **write things down in your own words,** and if possible, **summarize them in an outline, concept map, or diagram.** (Concept maps are discussed in Chapter 3 of the Study Guide.) If you write things down in your own words, you can't just repeat what it says in your notes or in the text. Instead, you have to **think about what the text or lecture really means.** The process of writing helps you to organize your thoughts, and looking at what you have written helps you evaluate your understanding. Outlining, mapping, and making sketches help you to see how ideas connect.

Can you describe how a bicycle works? You can do it because you remember what a bicycle looks like, not because you have memorized a verbal description of how a bicycle works. This same approach works in biology. Biology is a very concrete, tangible subject. **You can really see most of the things biologists talk about,** even though some of them are very small (molecules) or take a long time (evolution). A lot of effort has gone into making the textbook illustrations clear and complete. Study the illustrations carefully, and **try to learn in visual images,** not just in words. In fact, usually it is easier to learn in pictures. Note that the Web/CD animations are immensely helpful in this regard. Remember that even tiny things like molecules and genes are real, just as real as a bicycle, and thinking in pictures is an effective way to remember how they work.

Even if you have never taken a biology course, you already know a lot about the subject. I find that students who enjoy biology and do well in the course are curious about how things work. They devote sufficient time to studying, look at the big picture, and manage to avoid getting get bogged down in unnecessary details. And they realize that many of the things covered in the text and in the course are things they already know about. Don't set your experience aside because it seems "unscientific." Instead, embrace it; trust your experience and the knowledge that you bring to the subject of biology. Biology is not some esoteric subject "out there." In fact, there are many points of contact between biology and the rest of your life. Consider the function of your body (breathing, seeing, walking) and your connection with other organisms and the environment (the food you eat, the air you breathe, the microorganisms that make you sick). In addition, a multitude of important issues, including global warming, bioterrorism, cloning, health care, and population growth, have biological implications.

Many students find it useful to review that material together in pairs or small groups. Study on your own to get the basics, then get together, talk about it, and quiz each other. Discussing biology aloud gets another "learning channel" involved, and other members of your study group can evaluate your understanding. Compare your answers to the end-of-chapter and Study Guide questions. Make up questions of your own. **Don't just recite what you know, but make up new questions and problems. See if you can anticipate the kinds of questions that might show up on an exam.** If you disagree on an answer, draw in another student to break the deadlock. What you need is a reality check. Do you really understand how blood circulates? What is the connection between genes, DNA, and chromosomes? When my students miss exam questions, they seldom complain that the material was too difficult to understand. More often, they didn't realize certain topics were important, and neglected to go over them before the exam. Your study group is a great place to fill in knowledge gaps so they don't slow you down. A study group is also a great place to **get different points of view, and new ways to study and approach the subject.**

Finally, **if you have questions or difficulties, ask for help.** Make an appointment to talk with your professor during office hours. Don't be embarrassed if you don't understand something. If you knew everything about biology, you wouldn't have to take the course! Teaching assistants and laboratory instructors are also valuable resources. Many colleges and universities have tutorial assistance, learning assistance or study skills centers, and/or learning labs that provide additional help. Use all the resources that are available to you.

I hope this Study Guide contributes to your understanding of biology and to success in your biology course. If you have comments or suggestions, I would like to hear from you. You can e-mail me at: rilie1@comcast.net.

Acknowledgments. I would like to thank the staff at Benjamin Cummings—especially Editor-in-Chief Beth Wilbur and editorial project manager Ginnie Simione Jutson. A special thank you to production editor Joan Keyes of Dovetail Publishing Services, who expertly steered the Study Guide through its fifth revision. Thank you, Joan! Many thanks to the authors of *BIOLOGY: Concepts & Connections*, who had confidence in my ability to write the Study Guide and whose writing sets a high standard. I would also like to thank my friend and mentor Bob Thornton, my most inspiring teacher, and Cat Shelby, my co-conspirator and accomplice. A number of reviewers made important suggestions; I hope the Study Guide meets their expectations. Thanks also to my students and colleagues at Linn-Benton Community College, in Albany, Oregon, who have made teaching and learning a joy.

Richard Liebaert

Reviewers of the first through fifth editions

Jane Aloi Horlings, *Saddleback College*

Loren K. Ammerman, *University of Texas, Arlington*

M.C. Barnhart, *Southwest Missouri State University*

William E. Barstow, *University of Georgia, Athens*

Stephen Dina, *Saint Louis University*

Charles Duggins, Jr., *University of South Carolina*

Martin Hahn, *William Peterson College*

Lazlo Hanzley, *Northern Illinois University*

Yvonne Harris, *Truman College*

Jean Helgeson, *Collin County Community College*

George A. Hudock, *Indiana University*

Kris Hueftle, *Pensacola Junior College*

Ursula Jander, *Washburn University of Topeka*

Alan Jaworski, *University of Georgia, Athens*

Steve Lebsack, *Linn-Benton Community College*

Melanie Loo, *California State University, Sacramento*

Joseph Marshall, *West Virginia University*

Presley F. Martin, *Drexel University*

James E. Mickle, *North Carolina State University*

Henry L. Mulcahy, *Suffolk University*

Kathryn Stanley Podwall, *Nassau Community College*

Frank Romano, *Jacksonville State University*

Robert Salisbury, *State University of New York College at Oswego*

Roger Sauterer, *Jacksonville State University*

Gary Smith, *Tarrant County Junior College*

Mary Colavito, *Santa Monica College*

Gerald Summers, *University of Missouri, Columbia*

Hilda Taylor, *Acadia University*

Sandra Winicur, *Indiana University, South Bend*

Edmund Wodehouse, *Skyline College*

Uko Zylstra, *Calvin College*

Biology: Exploring Life

We are living in the golden age of biology. More biologists (and people who have taken a biology course) are alive today than have lived in all of history. Biology has always been important to human life—from the earliest times we have had to understand our bodies, the plants and animals in our surroundings, and how to obtain food—but in recent years biological issues have taken on a new urgency. Problems such as global warming, over-population, emerging diseases, bioterrorism, the challenges of genetic engineering, and de-pletion of global biodiversity demand that we apply more sophisticated knowledge to biological questions and issues. Just as "war is too important to be left to the generals," bi-ology is too important to be left to the biologists. This chapter introduces the science of biology and its importance in all our lives.

Organizing Your Knowledge

Exercise 1 (Module 1.1)

Web/CD Activity 1A   *The Levels of Life Card Game*

This module discusses the hierarchy of structural levels into which life is organized. Each of these levels has unique properties, which arise from the organization of its component parts. Review this structural hierarchy by completing the chart below.

Level	Description
Ecosystem	1. All the organisms in a particle area
2. community	All the organisms in a particular ecosystem
Population	3. an interacting group of individuals of one species
4. Organism	An individual living thing
5. organ system	Organs that work together to perform particular functions
Organ	6. Performs a specific function
7. Tissues	A group of similar cells with a specific function
8. cells	A unit of living matter, separated from its environment by a membrane
Organelle	9. a structure that performs a specific function in a cell
10. molecule	A cluster of atoms held together by bonds

Exercise 2   (Module 1.2)

This module describes how organisms interact with their environments. Review the web of interactions by completing this crossword puzzle.

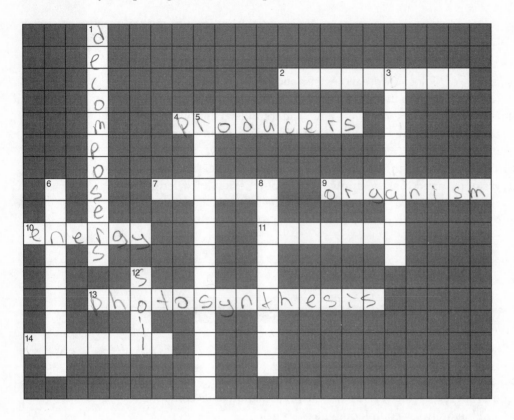

Across

2. Chemical ____ cycle within the ecosystem's web.

4. ____ make food that supports the ecosystem.

7. ____ are the main producers in most ecosystems.

9. A rabbit is a ____ in the ecosystem.

10. ____ comes into an ecosystem in the form of sunlight.

11. Decomposers ____ chemical nutrients.

13. Plants and other ____ organisms trap energy and make sugar.

14. ____ eat plants and each other.

Down

1. ____ such as prokaryotes and fungi convert dead matter to usable nutrients.

3. An ____ is the highest level in nature's hierarchy.

5. A web of ____ gives an ecosystem its structure.

6. In an ecosystem, there are many interactions among living organisms, and between living organisms and their ____ environment.

8. In an ecosystem or organism, ____ correlates with function.

12. Plants absorb mineral nutrients from the ____ .

Exercise 3   (Modules 1.3 – 1.4)

All living things are built from complex systems called cells, and all cells are controlled by the information in DNA molecules. These two modules discuss these and other common features shared by all living things. Write a paragraph describing how a living thing familiar to you—a college student—displays these features of life.

Exercise 4   (Module 1.5)

Web/CD Activity 1B   *Classification Schemes*

Review the three domains of life by matching each statement on the right with the correct domain. Write your answer in the first column. In addition, name the kingdom for each of the organisms in Domain Eukarya, and write your answer in the second column. Choose from:

Domain Bacteria
Domain Archaea
Domain Eukarya
      protists (several kingdoms)
      Kingdom Plantae
      Kingdom Fungi
      Kingdom Animalia

Domain	Kingdom	
Eukarya	*Plantae*	1. Tree and fern
Bacteria		2. Prokaryotes
Archaea		3. Another domain of prokaryotes
Eukarya	*protists*	4. Multicellular eukaryotes that eat other organisms
Eukarya	*fungi*	5. Molds, yeasts, and mushrooms
Eukarya	*protists*	6. Algae and protozoa
Bacteria		7. Organisms whose cells lack a nucleus
Eukarya	*Animalia*	8. Brown pelican, sloth, and spider
Eukarya	*Plantae*	9. Multicellular photosynthetic organisms with rigid cellulose cell walls
		10. Single-celled eukaryotes such as *Amoeba*

Exercise 5   (Module 1.6)

Web/CD Thinking as a Scientist   *How Do Environmental Changes Affect a Population?*

Evolution is biology's core theme. It explains both the diversity of life and the adaptation of living things to their environment. Fill in the blanks in the following story to review the concepts of evolution.

      While investigating the insect life of the rainforest canopy, a zoologist captured several specimens of a previously unknown species of butterfly. The butterfly was mostly black but had conspicuous red and yellow stripes on its wings. It rested on bare tree limbs in plain view; the zoologist was surprised she had not seen it before. The butterfly was very similar in structure to that of a much less conspicuous all-black species found in the same general area so the zoologist figured that the two species were closely
¹_____ members of the same family.

      Biologists have long marveled at the diversity of insect life in the tropics.
²_____, the English biologist who wrote *The Origin of Species*, was surprised by the large number of insect species he encountered in the rain forests of South America. In fact, biologists estimate that most species of living things are rain forest insects.

Like Darwin, the zoologist concluded that the black butterfly and the new species looked alike because they were both descended from a common ³_____ species. But why the difference in color pattern? When she first encountered the striped butterflies, she speculated that the red and yellow stripes were an ⁴_____, a beneficial feature that evolved by means of natural selection. But how could a bright color pattern be of any possible benefit? Wouldn't brightly colored butterflies be attacked by predators?

Her suspicions intensified when the zoologist saw the red and yellow winged butterfly resting on a tree limb. A predatory bird landed nearby and peered at the butterfly. The butterfly responded by rapidly flapping its wings, displaying their striped pattern, and the bird flew off.

This is what first caused the zoologist to suspect that the bright wing pattern was an example of "warning coloration," often seen in harmful or bad-tasting animals—for example, the conspicuous yellow and black stripes of bees and wasps. How could such a color pattern have evolved in this species of butterfly? The zoologist speculated that at one time a ⁵_____ of black butterflies existed in this area, breeding among themselves but not with other members of their species. These butterflies exhibited ⁶_____ traits—slightly different wing shapes, sizes, behaviors, and so on. They also may have tasted different. Perhaps some were able to make a bad-tasting substance or store a bad-tasting substance obtained from food plants. Just as Darwin reasoned, the zoologist realized that ⁷_____ variation must be present in the population for natural selection to operate. In the population of butterflies, good-tasting butterflies were more likely to be eaten by ⁸_____ than bad-tasting ones before they had a chance to ⁹_____. The surviving bad-tasting butterflies were able to reproduce, and they passed their ability to make the bad-tasting chemical on to their ¹⁰_____. Over time, the bad-tasting butterflies became more numerous. Among the bad-tasting butterflies, there may have been variation in wing coloration. Butterflies with bright colors on their wings were easier for predators to remember and avoid. They had more offspring than less-conspicuous individuals, and eventually bright-colored, bad-tasting butterflies became the norm in the population. In this situation, as in others explained by Charles Darwin, ¹¹_____ occurs as heritable ¹²_____ are exposed to ¹³_____ factors that favor the ¹⁴_____ success of some individuals over others.

The zoologist speculated that, over a long period, the changes in palatability, wing pattern, and other characteristics must have combined, and a new ¹⁵_____ of butterfly came into existence. According to Darwin, the ¹⁶_____ of new species results from the accumulation of minute changes resulting from natural selection over ¹⁷_____. This short story is just one illustration of evolution, biology's core ¹⁸_____.

Exercise 6   (Modules 1.7 – 1.8)

Web/CD Thinking as a Scientist   *The Process of Science: How Does Acid Precipitation Affect Trees?*

Review the scientific method by filling in the blanks in the following story. Choose from **variable, hypothesis, question, induction, control, prediction, deduction, discovery, observation, experiment, experimental, and scientific method.** Answers may be used more than once.

In Exercise 5, you read how a zoologist identified a previously unknown species of butterfly. This species was mostly black, but had conspicuous red and yellow stripes on its wings. It was very similar in appearance and structure to an all-black species found in the same area. Carefully comparing the two species, the zoologist concluded that they were closely related—members of the same family. Looking at many examples (of butterflies) and deriving a general principle (the characteristics of the butterfly family) employs a kind of logic called [1]_____. This kind of thinking is often involved when a scientist does [2]_____ science, which is mostly concerned with describing nature.

A different process is followed when a scientist seeks to explain natural events. The second method of inquiry is called [3]_____-driven science. It employs a process of inquiry sometimes called the [4]_____. Although it is not a single method, its key element is the kind of logic called [5]_____.

In our example, like many examples of hypothesis-driven science, the process started with a simple [6]_____: The zoologist noticed that predatory birds avoided the brightly colored butterflies even though they rested in tree branches in plain sight. This evoked a [7]_____: Is there something about the butterflies that the birds don't like? The researcher had a hunch; she suspected that the striped butterflies tasted bad, and that their bright colors acted as a sort of "warning" to predators to stay away. This kind of tentative explanation is called a [8]_____.

The zoologist decided to test this in the laboratory, under conditions that she could manipulate and monitor. Such a test is called a controlled [9]_____. She captured insect-eating birds native to the area and put them in cages at a nearby research station. Then she netted a number of brightly striped butterflies and their black cousins. For her first experiment, she allowed the birds to choose between a black butterfly and a striped one. The birds invariably chose the black butterflies and avoided the striped ones. This confirmed her field observations.

But did the striped butterflies taste bad? The researcher set up another controlled experiment, designed to compare a [10]_____ group—striped butterflies with their wings painted black—with a [11]_____ group of "normal" striped butterflies. (Actually, the "normal" butterflies were also handled and painted with clear paint, so that only one factor, or [12]_____, would differ between the two groups.) Her hypothesis led the zoologist to make a [13]_____ about how she thought the experiment would turn out: *If* the stripes really acted as a warning, *then* the birds would be fooled and eat the butterflies when the stripes were covered—and that *if* the striped butterflies tasted bad, *then* the birds would spit them out. Such "if-then" thinking is called [14]_____, and is an important feature of hypothesis-driven science.

Just as the researcher hypothesized, the birds chose the black-painted butterflies in every trial. Also, most of the birds quickly spat out the black-painted butterflies, and those that swallowed the butterflies became ill. Just to cover things, the zoologist performed another experiment in which she painted the wings of the edible black-winged butterflies. The birds ate them with gusto, demonstrating that the paint itself was not distasteful, and produced no ill effects.

After repeating the experiments several times, the researcher wrote a paper describing her hypothesis, experiments, results, and conclusions. It was published in the *Journal of Tropical Entomology*. There other scientists could read about the experiments, repeat and expand upon them, even challenge the results—all part of the process of science.

List six ways in which you think the science of biology and its technological applications may affect society in the next decade. Which of these are primarily scientific? Which are technological? Which do you think will have the most effect on you personally?

Testing Your Knowledge

Multiple Choice

1. All organisms have which of the following in common?
   a. They exhibit complex organization.
   b. They store genetic information in DNA molecules.
   c. They utilize energy.
   d. They reproduce.
   e. all of the above

2. Extrapolating from general premises to specific results is a kind of logic called
   a. induction.
   b. synthesis.
   c. deduction.
   d. experimentation.
   e. observation.

3. Biologists group living things into ___ domains.
   a. 2
   b. 3
   c. 4
   d. 5
   e. about 10

4. A bacterium and an amoeba are placed in different domains because
   a. a bacterium is single-celled.
   b. an amoeba is photosynthetic.
   c. an amoeba can move.
   d. a bacterial cell is much simpler.
   e. an amoeba is single-celled.

5. Most of the organisms in Kingdom ____ are photosynthetic.
   a. Animalia
   b. Protista
   c. Plantae
   d. Fungi
   e. none of the above

6. At the most fundamental level in life's hierarchy, all living things contain the same basic kinds of
   a. cells.
   b. organs.
   c. molecules.
   d. tissues.
   e. systems.

7. An educated guess posed as a tentative explanation is called a
   a. theory.
   b. control.
   c. variable.
   d. prediction.
   e. hypothesis.

8. The information in ____ underlies all of the properties that distinguish life from nonlife.
   a. carbon
   b. DNA
   c. proteins
   d. populations
   e. chemical nutrients

9. The ____ is the highest level in life's structural hierarchy.
   a. ecosystem
   b. cell
   c. organism
   d. population
   e. molecule

10. There are many interdependencies in an ecosystem. Prokaryotes and fungi play an important role in the ecosystem primarily because they
    a. cause diseases that keep populations in check.
    b. trap water, which is then used by other organisms.
    c. decompose the remains of dead organisms.
    d. are responsible for producing energy.
    e. do photosynthesis, which makes the food for other species.

Essay

1. Name the three domains of life and briefly describe their distinguishing characteristics.

2. Name the kingdoms of Domain Eukarya, identify the organisms in each, and briefly describe the criteria that separate each kingdom from the others.

3. What kinds of questions can be answered and what kinds of problems can be solved by science? What kinds of questions are outside the realm of science?

4. Explain how the information in DNA relates to the common features that characterize life.

Applying Your Knowledge

Multiple Choice

1. A crop scientist noted that, over a period of 10 years, a beetle species that feeds on rice gradually became resistant to insecticide. Which of the following best explains this in terms of natural selection?
   a. The insecticide mutated the beetles exposed to the biggest doses.
   b. Some beetles learned to tolerate the insecticide and passed this ability to their offspring.
   c. Beetles learned to avoid the spray and passed the knowledge to their offspring.
   d. The insecticide caused the beetles to reproduce more quickly than normal.
   e. Those beetles with natural resistance to the insecticide had the most offspring.

2. Researchers testing new drugs usually give the drug to one group of people and give placebos, "sugar pills," to another group. The group receiving the sugar pills
   a. constitutes the experimental group.
   b. is needed so that the test will be repeated enough times.
   c. is the control group.
   d. is a backup in case some of the people getting the drug drop out of the test.
   e. is the experimental variable.

3. ____ has characteristics that result from the organization of its component ____ .
   a. A population . . . ecosystems
   b. A tissue . . . organs
   c. A cell . . . tissues
   d. An organism . . . organ systems
   e. A molecule . . . cells

4. An ecologist studied the effect of nutrients and predators on the population growth of bacteria on the bottom of a pond. His study of bacteria could not involve which of the following levels in life's structural hierarchy?
   a. ecosystem
   b. organ
   c. organism
   d. population
   e. molecule

5. A rain forest primate called an aye-aye has a long middle finger that it uses to probe for insects in cracks and crevices in tree bark. This connection between structure and function developed gradually as a result of
   a. reproduction.
   b. population growth.
   c. natural selection.
   d. DNA replication.
   e. energy exchange.

6. Which of the following does *not* illustrate technology?
   a. DNA research is used to cure an inherited disease.
   b. Scientists develop a bacterium that destroys toxic wastes.
   c. A biologist identifies a new species of monkey.
   d. A chemical that slows cell division is used to treat cancer.
   e. Biologists breed a disease-resistant kind of corn.

7. Which of the following illustrates *discovery* science?
   a. testing a new blood-pressure drug
   b. trying to find a connection between dietary fat and heart disease
   c. cataloging all the species in a tide pool
   d. determining whether higher temperatures slow or speed plant growth
   e. all of the above

Essay

1. Choose a familiar wild animal—a squirrel, a toad, or a duck, for example—and describe some of the web of relationships that connect it with other organisms in its ecosystem.

2. Beavers are descendants of land-dwelling rodents similar to rats. Explain how natural selection could have shaped the beaver's flat tail and webbed feet, which it uses for swimming.

3. Jason tried a new fertilizer called MegaGro on his garden. He said, "I used it on all my tomato plants this year, and they grew much better than they did last year! MegaGro is fantastic!" Was Jason's test of MegaGro scientifically valid? Why or why not?

4. Tropical birds called oilbirds nest in caves and emerge at night to forage for seeds. Biologists think that oilbirds might be able to avoid obstacles in their caves and in the tangled growth of the rain forest much the way bats do—by making sounds and listening to the echoes. Describe a controlled experiment to test this hypothesis.

5. Explain how each of the following shows the connection between biological structure and function: your hand, a leaf, a hawk's beak, a frog's hind legs.

6. Describe the hierarchy of structural levels of which your body is composed. Give a specific example of a feature at one level that is not seen in the parts that make it up.

7. A camera sends back pictures of purple gelatinous blobs in near-boiling water near volcanic vents on the ocean floor. What properties should scientists look for to determine whether the blobs represent life?

8. How is human DNA different from the DNA of a chimpanzee? The DNA of a goldfish?

Extending Your Knowledge

1. Take a short walk outside, or look out a window, and describe how what you see relates to life's domains and kingdoms, life's hierarchy of structural organization, the common features that characterize life, and the interconnectedness of living things.

2. For one week, keep a list of biological issues in the news—the newspaper, magazines, television, radio, and conversations with friends and family. How many different issues did you encounter during the week? What were the major categories of biological issues in the news? Which of the news items are related to science, and which to technology?

The Chemical Basis of Life

2

You may have read something like "The value of all the chemicals in the human body is about 97 cents." This sounds pretty cheap, but putting a price tag on our chemicals misses an important point. This book was written on a computer. If you hit the computer with a sledgehammer, the chemicals in the pile of junk would be worth a lot less than I paid for the computer. But the value of the computer has less to do with the *kinds* of materials that go into it than *how* the materials are *arranged*. Like computers, living things are made from a few kinds of atoms obtained from water, soil, and air. These atoms, following a few basic principles, form the molecules and cells of a human being or a pine tree. What is special about life is not the chemicals themselves but the way they are organized to do what living things do—from the DNA that controls every cell to the valves that regulate the flow of blood through the heart. It is easy to believe that a human being is made up of 97 cents' worth of chemicals; the truly amazing thing is that 97 cents' worth of chemicals can make a human being. This chapter is about the chemicals of life and the rules they follow in shaping a living thing.

Organizing Your Knowledge

Exercise 1   (Introduction)

Thomas Eisner's research has revealed some of the roles of chemical messages in the lives of plants and animals. Review some of them by stating whether each of the following statements is true or false.

_____   1. The rattletrap moth is protected from predators by a chemical it obtains from a plant.

_____   2. Chemicals secreted by ovulating human females attract human males.

_____   3. A plant might secrete a chemical that discourages a butterfly from laying its eggs.

_____   4. A mating male moth might give a female a chemical that protects her and her offspring.

_____   5. A plant might use a chemical to attract a pollinating insect.

_____   6. Beetles are the only insects that do not use chemical signals.

_____   7. Human males secrete a chemical that affects ovulation in females.

_____   8. A female moth might choose a mate based on a male's chemical abilities.

_____   9. Chemical signals within an organism are called hormones.

_____   10. Insects embryos obtain protective chemicals from their mothers.

Exercise 2   (Modules 2.2 – 2.3)

Web/CD Activity 2A   *The Levels of Life Card Game*
Web/CD Thinking as a Scientist   *Connection: How Are Space Rocks Analyzed for Signs of Life?*

Write the chemical symbol for each of the following elements, and state whether it is one of the four elements used by living things in large amounts (L), whether it is used in moderate amounts (M), or whether it is a trace element (T) required in small amounts.

Symbol	Amount	Element	Symbol	Amount	Element
		1. Magnesium			7. Carbon
		2. Oxygen			8. Calcium
		3. Zinc			9. Phosphorus
		4. Hydrogen			10. Nitrogen
		5. Copper			11. Sodium
		6. Iodine			12. Iron

Exercise 3   (Modules 2.1 – 2.3)

Web/CD Activity 2A   *The Levels of Life Card Game*
Web/CD Thinking as a Scientist   *Connection: How Are Space Rocks Analyzed for Signs of Life?*

A compound is a substance that contains two or more elements in a fixed ratio. Indicate with a check mark which of the following are elements and which are compounds. (You may have to guess on some!)

	Element	Compound
1. Table salt		
2. Calcium		
3. Water (H_2O)		
4. Vitamin A		
5. Carbon		
6. Sulfur		
7. Carbon dioxide (CO_2)		
8. DNA		
9. Iodine		
10. Protein		

Exercise 4   (Modules 2.4 – 2.6)

Web/CD Activity 2B   *Structure of the Atomic Nucleus*
Web/CD Activity 2C   *Electron Arrangement*
Web/CD Activity 2D   *Build an Atom*

These modules introduce atoms. It is most important to know what the subatomic particles are, where they are located in an atom, and that atoms of different elements differ because they contain different numbers of protons. Some atoms not covered in these modules are compared below. You can figure out the subatomic particles they contain based on the concepts in the modules. First, fill in the blanks. Then sketch each atom, labeling and coloring **protons** red, **neutrons** gray, and **electrons** blue.

	Element	Symbol	Atomic Number	Mass Number	Number of Protons	Number of Neutrons	Number of Electrons
1.	Carbon-12	C	6	12	6	6	6
2.	Nitrogen-14	___	7	14	___	___	___
3.	Chlorine-35	___	___	35	17	___	___
4.	Oxygen-16	___	___	___	___	___	8
5.	Oxygen-17	___	___	___	___	___	___

Web/CD Activity 2E   *Ionic Bonds*
Web/CD Activity 2F   *Covalent Bonds*

The atoms of four elements important to life are diagrammed below. Pay particular attention to their electron shells. Remember that atoms with incomplete outer electron shells
participate in chemical reactions that allow them to attain complete outer shells: 2 electrons
for a hydrogen atom, 8 electrons for most other elements important to life.

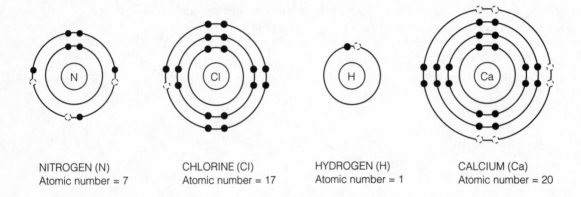

| NITROGEN (N) | CHLORINE (Cl) | HYDROGEN (H) | CALCIUM (Ca) |
| Atomic number = 7 | Atomic number = 17 | Atomic number = 1 | Atomic number = 20 |

1. Given the information and diagrams above, show how electrons would be trans-
    ferred between calcium and chlorine atoms to form calcium and chloride ions, which
    would then attract each other to form calcium chloride, $CaCl_2$. (Hint: An atom can
    gain or lose more than one electron.)

2. Using the information and diagrams above, show how nitrogen could form covalent
    bonds with several hydrogen atoms, forming a molecule of ammonia. What would
    be the molecular formula for ammonia?

Exercise 6 (Modules 2.9 – 2.14)

Web/CD Activity 2G   *Nonpolar and Polar Molecules*
Web/CD Activity 2H   *Water's Polarity and Hydrogen Bonding*
Web/CD Activity 2I   *Cohesion of Water*

Review the properties of water by filling in the blanks in the following story.

When Amy came through the door, she found Liz poised over a glass of water, ready to drop a needle into the glass. Amy asked, "Liz, what are you trying to do?"

"We're studying the [1]_____ basis of life in my biology class," Liz replied. "I don't believe some of this stuff she's teaching us, so I need to do some experiments to figure it out."

Liz gently placed the needle on the water surface. "Watch this," she said. The needle rested in a dimple on the surface of the liquid.

"How did you do that?"

"I didn't. The water did. Water molecules have a tendency to stick together, which is called [2]_____. The water molecules are stuck together so tightly at the surface that they form a film that can support the weight of the needle. Bugs can walk on it. It's called [3]_____."

Amy was getting interested. She asked, "So how do the water molecules do it? What's so special about water?"

Liz explained, "A water molecule is H_2O, right? It is made up of one [4]_____ atom and two [5]_____ atoms. The atoms stay together because they [6]_____ electrons. This holds them together. A shared pair of electrons forms a chemical bond called a [7]_____ bond between each hydrogen atom and the oxygen atom. Now, if the electrons were shared evenly, the bond would be called a [8]_____ covalent bond. But they are not shared evenly. The oxygen tends to 'hog' the electrons away from the hydrogens. It has a greater attraction for electrons; it is more [9]_____ than hydrogen."

"So what does this have to do with floating needles?"

"Well, because the oxygen atom attracts the electrons more strongly, the shared electrons are closer to the oxygen than to the hydrogens, giving the oxygen a slight [10]_____ charge. Because the electrons are pulled away from the protons in the nuclei of the hydrogen atoms, the hydrogens are left with slight [11]_____ charges. So the bonding electrons are shared unevenly, producing a [12]_____ covalent bond between each hydrogen atom and the oxygen atom. In fact, the whole water molecule is polar, even though the molecule as a whole is electrically [13]_____."

Amy was getting impatient. "So what does that have to do with surface tension? And what's the biology connection?"

Liz went on, "Well, it is their polarity that causes water molecules to stick together. The [14]_____ charged oxygen of one water molecule is attracted to the [15]_____ charged hydrogens of other water molecules. These special bonds between water molecules are called [16]_____ bonds. These bonds form a network at the water's surface, creating surface tension strong enough to support the needle. Each water molecule can connect with [17]_____ others. Hydrogen bonds are weak, but important. For example, they are responsible for holding the two strands of a [18]_____ molecule together, and for keeping [19]_____ molecules folded.

Hydrogen bonds give water some peculiar properties. For example, water is the only common substance on earth that naturally exists in all three states of matter—[20]_____, [21]_____, and [22]_____. And lots of things will dissolve in water; it is a versatile [23]_____. Blood plasma, for example, is an [24]_____ solution containing many different [25]_____, or dissolved substances, such as salt and blood sugar."

Amy got up and opened the bathroom door, looked inside, and said, "It's steamy in there. Are you going to take a bath?"

Liz replied, "No, that's just another experiment. I'm trying to figure out the difference between heat and temperature."

"Are they different?"

"Yes. [26]_____ is the total amount of energy resulting from the movement of molecules in a body of matter, like a bathtub full of water. [27]_____ measures the intensity of movement. I compared the amount of heat in a cup of water at 98°C and a bathtub of water at 45°C. In the [28]_____, the intensity of movement of water molecules was greater, but the [29]_____ held more heat energy. I knew it did because the bathtub of water added more heat to the room as it cooled, warming up the room more than the cup of hot water did.

"Water has a great capacity to store heat, by the way. When water is heated, a lot of the energy goes into breaking the [30]_____ between water molecules before the molecules can move faster. For instance, if you had a kilogram of water and a kilogram of rock, the same amount of heat would raise the temperature of the water [31]_____ than the temperature of the rock. This means water can soak up a lot of heat, and its temperature will go up only a few degrees."

"And when water cools a few degrees, it [32]_____ a lot of heat."

"Correct. And since animals are mostly water, this helps us control our body temperature. It also stabilizes the temperatures of the ocean and coastal areas. In the summer, the ocean [33]_____ heat, and in the winter, it [34]_____ heat."

Amy's eyes narrowed. "So why do we sweat when we are hot? Wouldn't we want to hang onto all that good water?"

Liz was ready with an answer. "No, not necessarily. Because of their strong hydrogen bonds, it takes a lot of heat energy to get a water molecule moving [35]_____ enough to [36]_____, to separate from its neighbors. This gives water an unusually high [37]_____, but it also makes [38]_____ cooling possible. The hottest—or fastest moving—water molecules evaporate first, taking a lot of heat energy with them and leaving the cooler—slower—molecules behind. So sweating cools you off on a hot day."

Amy looked at the clock and said, "It's 3:30. I told Sara I'd meet her at the ice rink at 3:30."

This inspired Liz anew. "Ice. Now, ice is very interesting. In ice the water molecules are locked into a crystal, linked by hydrogen bonds, but farther apart than they are in liquid water. This means that ice is [39]_____ dense than liquid water, so it [40]_____. This is important to life, because . . ."

But Amy was already out the door. Liz had a puzzled expression on her face as she got up and slid an ice tray out of the freezer.

Exercise 7   (Modules 2.15 – 2.16)

Web/CD Activity 2J   *Acids, Bases, and pH*
Web/CD Thinking as a Scientist   *Connection: How Does Acid Precipitation Affect Trees?*

Practice using the pH scale by giving the approximate pH of each of the following. Some are listed in the modules; others you can estimate from the information given.

____ 1. Tomato juice
____ 2. Human blood
____ 3. Vinegar (moderately acidic)
____ 4. Pure water
____ 5. Cola (moderately acidic)
____ 6. Household ammonia

____ 7. Concentrated nitric acid (very acidic)
____ 8. Acid precipitation
____ 9. Drain cleaner (very basic)
____10. Antacid pills (mildly basic)
____11. Urine
____12. Gastric juice

Exercise 8   (Module 2.17)

This module introduces chemical reactions, chemical processes that change matter. A common chemical reaction in many cells is one that changes hydrogen peroxide (H_2O_2) into water and oxygen gas:

$$2H_2O_2 \longrightarrow \underline{}H_2O + O_2$$

_____   _____

Hydrogen peroxide is a harmful by-product of many reactions. Cells get rid of it by carrying out the above reaction, converting it to harmless water and oxygen. What are the reactants in this reaction? What are the products? Label them in the blanks below the equation. Note that the equation for a chemical reaction must be "balanced." Since atoms cannot be created or destroyed in a chemical reaction—only rearranged—the numbers of atoms on both sides must be equal. In this example, there are four hydrogen atoms in the two hydrogen peroxide molecules on the left. After the reaction occurs, the hydrogen atoms reappear in the water on the right. Similarly, the four oxygen atoms in the hydrogen peroxide molecules on the left reappear in the water and oxygen molecule on the right. How many water molecules must be formed to account for all the atoms in the H_2O_2 molecules? Write the correct number in the small blank in front of H_2O.

Review basic chemical terminology by completing this crossword puzzle.

Across

2. ____ is the energy due to movement of molecules in a body of matter.

5. An ____ is a subatomic particle that circles an atom's nucleus.

7. The smallest particle of an element is called an ____.

9. An ____ is a charged atom or molecule.

12. Acid ____ is caused by pollutants that combine with water in the air.

16. ____ is anything that occupies space and has mass.

17. Two or more atoms held together by covalent bonds form a ____.

19. Neutrons and protons are found in an atom's ____.

22. The cohesion of water molecules is responsible for surface ____.

23. Variant forms of an element with different numbers of neutrons are called ____.

24. ____ is the tendency of water molecules to stick together.

Down

1. A ____ is a subatomic particle with no electrical charge.

3. ____ measures the intensity of heat.

4. Electrons are shared unequally in a ____ covalent bond.

5. There are 92 naturally occurring ____.

6. A ____ contains two or more elements in a fixed ratio.

8. Weak bonds between water molecules are called ____ bonds.

10. An ____ donates H^+ ions to solutions.

11. A ____ bond is formed when two atoms share electrons.

13. A ____ is a positively charged particle from the nucleus of an atom.

14. In a solution, the dissolving agent is called the ____.

15. A ____ is a liquid containing a uniform mixture of substances.

18. When two ions of opposite charges attract each other, an ____ bond forms.

20. The ____ is the substance dissolved in a solution.

21. A ____ accepts H^+ ions and removes them from solution.

Testing Your Knowledge

Multiple Choice

1. Which of the following is a trace element, required only in small amounts by most living things?
   a. oxygen
   b. iron
   c. nitrogen
   d. carbon
   e. hydrogen

2. A base is a substance that
   a. dissolves in water.
   b. forms covalent bonds with other substances.
   c. donates hydrogen ions to solutions.
   d. is a versatile solvent.
   e. removes hydrogen ions from solutions.

3. How an atom behaves when it comes into contact with other atoms is determined by its
   a. nucleus.
   b. size.
   c. protons.
   d. neutrons.
   e. electrons.

4. Most of water's unique properties result from the fact that water molecules
   a. are very small.
   b. tend to repel each other.
   c. are extremely large.
   d. tend to stick together.
   e. are in constant motion.

5. Atoms of different phosphorus isotopes
   a. have different atomic numbers.
   b. have different numbers of neutrons.
   c. react differently with other atoms.
   d. have different numbers of electrons.
   e. have different numbers of protons.

6. An ion is formed when an atom
   a. forms a covalent bond with another atom.
   b. gains or loses an electron.
   c. becomes part of a molecule.
   d. gains or loses a proton.
   e. gains or loses a neutron.

7. The smallest particle of water is
   a. an atom.
   b. a crystal.
   c. an element.
   d. a compound.
   e. a molecule.

8. Why are biologists so interested in chemistry?
   a. Chemicals are the fundamental parts of all living things.
   b. Most chemicals are harmful to living things.
   c. They know little about life except the chemicals it is made from.
   d. If you understand the chemistry of life, you can make a lot of money.
   e. Everything about life can be known by understanding its chemistry.

9. Molecules are always moving. Some molecules move faster than others; _____ is a measure of their average velocity of movement.
   a. polarity
   b. heat
   c. temperature
   d. electronegativity
   e. density

10. Which of the following holds atoms together in a molecule?
    a. ionic bonds between atoms
    b. transfer of protons from one atom to another
    c. sharing of electrons between atoms
    d. loss of neutrons by atoms
    e. sharing of protons between atoms

11. Ice floats because
    a. it is colder than liquid water.
    b. its molecules are moving faster than in liquid water.
    c. it is more dense than liquid water.
    d. its hydrogen molecules bond to the water surface film.
    e. its water molecules are farther apart than in liquid water.

12. Adding acid tends to ____ of a solution.
    a. increase the hydrogen ion concentration and raise the pH
    b. increase the hydrogen ion concentration and lower the pH
    c. decrease the hydrogen ion concentration and raise the pH
    d. decrease the hydrogen ion concentration and lower the pH
    e. c or d, depending on the original acidity

Essay

1. List the four elements needed by living things in large amounts, two others needed in moderate amounts, and two elements needed in trace amounts.

2. Explain why the smallest particle of iron is an atom, but the smallest particle of water is a molecule.

3. Explain the following statement: The temperature of the water in a teakettle is higher than the temperature of water in a swimming pool, but the swimming pool contains more heat.

4. How does acid precipitation form? How does it injure animals? Plants?

5. Explain why water molecules are polar, how this makes them tend to bond to each other, and how this causes water to have a large heat-storage capacity.

6. Explain how evaporation of water from your skin cools you on a hot day.

Applying Your Knowledge

Multiple Choice

1. Which of the following best states a reductionist point of view?
   a. You can understand something by taking it apart.
   b. Small things are more complex than large ones.
   c. A team can accomplish tasks individual members cannot.
   d. A system has functions more complex than its parts.
   e. If you look at something long enough, you will understand it.

2. An atom that normally has _____ in its outer shell would tend *not* to form chemical bonds with other atoms.
   a. 1 electron
   b. 3 electrons
   c. 4 electrons
   d. 6 electrons
   e. 8 electrons

3. Researchers studying the effects of toxic wastes knew that animals were poisoned by the heavy metal cadmium, but they wanted to know where cadmium accumulated in the body. They could find out by
   a. tracing the movement of cadmium isotopes in test animals.
   b. measuring the size of cadmium atoms.

c. finding out whether cadmium atoms form ionic or covalent bonds.
   d. finding out whether cadmium is acidic in water.
   e. determining the number of bonds formed by cadmium atoms.

4. Changing the number of _____ would change it into an atom of a different element.
   a. bonds formed by an atom
   b. electrons circling the nucleus of an atom
   c. protons in an atom
   d. particles in the nucleus of an atom
   e. neutrons in an atom

5. A glass of grapefruit juice, at pH 3, contains _____ H^+ as a glass of tomato juice, at pH 4.
   a. one-tenth as much
   b. half as much
   c. twice as much
   d. three times as much
   e. ten times as much

6. Fluorine atoms tend to take electrons from any atoms that come near. As a result, fluorine atoms
   a. tend to become positively charged.
   b. are nonpolar.
   c. do not react readily with other atoms.
   d. tend to form ionic bonds.
   e. are not very electronegative.

7. Tim added 10 milliliters (mL) of hydrochloric acid and 10 mL of water (pH 7) to a beaker containing 100 mL of water. The pH of the resulting solution was 4. Next he is going to add 10 mL of hydrochloric acid and 10 mL of pH 7 buffer to a different beaker containing 100 mL of water. What do you think will happen?
   a. The resulting pH will be less than 4.
   b. The resulting pH will be between 4 and 7.
   c. The resulting pH will be 7.
   d. The resulting pH will be between 7 and 11.
   e. The resulting pH will be greater than 11.

8. A sodium atom has a mass number of 23. Its atomic number is 11. How many electrons does it have (if it is not an ion)?
   a. 11
   b. 12
   c. 22
   d. 23
   e. 34

9. Which of the following is the smallest in volume?
   a. nucleus of an oxygen atom
   b. water molecule
   c. proton
   d. ice crystal
   e. electron cloud of an oxygen atom

10. Potassium chloride consists of potassium ions (K^+) and chloride ions (Cl^-) in a crystal. If potassium chloride is placed in water, what do you think happens?
    a. The K^+ ions are attracted to the oxygen atoms of water molecules.
    b. It will not dissolve.
    c. The Cl^- ions are attracted to the oxygen atoms of water molecules.
    d. It acts as an acid.
    e. The K^+ ions are attracted to the hydrogen atoms of water molecules.

Essay

1. A sulfur atom has 6 electrons in its outer shell. How many covalent bonds is it likely to form with other atoms? Why? What do you think the formula for hydrogen sulfide would be?

2. Plants carry out chemical reactions that make sugars, which contain carbon, hydrogen, and oxygen, from carbon dioxide (CO_2) and water (H_2O). Researchers want to know whether the oxygen atoms in sugar come from the carbon dioxide or the water. How could they use radioactive tracers to find out? What would they look for in terms of results?

3. If you drop a hot 10-kg rock into 10 kg of cold water, the rock cools off and the water warms up, but the final temperature of both is much closer to the starting temperature of the water than that of the rock. Why?

4. A sugar molecule contains carbon, hydrogen, and oxygen atoms. The oxygen atoms tend to steal electrons from the other atoms in the molecule, giving the oxygen atoms negative charges and leaving other parts of the sugar molecule positively charged. The molecules in oil, however, consist only of hydrogen and carbon, which share electrons equally. Oil molecules do not have areas of positive and negative charge. Using this information and what you know about water, explain why sugar mixes with water but oil does not.

5. Why are the ratios of elements in molecular formulas always fixed—H_2O instead of H_5O, and CH_4 instead of CH_6? Explain this in terms of the characteristics of the atoms making up the molecules.

6. On a typical July day in Seattle, Washington, the high temperature is around 75°F, the low around 55°F. In January, the average high is around 45°F, the low 35°F. In Minneapolis, Minnesota (at roughly the same latitude as Seattle), July highs average around 83°F, lows around 60°F; January highs average around 20°F, lows close to 0°F. Explain the differences between temperatures in Seattle and Minneapolis in terms of the concepts discussed in this chapter.

7. If you heat up a mixture of water and alcohol, the alcohol evaporates first, leaving most of the water behind. Why do you think this happens?

Extending Your Knowledge

1. Do you take vitamin supplements? A vitamin and mineral supplement contains most of the trace elements your body needs to function. Take a look at a vitamin bottle, and list these trace elements. What other elements does your body need besides these? In smaller or larger amounts? Why is it not necessary to take carbon or sodium pills?

2. Most acid precipitation results from pollutants produced by electric power plants. How might you reduce acid precipitation by recycling items like aluminum cans? Even if you live in a part of the country where electricity is generated in other ways, such as hydropower, how could your recycling reduce acid precipitation somewhere else?

3. Have you ever gone scuba diving or snorkeling? When you release air under water, the bubbles that form are always spherical, never square or pyramidal. Have you ever wondered why? Can you explain it in terms of hydrogen bonding and surface tension?

The Molecules of Cells

3

Sometime during the 21st century, people will walk on the planet Mars. One of the attractions of the red planet is the possibility that life exists there. The Viking missions of the 1970s tested Martian soil for the presence of life, but the results were at best inconclusive. In 2004, the *Spirit* and *Opportunity* rovers found convincing evidence that Mars was once wet enough to support life. Most scientists think there is good reason to expect that if we find life elsewhere—on Mars, or perhaps on Titan, a moon of Saturn, or maybe by scanning planets in other star systems—the life we find will be based on the chemistry of carbon. On Earth, only carbon seems able to form the variety and complexity of stable compounds that can perform the myriad of activities needed to produce life. This chapter is about carbon and carbon compounds in life.

Organizing Your Knowledge

Exercise 1   (Module 3.1)

Web/CD Activity 3A   *Diversity of Carbon-Based Molecules*

The great variety of organic compounds results from the ability of carbon atoms to bond with four other atoms, forming branching chains of different lengths. Several hydrocarbon molecules, consisting only of carbon and hydrogen, are shown in Module 3.1. Practice seeing the versatility of carbon by sketching some hydrocarbon molecules of your own, as suggested below.

1. Sketch a hydrocarbon molecule that is a straight chain, containing five carbon atoms and twelve hydrogen atoms, molecular formula C_5H_{12}:

Question: Why does each carbon bond to four other atoms?

2. Now sketch a shorter hydrocarbon chain, with only four carbon atoms:

Question: What is the molecular formula ($C_?H_?$) of the above molecule?

3. Sketch another five-carbon hydrocarbon, but this time include one double bond:

Question: What is the molecular formula of this molecule?

4. Sketch a five-carbon hydrocarbon molecule that is branched (and contains no double bonds):

Question: What is the molecular formula of this molecule? What is the term for its relationship to molecule 1 (in this exercise)?

5. Sketch two five-carbon hydrocarbon molecules in the form of rings, one without double bonds and one with one double bond.

Question: How many hydrogen atoms are in each of these molecules?

Exercise 2   (Module 3.2)

Web/CD Activity 3B   *Functional Groups*

Functional groups participate in chemical changes and give each molecule unique properties. Circle the functional groups that are discussed in this module in the molecules below. Label an example of each of the following: **hydroxyl group, carbonyl group, carboxyl group, amino group,** and **phosphate group.** There are a total of ____ hydroxyl groups, ____ carbonyl groups, ____ carboxyl groups, ____ amino groups, and _____ phosphate groups. (The properties of the molecules are described at the right.)

Formaldehyde is the starting point for making many chemicals.

Formic acid gives ant venom its sting.

Lactic acid builds up as a waste product in exercising muscles and makes them feel tired.

Ethylene glycol is in automobile antifreeze.

Acrolein is produced when meat is heated; it is the barbecue smell.

Serine is part of many protein molecules.

Urea is a waste product in urine.

Putrescene's name is descriptive; it is produced in rotting flesh.

G3P is an intermediate step in plants' production of sugar.

Exercise 3   (Module 3.3)

Web/CD Activity 3C   *Making and Breaking Polymers*

There are four main classes of macromolecules. Most are polymers, assembled from smaller monomers in a process called a dehydration reaction. Hydrolysis breaks polymers back down to monomers. State whether each of the following relates to dehydration (D) or hydrolysis (H).

_____ 1. Connects monomers to form a polymer.

_____ 2. Produces water as a by-product.

_____ 3. Breaks up polymers, forming monomers.

_____ 4. Water is used to break bonds between monomers.

_____ 5. Joins amino acids to form a protein.

_____ 6. Glycerol and fatty acids combine this way to form a fat.

_____ 7. Occurs when polysaccharides are digested to form monosaccharides.

_____ 8. —H and —OH groups form water.

_____ 9. Nucleic acid breaks up to form nucleotides.

_____10. Water breaks up, forming —H and —OH groups on separate monomers.

Exercise 4   (Modules 3.3 – 3.7)

Web/CD Activity 3C   *Making and Breaking Polymers*
Web/CD Activity 3D   *Models of Glucose*
Web/CD Activity 3E   *Carbohydrates*

After reading these modules, review carbohydrates by filling in the blanks in the following story.

Carbohydrates are a class of molecules ranging from the simplest sugars, called [1]_____, to giant molecules called [2]_____, built of many sugars. Carbohydrates are the main fuel molecules for cellular work.

Plants make their own carbohydrates, but humans, like all animals, must obtain them from plants or other animals. Imagine eating a piece of whole-wheat bread spread with strawberry jam. It contains a mixture of carbohydrates, along with other macromolecules like [3]_____ and [4]_____. Much of the carbohydrate in the bread itself is in the form of a polysaccharide called [5]_____, which is simply a chain of [6]_____ monomers. The monomers were linked together in the wheat plant in a process called a [7]_____ reaction. As the glucose units joined, [8]_____ was produced as a by-product. When you swallow a bite of bread, digestive juices in the intestine separate the monomers in the opposite reaction, called [9]_____. In the intestine, this is actually a two-step process.

Secretions from the pancreas first break the starch down to maltose, a type of carbohydrate called a [10]_____, which consists of two glucose monomers. Secretions from the walls of the intestine complete the process, breaking each maltose molecule down to two individual glucose molecules. Each glucose is a [11]_____-shaped molecule, containing [12]_____ carbon atoms.

There are other carbohydrates in the bread and jam. Whole-wheat flour contains the tough coats of the wheat seeds. These contain a lot of [13]_____, the fibrous polysaccharide that makes up plant cell walls. Like starch, it is made of glucose monomers, but these monomers are [14]_____ in a different orientation. The human digestive tract is not capable of [15]_____ cellulose, so it passes through the digestive tract unchanged, in the form of [16]_____. Sucrose, a [17]_____ refined from sugar cane or sugar beets, may be used to sweeten the strawberry jam. Each sucrose molecule is hydrolyzed in the small intestine to form one molecule of [18]_____ and one molecule of [19]_____. The jam naturally also contains a small amount of fructose, a [20]_____ that is produced by strawberries and is considerably sweeter than sucrose. (If the jam is artificially sweetened, it might contain other molecules whose [21]_____ are similar to natural sugars. These molecules bind to "sweet" [22]_____ on the tongue, producing the sensation of sweetness.)

Once all the carbohydrates have been hydrolyzed to small monosaccharides, they can be absorbed by the body. Glucose and fructose pass through the wall of the intestine and into the bloodstream, which carries them to the liver. Like all carbohydrate molecules, these sugars are [23]_____, so they easily dissolve in the water of blood plasma. In the liver, the fructose is converted to glucose. This process is relatively easy because glucose and fructose are [24]_____, having the same molecular formula, [25]_____, but slightly different structures. Glucose circulates around the body as "blood sugar" and is taken up by the cells for fuel as needed. Extra glucose molecules are taken up by liver and muscle cells and linked together by [26]_____ synthesis to form a polysaccharide called [27]_____. This molecule is similar to [28]_____, except it is more branched. Later the glycogen can be hydrolyzed to release [29]_____ into the blood.

Exercise 5 (Modules 3.8 – 3.10)

Web/CD Activity 3F   *Lipids*

Review the structures and functions of lipids by completing the following crossword puzzle.

Across

3. ____ means that hydrogen has been added to unsaturated fats.

4. ____ is a steroid common in cell membranes.

6. A ____ is similar to a fat; found in cell membranes.

9. A fat molecule is composed of ____ and three fatty acids.

10. Glycerol and 3 ____ acids make a triglyceride.

11. ____ is another name for "fat."

12. A ____ forms a waterproof coat that keeps a fruit or insect from drying out.

16. Olive and corn ____ are examples of unsaturated fats.

17. Fats with double bonds are said to be ____.

18. A ____ is a lipid-containing deposit in a blood vessel.

19. ____ are grouped together because they do not dissolve in water.

Down

1. ____ is when lipid-containing deposits block blood vessels.

2. Female and male sex hormones are examples of ____.

5. ____ is an illegal steroid recently banned by the International Olympic committee and professional sports.

7. Animal fats are said to be ____.

8. Lipids are water-avoiding, or ____ substances.

13. Unsaturated fats contain more ____ bonds than saturated fats.

14. A ____ is a large molecule whose main function is energy storage.

15. ____ steroids are dangerous variants of testosterone.

Exercise 6   (Module 3.11)

Web/CD Activity 3G   *Protein Functions*

Everything a cell does involves proteins. Seven classes of proteins are discussed in Module 3.11. Match each of the classes with one of the descriptions below.

_____ 1. Hemoglobin carries oxygen in the blood.

_____ 2. A protein in muscle cells enables them to move.

_____ 3. Antibodies fight disease-causing bacteria.

_____ 4. Collagen gives bone strength and flexibility.

_____ 5. Insulin signals cells to take in and use sugar.

_____ 6. Proteins in seeds provide food for plant embryos.

_____ 7. A protein called sucrase promotes the chemical conversion of sucrose into monosaccharides.

Exercise 7   (Modules 3.12 – 3.13)

Three amino acids not shown in the modules are diagrammed below.

1. Draw a box around the unique R group of each, and label it **R group**.

2. Draw a red circle around the amino group of each, and label it **amino group**.

3. Draw a blue triangle around the acid group of each, and label it **acid group**.

Alanine   Threonine   Asparagine

4. In the space below, sketch the three amino acids to show how they would join to form a tripeptide. What is this chemical reaction called? How many molecules of water would be formed? Show where the water would come from.

Exercise 8   (Modules 3.14 – 3.15)

Web/CD Activity 3H   *Protein Structure*

Identify each of the levels of protein structure in the diagrams. Then choose the descriptions from the list below that go with each of the levels.

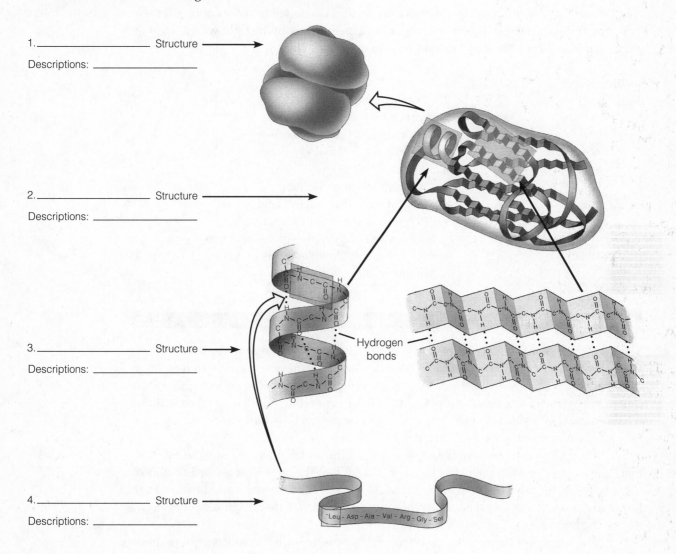

1._____ Structure ⟶

Descriptions: _____

2._____ Structure ⟶

Descriptions: _____

3._____ Structure ⟶

Descriptions: _____

4._____ Structure ⟶

Descriptions: _____

Hydrogen bonds

Leu – Asp – Ala – Val – Arg – Gly – Ser

Choose from these descriptions:

A. Overall three-dimensional shape
B. Amino acid sequence
C. Even a slight change in this can alter tertiary structure.
D. This level occurs in proteins with more than one polypeptide subunit.
E. Coiling and folding produced by hydrogen bonds between —NH and C=O groups
F. Not present in all proteins
G. Level of structure that is held together by peptide bonds
H. Alpha helix and pleated sheet
 I. Stabilized by clustering of hydrophobic R groups, hydrogen bonds, ionic bonds, and sometimes even covalent bonds
 J. "Globular" or "fibrous" might describe this level of structure.

Exercise 9   (Module 3.16)

Web/CD Activity 3I   *Nucleic Acid Functions*
Web/CD Activity 3J   *Nucleic Acid Structure*
Web/CD Thinking as a Scientist   *Connection: What Factors Determine the Effectiveness of Drugs?*

Nucleic acids are the fourth group of macromolecules discussed in this chapter. Review their structures and functions by matching each of the phrases on the right with a word or phrase from the list on the left. Answers may be used more than once.

A. Phosphate group
B. Deoxyribose
C. A, T, C, G
D. DNA
E. Nucleotide
F. A, U, C, G
G. Double helix
H. Ribose
I. Nitrogenous base
J. RNA

_____   1. Sugar in RNA
_____   2. Overall structure of DNA
_____   3. Short for ribonucleic acid
_____   4. Passed on from parent to offspring
_____   5. Nitrogenous bases of RNA
_____   6. Sugar in DNA
_____   7. Nitrogenous bases of DNA
_____   8. Short for deoxyribonucleic acid
_____   9. DNA works through this intermediary.
_____   10. Nucleotide is sugar, phosphate, and this.
_____   11. Sugar of one nucleotide bonds to this of next nucleotide.
_____   12. Monomer of nucleic acids

Exercise 10   (Summary)

You may find that making a concept map is a useful way to organize your knowledge. Such a map for the topic of carbohydrates is shown at the top of the next page. A concept map shows how key ideas are connected. Making a concept map can help you learn because it causes you to focus on main concepts and how they are related. It helps you to sort out what is important from unimportant details, and helps you tie your knowledge together into a more meaningful and useful whole.

To make a concept map, you must first decide which ideas are most important. Place the biggest, or most inclusive, concept at the top of the page. Just a word or phrase is enough. Cluster subconcepts around it, and cluster sub-subconcepts around them. Draw lines between the concepts to show how they are connected, and describe these connections next to the lines. Again, use only a word or two.

If the topics or connections are not clear, perhaps they are unimportant, or perhaps you are not clear on how they connect, or perhaps they do not really connect. Remember, clarifying relationships is the purpose of making the map. Generally, maps that are more "branched" are more useful than ones with many long straight "chains" of boxes, but there is no one "correct" map for a particular topic.

Focus on the process of making the map, rather than on the map itself. More learning will take place while you are making the map than when you look at the finished product. You might want to "tune up" your maps by comparing them with maps made by other students. After reviewing the following concept maps, on separate paper, try making your own concept maps for proteins and nucleic acids. Keep them simple at first. Remember, "Practice makes perfect!" Also, keep the concept map idea in mind for upcoming chapters.

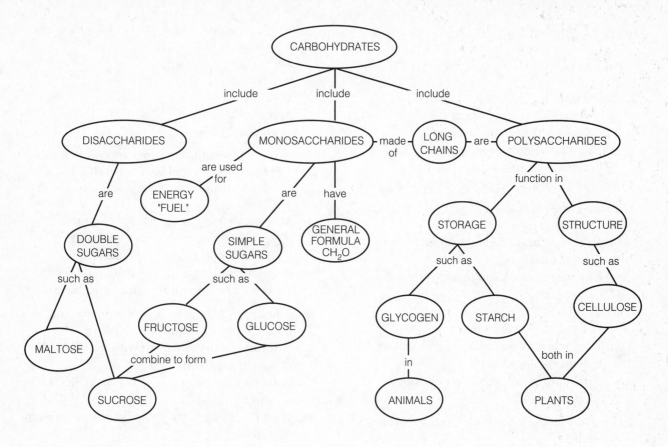

Practice working with a concept map by filling in the blanks on this map for lipids.

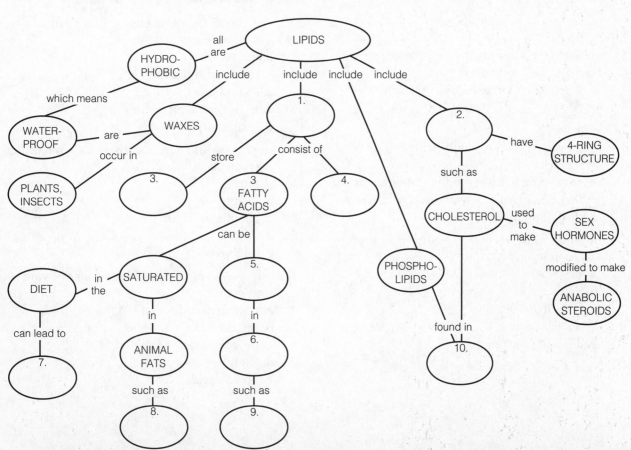

Testing Your Knowledge

Multiple Choice

1. Cellulose is a _____ made of many _____.
   a. polypeptide . . . monomers
   b. carbohydrate . . . fatty acids
   c. polymer . . . glucose molecules
   d. protein . . . amino acids
   e. lipid . . . triglycerides

2. In a hydrolysis reaction, _____, and in this process water is _____.
   a. a polymer breaks up to form monomers . . . consumed
   b. a monomer breaks up to form polymers . . . produced
   c. monomers are assembled to produce a polymer . . . consumed
   d. monomers are assembled to produce a polymer . . . produced
   e. a polymer breaks up to form monomers . . . produced

3. The four main categories of macromolecules in a cell are
   a. proteins, DNA, RNA, and steroids.
   b. monosaccharides, lipids, polysaccharides, and proteins.
   c. proteins, nucleic acids, carbohydrates, and lipids.
   d. nucleic acids, carbohydrates, monosaccharides, and proteins.
   e. RNA, DNA, proteins, and carbohydrates.

4. A major characteristic that all lipids have in common is
   a. they are all made of fatty acids and glycerol.
   b. they all contain nitrogen.
   c. none of them is very high in energy content.
   d. they are all acidic when mixed with water.
   e. they don't dissolve in water.

5. A flower's color is determined by the genetic instructions in its
   a. proteins.
   b. lipids.
   c. carbohydrates.
   d. nucleic acids.
   e. all of the above.

6. The most concentrated source of stored energy is a molecule of
   a. DNA.
   b. cellulose.
   c. fat.
   d. protein.
   e. glucose.

7. In some places the backbone of a protein molecule may twist or fold back on itself. This is called _____ and the coils or folds are held in place by _____.
   a. tertiary structure . . . hydrogen bonds
   b. primary structure . . . covalent bonds
   c. secondary structure . . . peptide bonds
   d. tertiary structure . . . covalent bonds
   e. secondary structure . . . hydrogen bonds

8. A hydrophobic amino acid R group would be found where in a protein?
   a. forming a peptide bond with the next amino acid in the chain
   b. on the outside of the folded chain, in the water
   c. on the inside of the folded chain, away from water
   d. forming hydrogen bonds with other R groups
   e. only at one end of a protein chain

9. The overall three-dimensional shape of a polypeptide is called the
   a. double helix.
   b. primary structure.
   c. secondary structure.
   d. tertiary structure.
   e. quaternary structure.

10. How many different *kinds* of protein molecules are there in a typical cell?
    a. four
    b. twenty
    c. about a hundred
    d. thousands
    e. billions

11. Estrogen, cholesterol, and other steroids are examples of
    a. polysaccharides.
    b. lipids.
    c. polypeptides.
    d. triglycerides.
    e. nucleic acids.

12. Functional groups called _____ groups are often used to transfer energy between organic molecules.
    a. hydroxyl
    b. amino
    c. carboxyl
    d. carbonyl
    e. phosphate

13. The "building blocks" of nucleic acid molecules are called
    a. polysaccharides.
    b. amino acids.
    c. fatty acids.
    d. nucleotides.
    e. DNA and RNA.

Essay

1. Briefly describe the various functions of proteins in the cell and body.

2. Animal fats tend to be solid at room temperature, plant oils more liquid. Explain how a difference in the chemical structure of their molecules causes this physical difference.

3. What forces and bonds maintain the three-dimensional folded shape of a protein molecule? How does this relate to the sensitivity of proteins to changes in their environment?

4. Using circles to represent monosaccharides, show the difference between glucose, maltose, and starch. Maltose is an example of what kind of carbohydrate? Starch is an example of what kind of carbohydrate?

5. Sketch a protein molecule, using squares connected by lines to represent amino acids connected by peptide bonds. Does your protein display primary, secondary, and tertiary structure? Where?

Applying Your Knowledge

Multiple Choice

1. Citric acid makes lemons taste sour. Which of the following is a functional group that would cause a molecule like citric acid to be acidic?
   a. hydroxyl
   b. hydrocarbon
   c. amino
   d. carbonyl
   e. carboxyl

2. Which of the following do nucleic acids and proteins have in common?
   a. They are both made of amino acids.
   b. Their structures contain sugars.
   c. They are hydrophobic.
   d. They are large polymers.
   e. They each consist of four basic kinds of subunits.

3. A biochemist is analyzing a potato plant for the disaccharide sucrose. Where would he be most likely to find it?
   a. in cell membranes
   b. in grains in the cells of underground tubers (potatoes)
   c. in the nuclei of potato cells
   d. in the sap of the potato plant
   e. in the walls of the potato plant cells

4. Which of the following ranks the molecules in the correct order by size?
   a. water . . . sucrose . . . glucose . . . protein
   b. protein . . . water . . . glucose . . . sucrose
   c. water . . . protein . . . sucrose . . . glucose
   d. protein . . . sucrose . . . glucose . . . water
   e. glucose . . . water . . . sucrose . . . protein

5. How does glucose differ from sucrose, cellulose, and starch?
   a. It is a carbohydrate.
   b. It is larger.
   c. The others are polysaccharides.
   d. It is a monosaccharide.
   e. It contains carbon, hydrogen, and oxygen.

6. Steve noticed that his friend Jon had gained a little weight during the holidays. He commented, "Storing up some _____ for the winter, I see."
   a. polysaccharides
   b. triglycerides
   c. nucleotides
   d. polypeptides
   e. steroids

7. How does DNA differ from RNA?
   a. DNA is larger.
   b. One of their nitrogenous bases is different.
   c. They contain different sugars.
   d. DNA consists of two strands in a double helix.
   e. All of the above are differences.

8. Certain fatty acids are said to be essential because the body cannot make them itself; they must be obtained in the diet. If your diet were deficient in these essential fatty acids, you would not be able to make certain
   a. fats.
   b. glycerol molecules.
   c. monosaccharides.
   d. proteins.
   e. You would not be able to make any of the above.

9. Hydrolysis of a protein would produce
   a. amino acids.
   b. monosaccharides.
   c. polysaccharides.
   d. peptide bonds.
   e. nucleotides.

10. Glucose and hexanoic acid each contain six carbon atoms, but they have completely different properties. Glucose is necessary in food; hexanoic acid is poisonous. Their differences must be due to different
    a. monomers.
    b. macromolecules.
    c. hydrolysis.
    d. quaternary structures.
    e. functional groups.

11. Which of the following would probably *not* be affected when a protein is denatured?
    a. primary structure
    b. secondary structure
    c. hydrogen bonds
    d. tertiary structure
    e. All of the above must be affected for the protein to be denatured.

12. Palm oil and coconut oil are more like animal fats than other plant oils. Because they ____ than other plant oils, they can contribute to cardiovascular disease.
    a. contain fewer double bonds
    b. are less saturated
    c. contain more sodium
    d. are less soluble in water
    e. contain less hydrogen

Essay

1. Briefly explain why all starch molecules are pretty much the same, but there are millions of kinds of protein molecules.

2. Specific enzymes in your intestine enable you to break down starch and use the glucose molecules produced by this process. But you cannot break down cellulose. Explain why, in terms of both carbohydrate structure and protein shape.

3. Slight heating is often enough to render a protein nonfunctional. But a polysaccharide such as starch must literally be boiled in acid before it is significantly affected. Explain why.

4. Sketch the structural formulas of two hydrocarbon molecules that are isomers. Be sure the C and H atoms form the correct numbers of bonds. What are the molecular formulas of the molecules? What is identical about the molecules? How do the molecules differ?

5. Fred suffers from a disease that makes it difficult for his cells to produce glycogen. For him, three meals a day are not enough; he needs to snack constantly. Explain why.

6. A tripeptide, a molecule consisting of three amino acids, is a very small protein. Yet a huge variety of tripeptides is possible. Assume that the first, second, and third amino acids can be any of 20 choices. How many different tripeptides could there be? (Hint: Imagine you are stringing beads. If you have 20 colors to choose from, how many three-bead sequences are possible?) How could you calculate the number of possible polypeptides 100 amino acids long? (You probably won't want to actually work it out—it's a *very* large number.)

Extending Your Knowledge

1. Most labels on packaged foods and household products list the chemicals they contain. You can use your knowledge of basic biological chemistry to figure out what the ingredients are. For example, vitamin pills contain folic acid. What makes it an acid? A shampoo contains hydroxypropyl methylcellulose and hydrolyzed soy protein. What kinds of polymers are these? Pretzels are made with hydrogenated cottonseed oil. What does this mean? Poke around in your kitchen cupboards and bathroom cabinet for other examples.

2. Have you used (or been tempted to use) over-the-counter diet supplements (such as the "andro" used by baseball player Mark McGuire) to try to lose weight, boost energy, or build muscle? Is there any evidence that these supplements work? Could they be dangerous? How can you get this information?

A Tour of the Cell

4

A sampling of the diversity of life on Earth might include a redwood tree, a jellyfish, a bacterium, a tiger, and a mushroom. At one time, living things seemed so varied that the only characteristic they were thought to have in common was a mysterious "vital force" that made them all alive. Then, with the invention of the microscope, biologists discovered cells. By the late nineteenth century, they realized that all living things are made of cells and that an organism is alive because its cells are alive. Even though the life of a redwood tree and that of a jellyfish seem quite different, these two organisms look and function much the same on the cellular level. Now we have electron microscopes, and we can zoom in on the intricate structures within a single cell. We can take cells apart and analyze their chemistry, or probe them with radioactive isotopes, antibodies, lasers, or fluorescent dyes. This chapter describes what these techniques have revealed about the life of a cell.

Organizing Your Knowledge

Exercise 1   (Modules 4.1 – 4.2)

Web/CD Activity 4A   *Metric System Review*
Web/CD Thinking as a Scientist   *Connection: What Is the Size and Scale of Our World?*

Use the information in the two modules and the chart in Module 4.2 to complete the following table comparing microscopes and the unaided human eye.

	Unaided Eye	*Light Microscope*	*Electron Microscope (SEM or TEM)*
Kind of radiation (beam) used	1.	2.	3.
Parts that focus beam	4.	5.	6.
Maximum magnification	7.	8.	9.
Smallest objects visible	10.	11.	12.
Ability to separate close objects (resolution)	13.	14.	15.
Limitations	16.	17.	18.

Exercise 2   (Module 4.2)

Web/CD Thinking as a Scientist   *Connection: What Is the Size and Scale of Our World?*

We need to use a microscope to see cells because cells are so small. Why can't a cell be as big as a house, or at least as big as a baseball? Compare the two cells diagrammed below. **For each cell, calculate the surface area, volume, and ratio of surface area to volume. Then answer the questions.**

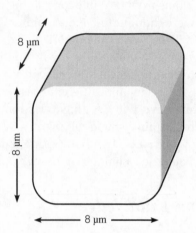

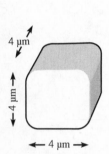

Cell 1	Cell 2
Surface area: $s = 6 \times (l \times l) =$	Surface area: $s = 6 \times (l \times l) =$
Volume: $v = l \times l \times l =$	Volume: $v = l \times l \times l =$
Surface/volume: $s/v =$	Surface/volume: $s/v =$

1. Which cell has the greater surface area?

2. Which cell has the greater volume?

3. Which cell has the greater ratio of surface area to volume?

4. In which cell would the surface area of the membrane most efficiently service the cytoplasm?

Exercise 3   (Module 4.3)

Web/CD Activity 4B   *Prokaryotic Cell Structure and Function*

Bacteria and archaea consist of small, simple prokaryotic cells. Label the following on this diagram of a prokaryotic cell: **capsule, cell wall, plasma membrane, nucleoid region, ribosome, prokaryotic flagella, pili.** Briefly state the function of each structure next to its label.

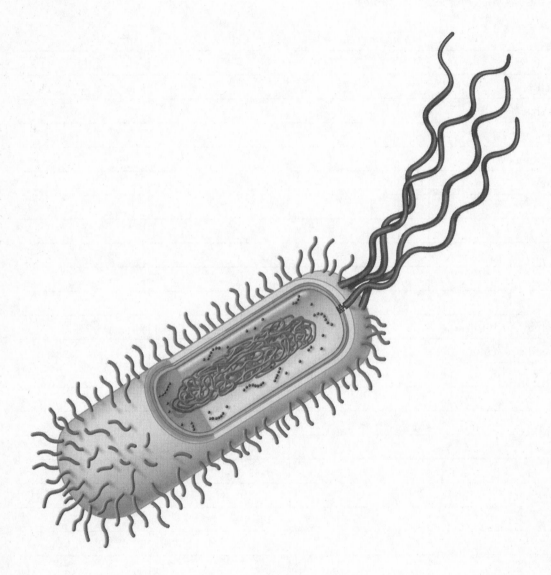

Exercise 4   (Modules 4.3 – 4.4)

Web/CD Activity 4B   *Prokaryotic Cell Structure and Function*
Web/CD Activity 4C   *Comparing Prokaryotic and Eukaryotic Cells*
Web/CD Activity 4D   *Build an Animal Cell and a Plant Cell*

Examine the diagrams and text, and then compare the structures of the cells of prokaryotes, plants, and animals by checking off their characteristics below. You may want to revise or refer to this checklist as you complete the chapter.

Characteristic	*Prokaryote Cell*	*Plant Cell*	*Animal Cell*
Prokaryotic structure			
Eukaryotic structure			
Relatively large size			
Relatively small size			
Membranous organelles			
Plasma membrane			
Cell wall			
Cytoplasm			
Ribosomes			
Bacterial flagellum			
Nucleus			
Rough endoplasmic reticulum			
Smooth endoplasmic reticulum			
Golgi apparatus			
Lysosome			
Peroxisome			
Mitochondrion			
Chloroplast			
Central vacuole			
Cytoskeleton			
Flagellum			
Centriole			

Exercise 5   (Modules 4.5 – 4.13)

Web/CD Activity 4E   *Overview of Protein Synthesis*
Web/CD Activity 4F   *The Endomembrane System*

Review the nucleus and the various structures that make up the endomembrane system by matching each phrase on the right with a structure from the list on the left. Answers can be used more than once.

A. Nucleus
B. Transport vesicle
C. Central vacuole
D. Smooth ER
E. Lysosome
F. Golgi apparatus
G. Rough ER
H. Contractile vacuole
 I. Ribosome

_____ 1. Lipids manufactured here
_____ 2. Small structure that makes protein
_____ 3. Contains chromatin
_____ 4. Sac of enzymes that digest things
_____ 5. Carries secretions for export from cell
_____ 6. Breaks down drugs and toxins in liver
_____ 7. Makes cell membranes
_____ 8. Cell control center
_____ 9. Numerous ribosomes give it its name
_____ 10. "Ships" products to plasma membrane, outside, or other organelles
_____ 11. May store water, needed chemicals, wastes, pigments in plant cell
_____ 12. Buds off from Golgi apparatus
_____ 13. Defective in Pompe's disease and Tay-Sachs disease
_____ 14. Proteins made and modified here for secretion from cell
_____ 15. Pumps out excess water from some cells
_____ 16. Nonmembranous organelle
_____ 17. Takes in transport vesicles from ER and modifies their contents
_____ 18. Digests food, wastes, foreign substances
_____ 19. Surrounded by double layer of membrane with pores
_____ 20. How proteins, other substances get from ER to Golgi apparatus
_____ 21. Stores calcium in muscle cells
_____ 22. Marks and sorts molecules to be sent to different destinations
_____ 23. Buds off lysosomes

Exercise 6   (Modules 4.6 – 4.13)

Web/CD Activity 4E   *Overview of Protein Synthesis*
Web/CD Activity 4F   *The Endomembrane System*

Sketch and label the endomembrane system on this diagram. Include **rough ER, smooth ER, ribosomes, Golgi apparatus, lysosome, nuclear envelope, transport vesicles,** and **plasma membrane.** (1) Trace the path of a protein from its site of manufacture to the outside of the cell with a red arrow. (2) Trace the path of a protein incorporated into a lysosome in blue. (3) Trace the path of a protein incorporated into the plasma membrane in green. (4) Trace the path of a lipid secreted from the cell in yellow.

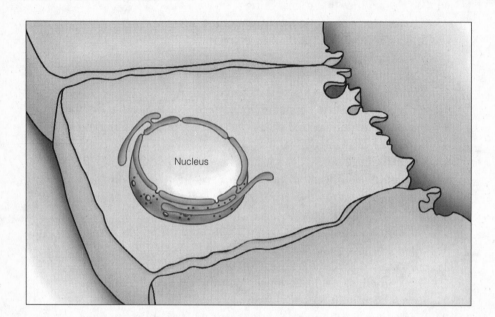

Nucleus

Exercise 7   (Modules 4.14 – 4.15)

Web/CD Activity 4G   *Build a Chloroplast and a Mitochondrion*

Both mitochondria and chloroplasts are energy converters, but their functions are quite different. Compare them by filling in the chart below.

	Chloroplast	*Mitochondrion*
Found in the following organisms . . .		
Carries out process of . . .		
Converts energy of . . .		
Into chemical energy in . . .		

Exercise 8   (Modules 4.16 – 4.17)

Web/CD Activity 4H   *Cilia and Flagella*

Compare the components of the cytoskeleton by indicating with a checkmark which of the following are characteristics of microfilaments, intermediate filaments, or microtubules.

	Microfilaments	Intermediate Filaments	Microtubules
Hollow tubes			
Solid rods			
Ropelike structure			
Made of tubulin			
Made of actin			
Made of fibrous proteins			
Help cell change shape			
Reinforcing rods, anchor organelles			
Act in muscle cell contraction			
Move chromosomes			
Act as tracks for organelle movement			
Give cell rigidity, shape			
In cilia			
In flagella			
In centrioles			
9 + 2 pattern			
Dynein arms cause bending movement			

Exercise 9   (Module 4.18)

Web/CD Activity 4I   *Cell Junctions*

Match each of the cell surface characteristics or structures on the left with a phrase on the right.

A. Tight junction
B. Plasmodesma
C. Anchoring junction
D. Cell wall
E. Gap junction
F. Extracellular matrix

_____ 1. Channel between animal cells
_____ 2. Rigid cellulose covering of plant cell
_____ 3. Link animal cells in leakproof sheet
_____ 4. Channel between plant cells
_____ 5. Connects animal cells into a strong sheet
_____ 6. Sticky layer holds animal cells together

Exercise 10   (Module 4.19 and Summary)

Web/CD Activity 4J   *Review: Animal Cell Structure and Function*
Web/CD Activity 4K   *Review: Plant Cell Structure and Function* •
Web/CD Thinking as a Scientist   *Connection: How Are Space Rocks Analyzed for Signs of Life?*

Label the organelles listed in Module 4.19 on these diagrams of animal and plant cells. (If you get stuck, refer to Module 4.4.) Try to group your labels according to the functional categories in Module 4.19 so that you can circle and label each category. Complete your diagrams by putting red boxes around the names of structures found in animal cells but not in most plant cells. Put green boxes around the names of structures found in plant cells but not in animal cells.

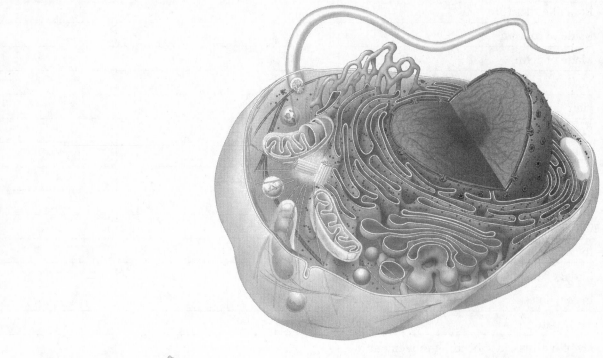

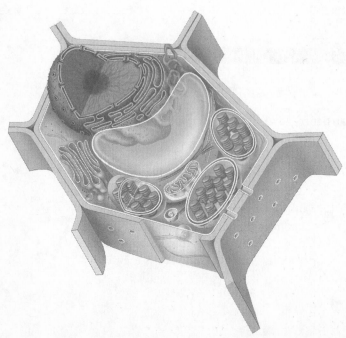

Testing Your Knowledge

Multiple Choice

1. To enter or leave a cell, substances must pass through
   a. a microtubule.
   b. the Golgi apparatus.
   c. a ribosome.
   d. the nucleus.
   e. the plasma membrane.

2. Which of the following would *not* be considered part of a cell's cytoplasm?
   a. a ribosome
   b. the nucleus
   c. a mitochondrion
   d. a microtubule
   e. fluid between the organelles

3. Which of the following consist of prokaryotic cells?
   a. plants and animals
   b. bacteria and archaea
   c. plants, fungi, bacteria, and archaea
   d. animals
   e. plants, bacteria, and archaea

4. Organelles involved in energy conversion are the
   a. rough ER and Golgi apparatus.
   b. nucleus and smooth ER.
   c. nucleus and chloroplast.
   d. lysosome and ribosome.
   e. mitochondrion and chloroplast.

5. The maximum size of a cell is limited by
   a. its need for enough surface area for exchange with its environment.
   b. the number of organelles that can be packed inside.
   c. the materials needed to build it.
   d. the amount of flexibility it needs to be able to move.
   e. the amount of food it needs to survive.

6. You would expect a cell with an extensive Golgi apparatus to
   a. make a lot of ATP.
   b. secrete a lot of material.
   c. move actively.
   d. perform photosynthesis.
   e. store large quantities of food.

7. Which of the following correctly matches an organelle with its function?
   a. mitochondrion—photosynthesis
   b. nucleus—cellular respiration
   c. ribosome—manufacture of lipids
   d. lysosome—movement
   e. central vacuole—storage

8. Cellular metabolism is
   a. a type of cell division.
   b. the process by which certain parts cause a cell to "self-destruct."
   c. the chemical activity of a cell.
   d. movement of a cell.
   e. control of the cell by the nucleus.

9. Which of the following stores calcium, important in muscle contraction?
   a. mitochondria
   b. smooth ER
   c. the Golgi apparatus
   d. contractile vacuoles
   e. rough ER

10. Of the following organelles, which group is involved in manufacturing substances needed by the cell?
    a. lysosome, vacuole, ribosome
    b. ribosome, rough ER, smooth ER
    c. vacuole, rough ER, smooth ER
    d. smooth ER, ribosome, vacuole
    e. rough ER, lysosome, vacuole

11. The internal skeleton of a cell is composed of
    a. microtubules, intermediate filaments, and microfilaments.
    b. cellulose and intermediate filaments.
    c. cellulose, microtubules, and centrioles.
    d. microfilaments.
    e. microfilaments and cellulose.

Essay

1. What are the advantages of an electron microscope over a light microscope? For what tasks would it be preferable to use a light microscope?

2. Briefly describe the major differences between prokaryotic and eukaryotic cells.

3. Name the structures present in plant cells but lacking in animal cells, and describe their functions.

4. Explain the advantages eukaryotic cells derive from being compartmentalized by many internal membranes.

5. Compare the functions of chloroplasts and mitochondria in a plant cell.

Applying Your Knowledge

Multiple Choice

1. A cell has mitochondria, ribosomes, smooth ER, and other parts. Based on this information, it could *not* be
   a. a cell from a pine tree.
   b. a grasshopper cell.
   c. a yeast (fungus) cell.
   d. a bacterium.
   e. Actually, it could be any of the above.

2. Dye injected into a cell might be able to enter an adjacent cell through a
   a. tight junction.
   b. microtubule.
   c. vacuole.
   d. plasmodesma.
   e. lysosome.

3. If a cell's chromatin were damaged, the cell would
   a. swell up and burst.
   b. run out of energy needed for its activities.
   c. go out of control.
   d. not be able to absorb light.
   e. divide immediately.

4. A researcher made an interesting observation about a protein made by the rough ER and eventually used to build a cell's plasma membrane. The protein in the membrane was actually slightly different from the protein made in the ER. The protein was probably changed in the
   a. Golgi apparatus.
   b. smooth ER.
   c. mitochondrion.
   d. nucleus.
   e. chloroplast.

5. If the nucleus is a cell's "control center," and chloroplasts its "solar collectors," which of the following might be called the cell's combination "food processor" and "garbage disposer"?
   a. lysosome
   b. Golgi apparatus
   c. flagellum
   d. ribosome
   e. nucleolus

6. When elongated, tube-shaped cells from the lining of the intestine are treated with a certain chemical, the cells sag and become round blobs. The internal structures disrupted by this chemical are probably
   a. cell junctions.
   b. microtubules.
   c. smooth and rough ER.
   d. mitochondria.
   e. microfilaments.

7. The electron microscope has been particularly useful in studying prokaryotes, because
   a. electrons can penetrate tough prokaryotic cell walls.
   b. prokaryotes are so small.
   c. prokaryotes move so quickly they are hard to photograph.
   d. they aren't really alive, so it doesn't hurt to "kill" them for viewing.
   e. their organelles are small and tightly packed together.

8. A cell possesses ribosomes, a plasma membrane, a cell wall, and other parts. It could *not* be
   a. a bacterium.
   b. a cell from a fungus.
   c. a cell from a mouse.
   d. an oak tree cell.
   e. a bacterium or a cell from a pine tree.

9. A mutant plant cell unable to manufacture cellulose would be unable to
   a. build a cell wall.
   b. divide.
   c. capture sunlight.
   d. move.
   e. store food.

10. A plant cell was grown in a test tube containing radioactive nucleotides, the parts from which DNA is built. Later examination of the cell showed the radioactivity to be concentrated in the
    a. rough ER.
    b. Golgi apparatus.
    c. smooth ER.
    d. central vacuole.
    e. nucleus.

Essay

1. Explain whether you would use a light microscope, a transmission electron microscope (TEM), or a scanning electron microscope (SEM) to perform each of the following tasks, and explain why: examining fine structural details within cell organelles, observing how a cell changes shape as it moves, studying tiny bumps on the cell surface, filming changes in the shape of the nucleus as a cell prepares to divide.

2. Imagine a cell shaped like a cube, 5 μm on each side. (Cells are not perfect cubes, but this assumption simplifies the question.) What is the surface area of the cell, in $μm^2$? What is its volume, in $μm^3$? What is the ratio of surface area to volume for this cell? (Sketches may help.) Now imagine a second cell, this one 10 μm on each side. What are its surface area, volume, and surface-to-volume ratio? Compare the surface-to-volume ratios of the two cells. How is this comparison significant to the functioning of cells? How could the surface-to-volume ratio of the larger cell be increased?

3. An enzyme (a type of protein) called salivary amylase is manufactured in the cells of your salivary glands and secreted as part of saliva. Explain how these parts of the cell cooperate to produce and secrete salivary amylase: transport vesicles, rough ER, plasma membrane, nucleus, Golgi apparatus, ribosomes.

4. A poison that acts specifically on mitochondria was found to interfere with the movement of cilia, slow down protein synthesis, reduce the frequency of cell division, and slow down the manufacture of lipids. Explain how one chemical could affect so many different cell activities.

5. When you work harder, your muscle cells work harder and increase in size. How might various organelles in a muscle cell increase in size, number, or activity to respond to the challenge of an increased workload?

Extending Your Knowledge

1. Analogies can be helpful in learning and understanding a new topic. For example, the cell can be thought of as a chemical factory, and the nucleus as its main office or control center. Can you compare the functions of other cell parts to the parts of a factory? What might be the factory doors? The power plant that provides energy for running the factory? The warehouse where raw materials or products are stored? Can you think of other comparisons? How does the cell differ from a factory?

2. Basic research is scientific investigation aimed at figuring out how something works, without thought for the immediate usefulness of the information obtained. The goal of applied research, on the other hand, is to put scientific knowledge to work. Much of cell biology is basic research—simply figuring out what the parts of a cell are and how they work. Nevertheless, the recent explosion of knowledge about cells has been immensely useful. For example, understanding lysosomes has helped us to understand disorders such as Tay-Sachs disease. Can you think of other examples where our understanding of cells has made our lives more healthy, comfortable, or productive?

The Working Cell

You can read these words because enzymes and membranes enable your cells to use energy. The light energy that bounces off the page enters your eyes and stimulates pigments held in special cell membranes. Enzymes make these pigments and convert them to a form that can absorb light. The eye cells can transmit signals through nerve cells to the brain because the membranes of these cells can selectively absorb and pump out charged particles. The energy for moving these particles comes from processes that make ATP. These processes take place through the action of enzymes on and between cell membranes. Every biological activity—not just reading, but walking, laughing, and thinking—depends on energy produced by processes that involve enzymes and membranes. Energy, enzymes, and membranes are the subjects discussed in this chapter.

Organizing Your Knowledge

Exercise 1 (Modules 5.1 – 5.5)

Web/CD Activity 5A *Energy Transformations*
Web/CD Activity 5B *Chemical Reactions and ATP*
Web/CD Activity 5C *The Structure of ATP*

After reading Modules 5.1–5.5, review energy, chemical reactions, and the function of enzymes by filling in the blanks in the following story.

If you were to stop eating, you would probably starve to death in weeks or months. If you were unable to breathe, you would die in minutes. Organisms need the energy that is released when food and oxygen combine. This energy is used not only to move the body but also to keep it from falling apart.

Energy is the ability to perform [1]_____. The sun is the source of the energy that sustains living things. Sunlight is pure [2]_____ energy, energy of movement that is actually doing work. In the process of photosynthesis, plants are able to use the energy of sunlight to produce food molecules. This process obeys the laws of [3]_____, the principles that govern energy transformations. Plants do not make the energy in food. According to the [4]_____ law of thermodynamics, energy can be [5]_____ or transferred, but it cannot be created or destroyed. In photosynthesis, no energy is created. Rather, the plant transforms the energy of sunlight into stored energy, called [6]_____ energy, stored in the chemical bonds of molecules of glucose.

No energy change is 100% efficient, and the changes that occur in photosynthesis are no exception to this rule. Some of the energy of sunlight is not stored in glucose, but rather is converted to [7]_____, which is random molecular motion. This energy is "lost" as far as the plant is concerned, and this random motion contributes to the disorder of the plant's surroundings. The [8]_____ law of thermodynamics says that energy changes are always accompanied by an increase in [9]_____, a measure of disorder. One of the reasons living things need a constant supply of energy is to counter this natural tendency toward disorder.

The products of photosynthesis contain [10]_____ potential energy than the reactants. This means that, overall, photosynthesis is an [11]_____ reaction. Such a reaction consumes energy, which in photosynthesis is supplied by the sun.

Photosynthesis produces food molecules, such as glucose, which store energy. An animal might obtain this food by eating a plant or an animal that has eaten a plant. The food molecules enter the animal's cells, where their potential energy is released in the process of cellular respiration. The products of this chemical reaction (actually a series of reactions) contain less potential energy than the reactants. Therefore, cellular respiration is an [12]_____ process; it [13]_____ energy. In fact, this is the same overall change that occurs when glucose in a piece of wood or paper burns in air. When paper burns, the energy escapes as the heat and light of the flames. In a cell, the reaction occurs in a more controlled way, and some of the energy is captured for use by the cell.

Energy released by the exergonic "burning" of glucose in cellular respiration is used to make a substance called [14]_____ . A molecule of [15]_____ and a [16]_____ group are joined to form each molecule of ATP. This is an endergonic reaction, because it takes energy to assemble ATP. The covalent bond connecting the phosphate group to the rest of the ATP molecule is unstable and easily broken. This arrangement of atoms stores [17]_____ energy. The [18]_____ of ATP is an exergonic reaction. When ATP undergoes hydrolysis, a [19]_____ is removed, ATP becomes [20]_____, and energy is released. Thus, ATP is a kind of energy "currency" that can be used to perform cellular [21]_____ . There are three kinds of cellular work: [22]_____, [23]_____, and [24]_____. Most cellular activities depend on ATP energizing other molecules by transferring its phosphate group to them—a process called [25]_____ . This happens in mechanical work, when ATP causes molecules in muscle cells to move. It should be noted that energy is not destroyed when ATP is used to do work. When an ATP molecule is hydrolyzed to make muscles move, some of its energy moves the body, and some ends up as random molecular motion, or [26]_____ . Similarly, ATP is used to move substances through [27]_____; this is called transport work.

A less obvious but important function of ATP is supplying the energy for fighting the natural tendency for a system to become disordered. A cell constantly needs to manufacture molecules to replace ones that are used up or damaged. This is chemical work. Building a large molecule from smaller parts is an [28]_____ reaction. Energy released by the exergonic hydrolysis of ATP is used to drive essential endergonic reactions. The linking of exergonic and endergonic processes is called energy [29]_____, and ATP is the critical connection between the processes that release energy and those that consume it.

What prevents a molecule of ATP from breaking down until its energy is needed? Molecules can break down spontaneously; that is why ATP energy is needed to repair them. Fortunately for living things, it takes some additional energy, called energy of [30]_____, to get a chemical reaction started. This creates an energy [31]_____ that prevents molecules from breaking down spontaneously. Energy barriers exist for both exergonic and endergonic reactions. Most of the time, most molecules in a cell lack the extra energy needed to clear the barrier, so chemical reactions occur slowly, if at all.

So what enables the vital reactions of metabolism to occur when and where they are needed, at a rate sufficient to sustain life? This is where enzymes come in. An enzyme is a special ³²_____ molecule that acts as a biological ³³_____. It ³⁴_____ the rate of a chemical reaction without being ³⁵_____ by it. An enzyme holds reactants in such a way as to ³⁶_____ the energy barrier that prevents them from reacting. Even though reactants would not normally possess the activation energy needed to start the reaction, the enzyme creates conditions that make the reaction possible. Enzymes enable the cell to carry out vital chemical changes when and where they are needed, enabling it to control the many chemical reactions that make up cellular ³⁷_____ .

Exercise 2   (Modules 5.1 – 5.5)

Web/CD Activity 5A   *Energy Transformations*
Web/CD Activity 5B   *Chemical Reactions and ATP*
Web/CD Activity 5C   *The Structure of ATP*

Briefly summarize the differences between the words or phrases in each of the following sets.

1.   Kinetic energy and potential energy

2.   Exergonic reactions and endergonic reactions

3.   Reactants and products

4.   ATP and ADP

5.   A reaction without an enzyme and a reaction with an enzyme

6.   Photosynthesis and cellular respiration

7.   First and second laws of thermodynamics

8.   Mechanical, transport, and chemical work

Exercise 3   (Modules 5.6 – 5.9)

Web/CD Activity 5D   *How Enzymes Work*
Web/CD Thinking as a Scientist   *How Is the Rate of Enzyme Catalysis Measured?*

Review enzyme action by completing the activities below.

1. Complete the diagram below so that it shows the cycle of enzyme activity. Imagine that the reaction carried out by this enzyme is splitting a substrate molecule into two parts. Color the diagram as suggested, and label the items in **boldface** type. Color the **enzyme** purple. Sketch the **substrate** as a dark pink shape. Sketch the **products,** and color them light pink. Also label the **active site.**

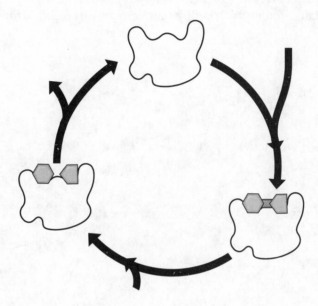

2. Make a sketch showing how heat or change in pH might change the above enzyme and alter its ability to catalyze its chemical reaction. Color and label the **enzyme,** its **active site,** and its **substrate,** as above.

3. On the left side of the space below, make a sketch showing how a competitive inhibitor might interfere with activity of the enzyme. Label the **competitive inhibitor,** and color it blue. On the right side, make a sketch showing how a noncompetitive inhibitor might interfere with activity of the enzyme. Label the **noncompetitive inhibitor,** and color it yellow.

Exercise 4   (Modules 5.10 – 5.13)

Web/CD Activity 5E   *Membrane Structure*
Web/CD Activity 5F   *Signal Transduction*
Web/CD Activity 5G   *Selective Permeability of Membranes*
Web/CD Thinking as a Scientist   *How Do Cells Communicate with Each Other?*

Review fluid mosaic membrane structure by coloring and labeling this diagram. It is a composite based on the figures in Modules 5.12 and 5.13. Label the items in **boldface** type: Start with the **cytoplasm, extracellular fluid,** and a **fiber of the extracellular matrix.** In the membrane, color **phospholipids** gray, protein molecules purple, **carbohydrate I.D. tags** on **glycoprotein** and **glycolipid** molecules green, and **cholesterol** molecules yellow. Also show the functions of certain proteins by labeling them **enzyme, receptor protein,** and **transport protein.**

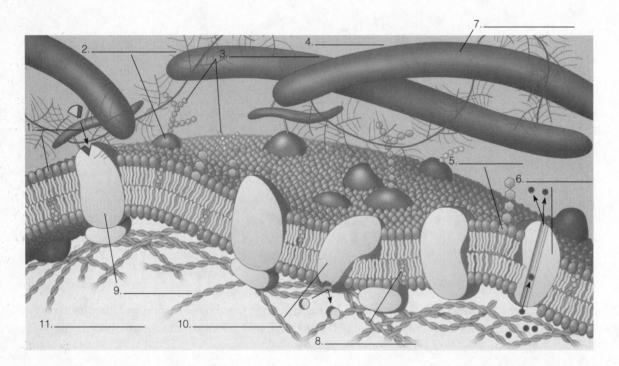

Exercise 5   (Modules 5.14 – 5.20)

Web/CD Activity 5H   *Diffusion*
Web/CD Activity 5I   *Facilitated Diffusion*
Web/CD Activity 5J   *Osmosis and Water Balance in Cells*
Web/CD Activity 5K   *Active Transport*
Web/CD Activity 5L   *Exocytosis and Endocytosis*
Web/CD Thinking as a Scientist   *How Does Osmosis Affect Cells?*

Review diffusion and the function of cell membranes by matching each of the phrases on the right with the appropriate mechanisms from the list on the left. Two questions require more than one answer.

A. Diffusion
B. Active transport
C. Osmosis
D. Phagocytosis
E. Passive transport
F. Facilitated diffusion
G. Pinocytosis
H. Receptor-mediated
   endocytosis
I. Exocytosis

_____ 1. Diffusion across a biological membrane
_____ 2. Moves solutes against concentration gradient
_____ 3. Any spread of molecules from area of higher concentration to area of lower concentration
_____ 4. Diffusion with the help of a transport protein
_____ 5. Three types of endocytosis
_____ 6. Engulfing of fluid in membrane vesicles
_____ 7. Diffusion of water across selectively permeable membrane, from hypotonic to hypertonic solution
_____ 8. Transport molecules need ATP to function
_____ 9. Enables cell to engulf bulk quantities of specific large molecules
_____10. How oxygen and carbon dioxide enter and leave cells
_____11. Two types of passive transport
_____12. Engulfing of particle in membrane vesicle
_____13. Fusion of membrane-bound vesicle with membrane, and dumping of contents outside cell
_____14. How a cell might capture a bacterium
_____15. Helped by aquaporins

Exercise 6 (Modules 5.15 – 5.16)

Web/CD Activity 5I   *Facilitated Diffusion*

Osmosis is an important process that has many effects on living things. Test your understanding of osmosis by predicting in each of the following cases whether water will enter the cell (*In*) or leave the cell (*Out*), or whether there will be no net movement of water (*None*). Assume that the plasma membrane is permeable to water but not solutes.

_____ 1. Cell is exposed to hypertonic solution.
_____ 2. Cell is placed in salt solution whose concentration is greater than cell contents.
_____ 3. Due to disease, solute concentration of body fluid outside cell is less than solute concentration of cells.
_____ 4. Cell is in isotonic solution.
_____ 5. Single-celled organism is placed in drop of pure water for examination under microscope.
_____ 6. Cell is immersed in solution of sucrose and glucose whose individual concentrations are less than concentration of solutes in cytoplasm, but whose combined concentration is greater than concentration of solutes in cytoplasm.
_____ 7. Solute concentration of cell is greater than solute concentration of surrounding fluid.
_____ 8. Cell is exposed to hypotonic solution.
_____ 9. Concentration of solutes in cytoplasm is equal to solute concentration of extracellular fluid.
_____10. Cytoplasm more dilute than surrounding solution.

Exercise 7   (Modules 5.10 – 5.20)

Web/CD Activity 5E   *Membrane Structure*
Web/CD Activity 5F   *Signal Transduction*
Web/CD Activity 5G   *Selective Permeability of Membranes*
Web/CD Activity 5H   *Diffusion*
Web/CD Activity 5I   *Facilitated Diffusion*
Web/CD Activity 5J   *Osmosis and Water Balance in Cells*
Web/CD Activity 5K   *Active Transport*
Web/CD Activity 5L   *Exocytosis and Endocytosis*
Web/CD Thinking as a Scientist   *How Do Cells Communicate with Each Other?*
Web/CD Thinking as a Scientist   *How Does Osmosis Affect Cells?*

Try to picture membranes and their functions close up by completing the following story.

Your first mission as a Bionaut requires you to enter a blood vessel and observe the structure and functions of cell membranes. You step into the water-filled chamber of the Microtron, which quickly shrinks you to a size much smaller than a red blood cell.

You tumble through the tunnel-like needle and into a blood vessel in the arm of a volunteer. Huge, rubbery red blood cells slowly glide past. Floating in the clear, yellowish blood plasma, you switch on your headlamp and examine the epithelial cells of the vessel wall. Their plasma membranes seem made of millions of small balloons. These are the polar "heads" of the [1]_____ molecules that make up most of the membrane surface. Through the transparent surface, you can see their flexible, [2]_____ tails projecting inward toward the interior of the membrane, and beyond them an inner layer of [3]_____ molecules with their tails pointing toward you. Here and there there are globular [4]_____molecules embedded in the membrane; some rest lightly on the surface, but most project all the way into the interior of the cell. The membrane is indeed a [5]_____ mosaic; the proteins are embedded like the pieces of a picture, but you can see that they are free to move around. You push on one of the proteins, and it bobs like an iceberg. Some of the phospholipids and proteins have chains of sugar molecules attached to them, forming [6]_____ and [7]_____. These are the molecules that act as cell [8]_____ tags. You notice that one of the proteins has a dimple in its surface. Just then a small, round molecule floating in the plasma nestles in the dimple. The molecule is a hormone, a chemical signal, and the dimpled protein is the [9]_____ that enables the cell to respond to it.

In your light beam, you can see the sparkle and shimmer of many molecules, large and small, in the blood and passing through the cell membrane. Oxygen is moving from the plasma, where it is more concentrated, to the cell interior, where it is less concentrated. This movement is [10]_____; when it occurs through a biological membrane, it is called [11]_____ transport. Similarly, carbon dioxide is flowing out of the cell, down its [12]_____ gradient, from the cell interior, where it is [13]_____ concentrated, to the blood, where it is [14]_____ concentrated.

You note that water molecules are passing through the membrane equally in both directions. The total concentration of solutes in the cell and in the blood must be equal; the solutions must be [15]_____. You signal the control team to inject a small amount of concentrated salt solution into the blood, making the blood slightly [16]_____ relative to the cell contents. This causes water to flow [17]_____ the cell, until the two solutions are again in equilibrium. This diffusion of water through a [18]_____ permeable membrane is called [19]_____.

Some sugar molecules floating in the blood are simply too large and polar to pass easily through the plasma membrane. The sugar molecules simply bounce off, unless they happen to pass through pores in special [20]_____ proteins. This is a type of passive transport, because the molecules move down a concentration gradient without the expenditure of [21]_____. Because transport proteins help out, it is called [22]_____ diffusion.

Your chemscanner detects a high concentration of potassium ions inside the cell. Transport proteins here and there in the membrane are able to move potassium into the cell against the concentration gradient. This must be [23]_____ transport; the cell expends [24]_____ to provide energy to "pump" the potassium into the cell.

Suddenly there is a tug at your foot. You look down to see your flipper engulfed by a rippling membrane. A white blood cell the size of a building quickly pins you against the vessel wall. The phospholipids of its membrane are pressed against your face mask. The cell is engulfing you, protecting the body from a foreign invader! Taking in a substance in this way is called [25]_____, more specifically [26]_____, if the substance is a solid particle. Suddenly the pressure diminishes, and you are inside the white blood cell, floating free in a membrane-enclosed bag, or [27]_____. Another sac is approaching; it is a [28]_____, full of digestive enzymes. You manage to get your legs outside of the vacuole and move it back toward the inner surface of the cell membrane. As the vacuole fuses with the membrane, you tear your feet free and swim away from the voracious cell, realizing that [29]_____ expelled you almost as fast as endocytosis trapped you!

You swim to the exit point, and the control team removes you by syringe. You are soon back in the lab, restored to normal size, and telling your colleagues about your close call.

Testing Your Knowledge

Multiple Choice

1. The movement of molecules from an area of higher concentration to an area of lower concentration is called
   a. diffusion.
   b. endocytosis.
   c. catalysis.
   d. active transport.
   e. osmosis.

2. Which of the following is *not* true of an enzyme? An enzyme
   a. is a protein.
   b. acts as a biological catalyst.
   c. supplies energy to start a chemical reaction.
   d. is specific.
   e. lowers the energy barrier for a chemical reaction.

3. Phospholipid molecules in a membrane are arranged with their____ on the exterior and their____ on the interior.
   a. hydrophobic heads . . . hydrophilic tails
   b. hydrophilic heads . . . hydrophobic tails
   c. nonpolar heads . . . polar tails
   d. hydrophobic tails . . . hydrophilic heads
   e. hydrophilic tails . . . hydrophobic heads

4. In osmosis, water always moves toward the ____ solution, that is, toward the solution with the ____ solute concentration.
   a. isotonic . . . greater
   b. hypertonic . . . greater
   c. hypertonic . . . lesser
   d. hypotonic . . . greater
   e. hypotonic . . . lesser

5. Which of the following enables a cell to pick up and concentrate a specific kind of molecule?
   a. passive transport
   b. diffusion
   c. osmosis
   d. receptor-mediated endocytosis
   e. pinocytosis

6. A cell uses energy released by ____ reactions to drive the ____ reaction that makes ATP. Then it uses the energy released by the hydrolysis of ATP, an ____ reaction, to do various kinds of work in the cell.
   a. exergonic . . . exergonic . . . endergonic
   b. endergonic . . . exergonic . . . endergonic
   c. exergonic . . . endergonic . . . exergonic
   d. endergonic . . . endergonic . . . exergonic
   e. exergonic . . . endergonic . . . endergonic

7. Energy of activation
   a. is released when a large molecule breaks up.
   b. gets a reaction going.
   c. is released by an exergonic reaction.
   d. is stored in an endergonic reaction.
   e. is supplied by an enzyme.

8. The laws of thermodynamics state that whenever energy changes occur, ____ always increases.
   a. disorder
   b. order
   c. kinetic energy
   d. potential energy
   e. energy of activation

9. Living things transform kinetic energy into potential chemical energy in the ____ , when ____ is made.
   a. mitochondrion . . . ADP
   b. chloroplast . . . ADP
   c. chloroplast . . . an enzyme
   d. mitochondrion . . . glucose
   e. chloroplast . . . glucose

10. Why does heating interfere with the activity of an enzyme?
    a. It kills the enzyme.
    b. It changes the enzyme's shape.
    c. It increases the energy of substrate molecules.
    d. It causes the enzyme to break up.
    e. It kills the cell, so enzymes can't work.

11. An enzyme is specific. This means
    a. it has a certain amino acid sequence.
    b. it is found only in a certain place.
    c. it functions only under certain environmental conditions.
    d. it speeds up a particular chemical reaction.
    e. it occurs in only one type of cell.

12. Diffusion of water across a selectively permeable membrane is called
    a. active transport.
    b. osmosis.
    c. exocytosis.
    d. passive transport.
    e. facilitated diffusion.

Essay

1. Describe the kinds of molecules that cannot easily diffuse through cell membranes. How do proteins facilitate diffusion of these substances?

2. Make a sketch showing why an enzyme acts only on a specific substrate.

3. Most enzyme-catalyzed chemical reactions in humans occur most readily around body temperature, 37°C. Why do these reactions slow down at lower temperatures? Why do they slow down at higher temperatures?

4. Which contains more potential energy, a large, complex molecule like a protein, or the smaller amino acid subunits of which it is composed? Is the joining of amino acids to form a protein an exergonic or endergonic reaction? Why must this be the case? Where does the cell obtain energy to carry out such reactions?

5. Describe the circumstances under which plant and animal cells gain and lose water by osmosis. Which of the following is the least serious problem: water gain by a plant cell, water loss by a plant cell, water gain by an animal cell, or water loss by an animal cell? Why?

Applying Your Knowledge

Multiple Choice

1. If a cell is like a factory, then enzymes are like
   a. the plans for the factory.
   b. the machines in the factory.
   c. the power plant for the factory.
   d. the raw materials used by the factory.
   e. the walls of the factory.

2. A molecule that has the same shape as the substrate of an enzyme would tend to
   a. speed metabolism by guiding the enzyme to its substrate.
   b. speed metabolism by acting as a cofactor for the enzyme.
   c. speed metabolism because it would also be a catalyst.

d. save the cell energy by substituting for the substrate.
   e. slow metabolism by blocking the enzyme's active site.

3. A plant cell is placed in a solution whose solute concentration is twice as great as the concentration of the cell cytoplasm. The cell membrane is selectively permeable, allowing water but not the solutes to pass through. What will happen to the cell?
   a. No change will occur because it is a plant cell.
   b. The cell will shrivel because of osmosis.
   c. The cell will swell because of osmosis.
   d. The cell will shrivel because of active transport of water.
   e. The cell will swell because of active transport of water.

4. A white blood cell is capable of producing and releasing thousands of antibody molecules every second. Antibodies are large, complex protein molecules. How would you expect them to leave the cell?
   a. active transport
   b. exocytosis
   c. receptor-mediated endocytosis
   d. passive transport
   e. pinocytosis

5. Which of the following would be *least* likely to diffuse through a cell membrane without the help of a transport protein?
   a. a large polar molecule
   b. a large nonpolar molecule
   c. a small polar molecule
   d. a small nonpolar molecule
   e. Any of the above would easily diffuse through the membrane.

6. Red blood cells shrivel when placed in a 10% sucrose solution. When first placed in the solution, the solute concentration of the cells is ____ the concentration of the sucrose solution. After the cells shrivel, their solute concentration is ____ the concentration of the sucrose solution.
   a. less than . . . greater than
   b. greater than . . . less than
   c. equal to . . . equal to
   d. less than . . . equal to
   e. greater than . . . equal to

7. A nursing infant is able to obtain disease-fighting antibodies, which are large protein molecules, from its mother's milk. These molecules probably enter the cells lining the baby's digestive tract via
   a. osmosis.
   b. passive transport.
   c. exocytosis.
   d. active transport.
   e. endocytosis.

8. Some enzymes involved in the hydrolysis of ATP cannot function without the help of sodium ions. Sodium in this case functions as
   a. a substrate.
   b. a cofactor.
   c. an active site.
   d. a noncompetitive inhibitor.
   e. a vitamin.

9. The relationship between an enzyme's active site and its substrate is most like which of the following?
   a. a battery and a flashlight
   b. a car and a driver
   c. a key and a lock
   d. a glove and a hand
   e. a hammer and a nail

10. In which of the following do both examples illustrate kinetic energy?
    a. positions of electrons in an atom—a ball rolling down a hill
    b. heat—arrangement of atoms in a molecule
    c. a rock resting on the edge of a cliff—heat
    d. a ball rolling down a hill—heat
    e. light—arrangement of atoms in a molecule

11. Which of the following is a difference between active transport (AT) and facilitated diffusion (FD)?
    a. AT involves transport proteins, and FD does not.
    b. FD can move solutes against a concentration gradient, and AT cannot.
    c. FD requires energy from ATP, and AT does not.
    d. FD involves transport proteins, and AT does not.
    e. AT requires energy from ATP, and FD does not.

12. An enzyme and a membrane receptor molecule are similar in that they
    a. are always attached to membranes.
    b. act as catalysts.
    c. require ATP to function.
    d. supply energy for the cell.
    e. bind to molecules of a particular shape.

Essay

1. The burning of glucose molecules in paper is an exergonic reaction, which releases heat and light. If this reaction is exergonic, why doesn't the book in your hands spontaneously burst into flame? You could start the reaction if you touched this page with a burning match. What is the role of the energy supplied by the match?

2. Seawater is hypertonic in comparison to body tissues. Explain what would happen to his stomach cells if a shipwrecked sailor drank seawater.

3. The laws of thermodynamics have imaginatively been described as the house rules of a cosmic energy card game: "You can't win, you can't break even (and to stay alive, you can't get out of the game)." State the law that says living things can't win the energy game. State the law that says they can't break even.

4. A farm worker accidentally was splashed with a powerful insecticide. A few minutes later he went into convulsions, stopped breathing, and died. The insecticide acted as a competitive inhibitor of an enzyme important in the function of the nervous system. Describe the structural relationship between the enzyme, its substrate, and the insecticide.

5. Lecithin is a substance used in foods such as mayonnaise as an emulsifier, which means that it helps oil and water mix. Lecithin is a phospholipid; a lecithin molecule has a polar "head" and a nonpolar "tail." How might the structure of lecithin allow water to surround fat droplets? Sketch a microscopic view of some fat droplets in mayonnaise, and show how you think the fat, surrounding water, and lecithin molecules might be arranged.

Extending Your Knowledge

1. Soap bubbles have many of the physical and chemical properties of cell membranes. Several books are available in bookstores and libraries that describe the physics and chemistry of bubbles. You might want to blow some bubbles, observe their behavior, and learn about the scientific principles that shape them.

2. Are you reading this under an electric light? Recall that energy cannot be created or destroyed, but only changed from one form into another. In what form was the light energy coming from the lamp before the light bulb changed it into light? Where did it come from? Trace the steps in the transformation of this energy as far back as you can. What happens to the light energy after it shines on you and the book?

3. Enzyme names usually end in the suffix "-ase," and the name of the enzyme often describes its substrate and the kind of reaction the enzyme catalyzes. For example, DNA polymerase catalyses the building of a DNA polymer from smaller nucleotide building blocks. Lactase breaks down the sugar lactose. Look for other enzyme names (in the text and elsewhere—see questions 4 and 5) and try to imagine what the enzymes do.

4. Various household products contain enzymes—cleaners, laundry detergent, meat tenderizer, and so on. See if you can find products containing enzymes. Where are the enzymes in these products obtained? What are their substrates?

5. Many drugs, such as antidepressants, act as enzyme inhibitors. Certain heart and blood-pressure drugs act by blocking receptors in cell membranes. If you have access to any such medications, look at the accompanying information sheets to find out how they work.

How Cells Harvest Chemical Energy

6

You need to eat and breathe because your cells need food and oxygen for energy. In every cell in your body, organic molecules and oxygen interact in a complex process called cellular respiration. In this process, food molecules such as glucose are broken down and the energy contained in their chemical bonds is used to make ATP. The ATP made in cellular respiration is then used to drive cellular activities. Right now, ATP produced in cellular respiration is being used to generate the nerve impulses from your eyes to your brain, to move your muscles, and to drive your heartbeat. This chapter explains how your cells harvest the energy that keeps you alive.

Organizing Your Knowledge

Exercise 1   (Introduction – Module 6.4)

Review the basic terms and concepts of cellular respiration by filling in the blanks below.

Right now, you are breathing at a steady rate of 12 to 20 breaths per minute. Breathing, or [1]_____, is necessary for life, but why? Breathing allows the body to take in [2]_____ gas and expel waste [3]_____. Your breathing is closely related to [4]_____, the aerobic harvest of the energy in food molecules by cells. Most of the time, most cells acquire energy by taking in both [5]_____ and [6]_____ from the blood. These two substances interact, the glucose is broken apart, and [7]_____ and [8]_____ are produced. In the process, some of the energy is stored in molecules of [9]_____, which provide the energy for body activities. To make enough ATP for their needs, average human beings must take in food that provides about [10]_____ kilocalories (kcal) of energy per day.

OK. So your cells use sugar and O_2 to get energy to make ATP. But where do the sugar and O_2 come from? Ultimately, all energy used by living things comes from the [11]_____. Through the process of [12]_____, plants, algae, and some bacteria convert light energy into [13]_____ energy by making sugar. In a plant or alga, photosynthesis rearranges the atoms in [14]_____ and [15]_____ to make [16]_____ and O_2. Then we obtain glucose from our [17]_____ and O_2 from the [18]_____.

When using oxygen, your cells are said to function [19]_____. Cells making ATP this way capture about [20]_____ % of the energy in glucose. For short bursts, cells can make ATP [21]_____ — that is, without using oxygen. This process is much less efficient; it only banks about [22]_____ % of the energy in glucose. But it is useful during intense bursts of activity, such as sprinting or heavy lifting.

56

Muscles contain a mixture of two kinds of cells, or fibers, specialized either for aerobic or anaerobic ATP production. [23]_____ fibers can sustain repeated, long contractions, by continuously producing ATP via [24]_____cellular respiration. Slow fibers are long and [25]_____, maximizing their surface area and contact with nearby [26]_____ that deliver oxygen. They have many [27]_____, the structures where aerobic ATP breakdown occurs. And they are rich in [28]_____, a red protein related to hemoglobin that supplies O_2 molecules. The [29]_____ meat of a turkey leg consists mostly of myoglobin-rich slow muscle fibers. The white meat of a turkey breast, on the other hand, consists mostly of [30]_____ muscle fibers, which are specialized for quick, powerful bursts of flight. These fibers are [31]_____ and are more powerful than slow fibers, having [32]_____ mitochondria and [33]_____myoglobin than slow fibers. During intense activity, when the blood cannot deliver O_2 fast enough for aerobic cellular respiration, fast fibers can function anaerobically, making small amounts of ATP without oxygen. They don't completely break down [34]_____, and therefore do not capture all its energy, and instead of producing CO_2 they make [35]_____, which makes muscles [36]_____ and fatigue. This is why [37]_____ fibers are best at producing short bursts of power.

Human muscles contain both kinds of fibers. Their proportions vary from muscle to muscle, and person to person. A runner whose leg muscles are primarily composed of [38]_____ fibers would be more likely to excel in distance events, while an individual with an abundance of [39]_____ fibers might make a better sprinter.

Exercise 2 (Modules 6.5 – 6.6)

How does the cell capture the energy of organic molecules in ATP? Review the basic concepts by completing this crossword puzzle.

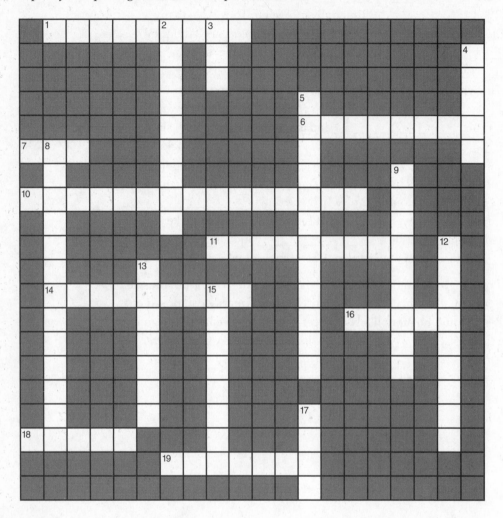

Across

1. The cell transfers energy by shuttling ____ from molecule to molecule.

6. Energy released in the electron transport chain is used to pump ____ ions (H^+) across a membrane.

7. ATP is made by adding ____ to an inorganic phosphate.

10. Oxidative _____ involves an electron transport chain and a process called chemiosmosis.

11. A sequence of electron carriers forms the electron ____ chain.

14. ____ is loss of an electron.

16. A molecule such as glucose is oxidized when it ____ an electron.

18. ____ is short for "oxidation-reduction."

19. Oxygen is ____ in cellular respiration.

Down

2. ____ is gain of an electron.

3. A coenzyme called ____ is used to carry electrons in redox reactions.

4. A molecule is reduced when it ____ an electron.

5. Most cells make most of their ATP via a process that involves an electron transport chain and a process called ____.

8. A ____ enzyme strips hydrogen atoms from organic molecules.

9. ATP ____ are protein complexes that use energy from a gradient of ion concentration to make ATP.

12. Phosphorylation is adding a ____ group to ADP to make ATP.

13. NADH delivers electrons to an electron ____ at the beginning of the electron transport chain.

15. Glucose is ____ in cellular respiration.

17. NAD^+ picks up electrons and hydrogen, forming ____.

Exercise 3   (Module 6.6)

Figure 6.6 introduces the three stages of cellular respiration. After studying it, see if you can label the diagram below without referring to the text. Include **oxidative phosphorylation, pyruvate, mitochondrion, CO$_2$, electrons carried by NADH, citric acid cycle, glycolysis, cytoplasm, ATP, glucose,** and **electrons carried by NADH and FADH$_2$.** (Note: 3, 6, and 7 are processes, 8 and 11 are places, and the rest are inputs and outputs.)

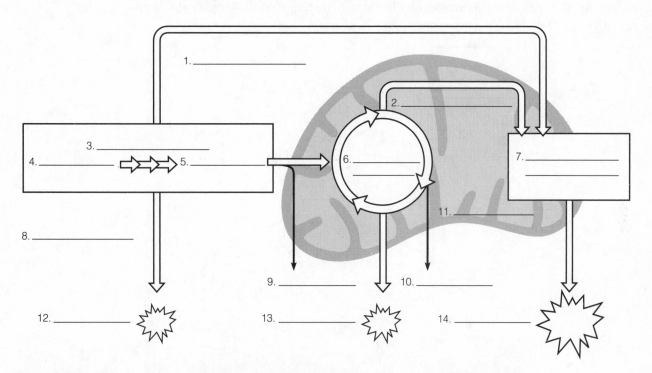

Exercise 4   (Module 6.7)

Web/CD Activity 6B   *Glycolysis*

Glycolysis is the first of three steps in cellular respiration. Review glycolysis by matching each phrase on the right with a term on the left. Some terms are used twice.

A. NADH
B. Pyruvate
C. ATP
D. NAD$^+$
E. Glucose
F. Glycolysis
G. ADP and P
H. Oxygen
I. Intermediate
J. Preparatory phase
K. Energy payoff phase
L. Substrate-level phosphorylation

_____ 1. Compound formed between glucose and pyruvate
_____ 2. Steps in glycolysis that produce ATP and pyruvate
_____ 3. Fuel molecule broken down in glycolysis
_____ 4. Produced by substrate-level phosphorylation
_____ 5. Invested to energize glucose molecule at start of process
_____ 6. Reduced as glucose is oxidized
_____ 7. Glucose converted to two molecules of this
_____ 8. Steps in glycolysis that consume energy
_____ 9. "Splitting of sugar"
_____ 10. Carries hydrogen and electrons from oxidation of glucose
_____ 11. When an enzyme transfers a phosphate from a substrate to ADP
_____ 12. Assembled to make ATP
_____ 13. Not involved in glycolysis

Exercise 5   (Modules 6.8 – 6.9)

Web/CD Activity 6C   *The Citric Acid Cycle*

Pyruvate from glycolysis is chemically altered and then enters the citric acid cycle, a series of steps that completes the oxidation of glucose. The energy of pyruvate is stored in NADH and FADH$_2$. To review these processes, fill in the blanks in the diagram below. (Try to do as many as you can without referring to the text.) Include the following: **NAD$^+$, pyruvate, CO$_2$, FADH$_2$, NADH, coenzyme A, ATP,** and **acetyl CoA.**

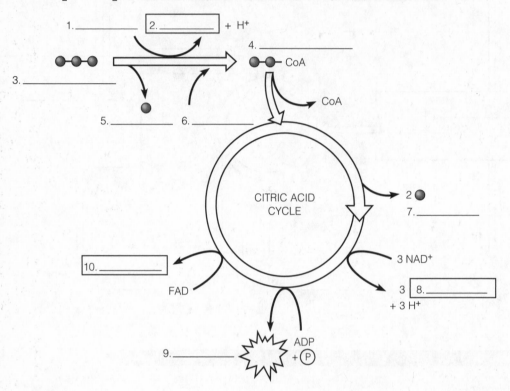

Exercise 6   (Module 6.10)

Web/CD Activity 6D   *Electron Transport and Chemiosmosis*

Circle the correct words or phrases in parentheses to complete each sentence.

The [1] *(final, second)* stage of cellular respiration is the electron transport chain and synthesis of [2] *(glucose, ATP)* by [3] (a process called *oxidative phosphorylation, active transport*). The electron transport chain is a sequence of [4] *(electron, proton)* carriers built into the [5] *(outer, inner)* membrane of the mitochondrion. Molecules of FADH$_2$ and [6] *(ADP, NADH)* bring high-energy electrons to the chain from glycolysis and [7] *(the citric acid cycle, chemiosmosis).* The electrons move along the chain from carrier to carrier in a series of redox reactions, finally joining with [8] *(H$_2$O, CO$_2$, O$_2$)* and H$^+$ from the surrounding solution to form [9] *(H$_2$O, CO$_2$, O$_2$).* Energy released by the electrons is used to move protons—[10] *(H$^+$ ions, ADP molecules)*—by [11] *(active transport, passive transport)* into the space between the inner and outer mitochondrial membranes.

The buildup of protons in the intermembrane space—a proton gradient—constitutes [12] *(kinetic, potential)* energy that the cell can tap to make [13] *(ATP, glucose).* The concentration of protons tends to drive them back through the membrane into the [14] *(inner compartment of the mitochondrion, cytoplasm of the cell),* but protons can cross the membrane only by passing through special protein complexes, called [15] *(coenzyme As, ATP synthases)*. As each of these complexes allows protons back through the membrane, a component of the complex rotates like a turbine, causing catalytic sites to phosphorylate [16] *(NAD, ADP)* and make [17] *(NADH, ATP)*. Thus, oxidative phosphorylation transforms energy extracted from glucose into the phosphate bonds of ATP.

Exercise 7   (Modules 6.10 – 6.11)

Web/CD Activity 6D   *Electron Transport and Chemiosmosis*

These diagrams will help you review oxidative phosphorylation (electron transport and chemiosmosis), and how poisons disrupt them. In the first diagram, show how the processes work normally. Trace movement of an electron with an orange arrow, movement of H+ ions (active transport and chemiosmosis) with green arrows, and formation of ATP with a black arrow.

In the second diagram, draw arrows showing the movement of electrons and H+ and the formation of ATP, as in the first diagram. Then draw a red line where each poison acts, to show how each of the poisons short-circuits the normal processes. Label the poisons **rotenone, cyanide, carbon monoxide, DNP,** and **oligomycin.**

1. Normal electron transport and chemiosmosis:

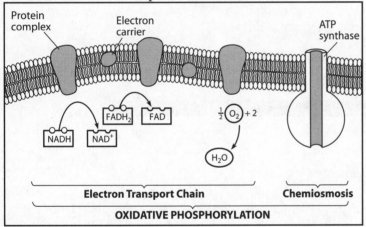

2. Effects of poisons:

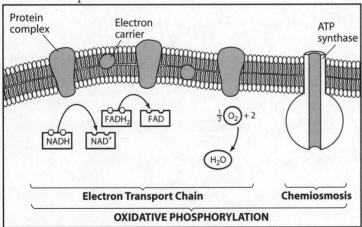

Exercise 8   (Modules 6.6 – 6.12)

Web/CD Activity 6A   *Glycolysis*
Web/CD Activity 6B   *The Citric Acid Cycle*
Web/CD Activity 6C   *Electron Transport and Chemiosmosis*
Web/CD Activity 6D   *Overview of Cellular Respiration*
Web/CD Thinking as a Scientist   *How Is the Rate of Cellular Respiration Measured?*

Check your overall understanding of cellular respiration by matching each of the phrases below with one of the three stages of the process. Use G for glycolysis, CA for the citric acid cycle, and OP for oxidative phosphorylation.

_____ 1. Generates most of the ATP formed by cellular respiration
_____ 2. Begins the oxidation of glucose
_____ 3. Occurs outside the mitochondrion
_____ 4. Produces 4 ATPs per glucose by substrate-level phosphorylation, but 2 ATPs per glucose are used to get it started
_____ 5. Oxidizes NADH and $FADH_2$, producing NAD^+ and FAD
_____ 6. Carried out by enzymes in the matrix (fluid) of the mitochondrion
_____ 7. Here electrons and hydrogen combine with O_2 to form H_2O
_____ 8. Occurs along the inner mitochondrial membrane
_____ 9. Generates most of the CO_2 produced by cellular respiration
_____ 10. $FADH_2$ and NADH deliver high-energy electrons to this stage
_____ 11. ATP synthase makes ATP
_____ 12. Reduces NAD^+ and FAD, producing NADH and $FADH_2$

Exercise 9   (Module 6.13)

Web/CD Activity 6E   *Fermentation*

Review fermentation by filling in the blanks below.

1. _____ anaerobes can make their ATP by fermentation or oxidative phosphorylation.
2. _____ are organisms that normally uses aerobic respiration to produce ATP, but can generate ATP without oxygen, via alcoholic fermentation.
3. Fermenters replenish their supply of NAD^+ by using NADH to oxidize _____ .
4. When oxygen is scarce, human _____ cells can make ATP by lactic acid fermentation.
5. Fermentation enables cells to make ATP in the absence of _____ .
6. For every molecule of glucose consumed, glycolysis produces two molecules of pyruvic acid, two molecules of ATP, and two molecules of _____ .
7. The waste products of alcoholic fermentation are _____ and carbon dioxide.
8. _____ acid fermentation is used to make cheese and yogurt.
9. Fermentation generates two energy-rich _____ molecules for every molecule of glucose consumed.
10. A cell can use _____ to generate a small amount of ATP, but it must somehow recycle its supply of NAD^+.
11. Like aerobic respiration, alcoholic fermentation produces _____ gas as a waste product.
12. Strict _____ require anaerobic conditions and are poisoned by oxygen.
13. Fermentation provides an _____ pathway that allows NADH to get rid of electrons and recycle as NAD^+.
14. The ATP yield of fermentation is much_____than that of aerobic respiration.
15. The buildup of _____from strenuous exercise can cause muscle fatigue and soreness.

Exercise 10   (Module 6.14)

Review the molecules that can be used as fuel for cellular respiration by writing their names in the blanks in this diagram. Include **glucose, amino acids, fats, fatty acids, proteins, sugars, carbohydrates,** and **glycerol.**

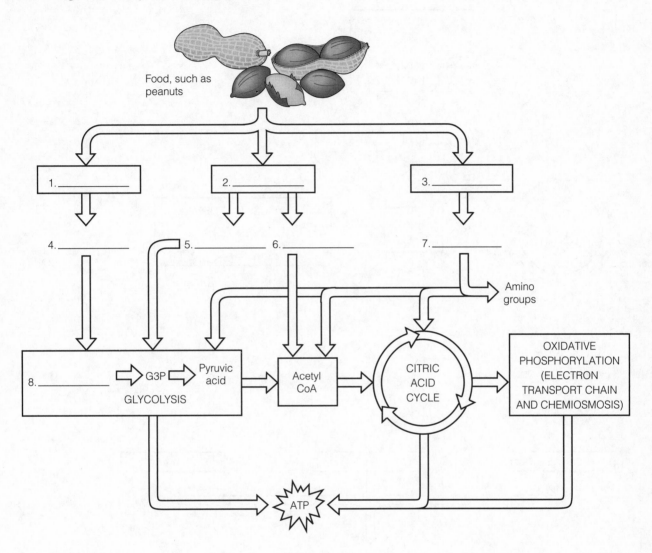

Show how a cell obtains organic molecules for biosynthesis of proteins, polysaccharides, and fats by drawing the missing arrows on this diagram.

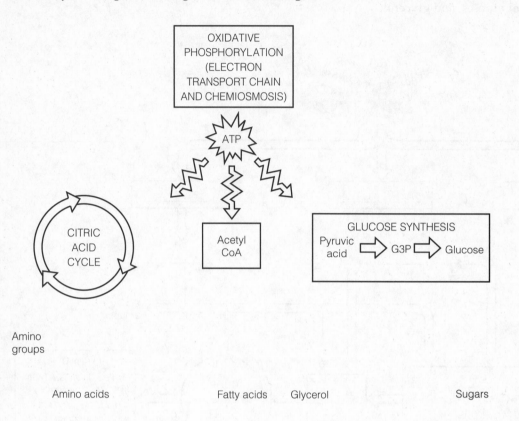

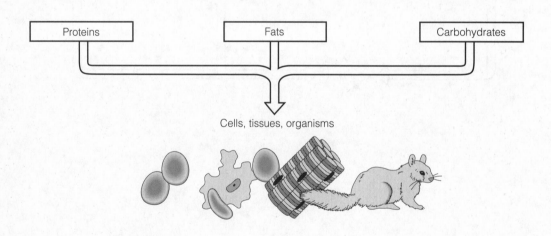

Cells, tissues, organisms

Testing Your Knowledge

Multiple Choice

1. A molecule is oxidized when it
   a. changes shape.
   b. gains a hydrogen (H^+) ion.
   c. loses a hydrogen (H^+) ion.
   d. gains an electron.
   e. loses an electron.

2. The main function of cellular respiration is
   a. breaking down toxic molecules.
   b. making ATP that powers cell activities.
   c. making food.
   d. producing chemical "building blocks" for cell structures.
   e. breaking down ATP, so that ADP and P can be reused.

3. In cellular respiration, _____ is oxidized and _____ is reduced.
   a. O_2 . . . ATP
   b. ATP . . . O_2
   c. glucose . . . O_2
   d. CO_2 . . . H_2O
   e. glucose . . . ATP

4. Most of the ATP produced in cellular respiration comes from
   a. glycolysis.
   b. oxidative phosphorylation.
   c. lactic acid fermentation.
   d. biosynthesis.
   e. the citric acid cycle.

5. _____ is used and _____ is produced in the overall process of cellular respiration.
   a. CO_2 . . . H_2O
   b. O_2 . . . glucose
   c. H_2O . . . ATP
   d. glucose . . . CO_2
   e. ATP . . . O_2

6. The energy given up by electrons as they move through the electron transport chain is used to
   a. break down glucose.
   b. make NADH and $FADH_2$.
   c. pump H^+ through a membrane.
   d. oxidize water.
   e. manufacture glucose.

7. Fermentation is essentially glycolysis plus an extra step in which pyruvate is reduced to form lactic acid or alcohol and CO_2. This last step
   a. removes poisonous oxygen from the environment.
   b. extracts a bit more energy from glucose.
   c. enables the cell to recycle NAD^+.
   d. inactivates toxic pyruvate.
   e. enables the cell to make pyruvate into substances it can use.

8. A small amount of ATP is made in glycolysis and the citric acid cycle
   a. by transfer of a phosphate group from a fragment of glucose to ADP.
   b. using energy from the sun to perform the process of photosynthesis.
   c. by transport of electrons through a series of carriers.
   d. when electrons and hydrogen atoms are transferred to NAD^+.
   e. as a product of chemiosmosis.

9. The ATP synthase in a human cell gets energy for making ATP directly from
   a. sunlight.
   b. flow of H^+ down a concentration gradient.
   c. oxidation of glucose.
   d. movement of electrons through a series of carriers.
   e. reduction of oxygen.

10. Which of the following describes glycolysis?
    a. It begins the oxidation of glucose.
    b. It produces a small amount of ATP.
    c. It generates NADH.
    d. It splits glucose to form two molecules of pyruvate.
    e. All of the above.

11. Most of the NADH that delivers high-energy electrons to the electron transport chain comes from
    a. chemiosmosis.
    b. the cytoplasm.
    c. glycolysis.
    d. biosynthesis.
    e. the citric acid cycle.

12. When protein molecules are used as fuel for cellular respiration, _____ are produced as waste.
    a. amino groups
    b. fatty acids
    c. sugar molecules
    d. molecules of lactic acid
    e. ethanol and CO_2

Essay

1. Describe the relationship between breathing and cellular respiration.

2. Compare the advantages and disadvantages of aerobic cellular respiration with the advantages and disadvantages of fermentation as methods of making ATP for cellular activities.

3. Compare the two mechanisms that generate ATP in cellular respiration—oxidative phosphorylation and substrate-level phosphorylation. In what stage(s) of cellular respiration does each occur? Where does each get the energy for making ATP? Which produces the most ATP under aerobic conditions? Under anaerobic conditions?

4. Describe three ways in which poisons can interfere with cellular respiration.

5. Explain the roles of glycolysis and the citric acid cycle in the biosynthesis of organic molecules.

Applying Your Knowledge

Multiple Choice

1. Which of the following illustrates oxidation?
    a. $CO_2 + H_2O \longrightarrow C_6H_{12}O_6$
    b. $C_6H_{12}O_6 \longrightarrow CO_2 + H_2O$
    c. $ADP + P \longrightarrow ATP$
    d. $ATP \longrightarrow ADP + P$
    e. $H_2O \longrightarrow O_2 + H$

2. In an experiment, mice were fed glucose ($C_6H_{12}O_6$) containing a small amount of radioactive oxygen. The mice were closely monitored, and in a few minutes radioactive oxygen atoms showed up in
    a. CO_2.
    b. NADH.
    c. H_2O.
    d. ATP.
    e. O_2.

3. In a second experiment, mice were allowed to breathe oxygen gas (O_2) laced with radioactive oxygen. In this experiment, the radioactive oxygen atoms quickly showed up in
    a. CO_2.
    b. NADH.
    c. H_2O.
    d. ATP.
    e. glucose, $C_6H_{12}O_6$.

4. A chemist has discovered a drug that blocks glucose phosphate isomerase, an enzyme that catalyzes the second reaction in glycolysis. He thought he could use the drug to kill bacteria in people with infections. But he can't do this because
    a. bacteria are facultative anaerobes; they usually don't need to do glycolysis.
    b. glycolysis produces so little ATP that the drug will have little effect.
    c. human cells also do glycolysis; the drug might also poison them.
    d. bacteria do not perform glycolysis.
    e. glycolysis can occur without the action of enzymes.

5. A glucose molecule is completely broken down in glycolysis and the citric acid cycle, but these two processes yield only a few ATPs. Where is the rest of the energy the cell obtains from the glucose molecule?
    a. in FAD and NAD^+
    b. in the oxygen used in the electron transport chain
    c. lost as heat
    d. in NADH and $FADH_2$
    e. in the CO_2 molecules released by the processes

6. Which of the following contains energy that a cell could use to make ATP?
    a. O_2
    b. CO_2
    c. NAD^+
    d. NADH
    e. H_2O

7. NADH is sometimes used by the cell in biosynthesis of needed organic molecules. Based on what you know about NADH, which of the following might be its function in biosynthesis?
    a. oxidizing organic molecules
    b. aiding in direct phosphorylation
    c. reducing organic molecules
    d. producing NAD^+
    e. breaking down ATP

8. Gram for gram, sugars are not as good as fats as a source of energy for cellular respiration, because sugars
   a. produce toxic amino groups when broken down.
   b. contain more hydrogen.
   c. usually bypass glycolysis and the Krebs cycle.
   d. contain fewer hydrogen atoms and electrons.
   e. are not as easily reduced.

9. A microbiologist discovered a new antibiotic that slowed the growth of bacteria by interfering with cellular respiration. She found that bacteria treated with the antibiotic produced about 15 ATP molecules for every glucose molecule they consumed. Which of the following hypotheses could explain the antibiotic's effect? The treated bacteria
   a. cannot perform glycolysis.
   b. have partially crippled electron transport chains.
   c. cannot produce NADH.
   d. have to rely at least partially on biosynthesis for their ATP.
   e. are forced to rely on fermentation for ATP.

10. A drug that blocks dehydrogenase enzymes would cause cells to
    a. run out of NADH.
    b. suffocate.
    c. deplete their supply of ADP.
    d. rely totally on electron transport and chemiosmosis for energy.
    e. be poisoned by lactic acid.

11. Unlike turkey breast, the breast of a duck is "dark meat." Why?
    a. Ducks fly longer distances, so their breast muscles consist of fast fibers.
    b. Ducks fly faster than turkeys, so duck breast muscles consist of fast fibers.
    c. The fibers of duck breast muscles contain less myoglobin.
    d. Ducks fly longer distances, so their breast muscles consist of slow fibers.
    e. Duck muscle fibers are specialized for anaerobic ATP production.

Essay

1. Fermentation is a much less efficient way to make ATP than aerobic cellular respiration via oxidative phosphorylation. This being the case, why do you think the fermenters have not been driven to extinction by competition with aerobes?

2. Without oxygen, cellular respiration grinds to a standstill, although glycolysis can continue to make some ATP anaerobically for a short time. When oxygen runs out, why does electron transport stop? Why do you think the citric acid cycle stops?

3. FAD and NAD$^+$ are made from the B vitamins riboflavin and niacin. Why do you think these substances are required in such tiny amounts in your diet? How would a deficiency in one of these vitamins interfere with cell function?

4. After a biochemical analysis of the victim's tissues, the brilliant forensics expert J. C. Mickleberry announced his findings: "Contrary to the conclusions of the police, the victim did not suffocate. The electron carriers in his mitochondria were all in the oxidized state. We will need to perform a second autopsy to determine the actual cause of death." Explain how the data led Mickleberry to his conclusion.

5. A microbiologist poured a test tube full of yeast into a flask of sugar water and periodically took samples from the flask. At first, the amount of sugar in the flask decreased gradually. Then there was a sharp drop in sugar, accompanied by the appearance of ethanol in the flask. Explain these results.

Extending Your Knowledge

1. Look around your home and find as many products as possible that either contain yeast or were made by yeast. Why were yeasts used to make these products?

2. Health and fitness experts recommend 20 minutes of aerobic exercise at least three times a week for peak cardiovascular conditioning and weight control. What kinds of exercise do you do? Do you exercise enough?

Photosynthesis: Using Light to Make Food 7

Seen from space, Earth is a blue-green jewel—blue light reflected from the oceans and green light reflected from the mantle of vegetation that covers much of the land. From orbit, astronauts can see the deep green that cloaks the equator and watch the advance and retreat of green as the seasons unfold in the temperate regions. Green is the color of chlorophyll, the most important chemical on Earth. In the process of photosynthesis, chlorophyll captures the energy of the sun. It enables plants and other producers to store solar energy in the food used by all living things. This chapter is about photosynthesis, the food-making process that starts with sunlight shining on green chlorophyll.

Organizing Your Knowledge

Exercise 1   (Module 7.1)

Autotrophs are able to produce their own organic molecules. All organisms that use light energy to make food are called photoautotrophs. Circle the organisms below that are photoautotrophs.

mushroom	pine tree	squirrel	green bacterium
rosebush	seaweed	moss	bread mold
parasitic bacterium	alga	sponge	grass

Exercise 2   (Modules 7.1 – 7.7)

Web/CD Activity 7A   *The Sites of Photosynthesis*
Web/CD Activity 7B   *Overview of Photosynthesis*
Web/CD Activity 7C   *Light Energy and Pigments*
Web/CD Thinking as a Scientist   *How Does Paper Chromatography Separate Plant Pigments?*

Review some of the basic terminology of photosynthesis by completing this crossword puzzle.

Across

5. ____ is oxidized in the process of photosynthesis.
6. The ____ is the cell organelle where photosynthesis takes place.
10. ____ are the light-catching membranes in a chloroplast.
11. Stacks of thylakoids in a chloroplast are called ____.
13. ____ energy travels through space as rhythmic waves.
14. When chlorophyll absorbs a photon, an ____ becomes excited.
16. Sugar is actually made in the ___ cycle.
18. The Calvin cycle occurs in the ____—the fluid of the chloroplast.
19. ____ is the green pigment in a leaf.
20. Carbon dioxide enters a leaf via holes called____.
22. Shorter wavelengths of light have____ energy than longer wavelengths.

Down

1. A ____ is a cluster of light-harvesting complexes in a thylakoid.
2. A ____ is a fixed quantity of light energy.
3. ____ is the process by which plants make food from carbon dioxide and water.
4. The color of light is related to its ____.
7. ____ is the source of energy for photosynthesis.
8. Carbon ____ is the incorporation of carbon dioxide into organic compounds.
9. The reaction ____ is the chlorophyll molecule that donates excited electrons.
12. ____ are yellow-orange pigments in a chloroplast.
15. ____ is the green tissue in the interior of a leaf.
17. A photosynthetic ____ is an organism that uses light to make food.
21. Photosynthesis occurs in ____ main stages.

Exercise 3 **(Modules 7.3 – 7.4)**

Write the overall equation for photosynthesis in the boxes below. Show the substances used on the left, and those produced on the right. Use different colors for carbon, hydrogen, oxygen in carbon dioxide, and oxygen in water, and then use your color code to show where atoms of C, H, and O on the left end up in the products on the right. On the lines under the substances used, state which is oxidized and which is reduced.

```
┌──────────┐     ┌──────────┐   LIGHT   ┌──────────┐     ┌──────────┐     ┌──────────┐
│          │  +  │          │  ──────▶  │          │  +  │          │  +  │          │
└──────────┘     └──────────┘           └──────────┘     └──────────┘     └──────────┘
──────────      ──────────
```

Exercise 4 **(Modules 7.3 – 7.5)**

Web/CD Activity 7B   *Overview of Photosynthesis*

Label this diagram summarizing the two stages of photosynthesis. Include **outer membrane of chloroplast, thylakoids, granum, stroma, light reactions, Calvin cycle, light, H₂O, O₂, electrons, NADPH, ATP, CO₂, sugar, ADP + P,** and **NADP⁺.** (Note: 5 and 7 are processes, 3, 9, 10, and 11 are places or structures, and the rest are inputs and outputs.)

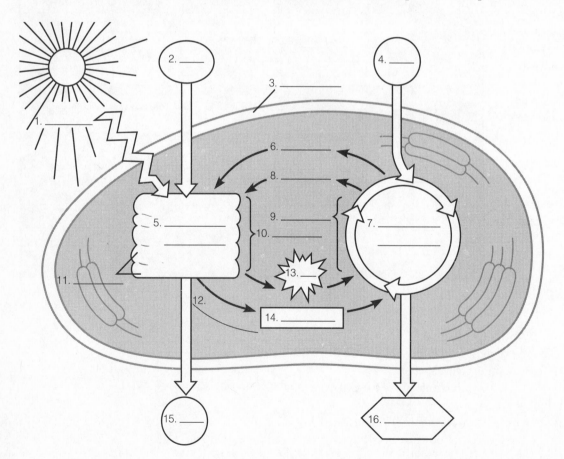

Exercise 5 (Modules 7.3 – 7.5)

Web/CD Activity 7B *Overview of Photosynthesis*

Refer to the equations and diagrams in the modules to match each of the phrases on the left with one of the ingredients or products of photosynthesis listed on the right.

_____ 1. Oxidized in the light reactions

_____ 2. Reduced in the Calvin cycle

_____ 3. Carries H and electrons from the light reactions to the Calvin cycle

_____ 4. Food produced by photosynthesis

_____ 5. Source of H and electrons that end up in glucose

_____ 6. Source of O atoms that end up in glucose

_____ 7. Where O atoms from water end up

_____ 8. Oxidized in the Calvin cycle

_____ 9. Reduced in the light reactions

_____ 10. Supplies energy to the Calvin cycle

_____ 11. Where C and O atoms in carbon dioxide end up

_____ 12. Recycled from the Calvin cycle to make ATP

_____ 13. Supplies energy to the light reactions

_____ 14. Gas produced by reactions in the thylakoids

_____ 15. Gas consumed by reactions in the stroma

_____ 16. Source of carbon for carbon fixation

_____ 17. Source of H for the Calvin cycle

_____ 18. Picks up energized electrons from reactions in the thylakoids

A. Carbon dioxide, CO_2

B. Water, H_2O

C. Glucose, $C_6H_{12}O_6$

D. Oxygen, O_2

E. ADP + P

F. ATP

G. $NADP^+$

H. NADPH

I. Light

Exercise 6 (Modules 7.6 – 7.7)

Web/CD Activity 7C *Light Energy and Pigments*

Order the following forms of electromagnetic energy from the shortest wavelength (1) to the longest (9). Which photons have the most energy? Which are used by plants in photosynthesis?

_____ a. Green light

_____ b. Radio waves

_____ c. X-rays

_____ d. Red light

_____ e. Ultraviolet light

_____ f. Infrared

_____ g. Microwaves

_____ h. Blue light

_____ i. Gamma rays

Web/CD Activity 7D   *The Light Reactions*
Web/CD Activity 7E   *The Calvin Cycle*
Web/CD Thinking as a Scientist   *How Is the Rate of Photosynthesis Measured?*

To review photosynthesis, fill in the blanks in the following story.

The next time you eat an apple, reflect on the apple tree's ability to make the sugars it contains, using sunlight to assemble simple substances from air and soil. This process is called [1]_____, and it takes place in structures called [2]_____ in cells of tissues called the [3]_____ inside the leaves of the apple tree. Photosynthesis actually consists of two processes: In the [4]_____ reactions, [5]_____ molecules in membranes called [6]_____ in the chloroplast capture light energy. In the [7]_____ cycle, which takes place in the [8]_____ surrounding the thylakoids, this energy is used to make sugar, a process called [9]_____ fixation.

Chlorophyll molecules absorb [10]_____, packets of light energy. Chlorophyll absorbs only certain wavelengths, or colors, of light, mainly in the [11]_____ and [12]_____ parts of the spectrum. It reflects [13]_____ light. Other pigments, such as [14]_____, can absorb colors that chlorophyll cannot use directly, and transfer this energy to chlorophyll. Chlorophyll and other pigments are clustered on the thylakoid membranes in groups called photosystems. All the pigment molecules in a photosystem pass their energy along to a structure called the [15]_____, in the middle of the photosystem. The reaction center consists of a special chlorophyll molecule and a primary electron acceptor. There are two kinds of photosystems, photosystem I and photosystem II, which absorb slightly different colors of light.

When light strikes a leaf, and pigment molecules absorb photons, they pass their energy to a photosystem II reaction center, and there the energy excites a chlorophyll [16]_____ to a higher energy level. This electron is passed to a primary electron acceptor and on to an electron transport chain. On their way down the electron transport chain, the electrons from photosystem II perform important work. One of the electron carriers in the chain uses the energy released by the electrons to transport [17]_____ ions from the [18]_____ into the space inside the [19]_____. This creates a buildup of H^+ ions, a concentration [20]_____ of H^+ across the membrane. The H^+ ions then diffuse through the membrane via a protein complex called [21]_____, which captures their energy to make [22]_____. In photosynthesis, this chemiosmotic production of ATP is called [23]_____. How does photosystem II replace its lost electrons? It gets them by splitting [24]_____. When the electrons of photosystem II are jarred loose, the reaction center develops a strong attraction for electrons. It obtains them by breaking apart a molecule of [25]_____. This leaves two H^+ ions (which pass into the thylakoid space) and an [26]_____ atom. This atom combines with one from another water molecule to form a molecule of [27]_____ gas, which diffuses out of the leaf—a product of photosynthesis important to us and other animals.

Meanwhile, [28]_____ energy excites an electron in a [29]_____ molecule in the reaction center of photosystem I. (This electron is replaced by one from the electron transport chain from photosystem II.) The primary electron [30]_____ captures the electron from photosystem I and passes it on to $NADP^+$, reducing it to a molecule of [31]_____.

At this point, the cells of the apple leaf have captured the energy of the sun in molecules of NADPH and ATP, and its leaf has released some O_2 gas, but so far no sugar has been produced. NADPH and ATP are used, and sugar is made, in the [32]_____ cycle, the second portion of [33]_____ that takes place in the [34]_____ of the chloroplast, around the thylakoids. Using carbon from [35]_____ obtained from the air, energy from [36]_____, and hydrogen and high-energy electrons carried by [37]_____, the enzymes of the Calvin cycle construct [38]_____, a high-energy three-carbon sugar molecule. In a series of steps, these molecules are combined to form the important six-carbon sugar [39]_____ and other organic compounds.

The cellulose that gives an apple its crunch and the sugar that gives it its sweet taste are made from the glucose made in photosynthesis. In your intestine, the sugars enter your blood and are transported to your body cells. There the chemical pathways of cellular [40]_____ release the energy in the sugar molecules and use it to build [41]_____, which is in turn used to power cellular work. Energy from the sun, captured by the apple and passed on to you, enables you to see, to move, and to contemplate this amazing story.

Exercise 8   (Module 7.12)

Web/CD Activity 7F   *Photosynthesis in Dry Climates*

Plants employ a variety of ways of fixing CO_2 and saving water. State whether each of the following statements relates to C_3 plants, C_4 plants, or **CAM** plants.

_____ 1. May waste energy on photorespiration on a hot day
_____ 2. Trap carbon in four-carbon compound, which donates it to Calvin cycle
_____ 3. Corn and sugarcane
_____ 4. Open stomata and trap CO_2 at night
_____ 5. Most plants
_____ 6. Soybeans, oats, wheat, rice
_____ 7. Can grow in hot, dry climates
_____ 8. Also can grow in hot, dry climates
_____ 9. Pineapple and many cacti
_____10. Calvin cycle uses CO_2 directly from the air

Exercise 9   (Module 7.13)

Test your understanding of this module by stating whether you think each of the following would be likely to **warm** or **cool** the Earth through alteration of the greenhouse effect.

_____   1. Increased burning of coal to produce electricity
_____   2. Increased rate of growth of algae in the oceans
_____   3. Increased cloud cover, which would reflect more sunlight
_____   4. Hybrid cars that go farther on a tank of gasoline
_____   5. Increased cutting of tropical rain forests
_____   6. Using nuclear power instead of coal and oil to make electricity
_____   7. Reforesting deforested and overgrazed land
_____   8. Increased rate of decomposition of organic matter
_____   9. Slowing the population growth rate
_____ 10. Using solar cells to generate electricity

Exercise 10   (Module 7.14)

Summarize this module by describing the relationship between CFCs and the Earth's protective ozone layer *in exactly 25 words*.

Testing Your Knowledge

Multiple Choice

1. The ultimate source of energy in the sugar molecules produced by photosynthesis is
   a. sugar.
   b. the sun.
   c. oxygen.
   d. ATP.
   e. chlorophyll.

2. Which of the following is produced by the light reactions of photosynthesis and consumed by the Calvin cycle?
   a. NADPH
   b. O_2
   c. H_2O
   d. sugar
   e. ADP + P

3. Which of these wavelengths is *least* useful for photosynthesis?
   a. green
   b. yellow
   c. blue
   d. orange
   e. red

4. When chloroplast pigments absorb light,
   a. they become reduced.
   b. they lose potential energy.
   c. their electrons become excited.
   d. the Calvin cycle is triggered.
   e. their photons become excited.

5. The light reactions of photosynthesis generate high-energy electrons, which end up in ____ . They also produce ____ and ____ .
   a. ATP . . . NADPH . . . O_2
   b. O_2 . . . sugar . . . ATP
   c. chlorophyll . . . ATP . . . NADPH
   d. water . . . sugar . . . O_2
   e. NADPH . . . ATP . . . O_2

6. The overall function of the Calvin cycle is
   a. capturing sunlight.
   b. making sugar.
   c. producing CO_2.
   d. splitting water.
   e. oxidizing glucose.

7. Which of the following correctly matches each of the inputs of the Calvin cycle with its role in the cycle?
   a. CO_2: high-energy electrons; ATP: energy; NADPH: oxidation
   b. CO_2: carbon; ATP: energy; NADPH: high-energy electrons
   c. CO_2: high-energy electrons; ATP: carbon; NADPH: energy
   d. CO_2: energy; ATP: carbon; NADPH: high-energy electrons
   e. CO_2: hydrogen; ATP: carbon; NADPH: energy

8. The main photoautotrophs in aquatic environments are
   a. plants and animals.
   b. plants and fungi.
   c. animals and algae.
   d. algae and bacteria.
   e. plants and bacteria.

9. Which of the following is *not* a product of the light reactions of photosynthesis?
   a. O_2
   b. sugar
   c. high-energy electrons
   d. ATP
   e. NADPH

10. Which of the following is oxidized in photosynthesis?
    a. O_2
    b. CO_2
    c. $C_6H_{12}O_6$
    d. ATP
    e. H_2O

11. In photosynthesis, plants use carbon from _____ to make sugar and other organic molecules.
    a. water
    b. the air
    c. chlorophyll
    d. the sun
    e. soil

Essay

1. Photosynthesis uses water and carbon dioxide to produce sugar and oxygen gas. Scientists long wondered whether the oxygen atoms in the oxygen gas produced in photosynthesis were obtained from carbon dioxide or water. Describe the experiments that enabled them to find out, and the results of these experiments.

2. Draw two squares, one labeled to represent the **light reactions** and the other to represent the **Calvin cycle**. Using arrows, show the inputs and outputs of each process. Include the following: **NADPH, ADP + P, O_2, light, CO_2, sugar, H_2O, ATP, $NADP^+$, electrons.**

3. Photosynthesis has been called "the most important chemical process on Earth." Explain why.

4. State two activities of humans that tend to intensify the greenhouse effect. Why are people concerned about this? State two actions we could take that would reduce our contribution to the greenhouse effect.

Applying Your Knowledge

Multiple Choice

1. A photon of which of these colors would carry the most energy?
   a. green
   b. yellow
   c. blue
   d. orange
   e. red

2. A plant is placed in a sealed greenhouse with a fixed supply of water, soil, and air. After a year, the plant weighs 5 kg more than at the start of the experiment, and the _____ weighs almost 5 kg less.
   a. soil in the pot
   b. water left in the room
   c. organic matter in the soil
   d. air in the room
   e. soil in the pot together with the water in the soil

3. In a rosebush, chlorophyll is located in
   a. chloroplasts, which are in mesophyll cells in the thylakoids of a leaf.
   b. mesophyll cells, which are in the thylakoids in chloroplasts in a leaf.
   c. thylakoids, which are in mesophyll cells in the chloroplasts in a leaf.
   d. chloroplasts, which are in thylakoids in the mesophyll cells of a leaf.
   e. thylakoids, which are in chloroplasts in the mesophyll cells of a leaf.

4. In an experiment, a plant was given ____ containing radioactive ^{18}O, and the radioactive oxygen atoms were used to make sugar.
   a. water
   b. chlorophyll
   c. oxygen gas
   d. ATP
   e. carbon dioxide

5. The *photo* part of the word *photosynthesis* refers to _____, whereas *synthesis* refers to _____.
   a. the reactions that occur in the thylakoids . . . carbon fixation
   b. the reactions in the stroma . . . the reactions in the thylakoids
   c. the Calvin cycle . . . carbon fixation
   d. the Calvin cycle . . . the reactions in the stroma
   e. the light reactions . . . reactions in the thylakoids

6. The energy used to produce ATP in the light reactions of photosynthesis comes from
   a. the "burning" of sugar molecules.
   b. splitting water.
   c. movement of H^+ through a membrane.
   d. carbon fixation.
   e. fluorescence.

7. The following (*P* through *U*) are the main steps of chemiosmotic ATP synthesis in the light-dependent reactions of photosynthesis. Which answer places them in the correct order?

   P. H^+ concentration gradient established.

   Q. H^+ diffuses through ATP synthase.

   R. Carriers use energy from electrons to move H^+ across membrane.

   S. Electrons from photosystem II pass along electron transport chain.

   T. Light excites electrons in photosystem II.

   U. Energy of H^+ flow used by ATP synthase to make ATP.
   a. PQTSRU
   b. STPQRU
   c. TSRPQU
   d. TSRUQP
   e. PQUSTR

8. The way ATP synthase captures the energy of H^+ ions in the light reactions appears to be similar to the way
   a. a turbine in a dam harnesses running water.
   b. an automobile engine uses hydrocarbons for fuel.
   c. heating of water forms steam in a nuclear reactor.
   d. electricity makes a lightbulb glow.
   e. a rotating wheel stores mechanical energy.

9. An oceanographer has suggested slowing the rate of greenhouse warming by fertilizing the ocean to increase the growth of algae. How would this reduce the greenhouse effect?
   a. It would produce oxygen, which reflects sunlight from the atmosphere.
   b. It would "repair" the Earth's ozone layer.
   c. It would use up CO_2, which traps heat in the atmosphere.
   d. It would change the color of the ocean, reflecting the sun's heat.
   e. It would trap sunlight that would otherwise warm the Earth.

10. Which of the following would not be capable of performing photosynthesis?
    a. bacterium
    b. pine tree
    c. mushroom
    d. seaweed
    e. algae

11. Ozone depletion might lead to
    a. warming of the climate.
    b. storms.
    c. cooling of the climate.
    d. increases in skin cancer.
    e. depletion of atmospheric CO_2.

Essay

1. In an experiment, plants were grown under colored filters that allowed equal amounts of light of different colors to strike different plants. Under which filter do you think plants grew the slowest? Why?

2. Carotenoids are yellow and orange pigments involved in photosynthesis. What colors of light must carotenoids absorb? Reflect? How is this useful to the plant?

3. Remember that the greater the concentration of hydrogen ions (H^+), the lower the pH of a solution. How do you suppose the pH of the solution in the thylakoid space compares with the pH of the solution in the stroma? What is responsible for the difference?

4. Compare chemiosmotic ATP production in photosynthesis (photophosphorylation) with ATP production in cellular respiration (oxidative phosphorylation): What is the source of high-energy electrons for each? What harnesses the energy of the electrons, and what is the energy used for? Where does each process take place? Where do the electrons end up? What actually manufactures the ATP? What is the immediate source of energy for ATP synthesis?

5. ATP is used to power the movement of your muscles as you turn the pages of this book. Where did the energy in the ATP come from? Trace the energy in the ATP molecules back to the sun.

6. In the mid-1600s, Belgian physician and chemist Jan Baptista van Helmont grew a small willow tree in a pot, adding only water to the soil. After five years, he found that the soil in the pot had lost only 60 grams, while the tree had grown by nearly 75 kilograms—more than 1000 times the material lost from the soil. Van Helmont concluded that the tree had gained most if its substance not from soil, but rather from the water he supplied. Was van Helmont right? Explain.

Extending Your Knowledge

1. What are some products you use or activities in which you participate that contribute to global warming? How might you modify your lifestyle to reduce your contribution to global warming?

2. Because photosynthesis uses carbon dioxide, deforestation increases global warming and planting trees reduces it. You can do your part to slow global warming by planting trees and encouraging others in your community to do so. Some groups that offer advice and information about planting trees:

   American Forests, P.O. Box 2000, Washington, DC 20013; www.americanforests.org

   Friends of Trees, 3117 NE ML King, Jr. Blvd., Portland, OR 97212; www.friendsoftrees.org

   National Arbor Day Foundation, 100 Arbor Ave., Nebraska City, NE 68410; www.arborday.org

   Plant-It 2020, PMB 310, 9457 S. University Blvd., Highlands Ranch, CO 80126; www.plantit2020.com

   Trees for the Future, P.O. Box 7027, Silver Spring, MD 20907; www.treesftf.org

   Trees, Water, & People, 633 Remington St., Fort Collins, CO 80524; www.treeswaterpeople.org

The Cellular Basis of Reproduction and Inheritance

8

In laboratories all over the world, biologists are studying cell division to try to understand why cells divide, what determines when they divide, what happens during cell division, and why some cells divide and others do not. The process of cell division is fundamental to inheritance, reproduction, growth, and development. Much of the inquiry into cell division is basic research; but this research has immense practical importance because understanding cell division will enable us to understand cancer, birth defects, heredity, and inherited diseases. This chapter explores the connections among cell division, reproduction, and inheritance.

Organizing Your Knowledge

Exercise 1   (Introduction – Module 8.3)

Review the concepts introduced in these modules by filling in the blanks.

"Like begets [1]_____." This old saying means offspring look like their parents. Technically, only offspring produced by [2]_____ reproduction look exactly like their parents, because they inherit all their [3]_____ from a single parent. For example, when an amoeba divides, its [4]_____ is duplicated, and identical sets of [5]_____ (the structures that contain most of the amoeba's DNA) are allocated to opposite sides of the cell. The parent amoeba splits, and the two daughter amoebas that are formed are genetically [6]_____ to each other and to the [7]_____ cell.

Prokaryotes also reproduce asexually, via a type of cell division called [8]_____. Most genes in a prokaryote are carried on a single [9]_____ DNA molecule, which is much [10]_____ and [11]_____ in structure than the multiple chromosomes of eukaryotes. The prokaryote replicates its DNA and the copies move toward opposite ends of the cell. As new [12]_____ grows between them, the chromosomes become separated. Finally, the plasma membrane and cell [13]_____ grow inward, separating the chromosomes and dividing the cell in two. Like the reproduction of an amoeba, binary fission produces daughter cells identical to the parent cell. Parent and offspring share identical [14]_____, or sets of genetic information.

The offspring produced by sexual reproduction resemble their parents, but they are not identical to their parents or to each other. Sexual reproduction begins with the production of an [15]_____ and a [16]_____, specialized cells that join to produce an offspring. The egg and sperm fuse—a process called [17]_____—and the fertilized egg inherits a unique combination of genes from both parents. Through repeated cell divisions, the fertilized egg develops into an organism with a unique combination of traits—for example, a cat with long, gray fur or a human with blue eyes and freckles. Thus, sexual reproduction produces [18]_____ among offspring. Through sexual reproduction "like begets like," but not exactly.

Exercise 2   (Module 8.4)

Test your knowledge of chromosomes by completing this crossword puzzle.

Across

4. ____ cells have more chromosomes and genes than prokaryotes do.

5. A typical ____ has about 3000 genes.

6. Chromosome duplication produces two sister ____.

8. A prokaryote has a ____ simple chromosome.

9. A chromosome contains one long ____ molecule.

10. Chromosomes are named for their affinity to ____ used in microscopy.

11. Eukaryotic chromosomes involve more ____ molecules than those of bacteria.

13. When a chromosome divides, one sister chromatid goes to each ____ cell.

14. Proteins help determine chromosome structure and control ____ activity.

Down

1. Sister chromatids are joined at the ____.

2. The number of chromosomes in a eukaryotic cell depends on the ____.

3. When a cell is not dividing, chromosomes form long, thin fibers called ____.

7. Genes in eukaryotic cells are grouped into ____ chromosomes.

9. Chromosomes are clearly visible only when a cell is ____.

12. ____ cells carry about 35,000 genes in 46 chromosomes.

Web/CD Activity 8A   *The Cell Cycle*

Review the cell cycle: First identify the parts of the cycle and place them in order by writing the name of each phase or process on the diagram. Choose from **S, interphase, mitosis, G₁, mitotic phase, cytokinesis,** and **G₂.** Then add a brief description of what is happening during that portion of the cycle. Choose from **growth and DNA synthesis, cell growth following division, division of cytoplasm, activity between divisions, division of nucleus and chromosomes, growth and activity between DNA replication and division,** and **mitosis plus cytokinesis.**

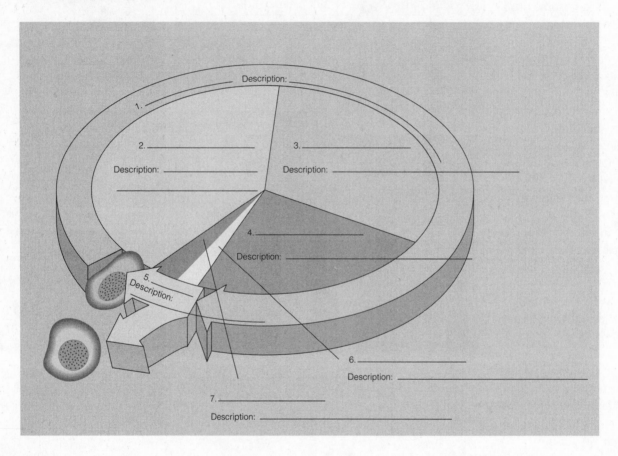

Exercise 4   (Module 8.6)

Web/CD Activity 8B   *Mitosis and Cytokinesis Animation*
Web/CD Thinking as a Scientist   *How Much Time Do Cells Spend in Each Phase of Mitosis?*

Summarize mitotic cell division. Briefly describe the appearance and activities of each of these cell parts during interphase and the four stages of mitosis. Include a simple sketch for each phase (just cell outlines and chromosomes).

	Interphase	*Prophase*	*Metaphase*	*Anaphase*	*Telophase*
Nucleus and nuclear envelope					
Mitotic spindle					
Chromosomes					
Cell size and shape					
Sketch					

Exercise 5   (Module 8.6)

Web/CD Activity 8B   *Mitosis and Cytokinesis Animation*
Web/CD Thinking as a Scientist   *How Much Time Do Cells Spend in Each Phase of Mitosis?*

This is a photograph of cells in an onion root tip, an area of rapid cell division. In which stage of mitosis (or interphase) is each of the numbered cells?

1. _____

2. _____

3. _____

4. _____

5. _____

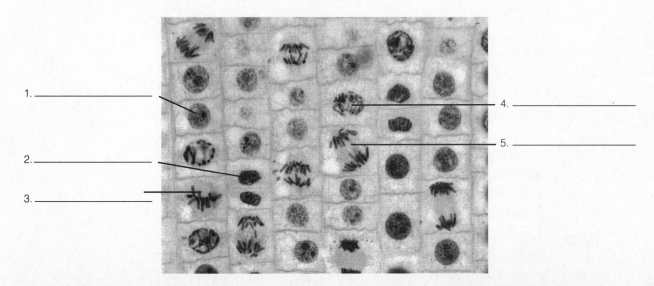

Exercise 6   (Module 8.6)

Web/CD Activity 8B   *Mitosis and Cytokinesis Animation*

Match the word or phrase on the right with the correct role in mitosis in an animal cell on the left.

A. Where spindle microtubules attach to chromosomes
B. Move chromosomes
C. Pulled apart by spindle microtubules
D. Material around centrioles from which mitotic spindle grows
E. Chromosomes come to rest here during metaphase

_____ 1. Metaphase plate
_____ 2. Kinetochores
_____ 3. Sister chromatids
_____ 4. Spindle microtubules
_____ 5. Centrosome

Exercise 7   (Module 8.7)

Web/CD Activity 8C   *Mitosis and Cytokinesis Video*

Read this module and then write a statement containing exactly 30 words (no more, no less!) comparing cytokinesis in plant and animal cells. (Writing *exactly* 30 words will help you to think about the processes and choose your words carefully. It's fun to try it.)

Exercise 8   (Modules 8.8 – 8.11)

Web/CD Activity 8D   *Connection: Causes of Cancer*

Review the functions of cell division and the factors that control it by filling in the blanks below.

Mitotic cell division has several important functions. Some animals rely on cell division for [1]_____ reproduction. *Hydra*, for example, produces buds that detach from the parent and take up life on their own. Cell division is also responsible for [2]_____, as seen in human embryos and plant roots. In an adult human, some cells, such as most nerve and muscle cells, cease to divide. Others, such as cells of the [3]_____, divide only if the organ is damaged. This process [4]_____ wounds. Some cells, such as those on the surface of the [5]_____ and the lining of the [6]_____, are constantly being abraded and lost. These cells are [7]_____ by cell division. In each of these cases, the new cells have exactly the same [8]_____ and [9]_____ of chromosomes as the parent cells, because of the way duplicated chromosomes divide in the process of [10]_____.

Growth, cell replacement, and reproduction require control of the rate and timing of cell division. Much has been learned by studying cells grown in laboratory [11]_____. Cells growing in a laboratory dish will divide only when in contact with a solid [12]_____. In the body, this [13]_____dependence may keep normal cells from dividing if separated from their normal surroundings. Cells will multiply only until they touch one another, a phenomenon known as [14]_____. Apparently, cells rely on proteins called [15]_____ for division, and will stop dividing when cells are crowded and these substances are depleted.

It appears that growth factors influence cell division by acting on the cell-cycle [16]_____ system, a set of molecules that triggers and coordinates events in the cell cycle. The system automatically [17]_____ cell division at several major checkpoints unless the "brakes" are overridden by go-ahead signals. There are checkpoints in the G_2 and M phases of the cell cycle, but the most important checkpoint for many cells is the [18]_____ checkpoint. If a cell receives a go-ahead signal, in the form of a growth factor, at the G_1 checkpoint, the cell will proceed into the [19]_____ phase of the cell cycle, replicate its DNA, and eventually divide. (A growth factor probably acts on a cell by attaching to a [20]_____ protein in the cell membrane. This protein in turn generates a signal that acts on the cell-cycle control system within the cell.) In the absence of a go-ahead signal, a cell will cease dividing. Many of our cells that can no longer divide—[21]_____ cells, for example—are stopped at the G_1 checkpoint.

Sometimes cells escape these control mechanisms, divide uncontrollably, and invade other body tissues. These [22]_____ cells can kill the organism. In cell culture, they can grow without being attached to a solid surface, are unaffected by density-dependent inhibition, and are less affected than normal cells by growth factors and [23]_____ signals. Cancer cells can go on dividing indefinitely (unlike normal cells, which can divide in culture for only [24]_____ generations). A mass of cancer cells is called a [25]_____. If the cells of a tumor remain at the original site, the tumor is called a [26]_____ tumor; if the cells spread, it is a [27]_____ tumor. The spread of cancer cells is called [28]_____. Carcinomas are cancers that arise in body coverings, such as the skin. [29]_____ arise in support tissues, such as muscle or bone. [30]_____ are cancers of blood-forming tissues.

Cancer treatments, such as [31]_____ and [32]_____, slow cancer by interfering with [33]_____. The anticancer drugs vinblastin and taxol prevent cell division by disrupting the mitotic [34]_____.

Exercise 9   (Modules 8.12 – 8.13)

Describe the relationship between the terms or items in each of the following pairs.

1. Sex chromosomes and autosomes

2. The two chromosomes of a homologous pair

3. The two sister chromatids of a single chromosome

4. A diploid cell and a haploid cell

5. A somatic cell and a gamete

6. An egg and a zygote

7. Fertilization and meiosis

8. Mitosis and meiosis

9. *X* and *Y* chromosomes

10. Gene and locus

Exercise 10   (Module 8.14)

Web/CD Activity 8E   *Asexual and Sexual Life Cycles*
Web/CD Activity 8F   *Meiosis Animation*

Review meiosis by drawing in the chromosomes to complete this sequence of diagrams. Some have been done for you. Label **meiosis I, meiosis II,** the **phases** of meiosis I and II, a pair of **homologous chromosomes,** two **sister chromatids,** and an example of **crossing over.** To make you think carefully about meiosis, the diploid number is 6 in this example. (It is 4 in Module 8.14.)

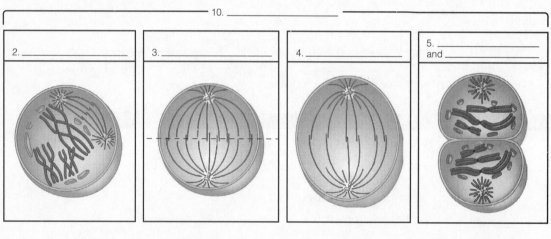

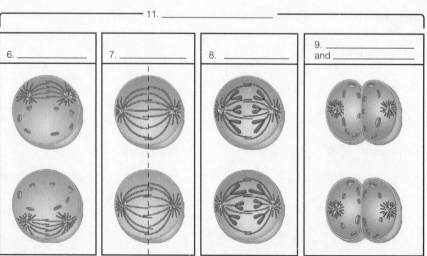

Exercise 11   (Module 8.15)

Compare mitosis and meiosis by completing this chart.

Mitosis	Meiosis
1.	Produces haploid daughter cells unlike parent cell
Involves one cell division	2.
Produces two daughter cells	3.
4.	Homologous chromosomes pair and then separate
Individual chromosomes line up at metaphase plate	5.
No crossing over occurs	6.
7.	Needed for sexual reproduction

Exercise 12   (Modules 8.16 – 8.18)

Web/CD Activity 8G   *Origins of Genetic Variation*
Web/CD Thinking as a Scientist   *How is Crossing Over Measured in the Fungus* Sordaria?

These modules discuss how independent orientation of chromosomes, random fertilization, and crossing over can lead to varied offspring. The diagram below shows the two homologous pairs of chromosomes in a cell with a diploid number of 4. Three different genes are also shown. On separate paper, complete the sketches described in questions 1 through 3.

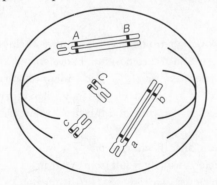

1. Show how two different orientations of the chromosomes during metaphase I of meiosis could lead to the four different combinations of genes in gametes (assuming crossing over does not occur). (You don't need to show meiosis step by step—just the outcome.)

2. Show how crossing over could recombine genes on the larger pair of chromosomes, producing different gametes.

3. How many different combinations of genes in gametes are possible if these two processes happen simultaneously? Try to sketch all of them.

Exercise 13   (Modules 8.19 – 8.22)

These modules discuss karyotyping—looking at chromosomes—and human abnormalities resulting from extra or missing chromosomes. Review them by completing this crossword puzzle.

Across

1. An abnormal number of chromosomes often results in ____.

4. A woman with only one X chromosome is said to have ____ syndrome.

5. The Y chromosome carries ____ genes than the X chromosome.

7. The incidence of Down syndrome increases with the ____ of the mother.

8. Down syndrome individuals are prone to ____ and Alzheimer's disease.

9. A single Y chromosome is enough to produce "____."

12. Nondisjunction results in extra or missing chromosomes in ____.

15. ____ is when members of a pair of chromosomes fail to separate during meiosis.

18. Most human offspring with abnormal numbers of chromosomes are spontaneously ____.

21. Down syndrome is the most serious common ____ defect in the U.S.

23. Pregnant women over 35 years of age are candidates for ____ testing.

25. In meiosis II, ____ chromatids may fail to separate.

26. A ____ is a photographic inventory of an individual's chromosomes.

Down

2. A person with Down syndrome has ____ number-21 chromosomes.

3. ____ who are lacking an X chromosome are designated XO.

4. XXY or XXYY males have small ____ and feminine body contours.

6. In meiosis I, a pair of ____ chromosomes may fail to separate.

10. If a sperm fertilizes an egg with an extra chromosome, the ____ will have an extra chromosome.

11. Nondisjunction can alter the number of ____ chromosomes, as well as autosomes.

13. Chromosomal defects usually result from errors in ____.

14. Individuals with Down syndrome are usually mentally ____.

16. People with Down syndrome live ____-than-normal life spans.

17. Meiosis begins in a woman's ovaries before she is ____ and is completed years later.

19. An extra chromosome 21 is called ____ 21.

20. A person with an extra number-21 chromosome has ____ syndrome.

22. ____ of the eggs of a Down syndrome woman will have an extra chromosome.

24. An unusual number of sex chromosomes has ____ effect than an unusual number of autosomes.

Exercise 14   (Module 8.23)

Chromosomes sometimes break, their parts can become scrambled, and abnormalities can result. Match each of the diagrams of chromosome alterations with its name and a description of its effects.

Diagram   *Name*   *Effects*

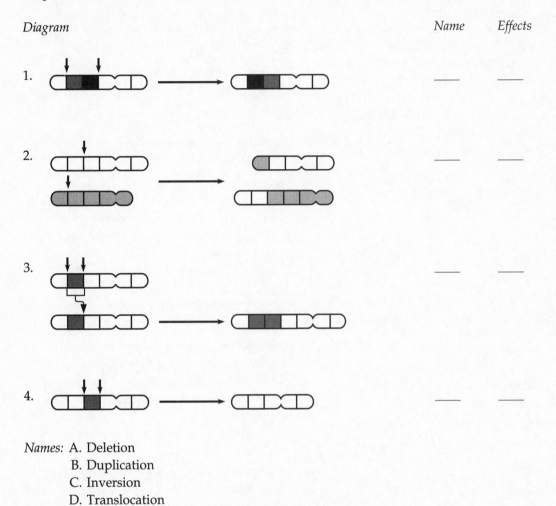

1. —— ——

2. —— ——

3. —— ——

4. —— ——

Names: A. Deletion
　　　　 B. Duplication
　　　　 C. Inversion
　　　　 D. Translocation

Effects: W. May cause chronic myelogenous leukemia in somatic cells
　　　　 X. Least likely to have serious effects, because genes are still present in normal numbers
　　　　 Y. Likely to have the most serious effects, as in *cri du chat* syndrome
　　　　 Z. A chromosome fragment breaks off and joins a homologous chromosome

Testing Your Knowledge

Multiple Choice

1. There are a number of differences between fission of a bacterium and human cell division. Which of the following is *not* one of them?
   a. A bacterium has only one chromosome.
   b. Human cells undergo mitosis and cytokinesis.
   c. Bacteria are smaller and simpler than human cells.
   d. Bacteria have to duplicate their DNA before dividing; human cells do not.
   e. Human chromosomes are larger and more complex.

2. You would be unlikely to see which of the following human cells dividing?
   a. muscle cell
   b. skin cell
   c. cancer cell
   d. cell from an embryo
   e. intestinal lining cell

3. Which of the following correctly matches a phase of the cell cycle with its description?
   a. M—replication of DNA
   b. S—immediately precedes cell division
   c. G_2—cell division
   d. G_1—immediately follows cell division
   e. All of the above are correctly matched.

4. Which of the following is *not* true of human somatic cells?
   a. They arise by mitotic cell division.
   b. They are haploid.
   c. They are body cells other than eggs and sperm.
   d. They are larger and more complex than bacterial cells.
   e. They contain 46 chromosomes.

5. In telophase of mitosis, the mitotic spindle breaks down and nuclear membranes form. This is essentially the opposite of what happens in
   a. prophase.
   b. interphase.
   c. metaphase.
   d. S phase.
   e. anaphase.

6. Sister chromatids
   a. cross over during prophase I of meiosis.
   b. separate during the first meiotic division.
   c. are produced during S phase between cell divisions.
   d. cross over during prophase II of meiosis.
   e. are also called homologous chromosomes.

7. Which of the following is *not* a function of mitotic cell division in animals?
   a. asexual reproduction
   b. growth
   c. repair of damaged organs
   d. production of gametes
   e. cell replacement

8. Meiosis
   a. is responsible for body growth and repair.
   b. halves the number of chromosomes in cells.
   c. is the process by which the body produces diploid cells.
   d. follows mitosis and splits the cytoplasm in two.
   e. is important in asexual reproduction.

9. Crossing over is
   a. important in genetic recombination.
   b. what makes a cell become cancerous.
   c. a key process that occurs during mitosis.
   d. an important mechanism of chromosome repair.
   e. what prevents cells from multiplying indefinitely in cell culture.

10. Human _____ are diploid, and human _____ are haploid.
    a. sex chromosomes . . . autosomes
    b. autosomes . . . sex chromosomes
    c. somatic cells . . . gametes
    d. gametes . . . somatic cells
    e. chromosomes . . . chromatids

11. Which of the following does *not* lead to genetic variability?
    a. random fertilization
    b. crossing over during meiosis
    c. division of chromosomes during anaphase of mitosis
    d. orientation of chromosomes during metaphase I of meiosis
    e. mutation

12. Most cells will divide if they receive the proper signal at a checkpoint in the ____ phase of the cell cycle.
    a. M
    b. G_1
    c. S
    d. G_2
    e. cytokinesis

13. Geneticists suspect that the extra chromosome seen in Down syndrome usually comes from the egg, rather than the sperm, because
    a. eggs are produced so rapidly that there is more chance for error.
    b. Down syndrome is due to a dominant gene in women, a recessive gene in men.
    c. most women inherit Down syndrome from their mothers.
    d. eggs are produced in much larger numbers than sperm.
    e. meiosis takes longer in the ovary, increasing the likelihood of error.

14. Which of the following chromosomal alterations would you expect to have the most drastic consequences?
    a. inversion
    b. duplication
    c. translocation
    d. deletion
    e. a and b are equally the most serious

15. Disorders involving unusual numbers of sex chromosomes show that "maleness" is caused by the
    a. presence of an *X* chromosome.
    b. presence of a *Y* chromosome.
    c. absence of an *X* chromosome.
    d. absence of a *Y* chromosome.
    e. absence of an *X* chromosome and presence of a *Y* chromosome.

Essay

1. Explain why, strictly speaking, the phrase "like begets like" applies only to asexual reproduction.

2. Briefly describe mitosis and cytokinesis and state their functions.

3. Describe how cancer cells differ from normal body cells.

4. Compare mitosis and meiosis. What are their functions? Which produces haploid cells? Diploid cells? What kinds of cells undergo mitosis and meiosis? What kinds of cells are produced by each? How many cells are produced?

5. How might the genes on the sister chromatids of a certain chromosome compare with each other and with the genes on the sister chromatids of the homologous chromosome?

6. Describe three aspects of sexual reproduction that lead to the production of varied offspring.

7. Explain how an error in meiosis can cause a baby to be born with an extra or missing chromosome.

8. About one in 700 babies born in the United States possesses an extra chromosome 21, resulting in Down syndrome. Why are few individuals seen who have extra copies of other chromosomes?

Applying Your Knowledge

Multiple Choice

1. In certain fungi and algae, cells undergo mitosis repeatedly without subsequently undergoing cytokinesis. What would result from this?
   a. a decrease in chromosome number
   b. inability to duplicate DNA
   c. division of the organism into many cells, most lacking nuclei
   d. large cells containing many nuclei
   e. a rapid rate of sexual reproduction

2. A human bone marrow cell, in prophase of mitosis, contains 46 chromosomes. How many chromatids does it contain altogether?
   a. 46
   b. 92
   c. 23
   d. 23 or 46, depending on when during prophase you look
   e. 46 or 92, depending on when during prophase you look

3. Which of the following is the most significant difference between mitosis and meiosis?
   a. Chromosomes are duplicated before mitosis.
   b. Meiosis is not followed by cytokinesis.
   c. Homologous pairs of chromosomes are split up in meiosis.
   d. A spindle formed of microtubules moves the chromosomes in mitosis.
   e. Crossing over occurs in mitosis.

4. If there are 22 chromosomes in the nucleus of a toad skin cell, a toad egg would contain ____ chromosomes.
   a. 22
   b. 44
   c. 11
   d. 33
   e. 88

5. Which of the following carry the same genetic information?
   a. sister chromatids
   b. *X* and *Y* chromosomes
   c. all autosomes
   d. homologous chromosomes
   e. all haploid cells

6. A cell biologist carefully measured the quantity of DNA in grasshopper cells growing in cell culture. Cells examined during the G_2 phase of the cell cycle contained 200 units of DNA. What would be the amount of DNA in one of the grasshopper daughter cells seen in telophase of mitosis?
   a. 50 units
   b. 100 units
   c. between 50 and 100 units
   d. 200 units
   e. 400 units

7. What would be the quantity of DNA in one of the grasshopper cells (question 6) produced by telophase II of meiosis?
   a. 50 units
   b. 100 units
   c. between 50 and 100 units
   d. 200 units
   e. 400 units

8. The two chromosomes of a homologous pair
   a. carry identical genetic information at corresponding locations.
   b. carry information for the same characteristics at different locations.
   c. carry identical genetic information at different locations.
   d. carry information for the same characteristics at corresponding locations.
   e. Any of the above is possible.

9. A picture of a dividing pigeon cell taken through a microscope shows that the cell contains 7 chromosomes, each consisting of 2 chromatids. This picture must have been taken during
   a. metaphase of mitosis.
   b. prophase I of meiosis.
   c. telophase II of meiosis.
   d. prophase II of meiosis.
   e. telophase of mitosis.

10. A culture of mouse cells is treated with a chemical that interferes with the activity of microfilaments. Which of the following will probably be affected the most?
    a. mitosis
    b. chromosome duplication
    c. pairing of homologous chromosomes
    d. cytokinesis
    e. joining of sister chromatids at the centromere

11. A zoologist examined an intestine cell from a crayfish and counted 200 chromosomes, each consisting of 2 chromatids, at prophase I of mitosis. What would he expect to see in each of the four cells at telophase II of meiosis if he looked in the crayfish ovary?
    a. 50 chromosomes, each consisting of 2 chromatids
    b. 50 chromosomes, each consisting of 1 chromatid
    c. 100 chromosomes, each consisting of 2 chromatids
    d. 100 chromosomes, each consisting of 1 chromatid
    e. 200 chromosomes, each consisting of 1 chromatid

12. One chromosome of a homologous pair carries the genes *J* and *K*. The other chromosome of the pair carries the genes *j* and *k* at corresponding loci. Crossing over results in exchange of chromosome segments and production of gametes with new combinations of genes. A "recombinant"-type gamete resulting from this crossover might contain
    a. genes *J* and *K*.
    b. genes *j* and *K*.
    c. genes *J* and *j*.
    d. genes *j* and *k*.
    e. genes *K* and *k*.

13. Humans have 23 pairs of chromosomes, while our closest relatives, chimpanzees, have 24. Chromosome studies indicate that at some point early in human evolution, two chromosomes simultaneously broke into a large portion and a small portion. The large parts combined to form a large chromosome, and the small parts combined to form a much smaller chromosome, which was subsequently lost. This important chromosomal change could best be described as
    a. nondisjunction followed by deletion.
    b. translocation followed by deletion.
    c. duplication followed by deletion.
    d. translocation followed by inversion.
    e. nondisjunction followed by inversion.

14. A karyotype would be least likely to show which of the following?
    a. an extra chromosome
    b. part of a chromosome duplicated
    c. a missing chromosome
    d. part of a chromosome turned around
    e. a translocation

Essay

1. Recall how chemotherapy slows the growth of cancerous tumors. Common side effects of chemotherapy are hair loss, nausea, and loss of appetite. Why do you think that chemotherapy has its strongest side effects on the skin and lining of the digestive tract?

2. Most of the grasshopper cells examined by the cell biologist in Multiple Choice questions 6 and 7 above contained either 100 or 200 units of DNA. Some of the interphase cells, however, contained between 100 and 200 units. Explain what was happening in these cells.

3. A slide of dividing cells in an onion root tip (Figure 8.11A in the text) is a "snapshot" in time. Each cell is stopped at the particular point in its cell cycle when the slide was made. A biology student examined such a slide under a microscope. Out of 100 cells she caught in the act of dividing, 52 were in prophase, 8 in metaphase, 11 in anaphase, and 29 in telophase. Assuming that the cells are growing and dividing independently, what do these data tell you about the phases of mitosis in onion cells?

4. A white blood cell from a female golden retriever was found to contain a total of 78 chromosomes. How many different kinds (sizes and shapes) of chromosomes would you expect to find in the cell? Why?

5. The somatic cells of a mosquito contain three pairs of chromosomes—two large ones, two medium-sized ones, and two small ones. One large chromosome bears the *A* gene; its homologue bears the *a* gene. One medium-sized chromosome bears the *B* gene; its homologue bears the *b* gene. One small chromosome bears the *C* gene; its homologue bears the *c* gene. Sketch cells at metaphase I of meiosis in the ovary of the mosquito, showing the different alignments of chromosomes that are possible. Then show how these lead to different combinations of genes in the gametes. How many different combinations are possible in the eggs? In the sperm of a male mosquito with the same genes, how many different gene combinations are possible? If the male and female mate and her eggs are fertilized with his sperm, how many different combinations are possible in the zygotes?

6. One chromosome of a homologous pair carries genes *Q* and *R*. Its homologue carries genes *q* and *r*. Show how crossing over during meiosis could produce gametes with new combinations of genes. What combinations of genes occur in parental-type gametes? In recombinant-type gametes?

7. Which of the following abnormalities in sex chromosomes would result in a predominantly male phenotype? A predominantly female phenotype? Explain why. *XXXX, XXY, XYYY, X, XXYY*

Extending Your Knowledge

1. What are some features that you (and your siblings, if you have any) appear to have inherited from your parents? Why don't you look exactly like your parents (or siblings)?

2. Do you know the warning signs of cancer? Has anyone close to you been affected by cancer? Cancer is more common in older people, but cancer can strike even college students in their teens and twenties. For more information on cancer research, diagnosis, treatment, and support, see these websites:
The American Cancer Society: www.cancer.org
The National Cancer Institute: www.nci.nih.gov

Patterns of Inheritance

9

At some point in your life you must have wondered about your inherited physical characteristics. Why is your hair dark or light, curly or straight? Do your eyes look like your mother's, your father's, or like neither of them? How do your eyes and hair, your height, and the color of your skin compare to those of your parents and your siblings? What made you female or male? People have asked these kinds of questions for thousands of years, but we only began to get some answers in the mid-1800s, when Gregor Mendel started to experiment with inheritance in plants in his abbey garden. This chapter concerns the principles and patterns of inheritance.

Organizing Your Knowledge

Exercise 1 (Modules 9.1 – 9.4)

Web/CD Activity 9A *Monohybrid Cross*

These modules discuss the basic principles of heredity and introduce the vocabulary of genetics. Read the modules carefully, and then practice using the vocabulary by matching each phrase on the left with a word or phrase on the right.

_____ 1. A unit that determines heritable characteristics
_____ 2. Organisms that always produce offspring identical to parents
_____ 3. The offspring of two different varieties
_____ 4. When two alleles of a pair differ, the one that is hidden
_____ 5. An incorrect idea that acquired characteristics are passed on
_____ 6. Parent organisms that are mated
_____ 7. A diagram that shows possible combinations of gametes
_____ 8. A breeding experiment that uses parents different in one characteristic
_____ 9. One of the alternative forms of a gene for a characteristic
_____ 10. Relative numbers of organisms with various characteristics
_____ 11. An organism that has two different alleles for a gene
_____ 12. Old idea that hereditary materials from parents mix in offspring
_____ 13. An organism's genetic makeup
_____ 14. Separation of allele pairs that occurs during gamete formation
_____ 15. Fertilization of a plant by pollen from a different plant
_____ 16. An organism that has two identical alleles for a gene
_____ 17. The science of heredity
_____ 18. A characteristic most commonly found in nature
_____ 19. What an organism looks like; its expressed traits
_____ 20. Offspring of the F_1 generation
_____ 21. When pollen fertilizes eggs from the same flower
_____ 22. Allele pairs separate during gamete production

A. Allele
B. Homozygous
C. Hybrid
D. Genotype
E. Segregation
F. F_2 generation
G. True-breeding
H. Heterozygous
I. Self-fertilization
J. Dominant
K. P generation
L. Monohybrid cross
M. Wild type
N. Phenotype
O. Cross
P. F_1 generation
Q. Recessive
R. Homologous chromosomes
S. Gene
T. Phenotypic ratio
U. Pangenesis
V. Cross-fertilization

____ 23. When two alleles of a pair differ, the one that determines appearance

____ 24. Where genes for a certain trait are located

____ 25. Offspring of the P generation

____ 26. A hybridization

W. Punnett square
X. Blending
Y. Genetics
Z. Law of segregation

Exercise 2   (Modules 9.3 – 9.4)

Web/CD Activity 9A   *Monohybrid Cross*

Test your knowledge of Mendel's principles by answering the following questions. You may want to test your ideas on scratch paper.

1. A pea plant with green pods is crossed with a plant with yellow pods. (Note: We're talking about the *pods* here, not the peas inside!) All their offspring have green pods.

   a. Which allele is dominant? Which allele is recessive?

   b. Using letters, what is the genotype of the green parent? The yellow parent?

   c. What are the genotypes of the offspring?

2. F_1 pea plants from the above cross are crossed. Use a Punnett square to figure out the genotypic and phenotypic ratios in the F_2 generation.

   a. Genotypic ratios:

   b. Phenotypic ratios:

3. Two black mice mate. Six of their offspring are black and two are white.

   a. What are the genotypes of the parents?

   b. For which offspring are you sure of the genotypes?

Exercise 3   (Module 9.5)

Mendel studied the inheritance of two characteristics at once and found that each pair of alleles segregates independently during the formation of gametes. In other words, if a tall pea plant with purple flowers is crossed with a short plant with white flowers, some of their descendants can be tall with white flowers. The tall and purple alleles do not have to stick together—they are independent.

So far, the textbook has discussed inheritance in peas and dogs. Just to be different, let's look at a genetic cross involving rabbits. In rabbits, the allele for brown coat is dominant, the allele for white coat recessive. The allele for short fur is dominant, the allele for long fur recessive. Imagine mating a true-breeding brown, short-haired rabbit with a white, long-haired rabbit. Using Module 9.5 as a model, write the genotypes of rabbits and gametes in the P, F_1, and F_2 generations in the blanks in the Punnett square. You may want to modify the drawings to show the phenotypes of the rabbits in the F_2 generation. Then use the Punnett square to figure out the phenotypic ratios in the F_2 generation—the proportion of rabbits that you can expect to be brown and short-haired, brown and long-haired, white and short-haired, and white and long-haired. Write their phenotype and their proportions in the blanks at the bottom.

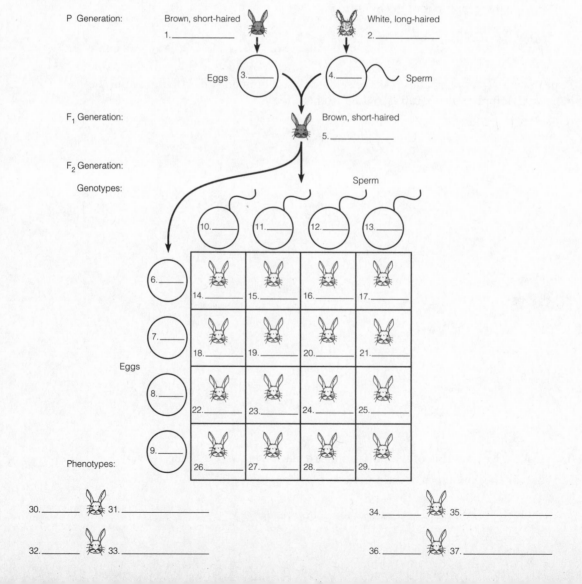

Exercise 4   (Module 9.6)

Web/CD Activity 9B   *Dihybrid Cross*

After reading this module on testcrosses, test your understanding by answering the following questions.

1. Recall that brown coat color in rabbits is dominant and white color is recessive. Suppose you have a group of rabbits—some brown and some white.
    a. For which phenotype(s) do you know the genotype(s)?

    b. For which phenotype(s) are you unsure of the genotype(s)?

2. Using *B* and *b* to symbolize the brown and white alleles:
    a. What are the possible genotypes of a white rabbit in your group?

    b. What are the possible genotypes of a brown rabbit?

3. Suppose you wanted to find out the genotype of a brown rabbit. What color rabbit would you mate it with?

4. A brown buck (male) is mated with a white doe (female). In their litter of 11 young, six are white and five are brown. Using a Punnett square to check your answer, what is the genotype of the buck?

5. Use a Punnett square to figure out the ratio of brown and white offspring that would have been produced by the above mating if the brown buck had been homozygous.

6. If half the offspring from a testcross are of the dominant phenotype and half are of the recessive phenotype, is the parent of the dominant phenotype (but unknown genotype) homozygous or heterozygous?

7. If all the offspring from a testcross are of the dominant phenotype, is the parent with the dominant phenotype (but unknown genotype) homozygous or heterozygous?

Exercise 5   (Module 9.7)

Web/CD Activity 9C   *Gregor's Garden*

The rules of probability can be used to predict the flip of a coin, the drawing of a card from a deck, or the roll of a pair of dice. They also govern segregation and recombination of genes. Read Module 9.7 carefully, and then fill in the blanks below.

The probability scale ranges from ^{1}_____ (an event that is certain not to occur) to ^{2}_____ (an event that is certain to occur). The probabilities of all possible outcomes for an event must add up to ^{3}_____. Imagine rolling a pair of dice, one die at a time. Each of the six faces of a die has a different number of dots, from one to six. If you roll a die, the probability of rolling a one is ^{4}_____. The probability of rolling any number other than one is ^{5}_____. The outcome of a given roll is unaffected by what has happened on previous rolls. In other words, each roll is a(n) ^{6}_____ event.

If you roll two dice simultaneously, what is the probability of "snake eyes" (both ones)? The roll of each die is an independent event. The probability of such a compound event (both dice coming up ones) is the [7]_____ of the separate probabilities of the independent events. Therefore, the probability of rolling two ones is [8]____ × [9]_____ = [10]_____. This is called the rule of [11]_____.

This rule also governs the combination of genes in genetic crosses. The probability that a heterozygous *(Pp)* individual will produce an egg containing a *p* allele is [12]_____. The probability of producing a *P* egg is also [13]_____. If two heterozygous individuals are mated, what is the probability of a particular offspring being [14]_____ recessive *(pp)*? The probability of producing a *p* egg is ½. The probability of producing a *p* sperm is also ½. The production of egg and sperm are independent events, so to calculate their combined probability we use the rule of [15]_____. Thus the chance that two *p* alleles will come together at fertilization to produce a *pp* offspring is [16]____ × [17]____ = [18]____.

Back to the dice for a moment. What is the probability that a roll of two dice will produce a three and a four? There are two different ways this can occur. One die can come up a three and the other a four, or one can come up a four and the other a three. The probability of the first combination is ⅙ × ⅙ = 1/36. The probability of the second is also ⅙ × ⅙ = 1/36. According to the rule of [19]_____, the probability of an event that can occur in two or more alternative ways is the [20]_____ of the separate probabilities of the different ways. The probability of rolling a three and a four is therefore [21]_____ + [22]_____ = [23]_____.

Similarly, what is the probability that a particular offspring of two heterozygous parents will itself be heterozygous? The probability of the mother producing a *P* egg is [24]_____. The probability of the father producing a *p* sperm is also [25]_____. Therefore, the probability of a *P* egg and a *p* sperm joining at fertilization is [26]_____ × [27]_____ = [28]_____. Or a *p* egg and a *P* sperm could join. The probability of this occurring is also [29]_____. According to the rule of addition, the probability of an event that can occur in two alternative ways is the sum of the separate probabilities. Therefore, the probability of heterozygous parents producing a heterozygous offspring is [30]____ + [31]____ = [32]_____.

Exercise 6   (Module 9.8)

After you read this module, use the information in the illustration to solve the following problems. You will probably want to work out Punnett squares on scratch paper.

1. A man and woman, both without freckles, have four children. How many of the children would you expect to have freckles?

2. Both Fred and Wilma have widow's peaks. Their daughter Shirley has a straight hairline. What are Fred and Wilma's genotypes?

3. A man and woman both have free earlobes, but their daughter has attached earlobes. What is the probability that their next child will have attached earlobes?

Exercise 7 (Module 9.8)

Family trees called pedigrees are used to trace the inheritance of human genes. The two pedigrees below show the inheritance of sickle-cell disease (described in Modules 9.14 and 9.15), which is caused by an autosomal recessive allele. In the first pedigree, the square and circle symbols are colored, as far as genotypes are known. Fill in the genotypes—*SS*, *Ss*, or *ss*—below the symbols. Use question marks to denote unknown genotypes. Complete the second pedigree by coloring in the symbols, following the rules described in Module 9.8. Again denote unknowns with question marks.

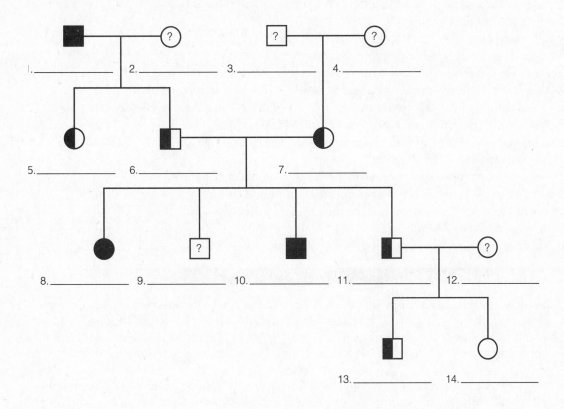

1. _____ 2. _____ 3. _____ 4. _____

5. _____ 6. _____ 7. _____

8. _____ 9. _____ 10. _____ 11. _____ 12. _____

13. _____ 14. _____

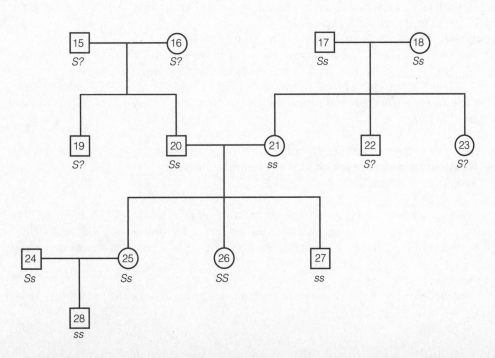

15 — *S?*   16 — *S?*   17 — *Ss*   18 — *Ss*

19 *S?*   20 *Ss*   21 *ss*   22 *S?*   23 *S?*

24 *Ss*   25 *Ss*   26 *SS*   27 *ss*

28 *ss*

Exercise 8   (Module 9.9)

This module discusses common human genetic diseases and their inheritance. Indicate whether each of the statements below is true or false, and change false statements to make them true.

_____ 1. Cystic fibrosis is the most common lethal genetic disease in the United States.

_____ 2. A genetic disorder is expected in half the children of two carriers of a recessive allele.

_____ 3. About 50 human genetic disorders are known to be inherited as Mendelian traits.

_____ 4. Most people afflicted with genetic disorders are born to afflicted parents.

_____ 5. Most serious human genetic disorders are recessive.

_____ 6. Cystic fibrosis is most common among Asian Americans.

_____ 7. Most genetic diseases are evenly distributed among ethnic groups.

_____ 8. Most societies have taboos and laws against marriage between close relatives.

_____ 9. Half of the offspring of two carriers of a recessive allele are likely to be carriers.

_____10. Dominant alleles are always more common in the population than recessive alleles.

_____11. Tay-Sachs disease is seen among Jews from Central Europe.

_____12. Lethal dominant alleles are much more common than lethal recessive alleles.

_____13. Achondroplasia is a form of dwarfism.

_____14. Geneticists agree that inbreeding always increases the risk of inherited diseases.

_____15. Huntington's disease is lethal, but it does not strike until middle age.

_____16. Symptoms of PKU include excess mucus in the lungs and digestive tract.

_____17. Extra fingers and toes is a dominant trait.

_____18. Sickle-cell disease is common among African Americans.

Exercise 9   (Modules 9.9 – 9.10)

Greg and Amy were excited and happy that she was pregnant, but their joy was mixed with anxiety. The couple had just received some bad news: Greg's sister had just given birth to a baby boy diagnosed with cystic fibrosis. Greg and Amy were at the clinic for genetic [1]_____ to discuss the possibility that Greg was a [2]_____ of cystic fibrosis and to determine their unborn child's chances of inheriting the disease.

Sharon, the genetic counselor, reviewed Amy's and Greg's family histories. She said, "Our first priority is to figure out whether the two of you are carriers. We knew that Amy could be, because her brother died of cystic fibrosis, but until Greg's nephew was diagnosed, we didn't know that there was CF in his family, too. Greg, if your sister is a carrier, you could be as well."

Amy interjected, "What does this mean for us and for our baby?"

"It means that you and Greg need to be tested for the cystic fibrosis allele. If you both are carriers, then we can talk about [3]_____ testing to determine whether your baby might have it."

Later that day, a technician withdrew blood for the tests, and the following week, Amy and Greg were back in the genetic counselor's office. Sharon breezed through the door. "You're in the clear for cystic fibrosis," she said matter-of-factly.

"What?"

"The CF tests were both negative: Neither of you are carriers. Plus, the [4]_____ of Amy's blood was fine—there doesn't appear to be much chance of a neural tube defect."

Greg and Amy both sighed with relief.

Then Sharon's expression became a bit more serious. "Unfortunately, we did find something else that concerns me. Besides testing for the CF allele, we did a routine screen for several other disorders, including [5]_____ disease—even though you are not Jewish—and PKU. Turns out you both are carriers for PKU."

Amy groaned, "Oh no."

Sharon quickly added, "Don't worry yet. Even though you are both carriers, the probability that the baby will have the disease is only [6]_____ ."

Amy asked, "What exactly is PKU? Is it a serious problem?"

Sharon explained that PKU, short for [7]_____ , is an inherited inability to break down an [8]_____ called phenylalanine. "The phenylalanine can build up in the blood and cause mental retardation. As I said, don't start worrying yet. We can test your fetus. If PKU is detected early, retardation can be prevented by putting the child on a special [9]_____ , low in phenylalanine."

Amy asked, "How will you test the baby?"

"We'll have to perform [10]_____—taking a sample of the [11]_____ fluid. We can check for PKU by testing for certain chemicals in the fluid itself. While we're at it we will culture some of the fetal [12]_____ from the fluid and do a [13]_____—take a picture of the chromosomes—to check for [14]_____ syndrome. It will take a couple of weeks to culture the cells. Or we could get the karyotype results right away by using a newer technique called [15]_____ sampling. Then —"

Greg interrupted. "Wait a minute. Do you have to get samples? Can't you just do [16]_____ imaging to look at the baby? Doctor Portillo did that before Kelly was born."

"We really can't check chromosomes or PKU by just looking at the fetus with ultrasound. Amy is over [17]_____ years old, so I think it is important to get a sample of amniotic fluid so we can check for Down syndrome. I'm sure everything will be okay, but it's best to be prepared. Plus, the karyotype will answer another question I'm sure you are eager to know the answer to—whether you are going to have a boy or a girl."

Exercise 10   (Modules 9.11 – 9.16)

Web/CD Activity 9D   *Incomplete Dominance*

These modules discuss examples of inheritance that are a bit more complex—and common—than the simple patterns of heredity observed by Mendel. After reading the modules, see if you can match each description with a pattern of inheritance. Choose from:

A. incomplete dominance
B. multiple alleles
C. codominance
D. pleiotropy
E. polygenic inheritance

_____ 1. There are three different alleles for a blood group—I^A, I^B, and i—but an individual has only two at a time.

_____ 2. Crosses between two cremello (off-white) horses always produce cremello offspring. Crosses between chestnut (brown) horses always result in chestnut offspring. A cross between chestnut and cremello horses produces palomino (a golden-yellow color somewhat intermediate between chestnut and cremello) offspring. If two palominos are mated, their offspring are produced in the ratio of 1 chestnut : 2 palominos : 1 cremello.

_____ 3. The sickle-cell allele, s, is responsible for a variety of phenotypic effects, from pain and fever to damage to the heart, lungs, joints, brain, or kidneys.

_____ 4. In rabbits, an allele for full color *(C)* is dominant over an allele for chinchilla *(c')* color. Both full color and chinchilla are dominant over the white allele *(c)*. A rabbit can be *CC, Cc', Cc, c'c', c'c,* or *cc.*

_____ 5. In addition to the A and B molecules found on the surface of red blood cells, humans also have M and N molecules. The genotype L^ML^M produces the M phenotype. The genotype L^NL^N gives the N phenotype. Individuals of genotype L^ML^N have both kinds of molecules on their red blood cells, and their phenotype is MN.

_____ 6. If a red shorthorn cow is mated with a white bull, all their offspring are roan, a phenotype that has a mixture of red and white hairs.

_____ 7. Independent genes at four different loci are responsible for determining an individual's HLA tissue type, important in organ transplants and certain diseases.

_____ 8. A recessive allele causes a human genetic disorder called phenylketonuria. Homozygous recessive individuals are unable to break down the amino acid phenylalanine. As a consequence, they have high levels of this substance in their blood and urine, reduced skin pigmentation, lighter hair than their normal brothers and sisters, and often some degree of mental impairment.

_____ 9. When graphed, the number of individuals of various heights forms a bell-shaped curve.

_____10. Chickens homozygous for the black allele are black, and chickens homozygous for the white allele are white. Heterozygous chickens are gray.

Exercise 11   (Module 9.17)

Genetic testing has revolutionized diagnosis and treatment of inherited conditions, and has become an important component of health care. But there are also problems and concerns related to genetic testing. List three of them.

1.

2.

3.

Exercise 12 (Module 9.18)

Genes are located on chromosomes. Genes undergo segregation and independent assortment because the chromosomes that carry them undergo segregation and independent assortment during meiosis. The illustration below is similar to that in Module 9.18. It shows how alleles and chromosomes are arranged in an F$_1$ rabbit and how meiosis sorts the alleles into their gametes. The diagram below shows only the chromosomes. Put a letter (*B*, *b*, *S*, or *s*) in each of the numbered boxes to show how segregation and independent assortment of chromosomes cause segregation and independent assortment of alleles.

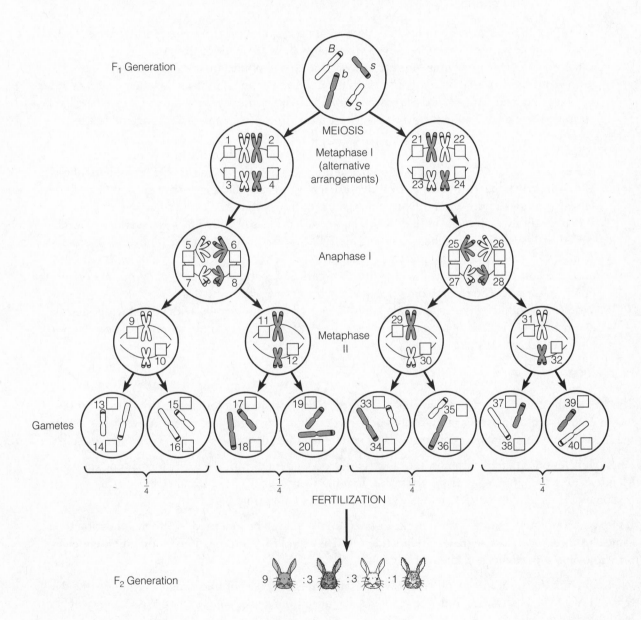

Exercise 13   (Modules 9.19 – 9.21)

Web/CD Activity 9E   *Linked Genes and Crossing Over*

These three modules discuss the inheritance of linked genes—genes that are on the same chromosome and therefore tend to be inherited together. Their pattern of inheritance is inconsistent with Mendel's "rules," but they illustrate important principles of chromosome structure and behavior. After reading the modules, match each of the observations below with the statement that explains the observation. Take your time; this exercise is not easy.

Observations

_____ 1. When two heterozygous round yellow peas are crossed, their offspring are produced in a 9:3:3:1 ratio (9 round yellow : 3 round green : 3 wrinkled yellow : 1 wrinkled green).

_____ 2. When two peas heterozygous for purple flowers and long pollen are crossed, the expected 9:3:3:1 ratio is not seen. The ratio is close to 3 purple long : 1 red round. Similarly, when a fruit fly with red eyes and long wings *(SsCc)* is crossed with a fly with scarlet eyes and curled wings *(sscc)*, offspring are not produced in the expected 1:1:1:1 ratio. Most offspring are red long and scarlet curled.

_____ 3. When two heterozygous purple long peas are crossed, most of their offspring are purple and long or red and round. But a very small number of offspring are purple and round or red and long. Similarly, when the *SsCc* and *sscc* fruit flies are mated, nearly all their offspring are *SsCc* and *sscc*. However, a small number of offspring (about 6% of the total) are *Sscc* and *ssCc*.

_____ 4. When a fruit fly with red eyes and long wings *(SsCc)* is crossed with a fly with scarlet eyes and curled wings *(sscc)*, 94% of their offspring are *SsCc* and *sscc*, and 6% are *Sscc* and *ssCc*. In other words, the recombination frequency between the *s* and *c* alleles is 6%. When a fly with red eyes and pale body *(SsEe)* is crossed with a fly with scarlet eyes and ebony body, 27% of their offspring are *Ssee* and *ssEe*. The recombination frequency between alleles *s* and *e* is 27%.

_____ 5. When a fly with long wings and pale body *(CcEe)* is crossed with a fly with curled wings and ebony body *(ccee)*, 21% of their offspring are *Ccee* and *ccEe*. The recombination frequency between alleles *c* and *e* is 21%.

Explanations

A. The greater the distance between two genes, the greater the opportunity for crossing over to occur between them. If crossing over is more likely, more recombinant offspring will result. If two genes are farther apart, the recombination frequency will be greater between them.

B. Pairs of alleles on different chromosomes segregate independently during gamete formation. They follow Mendel's principle of independent assortment. In other words, genes for different traits on different chromosomes do not tend to "stick together" when passed on to offspring.

C. If two genes are on the same chromosome, or linked, they tend to be inherited together. Alleles on the same chromosome do not segregate independently. They tend to "stick together," violating Mendel's principle of independent assortment.

D. Recombination frequencies can tell you how far apart genes are on a chromosome. If you know the distance from *a* to *b*, the distance from *a* to *c*, and the distance from *b* to *c*, you can map the sequence of genes on the chromosome.

E. Homologous chromosomes cross over during meiosis and exchange segments. This recombines linked genes into assortments not seen in the parents.

Exercise 14   (Module 9.22)

What determines an individual's sex? Sex is generally determined by genes and chromo-somes, but the process of sex determination works differently in different species. Match each group of organisms below with their system of sex determination.

____ 1. Most plants, including peas, corn
(earthworms and land snails too)
____ 2. Humans, fruit flies
____ 3. Date palms, marijuana
____ 4. Some butterflies, birds, fishes
____ 5. Grasshoppers, roaches
____ 6. Ants, bees
____ 7. Wild strawberries

A. Females are ZW, males ZZ.
B. Females are diploid, males haploid.
C. Females are XX, males XO (one X).
D. Females are XX, males XY.
E. Sexes not separate; all individuals produce both eggs and sperm.

Exercise 15   (Modules 9.23 – 9.24)

Web/CD Activity 9F   *Sex-Linked Genes*
Web/CD Thinking as a Scientist   *How Is the Chi-Square Test Used in Genetic Analysis?*

Genes located on the sex chromosomes—called sex-linked genes—determine many traits unrelated to maleness or femaleness. Red-green color blindness is a recessive sex-linked trait in humans. After reading Modules 9.23 and 9.24, see if you can describe the inheri-tance of color blindness by filling in the blanks below.

The genes for normal color vision and red-green color blindness, like most human sex-linked traits, are carried on the [1]_____ chromosome. A capital letter C represents the [2]_____ allele for normal vision; a small c represents the color-blindness allele. A male with normal color vision has the genotype [3]_____. (Because these genes are carried on the X chromosome, their sym-bols are shown as superscripts on the letter X.) A color-blind male has the genotype [4]_____.

A color-blind male will transmit the allele for color blindness to all his [5]_____ but none of his [6]_____. This is because only his daughters inherit his [7]_____ chromosome, and only his [8]_____ chromosome is passed to all his sons. All the children of a color-blind male and a homozygous dominant female will have normal color vision. Their sons will inherit only the normal vision allele, but their daughters will be [9]_____ of the color-blindness allele, thus possessing the genotype [10]_____.

A heterozygous female carrier transmits the color-blindness allele to [11]_____ of her offspring. If she and a male with normal vision have chil-dren, [12]_____of their sons will be normal and [13]_____ will be color blind. [14]_____ of their daughters will be normal, because they in-herit at least one dominant allele from their [15]_____. But half these daugh-ters will be [16]_____ of the color-blindness trait, because they inherit the color-blindness allele from their mother.

Color blindness is much more common in men than in women. If a man inherits a single color-blindness allele from his [17]_____, the gene will be expressed and he will be color blind. Because a man has only one ^{18}_____chromosome, whatever genes it carries are seen in the man's phenotype. If a woman inherits just one color-blindness allele, she has relatively normal vision, because the dominant normal allele on her other *X* chromosome masks most of the effects of the color-blindness allele. For a woman to be color blind, she would have to inherit ^{19}_____ alleles from both her mother and her father, which is much less likely.

Testing Your Knowledge

Multiple Choice

1. How did Mendel's studies in genetics differ from earlier studies of breeding and inheritance?
   a. Mendel worked with plants; earlier studies used animals.
   b. Mendel was able to explain the "blending" hypothesis.
   c. Mendel's work was more quantitative.
   d. Mendel worked with wild species, not domesticated ones.
   e. Mendel found that offspring inherit characteristics from both parents.

2. A true-breeding fruit fly would be _____ for a certain characteristic.
   a. homozygous dominant
   b. homozygous recessive
   c. heterozygous
   d. Any of the above can be true-breeding.
   e. a or b

3. When looking at the inheritance of a single characteristic, Mendel found that a cross between two true-breeding peas (between purple and white, for example) always yielded a _____ in the F$_2$ generation.
   a. 1:1 phenotypic ratio
   b. 3:1 genotypic ratio
   c. 1:2:1 phenotypic ratio
   d. 3:1 phenotypic ratio
   e. 1:1 genotypic ratio

4. Alternative forms of genes for a particular characteristic are called
   a. homologous chromosomes.
   b. alleles.
   c. linked genes.
   d. genotypes.
   e. phenotypes.

5. A fruit fly has two genes for eye color, but each of its sperm cells has only one. This illustrates
   a. independent assortment.
   b. linked genes.
   c. pleiotropy.
   d. polygenic inheritance.
   e. segregation.

6. Mendel made some crosses where he looked at two characteristics at once—round yellow peas crossed with wrinkled green peas, for example. He did this because he wanted to find out
   a. how new characteristics originated.
   b. whether different characteristics were inherited together or separately.
   c. how plants and animals adapt to their environments.
   d. whether the characteristics influence each other—whether color affects degree of roundness, for example.
   e. Actually, Mendel never had a clear purpose in mind.

7. A pea plant with purple flowers is heterozygous for flower color. Its genotype is *Pp*. The *P* and *p* alleles in the pea plant's cells are located
   a. next to each other on the same chromosome.
   b. at corresponding locations on homologous chromosomes.
   c. on the *X* and *Y* chromosomes.
   d. some distance apart on the same chromosome.
   e. at different locations on homologous chromosomes.

8. When an individual has both I^A and I^B blood group alleles, both genes are expressed and the individual has group AB blood. This is an example of
   a. codominance.
   b. a dihybrid.
   c. pleiotropy.
   d. incomplete dominance.
   e. linked genes.

9. How many genes are there on one chromosome?
   a. one
   b. two
   c. hundreds
   d. thousands
   e. millions

10. Which of the following is *not* true of linked genes?
    a. They tend to be inherited together.
    b. They violate Mendel's principle of independent assortment.
    c. They are on the same chromosome.
    d. They can form new combinations via crossing over.
    e. They are relatively rare; most genes are unlinked.

11. Morgan and his students were able to map the relative positions of genes on fruit-fly chromosomes by
    a. coloring chromosomes with dyes and observing them under a microscope.
    b. scrambling the chromosomes and observing how the flies changed.
    c. crossing various flies and looking at the proportions of offspring.
    d. transplanting chromosomes from one fly to another.
    e. looking at crosses that showed independent assortment.

12. The sex chromosomes of a human female are ____. The sex chromosomes of a human male are ____.
    a. $XX \ldots XY$
    b. $YY \ldots XX$
    c. $XX \ldots YY$
    d. $XY \ldots XX$
    e. $YY \ldots XY$

13. Most sex-linked traits in humans are carried on the ____ chromosome, and the recessive phenotypes are seen most often in ____.
    a. $X \ldots$ women
    b. $X \ldots$ men
    c. $Y \ldots$ women
    d. $Y \ldots$ men

14. The most common lethal genetic disease in the United States is
    a. sickle-cell disease.
    b. cystic fibrosis.
    c. Huntington's disease.
    d. hemophilia.
    e. PKU.

15. Which of the following human genetic disorders is sex linked?
    a. hemophilia
    b. PKU
    c. cystic fibrosis
    d. sickle-cell disease
    e. all of the above

16. There are various procedures that can be used to detect genetic disorders before birth. Among the tests discussed in this chapter, _____ is the least invasive, while _____ carries the highest risk.
    a. chorionic villus sampling . . . amniocentesis
    b. ultrasound imaging . . . fetoscopy
    c. fetoscopy . . . chorionic villus sampling
    d. fetoscopy . . . amniocentesis
    e. chorionic villus sampling . . . ultrasound imaging

Essay

1. Explain why Gregor Mendel was able to figure out the principles of heredity, while many other investigators before (and some after) Mendel failed to do so.

2. If you flip two coins, the probability that you will get two heads is ¼, but the probability that you will get one head and one tail is ½. Explain why.

3. Why are organisms such as peas and fruit flies better subjects for genetics studies than human beings?

4. What determines a human's sex? Describe two other systems of sex determination in different organisms.

Applying Your Knowledge

Multiple Choice

1. A brown mouse is mated with a white mouse. All of their offspring are brown. If two of these brown offspring are mated, what fraction of their offspring will be white?
   a. all
   b. none
   c. ¼
   d. ½
   e. ¾

2. Suppose you wanted to know the genotype of one of the brown F_2 mice in question 1. The easiest way to do it would be to
   a. keep careful records of the parent mice.
   b. mate it with a brown mouse.
   c. mate it with a mouse of its own genotype.
   d. mate it with a white mouse.
   e. It can't be done.

3. Some dogs bark while trailing; others are silent. The barker gene is dominant, the silent gene recessive. The gene for normal tail is dominant over the gene for screw (curly) tail. A barker dog with a normal tail who is heterozygous for both characteristics is mated to another dog of the same genotype. What fraction of their offspring will be barkers with screw tails?
   a. ¾
   b. ⁹⁄₁₆
   c. ³⁄₁₆
   d. ¼
   e. ¹⁄₁₆

4. Two heterozygous tall pea plants with purple flowers are crossed. The probability that one of their offspring will have white flowers is ¼. The probability that one of their offspring will be short is ¼. What is the probability that one of their offspring will be short with white flowers?
   a. 0
   b. ¹⁄₁₆
   c. ⅛
   d. ¼
   e. ½

5. A young unmarried woman had a baby and wished to collect child support from the father. Her blood group is AB. The baby's blood group is A. There are two possible fathers: Jim is group A, and Michael is group O. Which man could be the father?
   a. either
   b. Jim
   c. Michael
   d. neither
   e. impossible to tell given this evidence

6. Which of the following illustrates pleiotropy?
   a. In fruit flies, the genes for scarlet eyes and hairy body are located on the same chromosome.
   b. Matings between earless sheep and long-eared sheep always result in short-eared offspring.
   c. Wheat kernels can range from white to red in color, a trait controlled by several genes.
   d. The human cystic fibrosis gene causes many symptoms, from respiratory distress to digestive problems.
   e. An individual with both I^A and I^B alleles has blood group AB.

7. When two gray-bodied fruit flies are mated, their offspring total 86 gray-bodied males, 81 yellow-bodied males, and 165 gray-bodied females. The allele for yellow body is
   a. sex-linked and dominant.
   b. not sex-linked and dominant.
   c. sex-linked and recessive.
   d. not sex-linked and recessive.
   e. impossible to say on the basis of this information.

8. In fruit flies, the allele for red eyes is dominant, and the allele for purple eyes is recessive. Normal gray body is dominant, and black body is recessive. A geneticist mated a heterozygous red-gray male with a purple-black female. She predicted that there would be four phenotypes of offspring in equal numbers, but she was wrong. Instead, 48% of the offspring were red-gray, 46% were purple-black, 3% were red-black, and 3% were purple-gray. She concluded that in the male, the red and gray genes were linked, and in the female the purple and black genes were linked. If this is the case, how would you account for the red-black and purple-gray offspring?
   a. This is an example of pleiotropy.
   b. Body color and eye color are quantitative characteristics.
   c. Crossing over during meiosis recombined the genes.
   d. This cross shows incomplete dominance at work.
   e. Segregation of alleles occurred during meiosis.

9. Red-green color blindness is a human recessive sex-linked trait. A man and a woman with normal vision have a color-blind son. What is the probability that their next child will also be a color-blind son?
   a. 0
   b. ⅛
   c. ¼
   d. ½
   e. ¾

10. On a pedigree tracing the inheritance of PKU, a horizontal line joins a black square and a half-black circle. What fraction of this couple's children would you expect to suffer from PKU?
    a. none
    b. ¼
    c. ½
    d. ¾
    e. all

11. Duchenne muscular dystrophy is caused by a sex-linked recessive allele. Its victims are almost invariably boys, who usually die before the age of 20. Why is this disorder almost never seen in girls?
    a. Sex-linked traits are never seen in girls.
    b. The allele is carried on the *Y* chromosome.
    c. Nondisjunction occurs in males but not in females.
    d. Males carrying the allele don't live long enough to be fathers.
    e. A sex-linked allele cannot be passed on from mother to daughter.

12. Which of the following would be most useful for preventing a particular genetic disorder?
    a. knowing how the allele causes its phenotypic effects
    b. being able to identify carriers
    c. a test that can determine whether a fetus suffers from the disorder
    d. knowing which chromosome bears the allele that causes the disorder
    e. tracing the trait back through parents and grandparents

Essay

1. Two apparently normal parents have a daughter who suffers from agammaglobulinemia, an inherited defect of the immune system, which is supposed to protect the body from infection. Use a Punnett square to show how two normal parents could have a child afflicted with an inherited disease. What are the parents' genotypes? The daughter's genotype? What is the probability that their second child will also have agammaglobulinemia?

2. A pea plant with purple flowers and green pods is crossed with a plant that has white flowers and yellow pods. All the offspring have purple flowers and green pods. If two of these F_1 peas are crossed, what phenotypes will be seen in the F_2 generation, and in what proportions?

3. Freckles is dominant, no freckles recessive. A man with freckles and a woman with no freckles have three children with freckles and one with no freckles. What are the genotypes of the parents and children?

4. The inheritance of flower color in snapdragons illustrates incomplete dominance: When a red snapdragon is crossed with a white one, all their offspring are pink. What offspring would be produced, in what proportions, if two of these pink snapdragons were crossed? What offspring would be produced, in what proportions, if a pink snapdragon was crossed with a white one?

5. A man whose blood group is A and a woman whose blood group is B have a son whose blood group is O. What are their genotypes? What is the probability of the couple's next child having blood group B?

6. Recall that some characteristics, such as skin color, appear to be controlled by several genes. This creates a continuum of variation. If this polygenic explanation for the inheritance of human skin pigmentation is correct, how do the skin colors of the following four individuals compare? Which of the couples could have children with the widest range of skin colors? Why? Couple 1: *aaBbCC* and *aaBbCC*. Couple 2: *AaBbCc* and *AaBbCc*.

7. In fruit flies, the allele for red eyes is dominant over the allele for pink eyes. Straight wings is dominant over curled wings. Imagine that a red-eyed, straight-winged fly that is heterozygous for both characteristics is mated with a fly with pink eyes and curled wings. Predict the offspring that would be produced by this cross (genotypes, phenotypes, and fraction of each) if these two genes were on different chromosomes.
   When a geneticist actually carried out this mating, the offspring were as follows: 49% red eyes and straight wings, 49% pink eyes and curled wings, 1% red eyes and curled wings, and 1% pink eyes and straight wings. Does this agree with your prediction? How would you explain these results?

8. Numerous fruit-fly matings show that the *h* allele for hairy body, the *b* allele for spineless bristles, and the *s* allele for striped body are all located on the same chromosome. The recombination frequency between alleles *h* and *b* is 4%. The recombination frequency between alleles *s* and *b* is 15%. Why are the recombination frequencies between *h* and *b* and between *s* and *b* different? The recombination frequency between alleles *h* and *s* is 10%. What is the order of these three genes on the chromosome?

9. In humans, the presence of a fissure (gap) in the iris of the eye (called "coloboma iridis") is due to a sex-linked recessive gene. Show how an apparantly normal couple could have a child with this condition. Is the affected child more likely to be a boy or a girl?

10. Imagine that you are a genetic counselor, and a couple that is planning to have children comes to you for advice. Diane's brother has hemophilia. There is no history of hemophilia in Craig's family. What is the probability that their child will have hemophilia? (Recall that hemophilia is caused by a sex-linked recessive allele.)

Extending Your Knowledge

1. Gregor Mendel and Thomas Hunt Morgan were two of the more interesting figures in the history of biology. You may want to seek out their biographies in the library or look them up online.

2. If you live near your biological family, try to figure out the patterns of inheritance of some of the characteristics mentioned in Module 9.8 among your parents and siblings.

3. Is there any record of a genetic disorder in your family? Which one? How is it inherited? If you are able to collect data from your family, you might find it interesting to construct a pedigree for this genetic disorder. Alternatively, you could construct a pedigree for one of the simple human genetic traits illustrated in this chapter.

Molecular Biology of the Gene

10

Most biologists would agree that the most significant biological discovery of the twentieth century was the discovery of the structure of the gene. At the beginning of the century Mendel's rules were rediscovered, and genes were traced to the chromosomes. Soon it was possible to map the locations of genes, and scientists started wondering what exactly genes were made of and how they shaped an organism. By mid-century, it was clear that DNA is the genetic material and that genes act by directing the synthesis of proteins. Soon researchers discovered the double helix structure of DNA and deciphered the genetic code by which DNA shapes the body. Then scientists learned how to make genes and move them from one organism to another. As the twenty-first century begins, biologists map entire genomes and use their knowledge of genetics to reshape organisms, fight disease, and trace evolution. This chapter describes the molecular biology of the gene and the discoveries that continue to enlarge our understanding of genes.

Organizing Your Knowledge

Exercise 1   (Introduction and Modules 10.1 – 10.3)

Web/CD Activity 10A   *The Hershey-Chase Experiment*
Web/CD Activity 10B   *Phage T2 Reproductive Cycle*
Web/CD Activity 10C   *DNA and RNA Structure*
Web/CD Activity 10D   *DNA Double Helix*

Review the discovery that DNA is the genetic material, and the structures of DNA and RNA. Then match each phrase on the right with the correct term(s) on the left. Note that some answers are used more than once, and some questions have multiple answers.

A. Adenine (A)	_____	1. The basic chemical unit of a nucleic acid
B. Base	_____	2. The "transforming factor" that alters pneumonia bacteria
C. Cytosine (C)	_____	3. The two kinds of nucleic acids
D. DNA	_____	4. The three parts of every nucleotide
E. *E. coli*	_____	5. A pair of these forms a "rung" in the DNA ladder
F. Double helix	_____	6. Used to "label" DNA and protein in experiments
G. Guanine (G)	_____	7. The component of a bacteriophage that enters the host cell
H. Hydrogen bond	_____	8. Two alternating parts that form the nucleic acid "backbone"
I. Radioactive isotope	_____	9. The four bases in DNA
J. Covalent bond	_____	10. The DNA base complementary to T
K. Bacteriophage	_____	11. A virus that attacks bacteria
L. Protein	_____	12. The substance a phage leaves outside its host cell
M. Nucleic acid	_____	13. Ribose in RNA and deoxyribose in DNA
N. Nucleotide	_____	14. Watson and Crick deduced the structure of this molecule
O. Centrifuge	_____	15. Attacked by herpesvirus
P. Phosphate	_____	16. The DNA base complementary to G
Q. Polynucleotide	_____	17. A bacterium attacked by T2 phages
R. RNA	_____	18. The sequence of these encodes DNA information
S. Sugar	_____	19. Eukaryotic chromosomes consist of this and DNA
T. Thymine (T)	_____	20. The overall shape of a DNA molecule

(list continued on next page)

U. Uracil (U)
V. Herpesvirus
W. Nerve cell

_____ 21. Links adjacent nucleotides in a polynucleotide chain
_____ 22. Machine used to separate particles of different weights
_____ 23. Links a complementary pair of bases together
_____ 24. The four bases in RNA
_____ 25. RNA base that is not in DNA
_____ 26. A polymer of nucleotides
_____ 27. Causes cold sores, chickenpox, and other diseases

Exercise 2   (Modules 10.2 – 10.3)

Web/CD Activity 10C   *DNA and RNA Structure*
Web/CD Activity 10D   *The DNA Double Helix*

Review the structure of DNA by labeling these diagrams. Include **nucleotide, polynucleotide, sugar (deoxyribose), phosphate group, sugar-phosphate backbone, pyrimidine bases, purine bases, thymine (T), adenine (A), guanine (G), cytosine (C), hydrogen
bond, complementary base pair,** and **double helix.**

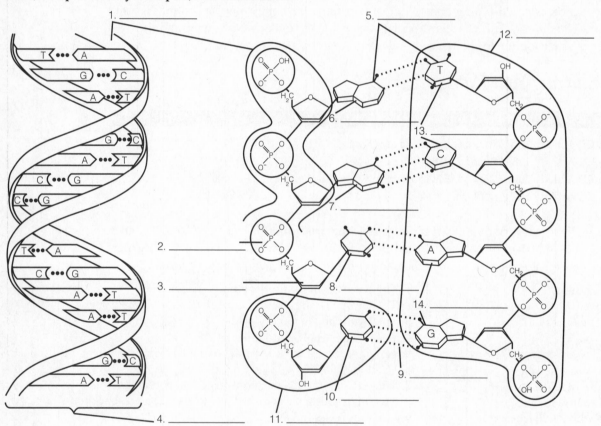

Exercise 3 (Module 10.4)

Reproduction and inheritance involve copying DNA instructions, so that they can be passed to the next generation. This process is carried out by DNA polymerases, enzymes that use each strand of the DNA helix as a template on which to build a complementary strand. Review DNA replication by completing the simplified diagrams below. The first diagram shows the parent DNA molecule; label the nucleotides in the right-hand strand. Complete and label the second diagram, so that it shows the parent strands separating and being used as templates. Label the third diagram, so that it shows two completed daughter molecules of DNA. Color the original DNA strands blue and the new strands red.

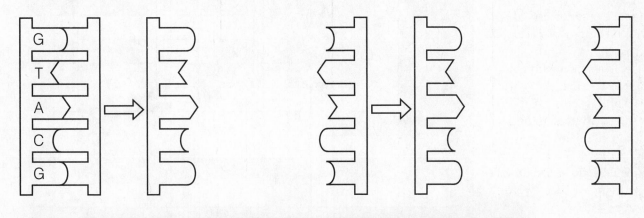

Exercise 4 (Module 10.5)

Web/CD Activity 10E   *DNA Replication: An Overview*

This module describes some of the ins and outs of DNA replication. Look at the diagrams carefully. Then see if you can match each of the numbers in the boxes on the diagram below with one of the lettered choices. Choices may be used more than once.

A. 5′ end of daughter strand
B. 3′ end of daughter strand
C. 5′ end of parental strand
D. 3′ end of parental strand
E. DNA polymerase
F. where DNA ligase will unite pieces

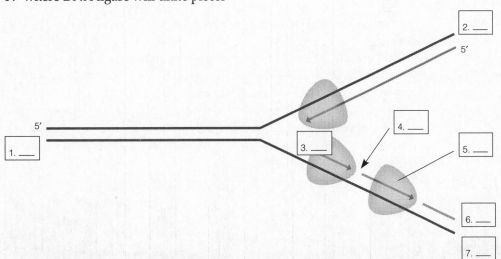

Exercise 5   (Modules 10.6 – 10.7)

Web/CD Activity 10F   *Overview of the Protein Synthesis*

In a cell, the genotype—genetic information in DNA—is expressed as phenotype in the form of proteins—structural proteins that shape the organism and enzymes that carry out metabolism. Review the relationship between genotype and phenotype by completing this crossword puzzle.

Across

3. A gene consists of hundreds or ____ of nucleotide bases.

6. The information in DNA specifies the synthesis of ____.

8. Garrod noted the gene-protein link in "inborn errors of ____."

9. Genetic instructions are written in 3-base "words" called ____.

10. An organism's expressed traits (what it looks like) make up its ____.

12. To make a protein, DNA information is first transcribed into ____.

13. The DNA language consists of a linear sequence of nucleotide ____.

14. The ____ is an organism's genetic makeup.

15. Phenotype is expressed via structural proteins, ____, and other proteins.

17. Making a polypeptide according to an RNA message is called ____.

Down

1. The base sequence of RNA is ____ to the DNA from which it is transcribed.

2. Genotype is the inheritable information encoded in ____.

4. Each codon in DNA and RNA specifies a certain ____ acid in a polypeptide.

5. Translation is conversion of an RNA message into a ____.

7. Transfer of information from DNA into an RNA molecule is called ____.

11. One ____ specifies how to build one polypeptide.

16. Using bread ____, Beadle and Tatum showed that a gene codes for an enzyme.

Web/CD Activity 10G   *Transcription*
Web/CD Activity 10H   *Translation*

These modules explain how the information in a gene is used to build a protein. Review the processes of transcription and translation by filling in the blanks below.

The first step in making a protein is transcription of a gene. This occurs in the [1]_____ of a eukaryotic cell. An enzyme called [2]_____ carries out the process of transcribing RNA from the DNA. It starts at a specific nucleotide sequence called a [3]_____, next to the gene. RNA polymerase attaches, and the two DNA strands separate. RNA polymerase moves along one strand, and as it does, RNA [4]_____ take their places one at a time along the DNA template. They hydrogen-bond with complementary bases, following the same pairing rules as in DNA—C with G, and U (replacing T in RNA) with A. As the RNA molecule elongates, it peels away from the DNA. Finally, RNA polymerase reaches the [5]_____, a base sequence that signals the end of the gene, and the enzyme lets go of the gene and the RNA molecule. In a prokaryote, the RNA transcribed from a gene, called [6]_____ (mRNA), can be used immediately in polypeptide synthesis. In a eukaryotic cell, the RNA is further modified, or [7]_____, before leaving the nucleus as mRNA. Extra nucleotides are added to the ends of the transcript, and noncoding regions called [8]_____ are removed. The remaining [9]_____ are spliced together to from a continuous coding sequence. The finished mRNA leaves the nucleus and enters the [10]_____, where translation takes place.

Translation of the "words" of the mRNA message into the [11]_____ sequence of a protein requires an interpreter—[12]_____ (tRNA)—which links the appropriate [13]_____ with each [14]_____ in the mRNA message. A tRNA molecule is a folded strand of RNA. At one end, a special [15]_____ attaches a specific amino acid. The other end of the tRNA molecule bears three bases called the [16]_____, which is complementary to a particular mRNA codon. During the translation process, the tRNA matches its amino acid with an mRNA codon.

[17]_____ are the "factories" where the information in mRNA is translated and polypeptide chains are constructed. A ribosome consists of protein and [18]_____ (rRNA). Each ribosome has a groove that serves as a binding site for mRNA. There are two binding sites for tRNA: The P site holds the tRNA carrying the growing [19]_____, while the A site holds a tRNA bearing the next amino acid.

Translation begins with initiation. An mRNA and a special [20]_____ tRNA bind to the ribosome and a specific mRNA codon, the [21]_____, where translation begins. The initiator tRNA generally carries the amino acid methionine (Met). Its anticodon UAC binds to the start codon, AUG. The initiator tRNA fits into the P site on the ribosome.

The next step in [22]_____ synthesis is elongation—adding amino acids to the growing chain. The anticodon of an incoming tRNA, carrying its amino acid, pairs with the mRNA codon at the open A site. With help from the ribosome, the polypeptide separates from its tRNA and forms a peptide bond with the [23]_____ attached to the tRNA in the A site. Then the "empty" tRNA in the P site leaves the ribosome, and the tRNA in the A site, with the polypeptide chain, is shifted to the P site. The mRNA and tRNA move as a unit, allowing the next codon to enter the A site. Another tRNA, with a complementary anticodon, brings its amino acid to the A site. Its amino acid is added to the chain, the tRNA leaves, and the complex shifts again. In this way, [24]_____ are added to the chain, one at a time.

Finally, a ^{25}_____ reaches the A site of the ^{26}_____, terminating the polypeptide. A stop codon causes the polypeptide to separate from the last tRNA and the ^{27}_____. The polypeptide folds up, and it may join with other polypeptides to form a larger ^{28}_____ molecule.

Exercise 7   (Module 10.15)

This module summarizes the key steps in the flow of genetic information from DNA to protein. Study the diagrams carefully, then label the numbered parts and processes.

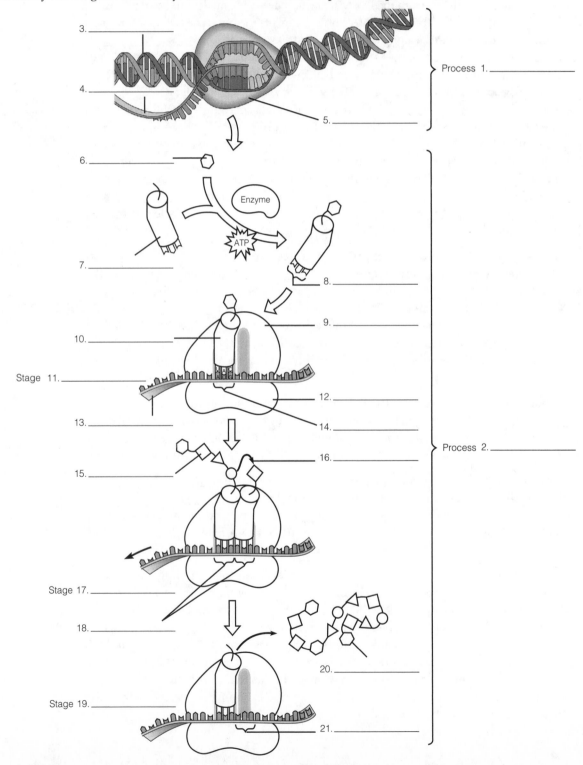

Exercise 8   (Modules 10.8 and 10.16)

These modules describe the genetic code, how biologists cracked the code, and how mutations change the meaning of the coded genetic message. Use the genetic code chart (Figure 10.8A in the textbook) to translate the following mRNAs into amino acid sequences and answer the questions.

mRNA nucleotide sequence:
(mRNA 1)

A U G C C A G A C A A U A U U A A G U G A

1. Amino acid sequence:

Mutation in mRNA:
(mRNA 2)

A U G C C A G A C C A U A U U A A G U G A

2. Amino acid sequence:
3. Number of bases changed in mRNA:
4. Type of mutation:
5. Number of amino acids changed:

Mutation in mRNA:
(mRNA 3; compare to 1)

A U G C C A G A C G A A U A U U A A G U G A

6. Amino acid sequence:
7. Number of bases changed in mRNA (look carefully!):
8. Type of mutation:
9. Number of amino acids changed (compared to mRNA 1):
10. Which mutation had the greatest effect and why?

Exercise 9   (Modules 10.17 – 10.21)

Web/CD Activity 10I   *Phage Lysogenic and Lytic Cycles*
Web/CD Activity 10J   *Simplified Reproductive Cycle of a DNA Virus*
Web/CD Activity 10K   *Retrovirus (HIV) Reproductive Cycle*
Web/CD Thinking as a Scientist   *Connection: Why Do AIDS Rates Differ Across the U.S.?*
Web/CD Thinking as a Scientist   *Connection: What Causes Infection in AIDS Patients?*

These modules describe the structures and life cycles of viruses. Match each phrase on the left with a term from the right. Some answers are used more than once.

____ 1. Consists of nucleic acid packaged in protein	A. RNA viruses
____ 2. Leads quickly to breaking open of host cell	B. Prophage
____ 3. Phage DNA inserted into bacterial chromosome	C. AIDS
____ 4. When virus "hides" as part of bacterial chromosome	D. Glycoprotein spikes
____ 5. Responsible for toxins of diphtheria, botulism	E. Virus
____ 6. Ineffective in fighting viruses	F. DNA
____ 7. This or DNA may be virus genetic material	G. Lytic cycle
____ 8. Cause of flu, colds, polio, mumps, AIDS	H. Vaccine
____ 9. Helps flu or mumps virus enter and leave host cell	I. Nucleus
____ 10. Used by mumps virus or HIV to attach to host receptors	J. Membranous
____ 11. Mumps virus reproduces here	envelope
____ 12. Mumps virus makes this and protein from RNA template	K. Bacteriophage
____ 13. Mumps virus gets envelope from this part of host cell	L. Provirus
____ 14. Herpesvirus reproduces here	M. HIV
____ 15. Genetic material of herpesvirus	N. Reverse transcriptase
____ 16. DNA of herpesvirus inserted into host cell DNA	O. Lysogenic cycle
____ 17. Can be used to prevent a viral disease	P. Retrovirus
____ 18. Virus protein coat	Q. White blood cell
____ 19. Genetic material of HIV	R. Prophage genes
____ 20. RNA virus that reproduces by means of DNA	S. Tobacco mosaic
____ 21. Enzyme that can make DNA from RNA template	T. Plasma membrane
____ 22. Form in which HIV "hides" in host cell	U. RNA
____ 23. Acquired immune deficiency syndrome	V. Cytoplasm
____ 24. Kind of cell infected by HIV	W. SARS
____ 25. Causes an African hemorrhagic fever	X. Ebola virus
____ 26. Virus like T2 that infects bacteria	Y. Antibiotic
____ 27. A deadly respiratory disease caused by an emerging virus	Z. Capsid
____ 28. Virus that causes AIDS	
____ 29. Rod-shaped plant virus	

Exercise 10   (Modules 10.21 – 10.22)

Bacteria have their own unique genetic characteristics and processes. They can transfer DNA via conjugation, transformation, and transduction. Match the following statements with one of the methods of bacterial DNA transfer. (Some statements are true of all methods of DNA transfer.)

____ 1. What happened in Griffith's experiment with pneumonia bacteria	A. Conjugation
____ 2. DNA may be integrated into chromosome of recipient cell	B. Transformation
____ 3. Taking up of DNA from the surrounding environment	C. Transduction
____ 4. Alters genetic makeup of recipient cell	D. All three of the
	above

_____ 5. Figure 5 below
_____ 6. Male and female cells joined by sex pili
_____ 7. Figure 7 below
_____ 8. Bacterial "mating"
_____ 9. Figure 9 below
_____ 10. Creates a recombinant cell
_____ 11. Transfer of genes by a bacteriophage
_____ 12. May involve transfer of genes by a plasmid
_____ 13. Usually controlled by a piece of DNA called an F factor
_____ 14. May transfer R plasmids which enable bacteria to resist antibiotics

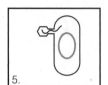
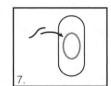
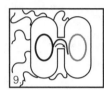

Testing Your Knowledge

Multiple Choice

1. In an important experiment, bacteriophages were allowed to infect bacteria. In the first trial, the phages used contained radioactive DNA, and radioactivity was detected in the bacteria. Next, other phages containing radioactive protein were allowed to infect bacteria, and no radioactivity was detected in the bacteria. When the experimenters compared the results of these two trials, they concluded that
    a. genes are made of DNA.
    b. bacteriophages can infect bacteria.
    c. DNA is made of nucleotides.
    d. genes carry information for making proteins.
    e. genes are on chromosomes.

2. An RNA or DNA molecule is a polymer made of subunits called
    a. bases.
    b. amino acids.
    c. nucleotides.
    d. nucleic acids.
    e. pyrimidines.

3. The information carried by a DNA molecule is in
    a. its amino acid sequence.
    b. the sugars and phosphates forming its backbone.
    c. the order of the bases in the molecule.
    d. the total number of nucleotides it contains.
    e. the RNA units that make up the molecule.

4. A gene is
    a. the same thing as a chromosome.
    b. the information for making a polypeptide.
    c. made of RNA.
    d. made by a ribosome.
    e. made of protein.

5. DNA replication occurs
    a. whenever a cell makes protein.
    b. to repair gene damage caused by mutation.
    c. before a cell divides.
    d. whenever a cell needs RNA.
    e. in the cytoplasm of a eukaryotic cell.

6. The flow of information in a cell proceeds
    a. from RNA to DNA to protein.
    b. from protein to RNA to DNA.
    c. from DNA to protein to RNA.
    d. from RNA to protein to DNA.
    e. from DNA to RNA to protein.

7. Which of the following is *not* needed for DNA replication?
    a. ribosomes
    b. DNA
    c. nucleotides
    d. enzymes
    e. All of the above are needed.

8. Which of the following processes occur(s) in the cytoplasm of a eukaryotic cell?
    a. DNA replication
    b. translation
    c. transcription
    d. DNA replication and translation
    e. translation and transcription

9. Beadle and Tatum showed that each kind of mutant bread mold lacked a specific enzyme. This experiment demonstrated that
   a. genes carry information for making proteins.
   b. mutations are changes in genetic information.
   c. genes are made of DNA.
   d. enzymes are required to repair damaged DNA information.
   e. cells need specific enzymes in order to function.

10. During the process of translation (polypeptide synthesis), ____ matches a nucleic acid codon with the proper amino acid.
    a. a ribosome
    b. DNA polymerase
    c. ATP
    d. transfer RNA
    e. messenger RNA

11. How does RNA polymerase "know" where to start transcribing a gene into mRNA?
    a. It starts at one end of the chromosome.
    b. Transfer RNA acts to translate the message to RNA polymerase.
    c. It starts at a certain nucleotide sequence called a promoter.
    d. The ribosome directs it to the correct portion of the DNA molecule.
    e. It looks for the AUG start codon.

12. When RNA is being made, the RNA base ____ always pairs with the base ____ in DNA.
    a. U . . . T
    b. T . . . G
    c. U . . . A
    d. A . . . U
    e. T . . . A

13. A mutagen is
    a. a gene that has been altered by a mutation.
    b. something that causes a mutation.
    c. an organism that has been changed by a mutation.
    d. the portion of a chromosome altered by a mutation.
    e. any change in the nucleotide sequence of DNA.

14. How do retroviruses, such as HIV, differ from other viruses?
    a. They are much simpler than other viruses.
    b. They contain DNA that is used as a template to make RNA.
    c. They can reproduce only inside of living cells.
    d. They contain nucleic acids that code for the making of proteins.
    e. They contain RNA that is used as a template to make DNA.

15. The primary difference between bacterial sex and sexual reproduction in plants and animals is that
    a. bacterial sex involves more than two individuals.
    b. bacterial sex does not involve genetic recombination.
    c. bacteria exchange RNA, not DNA.
    d. bacterial sex does not produce offspring.
    e. eggs and sperm are different, but bacterial gametes are all alike.

16. Sometimes a bacteriophage transfers a gene from one bacterium to another. This process is called
    a. transduction.
    b. conjugation.
    c. cloning.
    d. DNA splicing.
    e. transformation.

Essay

1. Sketch a short piece of a DNA molecule, five base pairs long. Use simple shapes to represent bases, sugars, and phosphates. Show proper base pairing, and label a nucleotide, a base, a phosphate group, a sugar, A, C, T, G, the double helix, and hydrogen bonds.

2. Explain why, in DNA, T pairs only with A and not with C or G.

3. Why does it take a group of three DNA nucleotides to specify one amino acid in a protein? Wouldn't it be simpler to have a one-to-one code, where one nucleotide specified one amino acid?

4. What is a mutation? What causes mutations? Why are most mutations harmful? Why aren't all mutations harmful?

5. Which type of mutation—a base substitution or a base deletion—is likely to have the greatest effect on the organism? Why?

6. Describe step by step, but in simple terms, the roles of mRNA, tRNA, ribosomes, and amino acids in making a polypeptide.

Applying Your Knowledge

Multiple Choice

1. Which of the following are arranged in the correct order by size, from largest to smallest?
   a. chromosome-gene-codon-nucleotide
   b. nucleotide-chromosome-gene-codon
   c. codon-chromosome-gene-nucleotide
   d. gene-chromosome-codon-nucleotide
   e. chromosome-gene-nucleotide-codon

2. A geneticist raised a crop of T2 bacteriophages in a medium containing radioactive phosphorus, so that the DNA of the bacteriophages was labeled with radioactivity. The labeled phages were then allowed to infect nonradioactive bacteria. In a few hours, these bacteria burst open, releasing many bacteriophages. Some of these phages contained labeled
   a. DNA.
   b. RNA.
   c. protein.
   d. all of the above.
   e. DNA and protein only.

3. A messenger RNA molecule for making a protein is made in the nucleus and sent out to a ribosome. The ribosome reads the mRNA message and makes a protein containing 120 amino acids. The mRNA consisted of at least how many codons?
   a. 30
   b. 40
   c. 120
   d. 360
   e. 480

4. The nucleotide sequence of a DNA codon is ACT. A messenger RNA molecule with a complementary codon is transcribed from the DNA. In the process of protein synthesis, a transfer RNA pairs with the mRNA codon. What is the nucleotide sequence of the tRNA anticodon? (Careful—this one is harder than it appears.)
   a. TGA
   b. UGA
   c. ACT
   d. TGU
   e. ACU

5. Imagine an error occurring during DNA replication in a cell, so that where there is supposed to be a T in one of the genes there is instead a G. What effect will this probably have on the cell?
   a. Each of its kinds of proteins will contain an incorrect amino acid.
   b. An amino acid will be missing from each of its kinds of proteins.
   c. One of its kinds of proteins might contain an incorrect amino acid.
   d. An amino acid will be missing from one of its kinds of proteins.
   e. The amino acid sequence of one of its kinds of proteins will be completely changed.

6. A cell is grown in a solution containing radioactive nucleotides, so that its DNA is labeled with radioactivity. It is removed from the radioactive solution and grown in a normal medium, so that any new DNA strands it makes will not be radioactive. In the normal medium, the cell replicates its DNA and divides. The two daughter cells also replicate their DNA and divide, producing a total of four cells. If a dotted line represents a radioactive DNA strand and a solid line represents a nonradioactive DNA strand, which of the following depicts the DNA of the four cells?

   a.

   b.

   c.

   d.

   e.

7. A particular ____ carry the information for making a particular polypeptide, but ____ can be used to make any polypeptide.
   a. gene and ribosome . . . a tRNA and an mRNA
   b. gene and mRNA . . . a ribosome and a tRNA
   c. ribosome and mRNA . . . a gene and a tRNA
   d. gene and tRNA . . . a ribosome and an mRNA
   e. tRNA and ribosome . . . a gene and an mRNA

8. A sequence of pictures of polypeptide synthesis shows a ribosome holding two transfer RNAs. One tRNA has a polypeptide chain attached to it; the other tRNA has a single amino acid attached to it. What does the next picture show?
   a. The polypeptide chain moves over and bonds to the single amino acid.
   b. The tRNA with the amino acid leaves the ribosome.
   c. The amino acid moves over and bonds to the polypeptide chain.
   d. The tRNA with the polypeptide chain leaves the ribosome.
   e. A third tRNA with an amino acid joins the pair on the ribosome.

9. A microbiologist analyzed chemicals obtained from an enveloped RNA virus (similar to a mumps virus) that infects monkeys. He found that the virus envelope contained a protein characteristic of monkey cells. Which of the following is the most likely explanation for this?
   a. The virus gets its envelope when it leaves its host cell.
   b. The virus forced the monkey cell to make proteins for its envelope.
   c. The virus has a lysogenic life cycle.
   d. The virus gets its envelope when it enters its host cell.
   e. The virus fools its host cell by mimicking its proteins.

10. At one point as a cell carried out its day-to-day activities, the nucleotides G A T were paired with the nucleotides C U A. This pairing occurred
    a. in a double-stranded DNA molecule.
    b. during translation.
    c. during transcription.
    d. when an RNA codon paired with a tRNA anticodon.
    e. It is impossible to say, given this information.

11. Which of the following does not take part in polypeptide synthesis?
    a. an exon
    b. mRNA
    c. an intron
    d. tRNA
    e. a ribosome

12. A microbiologist analyzed the DNA of *E. coli* before and after conjugation. She found that
    a. both cells lost some genes and gained others.
    b. both cells gained genes but lost none of their original genes.
    c. one cell lost genes, and the other gained genes.
    d. one cell gained genes, and the genes of the other were unchanged.
    e. the genes of both cells remained unchanged.

Essay

1. *E. coli* bacteria are used in many genetic studies. Type A *E. coli* can live on a simple nutrient medium, because they have all the genes necessary to produce the chemicals they need. Type V *E. coli* can live only on a nutrient medium to which a certain vitamin has been added, because they lack a gene that enables them to make this vitamin for themselves. It has been found that bacteria can absorb genes from other dead, ground-up bacteria. Describe an experiment using type A and type V *E. coli* to determine whether genes are made of protein or DNA.

2. It is possible to extract DNA from cells and analyze it to determine the relative amounts of the four DNA bases. The DNA of a goldfish contains more T and less G than human DNA, but in both goldfish and human DNA the amount of T is equal to the amount of A. Explain why.

3. Eric said to Renee, "The amino acid sequence of the proteins in your hair determines how curly or straight your hair will be." Renee replied, "I don't think that's right. Your genes determine whether your hair is curly or straight. That's why it's inherited." Who is right? Explain.

4. The DNA base sequence for a short gene is:

   TATGATACCTTGATAGCTATCTGATTG.

   What is the amino acid sequence of the polypeptide produced according to this DNA information? Use the genetic code chart (Figure 10.8A in the text) and your knowledge of transcription and translation to figure out the message.

5. A biochemist found that a bacterium produced an mRNA molecule consisting of 852 nucleotides and translated this mRNA into a polypeptide containing 233 amino acids. How many nucleotides in the mRNA message would actually be needed to carry the message for the polypeptide, and how many were "extras"? How would the bacterium know *which* nucleotides made up the message?

6. The virus that causes chickenpox can disappear for years and then reappear in a line of painful sores ("shingles") where a nerve cell passes through the skin. How can viruses go away and then reappear like this? Where are the viruses during the intervening period of time?

7. A gene can be removed from a eukaryotic cell and spliced into the DNA of a prokaryotic cell. The prokaryotic can transcribe the gene into mRNA and translate this mRNA into a polypeptide, but the polyeptide has an incorrect amino acid sequence, very different from the polypeptide normally produced by the eukaryotic cell. Why?

8. A mutant strain of *E. coli* bacteria will not grow unless they are supplied with the amino acid lysine. Another strain will not grow without a different amino acid, proline. When *E. coli* of the two strains are mixed, a few bacteria appear in the culture that are able to grow without either of the amino acids. Name and briefly describe three possible mechanisms that might account for this change.

Extending Your Knowledge

1. One of the most important things you can learn in your study of biology is how to keep from becoming infected with HIV, the AIDS virus. Information is available from college health centers, the CDC National STD AIDS Hotline, 1-800-342-AIDS, or this website: www.aidsinfo.nih.gov.

2. In recent years, there have been several major outbreaks of "childhood" viral diseases such as measles on college campuses. Most young children are vaccinated, so unvaccinated individuals are not likely to contract these diseases "naturally." Unfortunately, these viruses can have serious effects, especially on adults. Do you know whether you have been vaccinated for measles, mumps, chickenpox, and so on? Did you have any of these diseases when you were younger? Your family or family doctor may have information about this. If you are not immune to these diseases, you may want to ask your doctor or college health center about vaccination.

The Control of Gene Expression

11

One of the wonders of biology is how a fertilized egg, or zygote—a single, relatively generalized cell—can develop into an adult human being—a multicellular organism consisting of hundreds of kinds of specialized cells. The zygote contains all the genes needed to construct and direct the activities of an entire human being. As the zygote divides, different groups of cells differentiate, taking on different shapes and different functions, finally forming the cooperating tissues and organs of the adult body. Muscle cells, bone cells, nerve cells—experiments suggest that these different kinds of cells all contain identical sets of genes. What makes the cells different is that different subsets of genes are expressed in each kind of cell. A sophisticated regulatory system switches genes on and off, responding to environmental changes and shaping the differentiation of various cell lineages. Scientists are learning to control and reverse this differentiation process, producing clones for reproduction and therapy. Occasionally, genes escape from their normal controls, producing birth defects or cancer. This chapter considers gene regulation, a subject of much mystery and intensive research.

Organizing Your Knowledge

Exercise 1 (Module 11.1)

Web/CD Activity 11A   *The* lac *Operon in* E.Coli

A prokaryote can respond to changes in its environment by turning genes on and off. In bacteria, genes are grouped, with control sequences called operators and promoters, into clusters called operons. The *lac* and *trp* operons are two such gene clusters in the bacterium *E. coli*. Study the diagrams in Module 11.1, and then match each of the components of the *lac* and *trp* operon systems with its function.

lac operon:

_____ 1. Regulatory gene
_____ 2. Repressor protein + lactose
_____ 3. Repressor protein without lactose
_____ 4. RNA polymerase
_____ 5. Promoter
_____ 6. Operator
_____ 7. Operon genes
_____ 8. Enzymes

A. Keeps RNA polymerase from attaching to promoter and transcribing genes
B. Transcribes genes into RNA for protein synthesis
C. Repressor protein attaches here
D. Use lactose
E. Information for making repressor protein
F. Where RNA polymerase starts transcribing genes
G. Allows RNA polymerase to transcribe genes
H. Information for making enzymes that use lactose

trp operon:

_____ 1. Regulatory gene
_____ 2. Repressor protein + tryptophan
_____ 3. Repressor protein without tryptophan
_____ 4. RNA polymerase
_____ 5. Promoter
_____ 6. Operator
_____ 7. Operon genes
_____ 8. Enzymes

A. Keeps RNA polymerase from attaching to promoter and transcribing genes
B. Transcribes genes into RNA for protein synthesis
C. Repressor protein attaches here
D. Make tryptophan
E. Information for making repressor protein
F. Where RNA polymerase starts transcribing genes
G. Allows RNA polymerase to transcribe genes
H. Information for making enzymes that make tryptophan

Exercise 2   (Modules 11.2 – 11.3)

In eukaryotic organisms, cells become specialized—differentiated—for different functions. It has been demonstrated in several kinds of plants and animals that differentiated cells retain a complete set of genes for making all the specialized cells in the whole organism. Each differentiated cell *has* all the genes, but *different* genes are *active* in different kinds of cells. Check your understanding of this concept by coloring in the boxes below to show which genes you think would be active in each of the cell types.

	Stomach Gland Cell	Hair Follicle Cell	Stem Cell (Becomes Red Blood Cell)
Hemoglobin gene	☐	☐	☐
Keratin (hair protein) gene	☐	☐	☐
Citric acid cycle enzyme gene	☐	☐	☐
Digestive enzyme gene	☐	☐	☐
Insulin gene	☐	☐	☐

Exercise 3   (Modules 11.4 – 11.6)

Web/CD Thinking as a Scientist   *How Do You Design a Gene Expression System?*

Complete this crossword puzzle to review the roles of DNA packing and protein activators in gene expression.

Across

2. The color pattern of a ____ cat reflects the influence of chromosome inactivation.

4. Scientists think most eukaryotic regulatory proteins act as ____.

7. One X chromosome in each of a woman's cells is ____.

10. The DNA-histone beaded fiber is further wrapped into a tight ___ fiber.

12. Nucleosomes may control gene ____ by limiting access to DNA.

13. The DNA supercoil is further wrapped and folded into a ____.

15. In eukaryotes, many ____ proteins interact with DNA and one another to turn genes on and off.

17. DNA fits into a cell because of a system of folding, or ____.

18. In eukaryotes, structural and regulatory genes are ____.

19. The twisted DNA further coils into a ____ with a diameter of 200 nm.

Down

1. The DNA-histone complex looks like "____ on a string."

3. DNA is wound around small proteins called ____.

5. The first step in initiating gene transcription is binding of activators to sites called ____.

6. A transcription ____ may be involved in turning on a eukaryotic gene.

8. A ____ is a complex of DNA wrapped around eight histone molecules.

9. DNA packing seems to control gene expression at the ____ stage.

11. Activators trigger RNA ____ to begin transcription.

14. An activator may help position RNA polymerase on a gene's ____.

16. In most eukaryotic cells, most ____ are not expressed.

Exercise 4 (Modules 11.7 – 11.8)

Web/CD Activity 11B *Gene Regulation in Eukaryotes*

In eukaryotes, gene expression is also regulated after transcription of genes into mRNA and during and after translation of mRNA into protein. Review these processes by matching each of the processes on the left (listed in order of occurrence) with a description on the right.

_____ 1. First step in RNA splicing
_____ 2. Second step in RNA splicing
_____ 3. Alternative RNA splicing
_____ 4. Selective breakdown of mRNA
_____ 5. Inhibition of translation
_____ 6. Activation of finished protein
_____ 7. Selective breakdown of proteins

A. Altering a protein to form an active final product
B. Retaining or breaking down mRNA molecules, controlling how much they are translated
C. Action of inhibitors that may block synthesis of protein from mRNA message
D. Joining exons in different ways to produce more than one polypeptide from a single gene
E. Removal of noncoding introns from RNA
F. Joining of exons to produce mRNA
G. Retaining or breaking down proteins, depending on cell's needs

Exercise 5   (Module 11.9)

Web/CD Activity 11C   *Review: Gene Regulation in Eukaryotes*

This module summarizes the mechanisms that regulate gene expression in eukaryotes. After reviewing the module, match each of the mechanisms of regulation with the stage of gene expression at which it acts. Choose from **mRNA breakdown, DNA unpacking and changes, cleavage/modification/activation, protein breakdown, addition of cap and tail, transcription, splicing, translation,** and **flow through nuclear envelope.**

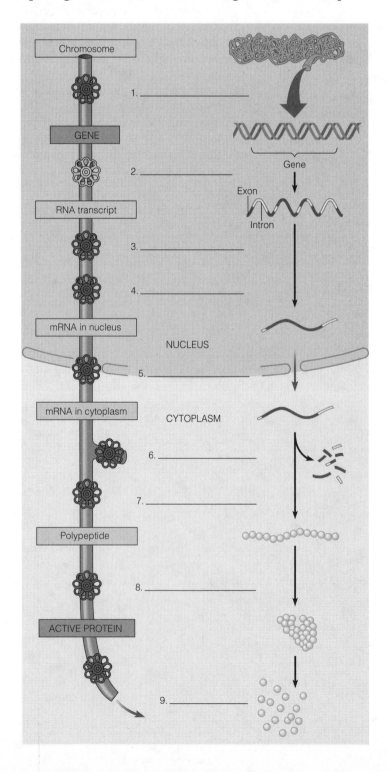

Exercise 6   (Modules 11.10 – 11.12)

Cloning experiments show that differentiated cells retain all of their genetic potential. Stem cells of embryos and adults are able to differentiate into many kinds of cells—useful for reproduction and therapy. Review cloning and stem cells by matching each phrase with a term from the list on the right.

_____ 1. Partially differentiated cells present in mature animals

_____ 2. Producing genetically identical organisms for agriculture, research, or saving endangered species

_____ 3. Cells that give rise to all specialized cells in the body

_____ 4. The process of cell specialization

_____ 5. Growing cells for replacement or repair of damaged or diseased organs

_____ 6. Genetically identical organisms

_____ 7. Replacing the nucleus of an egg or zygote with a nucleus from a differentiated cell

A. reproductive cloning
B. nuclear transplantation
C. differentiation
D. embryonic stem cells
E. therapeutic cloning
F. adult stem cells
G. clones

Exercise 7   (Modules 11.13 – 11.15)

Web/CD Activity 11D   *Development of Head-Tail Polarity*
Web/CD Thinking as a Scientist   *Can the "Head" Gene Be Regulated to Alter Development?*
Web/CD Activity 11E   *Signal-Transduction Pathway*

Powerful new methods of molecular biology have enabled scientists to explore how gene regulation controls animal development. Researchers have found that one of the first events in fruit-fly development is a sequence of changes that determine which end of an egg will develop into the fly's [1]_____ and which will develop into the [2]_____. One of the first [3]_____ that "turns on" in the egg cell codes for a protein that leaves the egg and signals nearby cells in its follicle, or egg chamber. The signal protein binds to a specific [4]_____ in the membrane of the target cell, which in turn activates a series of relay proteins in the target cell. The last relay protein activates a [5]_____ factor that triggers transcription of a specific target cell gene. The mRNA produced is then [6]_____ into a protein.

Via this mechanism, the egg cell signals the follicle cells. This [7]_____ genes in the follicle cells, and they produce proteins that signal back to the egg cell. One of the egg cell's responses is to localize a type of [8]_____ at one end of the egg cell. This marks where the fly's [9]_____ will develop. The other end of the egg will become the [10]_____. Similar processes establish the other body axes and thus the layout of the overall body plan of the fly.

After the egg is fertilized, the zygote is transformed into a multicellular embryo by repeated [11]_____. Translation of the "head" mRNA creates a regulatory [12]_____ that is concentrated mostly in the head of the developing fly. This protein in turn influences other genes, and these act to subdivide the embryo into repeating subunits called body [13]_____.

Protein products of the axis-forming and segment-forming genes now trigger another round of gene [14]_____ that shapes the details of the fly. Master control genes called [15]_____ genes determine what body parts—antennae, legs, and so on—will develop in each segment. Every homeotic gene contains a sequence of about 180 nucleotides called a [16]_____, which is translated into a protein segment of about 60 [17]_____. This homeobox polypeptide segment binds to specific [18]_____ base sequences, enabling the homeotic protein to turn groups of genes on and off during development.

[19]_____ in homeotic genes produce flies with spectacular changes in body structure, such as extra pairs of wings, or heads bearing legs instead of antennae. Such changes attracted the attention of researchers, which led to the discovery of homeotic genes and study of their important role in [20]_____. Further research has shown that homeoboxes are virtually identical in every eukaryotic organism studied so far—from yeast to plants to chickens to humans. For example, the chromosomes of a mouse and a fruit fly bear similar homeotic genes in the same [21]_____, and the genes are ordered from [22]"_____" to [23]"_____" on the chromosomes. These great similarities suggest that many of the base sequences in genes that control development arose [24]_____ in the history of life and have changed little since.

Exercise 8   (Modules 11.16 – 11.19)

Web/CD Activity 11F   *Connection: Causes of Cancer*

Review the causes and mechanisms of cancer by filling in the blanks in the following story.

In the United States, lung cancer kills about 174,000 people per year. Long the most common kind of cancer in men, lung cancer recently passed breast cancer to become the most frequent cancer in women as well. There has been a 136% increase in lung cancer deaths among women over a 20-year period.

Cancer is uncontrolled multiplication of cells. The [1]_____ control system is responsible for regulating cell [2]_____ and [3]_____. Sometimes cells escape from this control and multiply wildly. The cells form abnormal masses called tumors, which displace nearby normal tissues and can spread through the body.

Because the growing tumors block breathing passages, the first symptoms of lung cancer are usually coughing and difficulty breathing. The tumorous masses show up on chest X-rays, and usually a small sample of lung tissue is taken to examine the tumor cells.

What causes lung cancer? Cancer-causing agents are called [4]_____. Radiation, such as X-rays and UV light, are known to cause some cancers, but most are caused by chemicals. Carcinogens in tobacco smoke appear to be the major cause of lung cancer. An increase in cigarette smoking over the last century has been paralleled by a rise in lung cancer rates. Tobacco has also been linked to other forms of cancer, such as cancer of the throat and stomach.

Scientists have discovered much about the cellular mechanisms of cancer by studying cancers caused by viruses in humans and other animals. They were surprised to find that cancer-causing viruses carry cancer-causing genes, called [5]_____, as part of their genome. When the viruses insert their genes into the chromosomes of a host cell, the cancer-causing genes are inserted as well. Even more surprising, researchers found that oncogenes are simply altered versions of genes normally found in all cells. These genes, called [6]_____, usually code for proteins called [7]_____ factors—which normally stimulate cell [8]_____—or for other proteins that affect growth factors. What appears to cause most cancers is a [9]_____ in a proto-oncogene in a body (somatic) cell that changes it into an oncogene.

Changes in genes whose products *inhibit* cell division—so-called [10]_____ genes—can also contribute to the development of cancer. The normal products of proto-oncogenes and tumor-suppressor genes are involved in [11]_____-transduction pathways. Normally, the [12]_____ product of a proto-oncogene (such as one called *ras*) might act to conduct a signal from a [13]_____ factor to the interior of a cell, stimulating [14]_____ to occur. When the proto-oncogene mutates into an oncogene, it might produce a protein that increases cell division even in the [15]_____ of the growth factor.

A tumor-suppressor protein might act as a [16]_____ factor, which normally promotes production of a protein that blocks cell division. A mutation in a tumor-suppressor gene (such as *p53*) could produce a defective transcription factor. In this case, the inhibitory protein would not be transcribed, causing an [17]_____ rate of cell division.

Evidence suggests that more than one somatic [18]_____ is needed to produce cancer. These mutations are cumulative and passed to all of a cell's [19]_____. In most cases studied so far, changes in both proto-oncogenes and tumor-suppressing genes seem to be necessary for a full-fledged tumor to develop. Recent research has also brought to light a different kind of tumor-suppressor gene, like the *BRCA1* gene implicated in breast cancer, whose function is to [20]_____ damaged DNA. If this gene is defective, cancer is more likely to develop.

Exercise 9   (Module 11.20)

Web/CD Activity 11F   *Connection: Causes of Cancer*

After reading this module, match each of the following human cancers with the carcinogen(s) implicated as its major cause. (Answers may be used more than once or not at all.)

_____ 1. Lung
_____ 2. Colon and rectum
_____ 3. Breast
_____ 4. Prostate
_____ 5. Bladder
_____ 6. Kidney
_____ 7. Pancreas
_____ 8. Melanoma (skin)
_____ 9. Liver

A. Ultraviolet light
B. X-Rays
C. Cigarette smoke
D. Estrogen
E. Alcohol; hepatitis viruses
F. High dietary fat

Testing Your Knowledge

Multiple Choice

1. Your muscle and bone cells are different because
   a. they contain different sets of genes.
   b. they are differentiated.
   c. they contain different operons.
   d. different genes are switched on and off in each.
   e. they contain different histones.

2. Operons enable bacteria to
   a. function in frequently changing environments.
   b. increase their genetic diversity
   c. correct mutations that might interfere with their genetic instructions.
   d. differentiate.
   e. mutate and evolve more rapidly.

3. If the nucleus of a frog egg is destroyed and replaced with the nucleus of an intestine cell from a tadpole, the egg can develop into a normal tadpole. This demonstrates that
   a. intestine cells are fully differentiated.
   b. there is little difference between an egg cell and an intestine cell.
   c. an intestine cell possesses a full set of genes.
   d. intestine cells are not differentiated.
   e. frogs can regenerate lost parts.

4. DNA packing—the way DNA is folded into chromosomes—affects gene expression by
   a. controlling access to DNA.
   b. positioning related genes near each other.
   c. protecting DNA from mutations.
   d. enhancing recombination of genes.
   e. allowing "unpacked" genes to be eliminated from the genome.

5. The genes that malfunction in cancer normally
   a. control RNA transcription.
   b. are responsible for organizing DNA packing.
   c. code for enzymes that transcribe DNA.
   d. are not present in most body cells unless inserted by a virus.
   e. regulate cell division.

6. In most eukaryotic cells, most genes are not expressed. This suggests that most eukaryotic regulatory proteins act as
   a. exons.
   b. repressors.
   c. introns.
   d. enhancers.
   e. activators.

7. After an mRNA molecule is transcribed from a eukaryotic gene, portions called ____ are removed and the remaining ____ are spliced together to produce an mRNA molecule with a continuous coding sequence.
   a. operators . . . promoters
   b. exons . . . introns
   c. silencers . . . enhancers
   d. introns . . . exons
   e. promoters . . . operators

8. Which of the following mechanisms of gene regulation operates after mRNA transcription but before translation of mRNA into protein?
   a. mRNA splicing
   b. DNA packing
   c. repressors and activators
   d. protein degradation
   e. all of the above

9. Homeobox polypeptide segments
   a. serve as histones, facilitating DNA packing.
   b. bind to DNA and activate or repress gene transcription.
   c. are vastly different in different organisms.
   d. act as enzymes, carrying out important chemical reactions.
   e. carry out mRNA splicing in the cell nucleus.

10. In a eukaryote, a repressor protein may block gene expression by binding to a DNA site called
    a. an operon.
    b. a histone.
    c. an enhancer.
    d. a promoter.
    e. a silencer.

11. Which of the following has the most potential for use in therapeutic cloning?
    a. adult stem cells
    b. somatic cells
    c. gametes
    d. cancer cells
    e. embryonic stem cells

12. Gene expression in animal development seems to be regulated largely by
  a. controlling gene packing and unpacking.
  b. controlling the transcription of genes into mRNAs.
  c. controlling the translation of mRNAs into protein.
  d. selectively eliminating certain genes from the genome.
  e. selectively breaking down certain proteins so they cannot function.

13. Which of the following is a known or likely carcinogen?
  a. ultraviolet light
  b. cigarette smoke
  c. hormones
  d. X-rays
  e. all of the above

14. Which of the following is the first thing that happens when a signal molecule acts on a target cell?
  a. A transcription factor acts on the DNA.
  b. The signal molecule binds to the DNA.
  c. A new protein is made in the target cell.
  d. A specific gene is transcribed.
  e. The signal molecule binds to a receptor.

Essay

1. In the proper growth medium, a single cell from a Boston fern can be stimulated to grow into an entire plant. (This is how nurseries propagate many houseplants.) What does this signify with regard to cellular differentiation in plants?

2. What are introns and exons? Discuss three possible biological functions of introns.

3. What is a homeotic gene? Why does a mutation in a homeotic gene have a much more drastic effect on the organism than a mutation in other genes?

4. Very similar homeoboxes control gene transcription in the DNA of frogs, flies, and humans. Why do biologists consider this significant?

5. Briefly explain how genes control development of the head-to-tail axis of a fruit-fly embryo.

6. Describe the changes in a cell that can make the cell become cancerous.

7. What is the difference between reproductive and therapeutic cloning?

8. If a person wishes to avoid cancer, what factors in the environment should he or she try to avoid? What dietary and health habits would you recommend?

Applying Your Knowledge

Multiple Choice

1. When a certain bacterium encounters the antibiotic tetracycline, the antibiotic molecule enters the cell and attaches to a repressor protein. This keeps the repressor from binding to the bacterial chromosome, allowing a set of genes to be transcribed. These genes code for enzymes that break down the antibiotic. This set of genes is best described as
  a. an exon.
  b. regulatory genes.
  c. an operon.
  d. a homeobox.
  e. a nucleosome.

2. A genetic defect in humans results in the absence of sweat glands in the skin. Some men have this defect all over their bodies, but in women it is usually expressed in a peculiar way. A woman with the defect typically has small patches of skin with sweat glands and other patches where sweat glands are lacking. This pattern suggests the phenotypic effect of
  a. a mutation.
  b. chromosome inactivation.
  c. RNA splicing.
  d. an operon.
  e. the homeobox.

3. A bacterium makes the amino acid glycine or absorbs it from its surroundings. A biochemist found that glycine binds to a repressor protein and causes the repressor to bind to the bacterial chromosome, "turning off" an operon. If this is like other operons, the genes of this operon probably code for enzymes that
  a. control bacterial cell division.
  b. break down glycine.
  c. produce glycine.
  d. cause the bacterium to differentiate.
  e. manufacture the repressor protein.

4. In humans, the hormone testosterone enters cells and binds to specific proteins, which in turn bind to specific sites on the cells' DNA. These proteins probably act to
   a. help RNA polymerase transcribe certain genes.
   b. alter the pattern of DNA splicing.
   c. stimulate protein synthesis.
   d. unwind the DNA so that its genes can be transcribed.
   e. cause mutations in the DNA.

5. It is possible for a cell to make proteins that last for months; hemoglobin in red blood cells is a good example. However, many proteins are not this long-lasting. They may be degraded in days or even hours. Why do cells make proteins with such short lifetimes if it is possible to make them last longer?
   a. Most proteins are used only once.
   b. Most cells in the body live only a few days.
   c. Cells lack the raw materials to make most of the proteins they need.
   d. Only cancer cells, which can keep dividing, contain long-lasting proteins.
   e. This enables cells to control the amount of protein present.

6. Dioxin, produced as a by-product of various industrial chemical processes, is suspected of causing cancer and birth defects in animals and humans. It apparently acts by entering cells and binding to proteins, altering the pattern of gene expression. The proteins affected by dioxin are probably
   a. enzymes.
   b. DNA polymerases.
   c. transcription factors.
   d. enhancers.
   e. nucleosomes.

7. Which of the following would be most likely to lead to cancer?
   a. multiplication of a proto-oncogene and inactivation of a tumor-suppressor gene
   b. hyperactivity of a proto-oncogene and activation of a tumor-suppressor gene
   c. inactivation of a proto-oncogene and multiplication of a tumor-suppressor gene
   d. inactivation of both a proto-oncogene and a tumor-suppressor gene
   e. hyperactivity of both a proto-oncogene and a tumor-suppressor gene

8. A cell biologist found that two different proteins with largely different structures were translated from two different mRNAs. These mRNAs, however, were transcribed from the same gene in the cell nucleus. What mechanism below could best account for this?
   a. Different systems of DNA unpacking could result in two different mRNAs.
   b. A mutation might have altered the gene.
   c. Exons from the same gene could be spliced in different ways to make different mRNAs.
   d. The two mRNAs could be transcribed from different chromosome puffs.
   e. Different chemicals activated different operons.

9. Researchers have found homeotic genes in humans, but they are not yet certain how these genes shape the human phenotype. Considering the functions of homeotic genes in other animals, which of the following is most likely to be their function in humans?
   a. determining skin and hair color
   b. regulating cellular metabolic rate
   c. determining head and tail, back and front
   d. determining whether an individual is male or female
   e. regulating the rate and timing of cell division

Essay

1. Mutations sometimes affect operons. Imagine a mutation in the regulatory gene that produces the repressor of the *lac* operon in *E. coli*. The altered repressor is no longer able to bind to the operator. What effect will this have on the bacterium?

2. Describe how three different types of cells in your body are specialized for different functions. How do their differences reflect differences in gene expression? Suggest a gene that might be active in each of the cells but none of the others. Suggest a gene that might be active in all the cells. Suggest a gene that is probably not active in any of the cells.

3. A biochemist was studying a membrane-transport protein consisting of 258 amino acids. She found that the gene coding for the transport protein consisted of 3561 nucleotides. The mRNA molecule from which the transport protein was transcribed contained 1455 nucleotides. What is the minimum number of nucleotides needed to code for the protein? How can the protein be transcribed from an mRNA that is larger than necessary? How can this mRNA be made from a gene that is so much larger?

4. Explain how, in a eukaryotic cell, a gene on one chromosome might affect the expression of a gene on a different chromosome. How might a gene in a certain cell affect expression of a gene in a different cell?

5. A certain kind of leukemia can be caused by a virus, a chemical, or radiation. Explain how these different factors can all trigger identical forms of cancer in the same kind of tissue.

6. Researchers have suggested that it might be possible to clone an extinct woolly mammoth (an Ice-Age elephant) from a tissue sample obtained from a mammoth frozen in a glacier. Describe how this might be done.

Extending Your Knowledge

1. Pregnant women are advised to avoid unneeded exposure to chemicals from the environment, such as pesticides, alcohol, tobacco smoke, and medications. Does this chapter suggest how these substances might alter fetal development?

DNA Technology and Genomics

12

In recent years—within the lifetimes of most students reading this textbook—biologists have learned to construct genes and use them to alter organisms. A new toolkit consisting of enzymes, radioactive isotopes, cell culture techniques, and "gene machines" enables us to find genes, read them, compare them, insert them into new organisms, and harvest their products. Genetic engineering has revolutionized our understanding of genes and promises to transform our lives in many ways, from medicine to agriculture. The Human Genome Project has mapped our DNA, giving us new insights into human evolution, development, health, and disease. But our new understanding of genes and our ability to manipulate them are accompanied by some risks and numerous social and ethical questions. This chapter concerns DNA technology—its methods, promise, and challenges.

Organizing Your Knowledge

Exercise 1 (Modules 12.1 – 12.5)

Web/CD Activity 12A   *Restriction Enzymes*
Web/CD Activity 12B   *Cloning a Gene in Bacteria*
Web/CD Thinking as a Scientist   *How Can Antibiotic-Resistant Plasmids Transform* E. Coli?

"Clone," "sticky ends," "vector"—Gene engineers have their own lingo. Start building your genetic engineering vocabulary by matching each of the terms on the right with a description on the left. (Each answer is used only once.)

_____ 1. Used to cut DNA at a specific location for splicing
_____ 2. Using organisms or their components to make useful products
_____ 3. Direct manipulation of genes for practical purposes
_____ 4. Making multiple copies of gene-sized pieces of DNA
_____ 5. A small piece of bacterial DNA used for gene transfer
_____ 6. DNA transcribed from RNA
_____ 7. Used to "splice" pieces of DNA
_____ 8. A set of techniques for combining genes from different sources
_____ 9. A virus that attacks bacteria; used to clone genes
_____ 10. An organism used to clone genes
_____ 11. A collection of DNA fragments
_____ 12. Specific location where an enzyme cuts DNA
_____ 13. An enzyme used to make DNA from an RNA master
_____ 14. DNA in which genes from different sources are combined
_____ 15. A gene carrier

A. *E. coli*
B. Genetic engineering
C. Reverse transcriptase
D. Recombinant DNA technology
E. Vector
F. Restriction enzyme
G. Complementary DNA (cDNA)
H. Biotechnology
I. Plasmid
J. DNA ligase
K. Recombinant DNA
L. Gene cloning
M. Bacteriophage
N. Restriction site
O. Genomic library

Exercise 2   (Modules 12.1 – 12.9)

Web/CD Activity 12A   *Restriction Enzymes*
Web/CD Activity 12B   *Cloning a Gene in Bacteria*
Web/CD Thinking as a Scientist   *How Can Antibiotic-Resistant Plasmids Transform* E. Coli?

We can engineer bacteria to produce desired genes or proteins. Continue your review of techniques used to cut, splice, clone, and locate genes by filling in the blanks.

Gene engineers use plasmids as [1]_____ to insert genes into bacteria or eukaryotic cells. Imagine that you wanted to build a bacterium capable of making large quantities of human growth hormone (HGH), which is a protein. Your first step would be to obtain the [2]_____ that codes for HGH. One way to do this is to use a [3]_____ enzyme to cut up all the DNA in a human cell. The enzyme recognizes short nucleotide [4]_____ within DNA molecules and cuts the DNA at these [5]_____ sites. Restriction enzymes cut the two DNA strands unevenly, leaving single-stranded ends that can hydrogen-bond with complementary single-stranded [6]"_____ends." A restriction enzyme can chop up a cell's DNA into thousands of restriction fragments, each consisting of a few genes or parts of genes.

The next step in making human growth hormone is isolating a supply of [7]_____ to use as vectors, for carrying the DNA fragments into bacteria. These are treated with the same restriction enzyme that was used to cut up the human DNA, producing plasmids with sticky ends that are [8]_____ to sticky ends of the human DNA fragments.

Now the human DNA fragments are mixed with plasmids. The sticky ends on the fragments base-pair with the sticky ends on the plasmids, but these connections are weak and temporary. An enzyme called DNA [9]_____, which naturally functions in DNA [10]_____, is used to catalyze the formation of covalent bonds between adjacent nucleotides in the DNA fragments and plasmids. This forms [11]_____ DNA, a DNA molecule with a new, human-made combination of genes.

In the next step, each recombinant plasmid is added to a bacterium. Under specific conditions, a bacterium will take up the plasmid DNA from solution by the process of [12]_____. The bacterium, with its recombinant plasmid, is allowed to grow and reproduce on a nutrient medium. Each bacterium replicates its own DNA and the plasmid DNA and then divides repeatedly. Each bacterium grows into a colony of identical cells, all containing the recombinant DNA. This production of multiple copies of the genes is called gene [13]_____. Cloning all the different DNA fragments obtained from the human cell produces a genomic [14]_____ of DNA segments. Because this procedure does not target a particular gene (at least so far), it is called the [15]"_____ approach" to gene cloning. (DNA fragments can also be spliced into phages, [16]_____ that infect bacteria. The phages reproduce in bacteria to produce libraries of cloned DNA pieces.)

There are a lot of genes to sort through in a library produced from an entire eukaryotic genome. Plus, eukaryotic genes contain noncoding [17]_____, which must be removed before bacteria can read them. Why not simplify things by looking only at genes expressed in the particular cell you are interested in? You can identify the genes expressed in a particular kind of cell, by using the enzyme [18]_____ transcriptase to produce intron-free genes from the mRNA produced in the cell. If you wanted to obtain a human growth hormone gene, the place to start would be a cell from the pituitary gland, where HGH is made. In the cell, the HGH gene (and others) is transcribed into RNA. Enzymes then remove the introns from the RNA and splice the remaining [19]_____ together to make mRNA. If you extract the mRNA from a cell, and add reverse transcriptase (obtained from a [20]_____), the enzyme transcribes a strand of DNA along the mRNA. The RNA is then broken down, and a second DNA strand is synthesized, producing double-stranded DNA. The artificial genes produced this way lack introns, so they are more manageable than the original genes. They also can be transcribed and translated by [21]_____, which lack the ability to deal with introns. Complementary DNA produced in this fashion is cut and pasted into plasmids, using restriction enzymes and ligase, and then cloned in bacteria. There are many mRNA molecules in a pituitary gland cell, so this method also produces a [22]_____ library; however, this library is smaller than a library produced by cutting up the entire genome, because it is limited to the genes actually [23]_____ in a pituitary gland cell.

At this point you have isolated and cloned the HGH gene, but where is it? A genomic library can consist of thousands of bacterial colonies. The bacteria of one of the colonies contains the HGH gene, but which one? One way to look for the gene is to look for HGH, its [24]_____ product. But usually you look for the gene itself, a search that is made easier by using a nucleic acid [25]_____. If you know part of the amino acid sequence of HGH, you can work backward to figure out the probable nucleotide sequence of part of the HGH gene—AAGTGTAG, for example. Now you can produce an artificial RNA (or DNA) molecule with a complementary base sequence—[26]_____, in this case. This complementary molecule is labeled with a [27]_____ isotope or a fluorescent dye, and is called a probe because it can be used to find the gene. To find the bacterial clone that holds the gene, DNA is obtained from each colony of bacteria and treated to separate the DNA strands. The probe is then mixed with the DNA strands, and it hydrogen-bonds only with the recombinant DNA with a complementary base sequence—the HGH gene. Once you have figured out which bacterial colony in the library contains the HGH gene, you can grow these bacteria in larger amounts.

In actual practice, finding particular base sequences is much simplified by using a DNA [28]_____—a grid of thousands of DNA sequences. If cDNA from a sample is [29]_____ to a piece of DNA on the microarray, it sticks, causing a change that makes the DNA glow. Microarrays are used in medical diagnosis—to analyze the pattern of gene [30]_____ in breast cancer cells, for example.

The final step in engineering bacteria to produce human growth hormone is to grow the bacteria in large quantities (usually done in large vats) and extract the protein. The bacteria will manufacture the protein on command if you have spliced the proper control sequences into your recombinant plasmids. Now it is only necessary to collect and purify the protein (and get approval from the Food and Drug Administration!) to start treating patients with recombinant DNA HGH.

Gene-cloning procedures like these have been used to modify organisms to produce a variety of protein products. For example, *E. coli* can be grown in large quantities and induced to [31]_____ protein products into the medium where they live. They have been engineered to make [32]_____ such as insulin, taxol, a substance used to treat ovarian [33]_____ , and proteins from pathogens that are put into [34]_____ that protect us from infectious diseases. (Also, in medicine, recombinant DNA techniques are used to track down and identify [35]_____ , such as HIV, and to diagnose and target treatment for cancers. Genetically modified viruses can be used in vaccines to protect from disease without causing harm.)

Sometimes it works better to use eukaryotic cells rather than bacteria to produce a particular protein. [36]_____ cells are the easiest eukaryotes to grow in large quantities, and they also have the ability to take up DNA from the environment and [37]_____ it into their genomes. Yeast cells are currently being used to produce interferons (used to treat cancer and infections) and hepatitis B vaccine. Finally, some proteins can only be produced by modifying cells from mammals. Many proteins secreted by mammalian cells are glycoproteins—proteins with chains of [38]_____ attached to them. Only mammalian cells can make them properly. Researchers are starting to explore modifying the gene products of whole animals or plants rather than cells grown in cell cultures. For example, adding a human gene to the genome of a sheep might enable us to obtain the gene's protein product from the animal's [39]_____ !

Exercise 3 (Modules 12.1 – 12.5, 12.8 – 12.10)

Web/CD Activity 12A *Restriction Enzymes*
Web/CD Activity 12B *Cloning a Gene in Bacteria*
Web/CD Activity 12C *Gel Electrophoresis of DNA*
Web/CD Thinking as a Scientist *How Can Antibiotic-Resistant Plasmids Transform* E. Coli?

Can you visualize how recombinant DNA techniques are used? Test your visual memory by matching each of the diagrams (or parts of diagrams) below with one of the following processes: **isolating plasmid from** *E. coli;* **extracting DNA from a eukaryotic cell; obtaining copies of gene and protein from cloned bacteria; separating DNA fragments via gel electrophoresis; cutting DNA with restriction enzyme; using a DNA microarray to find a base sequence; joining plasmid and DNA fragment using DNA ligase; cloning recombinant DNA; using reverse transcriptase to make cDNA; using a nucleic acid probe to find a gene; inserting a plasmid into a bacterium via transformation; and mixing plasmids and DNA fragments with sticky ends.**

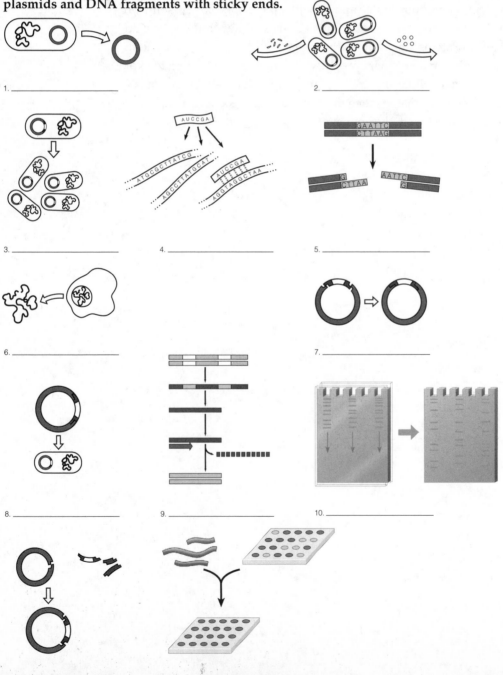

1. _____

2. _____

3. _____

4. _____

5. _____

6. _____

7. _____

8. _____

9. _____

10. _____

11. _____

12. _____

Exercise 4   (Modules 12.10 – 12.12)

Web/CD Activity 12C   *Gel Electrophoresis of DNA*
Web/CD Activity 12D   *Analyzing DNA Fragments Using Gel Electrophoresis*
Web/CD Thinking as a Scientist   *How Can Gel Electrophoresis Be Used to Analyze DNA?*
Web/CD Activity 12E   *Connection: DNA Fingerprinting*

Gel electrophoresis and restriction fragment analysis are powerful techniques used to separate, identify, and compare patterns of gene fragments from DNA samples, producing what are popularly called DNA fingerprints. These methods are widely used in forensics to compare crime scene evidence with tissues of victims and suspects. They are also used to trace family relationships (as in paternity cases), to spot genetic diseases, and to identify the remains of soldiers and victims of disasters (such as the World Trade Center 9/11 victims).

Imagine that you have been called upon to investigate a rape/homicide. Samples of the rapist's DNA have been recovered from the victim. Meanwhile, the police have taken two suspects into custody— Sam and Joe. Both were seen with the victim the night of the crime, but both say they are innocent. Your job is to compare the perpetrator's DNA (obtained from the victim) with samples of Sam's and Joe's DNA, to determine whether either of them should be tried for the crime. In this highly simplified example, you will chop up each of the DNA samples with a restriction enzyme that cuts at the CCGG restriction site. Start by drawing a line showing where each DNA molecule is cut. (Hint: Remember, DNA strands have 5′ and 3′ ends. Read the top DNA strand left to right and the bottom strand right to left. Note that CCGG is *not* the same as GGCC!)

Perpetrator: ATGCCGGTACATTAGTAGCCGGCATTTGAACGATCGTAATAAATGGCA
(Crime   TACGGCCATGTAATCATCGGCCGTAAACTTGCTAGCATTATTTACCGT
scene)

Sam: ATGCTGGTACATTAGTAGCCGGCATTTGAACGATCGTAATAAATGGCA
              TACGACCATGTAATCATCGGCCGTAAACTTGCTAGCATTATTTACCGT

Joe: TGCCATTTATTACGATCGTTCAAATGCCGGCTACTAATGTACCGGCAT
              ACGGTAAATAATGCTAGCAAGTTTACGGCCGATGATTACATGGCCGTA

1. Where does the restriction enzyme cut each piece of DNA? Show above.

2. In how many places does the enzyme cut each sample?

3. How many restriction fragments are produced from each sample?

4. Now show how each collection of restriction fragments will move through the electrophoresis gel. (Do the largest or smallest fragments move fastest and farthest?)

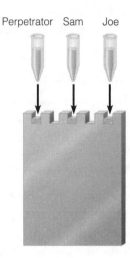

Perpetrator   Sam   Joe

In actual practice, only a few selected portions of the DNA from each individual would be examined using radioactive probes to identify specific nucleotide sequences, or markers, among the bands on the gel. But in our example, we will just look at the overall pattern of bands. This is the moment of truth.

5. How would you describe the three DNA fingerprints?

6. Which, if either, of the two suspects should be tried for rape and murder? Or should both suspects be set free?

Exercise 5   (Module 12.13)

Why not simply correct genetic defects by altering an affected individual's defective genes? Gene therapy is a good idea, but it has proved difficult to apply. List three technical and three ethical issues raised by gene therapy.

1.   Three technical issues:

2.   Three ethical issues:

Exercise 6 (Module 12.14)

PCR is used to amplify small DNA samples, to produce larger amounts of DNA for study or analysis. State four sources of DNA that might need to be amplified by the PCR method.

1.

2.

3.

4.

Exercise 7 (Module 12.15 – 12.17, 12.20)

Web/CD Activity 12F   *The Human Genome Project: Human Chromosome*

The science of genomics is giving us new information about the genomes of humans and other species. This understanding will help us answer questions about genome organization, regulation of genes, growth and development, and evolution. In addition, mapping the human genome is yielding new insights into genetic defects and disease. Review genomics and the Human Genome Project by matching each phrase on the left with a term from the list on the right. (Each answer is used only once.)

_____	1. All the proteins encoded by a genome	A. Genome
_____	2. The company that mapped the human genome using a "shotgun" approach	B. Repetitive DNA
		C. HGP
_____	3. Protective segment of repetitive DNA at the end of a chromosome	D. Transposon
_____	4. About 97% of human nucleotide sequences	E. Proteome
_____	5. A map that figures out the overlap of fragments between markers	F. Celera
_____	6. This animal has a genome roughly the same size as a human	G. Prokaryote
_____	7. Pedigree analysis helps produce this low-resolution map	H. Physical map
_____	8. This plant has about the same number of genes as a human	I. Noncoding DNA
_____	9. Eukaryote with a rather small, simple genome	J. *Drosophila*
_____	10. Nucleotide sequence that can move around the genome	K. *Arabidopsis*
_____	11. Many genomes are known for this kind of organism	L. Yeast
_____	12. Nucleotide sequences present in many copies in the genome	M. Linkage map
_____	13. All the genes of an organism	N. Mouse
_____	14. An effort to map the entire human genome	O. Telomere
_____	15. Its genome makes it seem like "a little person with wings"	P. Human
_____	16. Its gene count was much smaller than was expected at the beginning of the HGP	

Exercise 8   (Modules 12.18 – 12.19)

Web/CD Activity 12G   *Connection: Application of DNA Technology*

Web/CD Activity 12H   *Connection: DNA Technology and Golden Rice*

Genetically modified organisms promise great benefits, but also present some hazards and ethical questions. Review some of the techniques, uses, and potential problems associated with GMOs by completing this crossword puzzle.

Across

2. Wild and genetically modified plants might exchange genes via their _____.

4. In matters regarding genetically modified organisms, _____ risk is probably unattainable.

5. Transgenic chickens can make foreign proteins in their _____.

11. Genetically modified crops are monitored by the U.S. Department of _____.

13. Genetically modified crop plants can resist pests or _____.

15. _____ rice makes a vitamin A precursor not normally found in rice.

18. Genetically modified microorganisms are _____ so they cannot survive outside the laboratory.

19. To prevent gene transfer between GM and wild plants, GM plants can be engineered so they cannot _____.

20. Researchers use a _____ from a soil bacterium to transfer genes into plant cells.

Down

1. There is concern that some people might be _____ to proteins in GM food plants.

3. An organism with a gene from another species is called a _____ organism.

6. _____ is the abbreviation for "genetically modified organism."

7. Some people are concerned that genetically modified plants might give rise to _____ that are difficult to control.

8. Strict government _____ are designed to protect us from genetically modified microorganisms.

9. A transgenic animal can be created by _____ DNA into a fertilized egg.

10. Transgenic mammals can secrete useful proteins in their _____.

12. In the year 2000, 130 countries agreed on a _____ Protocol that requires identification of GM organisms.

14. The U.S. _____ and Drug Administration regulates biotechnology in medicine.

16. A genetically _____ organism has aquired one or more genes by artificial means.

17. A transgenic animal has genes from at least _____ parents.

Testing Your Knowledge

Multiple Choice

1. Comparing whole sets of genes, especially among different organisms, is called
   a. transgenics.
   b. genomics.
   c. recombinant DNA technology.
   d. molecular biology.
   e. genetics.

2. Which of the following has been produced by genetically modified microorganisms?
   a. human insulin
   b. human growth hormone
   c. cancer drugs
   d. growth factor for burn treatment
   e. all of the above

3. There are thought to be about ____ genes in a human cell.
   a. 23
   b. 46
   c. 2000
   d. 30,000
   e. 3 billion

4. The genome of a human is
   a. very similar to the genome of a mouse.
   b. much more complex than the genome of a mouse.
   c. much larger than the genome of a mouse.
   d. completely different from the genome of a mouse.
   e. much better understood than the genome of a mouse.

5. A genetic marker is
   a. a place where a restriction enzyme cuts DNA.
   b. a chart that traces the family history of a genetic trait.
   c. a nucleotide sequence near a particular gene.
   d. a radioactive probe used to find a gene.
   e. an enzyme used to cut DNA.

6. In recombinant DNA experiments, ____ is used to cut pieces of DNA, and ____ joins these segments to form recombinant DNA.
   a. a restriction enzyme . . . DNA ligase
   b. a transposon . . . a restriction enzyme
   c. a plasmid . . . DNA ligase
   d. DNA ligase . . . a restriction enzyme
   e. a transposon . . . a plasmid

7. A genomic library is
   a. where you look to find out how to make recombinant DNA.
   b. a listing of the known nucleotide sequences for a particular species.
   c. all the genes contained in one kind of cell.
   d. a collection of cloned DNA pieces from an organism's genome.
   e. a place where one can obtain DNA samples from various species.

8. A nucleic acid probe might be used to
   a. insert genes into a host cell.
   b. make DNA for gene cloning.
   c. splice pieces of DNA.
   d. cut pieces of DNA down to manageable size.
   e. find a nucleotide sequence.

9. It is sometimes necessary to genetically engineer mammalian cells to produce proteins because they
   a. can produce larger quantities of protein than bacteria.
   b. can read eukaryotic genes, and bacteria cannot.
   c. can add sugars to make glycoproteins, and bacteria cannot.
   d. are easier to grow than bacteria.
   e. can be induced to secrete proteins into their environment.

10. Which of the following is cited as a possible risk of genetically modified crop plants?
    a. allergic reactions
    b. hybridization with wild relatives
    c. creation of new pests that might be hard to control
    d. all of the above
    e. GM crops present no risks

11. Electrophoresis is used to
    a. separate fragments of DNA.
    b. clone genes.
    c. cut DNA into fragments.
    d. match a gene with its function.
    e. amplify small DNA samples to obtain enough for analysis.

Essay

1. How might transposons harm an organism? How might they contribute to the evolution of a species?

2. Explain how DNA segments can be cut and spliced together to produce recombinant DNA. How do the segments "find" each other and stick together? How is recombinant DNA then cloned to produce multiple copies of the gene?

3. Explain why bacteria and yeast are often used as "factories" for gene products.

4. Describe some uses of recombinant DNA technology in agriculture.

5. Describe some uses of recombinant DNA technology in medicine.

6. What are some potential risks of genetically engineered organisms being accidentally or purposefully released into the environment? What kinds of safety measures guard against accidental release? Are you concerned about these risks? Why or why not?

7. Discuss some of the ethical questions raised by recombinant DNA technology. What is the most difficult ethical question concerning human gene therapy?

8. Describe how restriction fragment analysis is done: How is DNA cut into fragments? How are the fragments separated? What does the DNA fingerprint look like, and why? Why do different people have different DNA fingerprints?

9. As we learn more about the human genome, how might individual DNA data be misused?

Applying Your Knowledge

Multiple Choice

1. A geneticist found that a particular nucleotide sequence was found on different chromosomes in different mouse skin cells. This suggested that
   a. transformation was occurring in some skin cells.
   b. transposons were moving around.
   c. the cells were engaging in conjugation.
   d. plasmids were transferring genes from cell to cell.
   e. the mice were transgenic animals.

2. Connection of "sticky ends" of restriction fragments is
   a. the opposite of separation of fragments via gel electrophoresis.
   b. similar to a nucleic acid probe locating a base sequence.
   c. carried out by DNA ligase.
   d. used to make complementary DNA (cDNA).
   e. allows the polymerase chain reaction (PCR) to copy DNA.

3. A mule is a cross been a horse and a donkey. Why do you think a mule is not considered a "genetically modified organism"?
   a. Because horses and donkeys are so much alike.
   b. Because a mule only inherits horse genes.
   c. Because a mule is produced naturally, not by artificial means.
   d. Because half the genes come from each parent.
   e. Actually, a mule is considered a genetically modified organism.

4. Because eukaryotic genes contain introns, they cannot be translated by bacteria, which lack RNA splicing machinery. If you want to engineer a bacterium to produce a eukaryotic protein, you can synthesize a gene without introns (if you know the nucleotide sequence) or
   a. alter the bacteria used so that they can splice RNA.
   b. use a molecular probe to find a gene without introns.
   c. work backward from mRNA to a piece of DNA without introns.
   d. use a phage to insert the desired gene into a bacterium.
   e. use a restriction enzyme to remove introns from the gene.

5. When recombinant plasmids are added to *E. coli*, not all the bacteria take them in. This makes it difficult to spot the bacteria that actually contain the plasmids. For this reason, R plasmids are often used as gene vectors in recombinant DNA experiments. Using R plasmids would allow experimenters to find the bacteria with recombinant DNA easily by
   a. using a molecular probe.
   b. killing off the bacteria without the R plasmids.
   c. feeding them a special diet.
   d. exposing the bacteria to restriction enzymes.
   e. watching what happens when the bacteria reproduce.

6. Scientists wished to create an organism capable of breaking down several kinds of toxic wastes, so they combined the genes of several bacteria to create a single "superbacterium." They probably did not need to use which of the following in creating the superbacterium?
   a. nucleic acid probes
   b. gel electrophoresis
   c. plasmids
   d. restriction enzymes
   e. DNA ligase

7. A crop scientist spliced genes for disease resistance into Ti plasmids and then treated tomato plants with the plasmids. Some parts of some plants resisted the disease, but most of the plants eventually died. The researcher could increase his chances for success by
   a. treating single cells and cloning whole plants from the cells.
   b. using molecular probes to figure out where to put the genes.
   c. using bacteriophages rather than Ti plasmids to introduce the genes.
   d. inserting the genes into the cells of the tomato plants with a needle.
   e. employing reverse transcriptase to get the genes into the plants.

8. A molecular biologist used a virus to introduce a gene coding for a certain enzyme into mouse cells. Most of the mouse cells were able to make the enzyme, but most of them lost the ability to make some other protein (different ones in different cells), and many died. Which of the following best explains these results?
   a. The viruses caused the mouse cells to become diseased.
   b. The viruses transferred genes from one mouse cell to another.
   c. The viruses inserted the enzyme gene into mouse cell genes.
   d. The viruses activated transposons, which disrupted other genes.
   e. The enzyme acted as a restriction enzyme, cutting up mouse DNA.

9. DNA fingerprints were used to determine whether Sam could be the father of Becky's baby. Which of the following would show that Sam is not the father? If ____ genetic fingerprint showed some bands not in ____ genetic fingerprint.
   a. Sam's . . . the baby's
   b. Becky's . . . the baby's
   c. the baby's . . . Sam's
   d. the baby's . . . Becky's
   e. the baby's . . . Sam's or Becky's

10. Archaeologists unearthed a human skull with a small dried fragment of the scalp still attached. They extracted a tiny amount of DNA from the scalp tissue. How could they obtain sufficient DNA for an analysis of the ancient man's genes?
    a. subject the DNA to electrophoresis
    b. use a molecular probe
    c. use reverse transcriptase
    d. use the polymerase chain reaction
    e. subject the DNA to restriction enzymes

11. There is about 6,000 times as much DNA in a human cell as in an *E. coli* cell but only about 6 times as many genes. Why?
    a. A human cell has much more noncoding DNA.
    b. The DNA is much more tightly coiled in a human cell.
    c. Most of the genes in a human cell are turned off.
    d. *E. coli* genes are less efficient than human genes.
    e. Human genes are much smaller than *E. coli* genes.

12. Researchers wanted to find out which genes were expressed in muscle tissue, so they ground up several pounds of hamburger (muscle!) and extracted the DNA and RNA from the cells. Their next step should be to
    a. amplify the DNA sample via the polymerase chain reaction (PCR).
    b. cut up the DNA with a restriction enzyme.
    c. amplify the RNA sample via the polymerase chain reaction (PCR).
    d. locate the desired genes with a nucleic acid probe.
    e. use the RNA and reverse transcriptase to make complementary DNA (cDNA).

Essay

1. Describe how you would go about genetically engineering a bacterium to produce human epidermal growth factor (EGF), a protein used in treating burns. Use the following: DNA ligase, *E. coli*, plasmids, genetic code chart, restriction enzyme, machine for synthesizing a gene, description of the amino acid sequence of EGF, and glassware and equipment for growing and handling bacteria and extracting protein.

2. In a "shotgun" experiment, all the DNA in a cell is cut up and cloned, producing a genomic library of DNA fragments, each fragment in a different bacterial colony. Why do you think some researchers question the safety of the shotgun approach?

3. A researcher is searching for the bacterial clone containing a particular cloned gene. She knows that part of the nucleotide sequence of the gene is ATGGCTATC. Explain how she might find the bacteria that contain the gene.

4. A microbiologist developed a strain of *E. coli* that were easily killed by sunlight and whose diet required two unusual amino acids not normally found outside the laboratory. Why would such a bacterium be useful in recombinant DNA work?

5. Manuel, a political activist, "disappeared" during the reign of a dictator. After the dictator was overthrown, human remains thought to be those of Manuel were found buried in a prison compound. A sample of DNA was extracted from the remains and subjected to restriction fragment analysis, along with DNA samples from Manuel's parents. The patterns of restriction fragments are shown below. Could the remains be those of Manuel, the missing activist? Explain your answer.

6. A certain genetic disorder results from the lack of a blood enzyme that is secreted by bone marrow cells. A second disease occurs when nerve cells are unable to produce a particular enzyme. Which of these disorders would be a better candidate for gene therapy and why?

Extending Your Knowledge

1. Have you used any products produced by means of DNA technology? Think about drugs (HGH, insulin), foods made from GM organisms, vaccines, and hospital diagnostic tests.

2. Hardly a day goes by without a story about DNA technology appearing in the newspaper. For the next few days, look for examples of DNA technology in the news. You may see articles about new drugs, cancer treatment, genetic diseases, transgenic crop plants, mapping of the human genome, clarification of evolutionary relationships, and crimes solved (or prisoners exonerated) using recombinant DNA techniques. Do you understand these articles better after studying recombinant DNA technology? What did the authors omit that you would like to know?

How Populations Evolve

13

Charles Darwin made sense out of biology. He looked at what many others had seen and had an idea that no one had ever thought of—evolution through the mechanism of natural selection. This idea is as important to biology as the atom is to chemistry. Natural selection explains both the diversity of life and its adaptation to the environment. Copernicus and Galileo discovered where we fit in the universe. Einstein showed us our place in space and time. Darwin revealed our relationship to the living world. It all started on an ocean voyage, when Darwin was only 22 years old. This chapter starts with Darwin's voyage of discovery and explains where that voyage has since taken us in our understanding of life.

Organizing Your Knowledge

Exercise 1   (Module 13.1 – 13.2)

Web/CD Activity 13A   *Darwin and the Galápagos Islands*
Web/CD Activity 13B   *The Voyage of the* Beagle: *Darwin's Trip Around the World*
Web/CD Thinking as a Scientist   *How Do Environmental Changes Affect a Population?*

Charles Darwin was not the first person to ponder the origin of species. Match each of the following with his place in unraveling the history of life. Don't focus on names and dates, but rather on how ideas about the origin and history of life have changed over the centuries.

A. Darwin   _____ 1. Ancient Greek who believed living things have changed

B. Lyell   _____ 2. Greek philosopher who believed species to be fixed

C. Wallace   _____ 3. Fossils led this Frenchman to believe the Earth is old

D. Anaximander   _____ 4. Proposed that acquired characteristics may be inherited

E. Aristotle   _____ 5. Believed in gradual geological change, not catastrophes

F. Lamarck   _____ 6. Asserted that populations grow faster than resources

G. Buffon   _____ 7. Wrote *The Origin of Species*, explaining "descent with modification"

H. Malthus   _____ 8. Conceived a theory of evolution almost identical to Darwin's

Exercise 2   (Modules 13.3)

Web/CD Activity 13B   *The Voyage of the* Beagle: *Darwin's Trip Around the World*

Review fossils by completing this crossword puzzle.

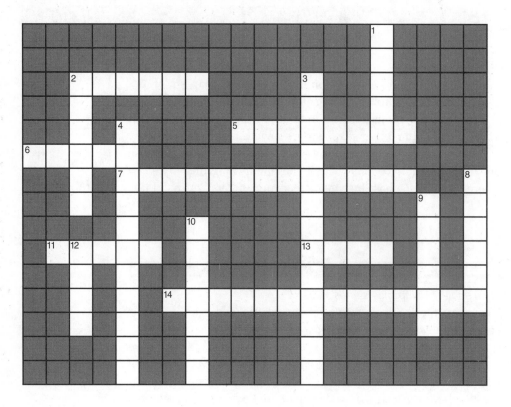

Across

2. Hard parts, such as bones, teeth, and ____, fossilize most easily.
5. A series of fossils shows how mammals evolved from ____.
6. Paleontologists recently found fossils of an ancient ____ with hind legs.
7. Sometimes a whole organism is preserved, if ____ does not occur.
11. Insects and other small animals are often preserved in ____, fossilized tree resin.
13. An impression left by a seashell may fill with minerals to form a ____.
14. A ____ is a scientist who studies fossils.

Down

1. Fossil footprints and burrows are called ____ fossils.
2. Layers of sedimentary rock are called ____.
3. ____ occurs when remains of dead organisms are turned to stone.
4. Fossils are usually found in ____ rocks.
8. The ____ fossils are found in the deepest strata.
9. A ____ is a preserved imprint or remnant of an ancient organism.
10. Some fossils retain ____ materials such as chlorophyll or proteins.
12. A seashell might make an imprint in mud, which forms a ____.

Exercise 3   (Modules 13.3 – 13.4)

Web/CD Activity13C   *Reconstructing Forelimbs*

In addition to fossils, there are other kinds of evidence for evolution. Name the category of evidence to which each of the following examples belongs.

Category

Example

_____

1. Fertilized eggs of earthworms, insects, and snails all go through the same pattern of cell division.

_____

2. The genes of humans and chimpanzees are about 99% identical.

_____

3. Remains of upright-walking but small-brained apes have been found in Africa.

_____

4. All animals with backbones have 12 pairs of nerves extending from the brain.

_____

5. A protein called albumin is very similar in dogs and wolves, less similar in dogs and cats.

_____

6. The farther an island is from the mainland, the more different its plants and animals are from those on the mainland.

_____

7. Animals called trilobites were common in the oceans 300 million years ago, but they have been extinct for millions of years.

Exercise 4   (Modules 13.2, 13.5 – 13.6)

Web/CD Thinking as a Scientist   *How Do Environmental Changes Affect a Population?*

Read these modules, and then review selection and population genetics by completing the following story.

If you think that the more you mow your lawn, the meaner the weeds get, you may be right. Researchers have found that in lawns that are mown regularly, the dandelions fight back! Of course, dandelions don't "know" what they are doing. But the dandelions in a regularly mown lawn reproduce faster than their ancestors in more "natural" environments.

^{1}_____, the English scientist who first devised the theory of ^{2}_____, would have explained it this way: Not all dandelions are alike; they ^{3}_____ in color, size, and rate of maturation. Many of these characteristics are ^{4}_____, or passed on to offspring. Every dandelion flower is capable of producing hundreds, perhaps thousands, of white-tufted seeds in a season. This constitutes an ^{5}_____ of offspring, because ^{6}_____ are limited; not every dandelion seed will find just the right environment in which to grow. Darwin speculated that those individuals whose inherited characteristics ^{7}_____ them best to their environment would be more likely to ^{8}_____ and ^{9}_____ than less suited individuals. Their type would become ^{10}_____ common in the next generation. Darwin called this phenomenon ^{11}_____. Many examples of natural selection are known, including the evolution of ^{12}_____ resistance in insects. Those whose genes protect them best from being poisoned leave more descendants. Resistant insects are seen in greater and greater numbers in succeeding generations. Less resistant individuals die out.

Reproduction is central to natural selection; in fact, natural selection can be defined as [13]_____ success in reproduction.

Natural selection is at work when you mow the lawn. It might be helpful to discuss the dandelions in terms of [14]_____ genetics, the study of how genes affect population changes. The dandelions in your lawn make up a [15]_____, a group of individuals of the same species. The species is all dandelions, the group of populations whose individuals can interbreed. Depending on where you live, your lawn may be more or less [16]_____ from other dandelion populations. Through some quirk of nature, dandelions only reproduce asexually, even though they continue to produce nectar and pollen. So unlike most other flowers, dandelions in your lawn cannot cross-pollinate with other populations, although seeds may blow in from elsewhere, and seeds from your lawn might blow across town.

The characteristics of dandelions—color, height, shape of root system, and so on—are dictated by their genes. All the genes in the dandelion population make up the [17]_____. Apparently, the height of dandelions when they mature and produce seed and how long it takes to do this are controlled by genes, and these traits can vary from dandelion to dandelion. If you mow the lawn often enough, the slower-growing, taller dandelions get lopped off before they can produce any seeds. In terms of reproductive success, these dandelions are not as well [18]_____ as the faster-growing individuals in the population. Because they don't produce as many offspring, this slow type will not be as numerous in the next generation, and their [19]_____ will make up a smaller proportion of the gene pool. On the other hand, dandelions that don't have to grow as tall or take as long to produce seed can reproduce between mowings. Their genes, and their fast-growing traits, will be better represented in the next generation. Such a small change in the frequencies of alleles in the gene pool is called [20]_____.

This story illustrates a familiar example of natural selection. Over time, the shorter, faster-growing dandelions will predominate in the lawn. Note that natural selection involves differences between individuals, but individual dandelions do not evolve. An individual does not change its growth rate. But because there is variation in the survival and reproduction of individuals with different characteristics, the [21]_____ of dandelions evolves.

Exercise 5   (Modules 13.7 – 13.8)

Web/CD Thinking as a Scientist   *How Can Frequency of Alleles Be Calculated?*

Imagine a population of 100 annual wildflowers, some red and some yellow. The red allele, *R*, is dominant; the yellow allele, *r*, is recessive. There are 36 *RR* plants in the population, 48 *Rr* plants, and 16 *rr* (yellow) plants. If the population is at Hardy-Weinberg equilibrium, what will be the frequencies of the various genotypes and the frequencies of the two alleles, *R* and *r*, in the next generation? Follow the example in Module 13.7 as a guide, and fill in the blanks below to figure out the frequencies for this example.

First, figure out genotype frequencies for the current generation:

A. Phenotypes   Red   Red   Yellow
B. Genotypes   ____   ____   ____
C. Number of plants
    (total 100)   ____   ____   ____
D. Genotype frequencies
    (number of genotypes/100) ____   ____   ____

Next, figure out the frequencies of *R* and *r* alleles in the gene pool:

E. Number of *R* alleles in gene pool ____   ____   ____
F. Number of *r* alleles in gene pool ____   ____   ____
G. Allele frequencies
    (number of *R* alleles/200
    or number of *r* alleles/200)   Frequency of *R: p* = ____   Frequency of *r: q* = ____

Now you know the frequency of *R* and *r* gametes these plants will produce:

H. Gamete frequencies
    (= allele frequencies)   Frequency of *R:* ____   Frequency of *r:* ____

Now you can use the rule of multiplication to calculate the frequencies of the three possible genotypes of plants in the second generation:

	R pollen $p = 0.6$	*r* pollen $q = 0.4$
R egg $p = 0.6$		
r egg $q = 0.4$		

I. Phenotype   Red   Red   Yellow
J. Genotype   ____   ____   ____
K. Genotype frequencies ____   ____   ____

Now you can figure out the frequencies of R and r alleles in the gene pool for the second generation (assuming the population stays at 100 individuals):

L. Number of R alleles in gene pool   ____   ____   ____

M. Number of r alleles in gene pool   ____   ____   ____

N. Allele frequencies
(number of R alleles/200
or number of r alleles/200)   Frequency of R: p = ____   Frequency of r: q = ____

O. What happened to the genotype and allele frequencies in the second generation?

What would you predict for the third generation? Why?

Exercise 6   (Modules 13.7 – 13.9)

Web/CD Activity 13D   *Causes of Microevolution*
Web/CD Thinking as a Scientist   *How Can Frequency of Alleles Be Calculated?*

The Hardy-Weinberg equilibrium is an idealized model. Equilibrium is maintained only if five conditions are met. This happens only in the fertile imaginations of biologists, not in real populations. Real populations always deviate from one or more of the conditions, and their gene pools change over time. The Hardy-Weinberg equilibrium is nevertheless a useful standard with which to compare real populations whose gene pools are changing slowly.

Let's continue to look at the wildflower population introduced in Exercise 5. If it is like other real populations, its gene pool is changing. For each of the scenarios below, state which of the Hardy-Weinberg conditions the population deviates from, and explain what agent of microevolution causes the gene pool to change. Also state which of these deviations would cause the flowers to adapt to their environment.

1. A windstorm blows in hundreds of seeds from a nearby meadow, where nearly all the flowers are yellow.

2. A cosmic ray hits one of the red flowers just as a developing egg cell is replicating its DNA. Quite by chance, a red allele is transformed into a yellow allele.

3. The flowers tend to grow in red or yellow patches. A landslide buries and kills most of the red flowers.

4. The red pigment in the petals of the red flowers is poisonous and protects them from beetles that eat the developing seeds. The yellow flowers are not protected in this way.

5. The bees that pollinate the flowers tend to develop a "search image." Once they start visiting flowers of a certain color, they stick to that color. So pollen from red flowers is more likely to be delivered to other red flowers, and pollen from yellow flowers is more likely to fertilize other yellow flowers.

Exercise 7   (Modules 13.10 – 13.14)

Web/CD Activity 13E   *Genetic Variation from Sexual Recombination*
Web/CD Thinking as a Scientist   *Connection: What Are the Patterns of Antibiotic Resistance?*

Review the various terms and concepts relating to variation by completing this crossword puzzle.

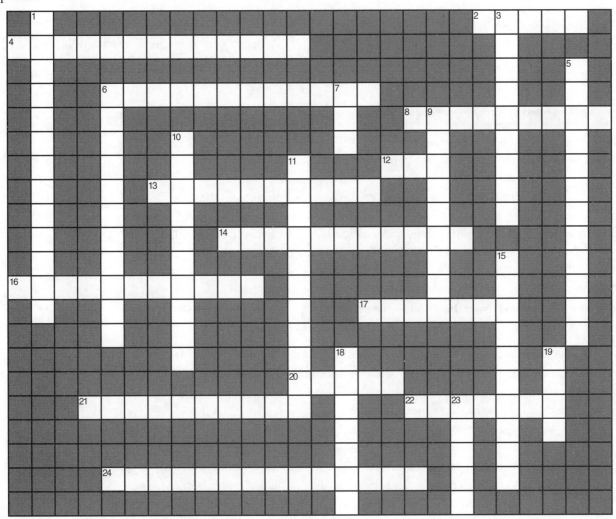

Across

2. Genetic variation allows a population to ____ to environmental change.
4. Mutation and sexual ____ generate variation.
6. Often characteristics that are controlled by several genes vary ____ .
8. ____ selection occurs when natural selection maintains two or more phenotypes.
12. Mutations allow ____ to resist antiviral drugs.
13. Cheetahs passed through a ____, which left them with reduced variation.
14. Endangered species often have low genetic____ .
16. A population is ____ if two or more morphs are noticeably present.
17. Only the ____ component of variation leads to adaptation.

Down

1. Crossing over, independent assortment, and random ____ contribute to sexual recombination.
3. Frequency ____ selection is seen when a morph declines if it becomes too common.
5. Human ____ show a large amount of neutral variation.
6. ____ mutations that rearrange loci on a large scale are usually harmful.
7. Mutation rates tend to be ____ in animals and plants.
9. If *Bb* individuals have more reproductive success than *BB* or *bb* individuals, it is called heterozygote ____ .
10. ____ variation occurs when the environment varies from place to place.

(*Clues continued on page 154*)

Across

20. A____ is a gradual change in a characteristic over a geographic continuum.

21. The average fruit fly is ____ at about 86% of its gene loci.

22. Only mutations in cells that produce ____ can be passed on to offspring.

24. ____, which multiply asexually, can rapidly produce variation by mutation alone.

Down

11. Because they adapt rapidly, bacteria can quickly become resistant to ____ .

15. ____ diversity is determined by comparing the base sequences of DNA samples.

18. There are many "hidden" recessive alleles in ____ organisms.

19. Human beings show much ____ genetic variability than most species.

23. One form of a phenotypic characteristic is called a ____ .

Exercise 8   (Modules 13.15 – 13.18)

Returning one last time to the wildflowers discussed in Exercises 5 and 6, complete the following scenario regarding fitness and natural selection.

An early writer on evolution described natural selection as "nature red in tooth and claw." This may be true for lions and zebras on the Serengeti, but natural selection is usually more subtle. All living things are engaged in what Charles Darwin called a ^{1}"_____ for existence." Natural selection may be "survival of the ^{2}_____," but fitness is more than simply brute strength. Biologists define Darwinian ^{3}_____ as the relative contribution that an individual makes to the gene pool of the next generation. It has more to do with reproduction than strength or cunning.

For the wildflowers in our previous example, the struggle for existence involves physical traits such as color and shape of leaves and metabolic characteristics like efficiency in capturing sunlight and resistance to cold. The ^{4}_____ of the plant is a composite of all its characteristics. Only the ^{5}_____, not the ^{6}_____, is exposed to the environment. A red flower may be protected from predation, while a yellow flower is eaten. Genes for red color increase fitness, and these genes are more likely to be passed on to the next generation. The frequency of red genes in the ^{7}_____ increases, as does the frequency of red individuals in the ^{8}_____. Of course, the fitness of an allele for a particular characteristic depends on its genetic ^{9}_____. A red flower with a weak stem that snaps easily in the wind might be less fit than a yellow flower with a stronger stem.

Most flowers contain both male and female parts, but because males and females are distinct in most animal species, natural selection often shapes differences in appearance, such as antlers or bright plumage. Such difference is called sexual ^{10}_____. Among some species, these ^{11}_____ sexual characteristics might be used in fighting over females. A more common form of sexual selection involves ^{12}_____ choice. Usually males display their bright "plumage" and females do the choosing.

There are three different ways in which natural selection can affect a population. Many characteristics are not simple "either/or" alternatives like red and yellow flower color. Characteristics like height vary continuously and can be described by a ^{13}_____-shaped curve: There may be a few very short plants in the population, a majority of plants of medium height, and a few very tall plants. Imagine our wildflowers growing in a cold, windy environment. Very tall plants might freeze before their seeds mature. Shorter plants would stay warmer, but very short plants might have trouble dispersing their seeds to favorable environments. In this kind of situation, [14]_____ selection favors the intermediate variants, not too tall and not too short. Next imagine a situation where the environment is gradually becoming drier. In this case, [15]_____ selection might favor those individuals genetically programmed to grow the deepest roots. This kind of natural selection is most common during periods of environmental [16]_____. Finally, [17]_____ selection occurs when two different sets of environmental conditions favor the extreme phenotypes and act against [16]_____ types. For example, plants with shallow, spreading roots might be at an advantage in dry rocky soil, where water tends to penetrate quickly. At the same time, a deep taproot system might be favored in richer soil that holds water longer. Intermediate root systems would be at a disadvantage in both environments.

Will natural selection ever shape a perfect flower? Probably not. [19]_____ can only act on existing variations which depend on history and chance. And because a plant must do many different things, adaptations are often [20]_____.

Testing Your Knowledge

Multiple Choice

1. During his voyage around the world, Charles Darwin was inspired to think about evolution by
   a. books that he read.
   b. fossils he collected.
   c. studying adaptations of organisms to their environments.
   d. unique organisms he saw in the Galápagos Islands.
   e. all of the above.

2. _____ and _____ generate variation, while _____ results in adaptation to the environment.
   a. genetic drift . . . natural selection . . . mutation
   b. mutation . . . sexual recombination . . . natural selection
   c. overproduction of offspring . . . mutation . . . sexual recombination
   d. natural selection . . . mutation . . . sexual recombination
   e. sexual recombination . . . natural selection . . . mutation

3. Breeding of plants and animals by humans is called
   a. natural selection.
   b. sexual recombination.
   c. founder effect.
   d. artificial selection.
   e. neutral variation.

4. Microorganisms can adapt to changes in the environment by means of mutation alone because
   a. they are so small in size.
   b. their populations are very isolated from one another.
   c. most of their mutations are helpful, rather than harmful.
   d. they multiply so rapidly.
   e. their populations are so large.

5. The smallest unit that can evolve is a
   a. species.
   b. genotype.
   c. gene.
   d. population.
   e. morph.

6. "Differential reproduction" is just another way of saying
   a. natural selection.
   b. mutation.
   c. variation.
   d. recombination.
   e. genetic drift.

7. Which of the following changes in the gene pool results in adaptation to the environment?
   a. nonrandom mating
   b. genetic drift
   c. natural selection
   d. gene flow
   e. mutation

8. The ultimate source of all genetic variation is
   a. natural selection.
   b. genetic drift.
   c. sexual recombination.
   d. the environment.
   e. mutation.

9. In evolutionary terms, an organism's fitness is measured by its
   a. health.
   b. contribution to the gene pool of the next generation.
   c. mutation rate.
   d. genetic variability.
   e. stability in the face of environmental change.

10. Organisms that possess homologous structures probably
    a. are headed for extinction.
    b. evolved from the same ancestor.
    c. have increased genetic diversity.
    d. by chance had similar mutations in the past.
    e. are not related.

11. Sexual recombination occurs when chromosomes are shuffled in _____ and fertilization.
    a. mitosis
    b. genetic drift
    c. natural selection
    d. mutation
    e. meiosis

12. Darwin
    a. was the first person to conceive that organisms could change over time.
    b. believed that organisms could pass on acquired changes to offspring.
    c. was eager to publish his theory so that he could get all the credit.
    d. worked out the mechanism of evolution—natural selection.
    e. was the first to realize that fossils are remains of ancient organisms.

Essay

1. Explain how heritable variations, overproduction of offspring, and limited natural resources cause a species to adapt to its environment.

2. Briefly describe five categories of evidence for evolution.

3. What is the difference between a population and a species?

4. Horses look a lot like zebras. How would Darwin have explained this?

5. Sometimes a harmful allele may be present in the gene pool at a relatively high frequency. How might this be explained in terms of genetic drift? How might it relate to the fact that most organisms are diploid? What might it have to do with heterozygote advantage?

6. Describe how each of the following might alter the gene pool: genetic drift, nonrandom mating, gene flow, mutation, and natural selection.

Applying Your Knowledge

Multiple Choice

1. In a population of black bears, which would be considered the fittest?
   a. the biggest bear
   b. the bear having the largest number of mutations
   c. the bear that blends in with its environment the best
   d. the strongest, fiercest bear
   e. the bear that leaves the most descendants

2. A geneticist mixed together many different kinds of fruit flies—some with long wings, some with short wings, some with red eyes, some with brown eyes, and so on. He allowed the flies to feed, mate randomly, and reproduce by the thousands. After many generations, most of the flies in the population had medium wings and red eyes, and most of the extreme types had disappeared. This experiment appears to demonstrate
   a. stabilizing selection.
   b. geographical variation.
   c. disruptive selection.
   d. genetic drift.
   e. fitness.

3. Rabbits living farther and farther north tend to have smaller and smaller ears. This is an example of
   a. a cline.
   b. polymorphism.
   c. artificial selection.
   d. heterozygote advantage.
   e. genetic drift.

4. Blue poppies native to China are grown at a plant-breeding center in California, where those with the thickest leaves survive and reproduce best in the drier climate. This evolutionary adaptation of the poppies to their new environment is due to
   a. genetic drift.
   b. stabilizing selection.
   c. directional selection.
   d. neutral variation.
   e. disruptive selection.

5. Biologists have noticed that most human beings enjoy sex. How would they explain this, in evolutionary terms?
   a. If sex were not enjoyable, the human species would have died out.
   b. Early humans who enjoyed sex most had the most babies.
   c. Only body structures evolve, not behavior, so enjoyment cannot evolve.
   d. This was due to a random mutation, so it did not affect evolution.
   e. Like most people, biologists are baffled by the phenomenon of sex.

6. Which of the following would result in evolutionary adaptation of a mouse population to its environment?
   a. Half the mice are killed by an avalanche.
   b. A mutation for spotted fur occurs.
   c. Several mice leave the area and mate with individuals elsewhere.
   d. Mice with thicker fur best survive a cold winter.
   e. Mice are most likely to mate with close neighbors.

7. The relationship of genome to organism is the same as that of _____ to population.
   a. species
   b. gene
   c. gene pool
   d. mutation
   e. variation

8. Some critics of evolution believe that the theory of evolution is flawed because it is based on random changes—mutations. They say that a random change in an organism (or a car or a TV set) is likely to harm it, not make it function better. What logical statement could a defender of evolution make in reply to this criticism?
   a. Fossils prove without a doubt that mutations drive evolution.
   b. Mutation is random, but natural selection is not.
   c. Mutation has little to do with evolution.
   d. This is a weak spot in the theory that remains to be worked out.
   e. Mutations are not actually random.

9. A zoologist found that in the population of frogs in MacGregor's pond, half the genes for skin color in the gene pool were alleles for green spots, and half the genes were alleles for brown spots. Which of the following could cause these proportions to change?
   a. A drought shrinks the pond so that only five frogs remain.
   b. Females prefer to mate with brown-spotted males.
   c. Green-spotted frogs can hide more easily among the pond weeds.
   d. Filling in a nearby pond causes those frogs to move to MacGregor's pond.
   e. Any of the above could cause the proportions to change.

10. Each of us is part of the ongoing evolution of the human species. Which of the following occurrences would have the greatest impact on the future biological evolution of the human population?
    a. You work out every day so that you stay physically fit and healthy.
    b. A mutation occurs in one of your skin cells.
    c. You move to Hawaii, the state with the longest life expectancy.
    d. A mutation occurs in one of your sperm or egg cells.
    e. You encourage your children to develop their intellectual abilities.

11. We know a lot about fossil crabs, snails, and corals, but not much about ancient seaweeds. Why do you suppose this is the case?
    a. There were no seaweeds in ancient oceans.
    b. Seaweeds were too soft to fossilize well.
    c. Animal life was much more abundant than seaweeds in ancient times.
    d. Plants moved onto land, leaving only animals in the sea.
    e. A mass extinction wiped out the seaweeds, but animals survived.

Essay

1. A butterfly has a long tubelike proboscis that it uses to suck the nectar from a certain kind of deep, tubular flower. Its close relatives have much shorter proboscises. How might Lamarck have explained the existence of this long proboscis? How would Darwin have explained it?

2. A long section of *The Origin of Species* describes the breeding of pigeons and how pigeon breeders have produced many shapes, sizes, and colors of pigeons, all starting from a common wild pigeon. Why do you suppose Darwin thought this was important?

3. Jeff and Trina collected some wildflower seeds from a meadow and scattered them along the roadside near their home. As the weather got hotter, some of the seedlings began to wither. Jeff said, "Hey, those scrawny plants are going to have to adapt or they are going to die." Trina replied, "*They* may not adapt, but the wildflower *population* might." What did she mean?

4. A population of 100 fish in an aquarium consists of 49 homozygous dominant green individuals, 42 heterozygous green individuals, and 9 white individuals. What are the frequencies of the three genotypes in the population? What are the frequencies of the green and white alleles in the gene pool? If the population is at Hardy-Weinberg equilibrium, what will the genotype and gene frequencies be in the next generation of fish?

5. Describe five ways in which the fish population in question 4 could be caused to deviate from Hardy-Weinberg equilibrium. Is this likely? Which of these deviations might cause the fish to become better adapted to their environment?

6. An ecologist studying predators and their prey in White Sand Dunes State Park found that nocturnal pocket mice exist in dark and light morphs. In this area, 42% of the mice are dark and 58% are light. Because owls swallow their prey whole and then cough up "pellets" consisting of bones and fur, the ecologist was able to discover that 61% of the mice caught by owls were dark, and 39% were light. What is likely to happen to the pocket mouse population in the future, and why?

Extending Your Knowledge

1. There is an ongoing controversy in the United States about the teaching of evolution in public schools. Proponents of "scientific creationism" and "intelligent design" demand equal time for explanations other than evolution by natural selection. Has this been an issue in your community or state? What evidence and arguments are presented by each side? How does each side answer the other's criticisms? Has your study of biology altered your opinions about this controversy? If so, in what ways?

2. Take a walk around your neighborhood, preferably in a natural area or park where living things are abundant. For each organism that you see, note an adaptation to its environment. How did the evolutionary processes discussed in this chapter shape each organism to fit its environment?

3. Many liquid hand "soaps" and even dish washing liquids are now available with added antibacterial agents. Why are some medical and public health practitioners leery of such additives?

The Origin of Species

14

Waialeale, a mountain on the Hawaiian island of Kauai, is one of the rainiest places on Earth. Here 15 meters of rain a year is not unusual. The Alakai swamp lies at an elevation of 1300 meters on the flanks of the mountain. This boggy forest is one of the best places to see honeycreepers, birds unique to the Hawaiian Islands. Walking on a trail through the forest, you might see an apapane, a common red honeycreeper that uses its curved bill to eat nectar and caterpillars. The bright red iiwi laps up nectar with its long ivory-colored beak. You would be unlikely to spot the greenish akialoa, which has a curved, needle-sharp beak half as long as its body. It was last seen in 1973 and may be extinct. Other honeycreepers have stout hooked bills used for crushing twigs to get insects, or short heavy beaks for cracking seeds. There are twenty-odd species of honeycreepers on the various islands, and several extinct species are known. Why do honeycreepers live only in the Hawaiian Islands? How did they get here? Why are there so many kinds? All honeycreepers are thought to be descended from a single ancestral species—maybe a finch—that reached the islands perhaps 5 million years ago. Speciation—how new species such as the diverse forms of honeycreepers arise—is one aspect of macroevolution, dramatic and large-scale evolutionary change. Speciation and macroevolution are the subjects of this chapter.

Organizing Your Knowledge

Exercise 1   (Module 14.1 – 14.2)

There are several ways to identify species. The biological species concept is one of the most useful. Briefly explain why it might be difficult to apply the biological species concept in each of the following situations.

1. Fossils of "Java Man" and "Peking Man" are both thought to represent a single species—*Homo erectus*.

2. A tiger and a lion can interbreed in a zoo and produce a hybrid offspring called a tiglon.

3. Dogs come in many shapes and sizes, from Chihuahuas to Saint Bernards.

4. There are many strains and species of *Streptococcus* bacteria, which reproduce asexually.

5. Among *Clarkia* wildflowers in California, flowers of population A can interbreed and produce fertile offspring when crossed with flowers from population B. Similarly, B can interbreed with C. But A and C cannot successfully interbreed.

159

6. One bird guide calls flycatchers of the genus *Empidonax* "the bane of bird-watchers." Several species look so much alike that birders can distinguish them only by their songs.

7. A song sparrow population in Baja, California, is separated from other song sparrows by over a hundred miles of desert.

Exercise 2   (Module 14.3)

Review the reproductive barriers that separate species by categorizing the following examples. State whether each barrier is prezygotic (Pre) or postzygotic (Post), and then name the specific kind of barrier (such as temporal isolation or hybrid inviability) it exemplifies. The chart in Module 14.3 is a helpful summary.

Pre or Post	*Kind of Barrier*	*Example*
1. _____	_____	The salamanders *Ambystoma tigrinum* and *A. maculatum* breed in the same areas. *A. tigrinum* mates from late February through March. *A. maculatum* does not start mating until late March or early April.
2. _____	_____	Two species of mice are mated in the lab and produce fertile hybrid offspring, but offspring of the hybrids are sterile.
3. _____	_____	When fruit flies of two particular species are crossed in the lab, their offspring are unable to produce eggs and sperm.
4. _____	_____	A zoologist observed two land snails of different species that were trying to mate with little success because they apparently did not "fit" each other.
5. _____	_____	Male fiddler crabs (genus *Uca*) wave their large claws to attract the attention of females. Each species has a slightly different wave.
6. _____	_____	When different species of tobacco plants are crossed in a greenhouse, the pollen tube usually bursts before the eggs are fertilized.
7. _____	_____	Blackjack oak (*Quercus marilandica*) grows in dry woodlands, and scrub oak (*Q. ilicifolia*) grows in dry, rocky, open areas. Pollen of one species seldom pollinates the other.
8. _____	_____	The tiglon offspring of a lion and a tiger are often weak and unhealthy.

Exercise 3   (Module 14.4)

Formation of new species often begins with geographical isolation. For each of the organisms listed below, name two geographical barriers that might lead to allopatric speciation.

1. Daisy

2. Mouse

3. Trout

4. Oak tree

5. Sparrow

6. Sea star

Exercise 4   (Modules 14.6 – 14.7)

Web/CD Activity 14 A   *Polyploid Plants*
Web/CD Thinking as a Scientist   *How Do New Species Arise by Genetic Variation?*

New species can result if a change in chromosome number produces a reproductive barrier that isolates organisms from their parent populations. Study the examples given in the Web/CD Activity and Modules 14.6 and 14.7. Then state whether you think each of the organisms in italics below would be able to reproduce (yes or no), whether it could represent a new species (yes or no), and how many chromosomes it would have.

Repro-duction	New Species	No. of Chrom.	Example
1. ____	____	____	A flower has 18 chromosomes ($2n = 18$). Nondisjunction occurs, and diploid gametes are formed. The *zygotes formed* are tetraploids, capable of self-fertilization when mature.
2. ____	____	____	A tree has 22 chromosomes ($2n = 22$). Nondisjunction produces a diploid pollen grain, which fertilizes a normal haploid egg. The resulting polyploid zygote develops into a *full-grown tree.*
3. ____	____	____	Antelope of two different species mate in a zoo. Species A has 36 chromosomes; species B has 34 chromosomes. They produce a *hybrid offspring.*
4. ____	____	____	A lily of species A has 16 chromosomes. Species B has 20 chromosomes. A hybrid with 18 chromosomes undergoes nondisjunction, producing a *lily* that is capable of self-pollination.
5. ____	____	____	A botanist treats some tissue from a strawberry plant ($2n = 14$) with a chemical that causes nondisjunction. This doubles the number of chromosomes in a cell. This cell is then cultured on a special medium, until it eventually develops into a *strawberry plant* that can self-pollinate.
6. ____	____	____	Pollen from the cell-cultured strawberry in question 5 is placed on the flower of a plant of the parent species, producing a triploid *hybrid zygote.*

Exercise 5   (Module 14.8 – 14.9)

Web/CD Activity 14B   *Exploring Speciation on Islands*

Speciation and adaptive radiation often occur on islands. The Galápagos and Hawaiian Islands provide many examples. But for geographical isolation to occur, an island does not have to be a bit of land in the midst of an ocean. An evolutionary island could be, for example, a patch of grassland surrounded by forest. List four other examples and the kinds of species that could be isolated and evolve in each.

1.

2.

3.

4.

Exercise 6   (Module 14.10)

Contrast the gradualist model of speciation with the punctuated equilibrium model by completing the chart below.

	Gradualism	*Punctuated Equilibrium*
1. Long-term tempo of evolution smooth or jumpy?		
2. Darwin's view or more recent interpretation?		
3. Kind of evidence in fossil record?		
4. Speciation fast or slow?		
5. Continuous change through life of species, or quick change and stability?		

Exercise 7   (Modules 14.1 – 14.10)

This chapter describes many examples of speciation. Match each description below with the correct species.

_____ 1.   Diverged after a population was split by the Grand Canyon.
_____ 2.   Different species evolved in separated Death Valley springs.
_____ 3.   Mating between species may have sped the spread of West Nile virus.
_____ 4.   Different diets led to reproductive barriers in a laboratory experiment.
_____ 5.   Look much alike, but sing different songs.
_____ 6.   These birds are unique to the Hawaiian Islands.
_____ 7.   Different species have different beaks, but sometimes hybridize.

A. Meadowlarks
B. Honeycreepers
C. Antelope squirrels
D. Darwin's finches
E. Mosquitoes
F. Fruit flies
G. Pupfish

Exercise 8   (Module 14.1 – 14.10)

Use the concepts and terms from this chapter to complete the following story about (imaginary) butterflies and asters.

The yellowspot butterfly is found over hundreds of square miles of land in the delta of an African river. Its primary food source is a species of purple aster—a flower related to daisies and dandelions. Patches of asters are scattered in sunny meadows in the delta, some several miles apart. The butterflies do not usually venture far for food. Each patch of asters supports a separate [1]_____ of yellowspots, but because the butterflies sometimes wander and mate with butterflies in other areas, all the butterflies have been classified as members of the same [2]_____.

Insect taxonomists noted that one population of yellowspots was centered across the main river channel from the other populations. They suspected that the river might act as a [3]_____ to the butterflies, since they do not usually fly far over water. The researchers examined the butterflies from the isolated population and found that the butterflies from across the river were a bit smaller than most yellowspots and more orange in color, so the researchers nicknamed them "orangespots." The biologists found that the differences in appearance were inherited. They suspected these could reflect [4]_____ due to chance differences in the butterflies that founded the orangespot population. The researchers also noted that the environment was slightly warmer and drier on the far side of the river, so [5]_____ may have caused the orangespots to adapt to conditions there.

The scientists suspected that the two populations could represent different species. To test this hunch, they had to find out whether butterflies from the far side of the river could [6]_____ with individuals from the main population. The biologists captured some butterflies from both areas and placed them in a cage. Surprisingly, the orangespots and yellowspots ignored each other. The researchers found that the female butterflies rest on leaves and flash their wing spots to attract the males. Yellowspot females flash their wings at a much faster rate than orangespot females. Apparently the wing-flashing display acts as a [7]_____-zygotic reproductive barrier. Apparently, [8]_____ keeps the two populations of butterflies from interbreeding. In one instance, the eggs of an orangespot female were fertilized by a yellowspot male, but the embryos soon died. Apparently, there is also a [9]_____ -zygotic reproductive barrier between the butterflies. The particular type of barrier is [10]_____ in this case. The researchers realized that the yellowspot and orangespot butterflies are separate [11]_____, incapable of interbreeding.

A study of river sediments showed that the channel separating the two butterfly populations formed about a thousand years ago, when the river shifted course. It has indeed acted as a geographical barrier, [12]_____ the two populations from one another and eventually leading to [13]_____ speciation. Biologists were particularly interested in how quickly (in geological terms) the new butterfly species must have evolved. This rapidity would seem to support the [14]_____ model of species evolution.

While studying the habits of the butterflies, the biologists turned their attention to the flowers, and found a group of asters with oval-shaped leaves and slightly larger flowers than all the others. A study of their chromosomes showed that the unusual plants were [15]_____, having more than the usual two sets of chromosomes. All the other purple asters in the area had two sets of chromosomes and a $2n$ chromosome number of 26. The unusual plants were [16]_____, having four sets of chromosomes, for a total of 52. [17]_____ must have occurred during meiosis in one of the diploid asters, creating diploid gametes. Because the plant could self-pollinate, zygotes were formed with four sets of chromosomes. The plants that developed from these zygotes could interbreed with one another, but they were reproductively [18]_____ from the parent species. They represented a new [19]_____, produced in one generation through the process of [20]_____ speciation.

Exercise 9   (Modules 14.11 – 14.13)

Web/CD Activity 14C   *Mechanisms of Macroevolution*

The evolution of new species is the beginning of macroevolution—large-scale changes such as the development of unique structures and the appearance of whole new groups of organisms. Review macroevolution by matching each of the descriptions on the left with the best evolutionary example on the right.

_____ 1. Complex structures often evolve step-by-step.

_____ 2. A structure that evolves in one context and takes on a new role is called an exaptation.

_____ 3. Changes in homeotic genes can radically alter the timing of development and the shape of body parts.

_____ 4. Selection among different species may result in large-scale evolutionary trends.

_____ 5. Paedomorphosis is a change in timing of development, causing juvenile features to be retained by adults.

A. The horse family
B. Eyes
C. Human skull and jaw
D. Wings
E. Fins to legs

Testing Your Knowledge

Multiple Choice

1. Two animals are considered different species if they
    a. look different.
    b. cannot interbreed.
    c. live in different habitats.
    d. are members of different populations.
    e. are geographically isolated.

2. Which of the following is the first step in allopatric speciation?
    a. genetic drift
    b. geographical isolation
    c. polyploidy
    d. hybridization
    e. formation of a reproductive barrier

3. The science of naming and classifying organisms is called
    a. biology.   d. taxonomy.
    b. polyploidy.   e. gradualism.
    c. genetics.

4. Most of the time, species are identified by their appearance. Why?
    a. If two organisms look alike, they must be the same species.
    b. This is the criterion used to define a biological species.
    c. If two organisms look different, they must be different species.
    d. This is the most convenient way of identifying species.
    e. Most organisms reproduce asexually.

5. A new species can arise in a single generation
    a. through geographical isolation.
    b. in a very large population that is spread over a large area.
    c. if a change in chromosome number creates a reproductive barrier.
    d. if allopatric speciation occurs.
    e. according to the gradualist model of speciation.

6. The evolution of numerous species, such as Darwin's finches, from a single ancestor is called
    a. adaptive radiation.
    b. sympatric speciation.
    c. gradualism.
    d. nondisjunction.
    e. geographical isolation.

7. According to the _____ model, evolution occurs in spurts; species evolve relatively rapidly, then remain unchanged for long periods.
   a. nondisjunction
   b. gradualist
   c. adaptive radiation
   d. punctuated equilibrium
   e. geographical isolation

8. Evolutionary changes, such as the development of walking legs from fins, and the appearance of new groups of organisms, such as birds, are termed
   a. macroevolution.
   b. adaptation.
   c. hybridization.
   d. microevolution.
   e. paedomorphosis.

9. Individuals of different species living in the same area may be prevented from interbreeding by responding to different mating dances. This is called
   a. ecological isolation.
   b. hybrid breakdown.
   c. mechanical isolation.
   d. temporal isolation.
   e. behavioral isolation.

10. It is unlikely that the human population will give rise to a new species because
    a. the human population is too large.
    b. geographical isolation is unlikely to occur.
    c. a change in chromosome number would be fatal.
    d. the human population is too diverse.
    e. natural selection cannot affect humans.

Essay

1. Give three situations in which it might be difficult to use the test of interbreeding to determine whether two organisms are of the same or different species.

2. Describe step by step how geographical isolation could lead to speciation.

3. Describe what has to happen for two species with different numbers of chromosomes to interbreed and produce a fertile hybrid. Will this hybrid be able to interbreed with either of its parents? Why? How common is this in nature?

4. State whether a large, widely distributed population or a small, isolated population is more likely to undergo speciation, and explain why.

5. Compare the gradualist and the punctuated equilibrium models of species evolution. What would the fossil record of speciation look like if it supported the gradualist model? What would it look like if punctuated equilibrium was the case? Which model does the fossil record seem to support most often? Why?

Applying Your Knowledge

Multiple Choice

1. Three species of frogs—*Rana pipiens*, *Rana clamitans*, and *Rana sylvatica*—all mate in the same ponds, but they pair off correctly because they have different calls. This illustrates a ____ and ____ .
   a. prezygotic barrier . . . behavioral isolation.
   b. postzygotic barrier . . . hybrid breakdown.
   c. prezygotic barrier . . . temporal isolation.
   d. postzygotic barrier . . . behavioral isolation.
   e. prezygotic barrier . . . gametic isolation.

2. Drastic reductions in the number of body segments and pairs of legs may have been responsible for the evolution of the first insects from millipede-like ancestors. This example might illustrate
   a. species selection.
   b. microevolution.
   c. polyploidy.
   d. hybrid breakdown.
   e. changes in homeotic genes.

3. Bullock's oriole and the Baltimore oriole are closely related, but are they the same species? To find out, you could see whether they
   a. sing similar songs.
   b. look alike.
   c. live in the same areas.
   d. have the same number of chromosomes.
   e. successfully interbreed.

4. Sometimes two quite different populations interbreed to a limited extent, so that it is difficult to say whether they are clearly separate species. This does not worry biologists much because it
   a. is quite rare.
   b. is true for almost every species.
   c. supports the theory of punctuated equilibrium.
   d. may illustrate the formation of new species in progress.
   e. happens only among plants, not among animals.

5. Two species of water lilies in the same pond do not interbreed because one blooms at night and the other during the day. The reproductive barrier between them is an example of
   a. temporal isolation.
   b. gametic isolation.
   c. mechanical isolation.
   d. hybrid breakdown.
   e. ecological isolation.

6. Comparison of fossils with living humans seems to show that there have been no significant physical changes in *Homo sapiens* in 30,000 to 50,000 years. What might an advocate of punctuated equilibrium say about this?
   a. It is about time for humans to undergo a burst of change.
   b. That is about how long we have been reproductively isolated.
   c. It is impossible to see major internal changes by looking at fossils.
   d. You would expect lots of changes in the skeleton in that time period.
   e. Lack of change is consistent with the punctuated equilibrium model.

7. Which of the following is an example of a postzygotic reproductive barrier?
   a. One species of frog mates in April; another mates in May.
   b. Two fruit flies of different species produce offspring that are sterile.
   c. The sperm of a marine worm only penetrate eggs of the same species.
   d. One species of flower grows in forested areas, another in meadows.
   e. Two pheasant species perform different courtship dances.

8. Lake Malawi, in the African Rift Valley, is home to over a hundred species of cichlid fishes, each with a slightly different diet and habits. All these fishes probably evolved from one ancestor, an example of
   a. sympatric speciation.
   b. hybrid breakdown.
   c. adaptive radiation.
   d. gradualism.
   e. punctuated equilibrium.

9. A botanist found that a kind of white daisy had a diploid chromosome number of 16. In the same area, he found a yellowish daisy. Its cells contained 24 chromosomes. He found that the yellowish daisy was a polyploid descendant of the white daisies. Which of the following would describe this unusual plant? P: tetraploid, Q: triploid, R: probably sterile, S: a new species.
   a. P R S   d. P S
   b. Q R   e. Q S
   c. Q R S

10. A fossil expert finds an impression of an ancient marine creature called a trilobite in a layer of rock. In the adjacent layer is another species of trilobite, clearly related to the first but quite different in form. If the expert is a gradualist, how might he or she interpret this?
   a. This kind of change is exactly what gradualism would predict.
   b. Sympatric speciation must have occurred.
   c. Intermediate forms could have existed but were not fossilized.
   d. Internally, the creatures were identical; only the outer shell changed.
   e. This kind of abrupt transition is rare in the fossil record.

Essay

1. There are dozens of species of small rodents, such as rats and mice, in western North America, but relatively few species, ranging over much larger areas, in the east. Suggest a hypothesis related to speciation that might explain this.

2. The mating of a horse and a donkey produces a mule, which is strong and hard working, but sterile. A horse cell contains 64 chromosomes, a donkey cell 62 chromosomes. How many chromosomes would you expect to find in a cell from a mule? Why? Explain why a mule is sterile.

3. A dog (*Canis familiaris*) and a coyote (*Canis latrans*) will readily mate in captivity, and their offspring are healthy and fully fertile. Are we justified in saying they are distinct species? What might this tell us about the history of dogs and coyotes?

4. In terms of speciation, how might freshwater streams in a desert be like islands in the ocean?

5. Various sections of the Hawaiian Islands were formed by lava flows at different times. Kenneth Kaneshiro compared older populations of fruit flies living in older habitats with presumably younger populations living in more recently formed habitats. He found that male fruit flies in both areas are generally eager to mate, but females of older populations are much more selective about their partners than females of newer populations. Suggest a hypothesis to explain why this might be the case.

6. In the Hawaiian Islands, there are several native species of moths in the genus *Hedylepta* that feed exclusively on bananas. Other related species specialize in palms, grasses, or legumes. Bananas were brought to Hawaii by the native Polynesians about a thousand years ago. Why might the banana-eating species be of particular interest to biologists working out the details of the punctuated equilibrium theory?

7. Critics of evolution often say things like, "Sure, moths and bacteria can adapt to a changing environment. But peppered moths are still peppered moths, and *Streptococcus* is still *Streptococcus*. Nobody has ever seen a new *species* evolve." Is this criticism valid? How would you respond to this comment?

8. How might exaptations, paedomorphosis, and species selection have contributed to human evolution? (See Chapter 19 for ideas.)

Extending Your Knowledge

1. The Nature Conservancy is an environmental organization that buys land to protect rare and endangered species of plants and animals. Are there any Nature Conservancy preserves near your home or campus? What species are protected? How did these species come to be located there? Why are they rare?

2. The desire to name and categorize things seems to be a basic human characteristic. This may be why many people like to collect things. Some biologists believe that we have gotten carried away with this desire to classify when it comes to the idea of species. They think that the concept of a "species" is just that—only a concept. They believe that a species is not a real biological entity, only a convenient idea. There are just too many flaws and exceptions in the species concept for it to be valid. Populations, however, are real. Based on what you know and have read in this chapter, do you think species are real or not? What kinds of evidence or information might support each side of this discussion?

Tracing Evolutionary History

Imagine walking down a dusty trail along a riverbank. It is a hot summer day; insects circle and buzz lazily in the shade of overhanging trees. Here and there, the trail skirts high bluffs of gray and brown rock. The rock is laid down in layers—different colors, different thicknesses, and different textures. In places, the river has undercut the bank, and sheets of rock have peeled away and tumbled downhill. You round a bend and see a large slab up-ended beside the trail. There is a crack running all the way through the slab, so it is easy to pull the layers apart. You have to jump back as the surface layer falls away and shatters at your feet. The newly exposed surface is solid with the impressions of hundreds of fossil seashells. It is like finding a buried treasure. Nobody has ever seen these fossils before; they have been hidden for millions of years. Where did the shells come from? How did they get here? How old are they? What can they tell us about the history of life? These are some of the questions explored in this chapter.

Organizing Your Knowledge

Exercise 1 (Module 15.1)

Web/CD Activity 15A *The Scrolling Geologic Record*

Review the geological timeline by numbering each of the following events in sequence (1, 2, 3 . . .) and naming the geologic eon and era (if applicable) when each occurred.

Sequence	Eon	Era	Event
_____	_____	_____	A. Cone-bearing plants and dinosaurs dominant
_____	_____	_____	B. Humans appear
_____	_____	(na)	C. Diverse soft-bodied animals
_____	_____	_____	D. Diversification of fishes
_____	_____	(na)	E. First eukaryotes
_____	_____	_____	F. Radiation of mammals
_____	_____	_____	G. Invasion of land by plants and arthropods
_____	_____	_____	H. Origin of reptiles
_____	_____	(na)	I. First living things
_____	_____	_____	J. Permian mass extinction of marine and terrestrial life

Exercise 2 (Module 15.2)

Use the concept of half-life to answer these questions about the ages of fossils.

1. The half-life of carbon-14 is 5730 years. If a mammoth has ⅛ the ^{14}C-to-^{12}C ratio that it was thought to have when it was frozen in a Siberian glacier, how old is the mammoth? At an error level of plus or minus 10%, what are the maximum and minimum ages of the fossil?

2. The half-life of potassium-40 is 1.3 billion years. If a rock specimen contained 1 g of potassium-40 when it was formed, how much potassium-40 would be left if the rock is 2.6 billion years old?

Exercise 3 (Modules 15.3 – 15.5)

Web/CD Activity 15B   *Mechanisms of Macroevolution*

Review the concepts of geological processes—and their biological consequences—by matching each of the phrases on the left with a word on the right. The illustrations in the textbook will help.

_____ 1. The surface of the Earth, broken into plates	A. Pangaea
_____ 2. This ocean is surrounded by violent geological events	B. Plate tectonics
_____ 3. Southern land mass formed when Pangaea broke up	C. Cretaceous
_____ 4. A very small plate	D. Earthquake
_____ 5. This ocean grows as North America and Eurasia split apart	E. Himalayas
_____ 6. Island destroyed by volcanic eruption, then recolonized	F. Continental drift
_____ 7. Supercontinent formed 250 million years ago	G. Gondwana
_____ 8. Continent formed from the eastern part of Laurasia	H. Mantle
_____ 9. Any large, moving segment of the Earth's crust	I. Australia
_____ 10. "Fallout" from meteor impact	J. Nazca
_____ 11. Mountains formed by collision of Indo-Australian and Eurasian plates	K. Laurasia
_____ 12. Movement of continents over the Earth's surface	L. Crust
_____ 13. Mass extinction probably caused by a meteor impact	M. Fault
_____ 14. Northern land mass formed when Pangaea broke up	N. Atlantic
_____ 15. Many marsupials evolved here, in isolation from other continents	O. Krakatau
_____ 16. Geological forces that cause movements of crustal plates	P. Eurasia
_____ 17. Place where plates slide along one another	Q. Permian
_____ 18. Movement resulting from forces at plate edges	R. Plate
_____ 19. Mass extinction that occurred 250 million years ago	S. Pacific
_____ 20. Hot layer that lies beneath the crust	T. Mexico
_____ 21. Probable site of meteor impact	U. Iridium

Exercise 4 (Modules 15.6 – 15.9)

Web/CD Thinking as a Scientist   *How Is Phylogeny Determined Using Protein Comparisons?*

Review the principles, methods, and vocabulary of systematics by inserting the correct terms into the following essay.

1_____ is the field of biology devoted to the study of the diversity of life and the relationships among organisms. Biologists use several kinds of evidence to reconstruct the 2_____, or evolutionary history, of a group of organisms. Sometimes phylogeny can be reconstructed by looking at morphological (structural) similarities. The teeth and skeletons of tigers and bobcats show many 3_____ that indicate that these animals share a common ancestry. But anatomical comparisons can sometimes be misleading. A process called 4_____ evolution sometimes causes unrelated organisms to look alike because they live in similar environments. The Tasmanian "tiger" looks like a cat (see text, Chapter 18 introduction), but it is actually a marsupial, more closely related to a kangaroo! Such similarity due to convergence is called 5_____ ; it is not useful in systematics.

6_____ systematics goes beneath the surface appearance of organisms and compares nucleic acids or other molecules (such as 7_____ proteins) to explore relationships. Researchers use computers to search through and compare nucleic acid 8_____ sequences form different species. In general, the more similar the base sequences, the more 9_____ related the organisms in question. Some nucleic acids, such as the DNA in 10_____, evolve rather rapidly. Thus, mDNA can be used to trace recent evolutionary events, such as the divergence of various human groups. Other nucleic acids, such as 11_____ RNA, change more slowly, so they can track changes occurring over hundreds of millions of years. Because a good fossil record goes back only about 12_____ million years, molecules can give us insights into the evolution and relationships of organisms for which we have few good fossils, such as 13_____. And because some genes appear to change at a known rate, they allow us to calibrate a molecular 14_____ that can be used to date evolutionary branchpoints. But biologists need to be careful in comparing molecules because sometimes molecular similarities can arise by chance. Such molecular analogies are called 15_____.

Molecular systematics has given us some surprising insights into evolutionary relationships. Comparing all the genes—the 16_____ —of chimps and humans shows that on a molecular level, we are 99% identical. Humans and chimps share a fairly recent common ancestor—perhaps only 5 million years ago. The genes of humans and yeasts are 50% alike, showing that humans and yeasts are only distantly related, but also demonstrating the relatedness of all life.

An important goal of systematics is to name and classify organisms. Biologists called 17_____ use morphological and molecular comparison to name and group species. Each species is given a two-part name, called a 18_____— *Homo sapiens*, for example. But naming is only a starting point. The ultimate goal of taxonomy is to place each organism into a hierarchy of taxonomic categories from 19_____ (the smallest) to 20_____ (the largest and most inclusive). Ideally, these categories reflect evolutionary history. Biologists depict these relationships in the form of 21_____ trees. Species are the twigs of such a tree. The limbs of the tree are larger groupings such as orders, classes, and phyla. The newest approach for constructing phylogenetic trees is called 22_____, which seeks to identify 23_____—branches that include an ancestral species and all its descendants. Cladistics has shaken some branches of the evolutionary tree. For example, this approach places 24_____ within the reptile clade, and separates humans and chimps from other apes.

Exercise 5   (Module 15.7)

The system of taxonomic categories used by biologists is like a set of boxes into which organisms are sorted. A cocker spaniel—*Canis familiaris*—for example, is first placed in a small box, the specific name *familiaris* that separates it from all other species. This is placed in a slightly larger box, the genus *Canis*, which also holds *Canis lupus* (the wolf) and *Canis latrans* (the coyote). This genus box is placed in a larger box, along with other genera of doglike animals, and so on, all the way up to the last box that separates eukaryotes from prokaryotes. Imagine that the nested boxes below represent the taxonomic categories, starting with species (omitting subphylum). Label the boxes to show the relationships among the categories.

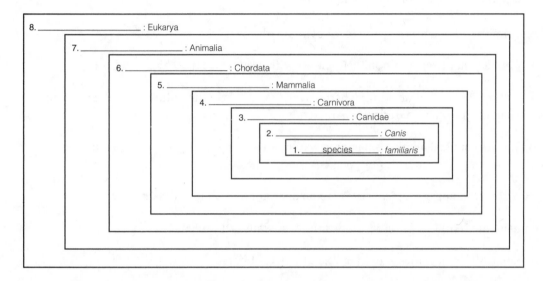

Exercise 6   (Module 15.8)

Cladistic analysis seeks to clarify evolutionary and taxonomic relationships by finding clades, groups of organisms made up of an ancestor and all its descendants, and constructing phylogenetic trees called cladograms. This simplified cladogram uses cladistic analysis (based on anatomy, but backed up by molecular data) to reconstruct the relationships among four groups of plants and their closest relatives, the green algae. Read Module 15.8, examine the trees in the module and below, and then answer the following questions. This exercise is rather difficult, so take your time.

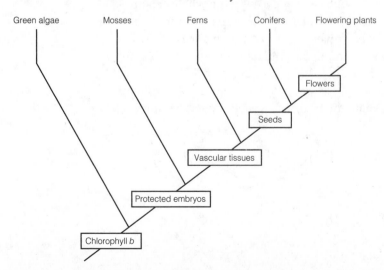

1. Which four groups of organisms above make up the ingroup?

2. Which organisms constitute the outgroup?

3. Are analogous or homologous features used in cladistic analysis?

4. Which characters are unique to a lineage of organisms, shared derived characters or shared primitive characters? Which are more useful in differentiating among (separating out) distinct lineages?

5. What is a shared primitive character common to all plants?

6. What is a shared derived character common to all plants?

7. What is a shared primitive character common to all plants with seeds?

8. What is a shared derived character common to all plants with seeds?

9. Which characters are most useful in deciding whether an organism is in the outgroup or the ingroup, shared primitive characters or shared derived characters?

10. If we are interested in focusing on all plants that have vascular tissues, which groups on the phylogenetic tree constitute the outgroup? The ingroup?

11. What is the name of a taxonomic group consisting of an ancestor and all its descendants?

12. What other organisms are in the clade that includes the first plants with seeds?

13. Name or describe nine different clades shown on the phylogenetic tree above.

Exercise 7   (Module 15.9)

Web/CD Thinking as a Scientist   *How Is Phylogeny Determined Using Protein Comparisons?*

Homologous structures—similar structures derived from the same structure in a common ancestor—tell us about phylogenetic relationships among organisms. But convergent evolution can make unrelated organisms look alike; their similarities may be analogous, not homologous. Fortunately, we can dig beneath surface similarities and compare biological molecules to measure relatedness between species. For example we can compare protein amino acid sequences, RNA base sequences, or DNA base sequences—even whole genomes. Imagine that you have sequenced mitochondrial DNA (mDNA) for six species of rodents, A through F. All the rodents are thought to have evolved from a common ancestor, X. The number of differences in mDNA sequence are compiled in the table below, which reads like a road map mileage chart. For example, there are 4 differences between A and C, and 9 differences between A and D. Use the differences in sequence to place Species A through F on the phylogenetic tree. (Hint: Don't get too mathematical; just "eyeball" overall numbers.)

A					
B	10				
C	4	11			
D	9	5	10		
E	14	16	15	15	
F	10	2	10	6	16
	A	B	C	D	E	F

1._____ 2._____ 3._____ 4._____ 5._____ 6._____

X

Exercise 8   (Module 15.10)

Web/CD Activity 15C   *Classification Schemes*

Since the 1960s, biologists have classified living things into five kingdoms. Recently, a three-domain classification system has gained favor among biologists. Compare these classification schemes by checking the statements that apply to a five-kingdom system, a three-domain system, or both.

	5-kingdom system	3-domain system
1. Four major groups of eukaryotes and one major group of prokaryotes	_____	_____
2. Two major groups of prokaryotes and one major group of eukaryotes	_____	_____
3. Two fundamentally different groups of organisms	_____	_____
4. Three fundamentally different groups of organisms	_____	_____
5. Classification based primarily on structure and nutrition	_____	_____
6. Classification based more on molecular studies	_____	_____
7. Implies that prokaryotes are more closely related to each other than to eukaryotes	_____	_____
8. Implies that one group of prokaryotes is closer to eukaryotes than to other prokaryotes	_____	_____
9. An attempt to classify life in a useful way that reflects evolution	_____	_____
10. Not a fact of nature, but a human construction	_____	_____

Testing Your Knowledge

Multiple Choice

1. Which of the following correctly matches an event in the history of life with the correct geologic eon?
   a. first mammals—Archaean
   b. first animals—Proterozoic
   c. dominance of dinosaurs—Archaean
   d. origin of life—Proterozoic
   e. first life on land—Phanerozoic

2. Systematics is concerned with
   a. naming organisms.
   b. studying biological diversity.
   c. taxonomy.
   d. tracing phylogeny.
   e. all of the above.

3. If you want to see a dinosaur, it would be best to set the controls of your time machine for the
   a. Mesozoic era.
   b. Paleozoic era.
   c. Cenozoic era.
   d. Paleozoic era.
   e. Precambrian.

4. Which of the following taxonomic categories contains all the others?
   a. genus
   b. class
   c. family
   d. subclass
   e. order

5. Until recently, most biologists classified living things into five kingdoms. A newer scheme recognizes three basic groups, called "domains":
   a. bacteria, plants, and animals
   b. prokaryotes, eukaryotes, and plants
   c. plants, animals, and fungi
   d. eukaryotes and two kinds of prokaryotes
   e. bacteria, protists, and eukaryotes

6. Organisms that survive mass extinctions
   a. often diversify, taking advantage of new opportunities.
   b. usually are so reduced in numbers that they soon go extinct.
   c. are often "living fossils" that have existed unchanged for long periods.
   d. usually cannot cope with the new conditions that follow.
   e. usually lack exaptations for their new environment.

7. Pangaea
   a. was a land mass that broke up to form the present-day continents.
   b. is the idea that all life on Earth is related.
   c. was an animal common in ancient seas but now extinct.
   d. is the evolutionary history of a species, family, or phylum.
   e. is the theory that crustal plates can move relative to one another.

8. The theory of plate tectonics helps us to explain all of the following except
   a. locations of volcanoes.
   b. formation of river systems.
   c. distribution of animals and plants.
   d. formation of mountain ranges like the Himalayas.
   e. earthquakes.

9. What evidence most strongly suggests that an impact by an asteroid or meteorite may have caused the extinction of the dinosaurs?
   a. Fossils show that dinosaurs suffered from cold and starvation.
   b. Sedimentary rocks contain a layer of mineral uncommon on Earth.
   c. There have been several near misses in recent years.
   d. The dinosaurs disappeared rather abruptly—virtually overnight.
   e. Fossils indicate that most dinosaurs were looking up when they died.

10. The oldest fossils usually
    a. contain more radioactive isotopes than younger fossils.
    b. are found in the deepest strata.
    c. have the longest half-lives.
    d. are found above younger fossils.
    e. are found in sediments from the Cenozoic era.

Essay

1. Explain how the formation of Pangaea may have led to mass extinctions at the end of the Permian period, about 250 million years ago.

2. Place the following events in the history of life in the proper order: appearance of humans, origin of eukaryotes, dominance of dinosaurs and cone-bearing plants, origin of animals, appearance of first vertebrates, diversification of mammals, origin of flowering plants, first prokaryotes, movement of plants and animals onto land.

3. Unlike hours, days, or weeks, geologic eons, eras, and periods are not of uniform lengths. For example, the Paleozoic era lasted more than 200 million years, the Cenozoic era only 65 million years. Explain why the units of the geologic time scale are uneven.

4. Describe two different ways in which plate tectonics and continental drift cause mountains to form.

5. The following list shows the classification of a human being, *Homo sapiens*. Name the category that corresponds to each of the taxa listed.

Eukarya	Primates
Animalia	Hominidae
Chordata	*Homo sapiens*
Vertebrata	
Mammalia	

Applying Your Knowledge

Multiple Choice

1. Which of the following would cast doubt on the asteroid-impact hypothesis for the extinction of the dinosaurs?
   a. finding a crater 200 million years old
   b. finding fossil dinosaur bones beneath a layer of iridium
   c. determining that birds are closely related to dinosaurs
   d. finding fossil dinosaur bones above a layer of iridium
   e. finding that many forms of marine life disappeared at the same time as the dinosaurs

2. The wings of birds and insects have the same function, but they do not have the same evolutionary origin. Bird and insect wings are
   a. homologous.
   b. phylogenetic.
   c. analogous.
   d. binomial.
   e. homoplasies.

3. Two animals in the same family would not have to be in the same
   a. genus.
   b. domain.
   c. order.
   d. phylum.
   e. class.

4. A phylogenetic tree of bird families would most clearly show which of the following?
   a. characteristics shared by all bird families
   b. evolutionary relationships among families
   c. families that look most alike
   d. analogous structures shared by various species
   e. relative ages of living species of birds

5. Fossils of an ancient reptile called *Lystrosaurus* have been found in Africa, India, and Antarctica. Which of the following best explains this distribution?
   a. They were able to move between continents before the oceans filled.
   b. Movement of India due to continental drift carried them from place to place.
   c. These areas were once next to each other and have since drifted apart.
   d. They were able to migrate over frozen seas during the Ice Ages.
   e. Changes in climate forced them to migrate from place to place.

6. A species of fruit fly is thought to have diverged from another species less than a thousand years ago. Study of which of the following would best clarify these fruit fly relationships?
   a. mitochondrial DNA
   b. the fossil record
   c. ribosomal RNA
   d. comparison of analogous structures
   e. radiometric dating

7. Which of the following taxa is *least* closely related to the others?
   a. Archaea
   b. Plantae
   c. Bacteria
   d. Eukarya
   e. Animalia

8. According to Figure 15.9, which of the following are most closely related?
   a. raccoon and lesser panda
   b. giant panda and raccoon
   c. giant panda and lesser panda
   d. sloth bear and spectacled bear
   e. brown bear and sloth bear

9. A microbiologist recently discovered a new single-celled eukaryotic organism. It is green and swims, propelled by two flagella. This organism would probably be placed in Kingdom
   a. Monera.
   b. Plantae.
   c. Archaea.
   d. Protista.
   e. Animalia.

10. In a detailed phylogenetic tree showing the evolutionary relationships within a class, the "trunk" of the tree might represent the _____, while the ends of individual "twigs" might represent _____.
    a. phylum . . . classes
    b. class . . . species
    c. genus . . . species
    d. species . . . families
    e. class . . . phyla

Essay

1. When a tree was buried in a swamp, it contained 400 mg of carbon-14. The half-life of carbon-14 is 5730 years. How long ago was the tree buried if it now contains 100 mg of carbon-14?

2. Australia is famous for its many marsupials—mammals that are born at an early stage in development and then mature in the mother's pouch. Koalas, kangaroos, wombats, and opossums are all marsupials. Marsupials are considered offshoots of an earlier type of mammal than the more familiar placental mammals, whose young remain longer in the uterus, connected to the mother by the placenta. Marsupials are found elsewhere, but in most places most marsupials have been displaced by the more "efficient" placentals. Speculate about why the marsupials managed to hang on in Australia.

3. The dinosaurs perished in a mass extinction at the end of the Cretaceous period, 65 million years ago. Briefly describe alternative hypotheses that have been suggested to explain the extinction of the dinosaurs. Did other forms of life disappear during the same period? Does this cast doubt on any of the hypotheses? Why?

4. Biologists like to look at fossils to trace evolutionary relationships. But birds are very delicate and unlikely to be fossilized; only a handful of old bird fossils are known. How might the molecules in the cells of living birds be compared to determine how ducks, sparrows, and hawks are related? Briefly name the molecules, what characteristic of the molecules would be compared, and how this would indicate how closely related the birds are.

5. Molecular comparisons indicate that the vultures of Africa and America are more closely related to separate ancestors than they are to each other. If the two groups of vultures are not closely related, why do you think they look so much alike?

Extending Your Knowledge

1. Have you ever collected fossils? It is a very interesting and rewarding hobby. In fact, several well-known paleontologists got started as amateurs. Are there fossil deposits in your area? If you are interested in looking for fossils, you can get information from state departments of geology or mining, geological surveys, departments of natural resources, bureaus of mines, university geology departments, or local rockhound clubs. Directions to many sites may be obtained from websites.

2. If you are interested in learning more about fossils and the history of life, many university museums have fossil collections. Some of the best dinosaur exhibits are located at the American Museum of Natural History in New York, the Smithsonian Institution in Washington, DC, the Royal Ontario Museum in Toronto, the National Museum of Canada in Ottawa, the Royal Tyrell museum in Drumheller, Alberta, the Peabody Museum at Yale University, the Carnegie Museum in Pittsburgh, the Field Museum of Natural History in Chicago, the Denver Museum of Natural History, and the Los Angeles County Museum. Assembled dinosaur skeletons can be seen at Dinosaur National Monument in Utah, and Dinosaur Provincial Park in Alberta. You can see dinosaur tracks at Rocky Hill, Connecticut, and near Glen Rose, Texas.

The Origin and Evolution of Microbial Life: Prokaryotes and Protists

16

A drop of pond water holds a beautiful and fascinating world. Under a microscope, pond scum becomes delicate filaments of green algae, their chloroplasts spiraling inside crystal cell walls. Other algae swim and spin lazily. Golden diatoms glide across the field of view, their intricately etched silica skeletons refracting rainbows of light. Necklaces of simpler cyanobacteria wave gently to and fro. Ciliated cells shaped like funnels suck algae and even smaller bacteria into their mouths, while formless amoebas crawl slowly in search of microscopic prey. Suddenly a huge *Paramecium* darts into view, its cilia beating in rhythm as it vacuums up small food particles. These creatures remind us of a time before there were any plants or animals—before there were any multicellular organisms at all. Some take us back to the very dawn of life on Earth. This chapter concerns the origin of life and the diversity of microbial life forms.

Organizing Your Knowledge

Exercise 1   (Module 16.1)

Web/CD Activity 16A   *The History of Life*

Summarize the early history of the Earth and the beginnings of life by numbering the following events in sequence. Some of the events are described in the text and some in Figure 16.1C.

_____ A. Formation of the seas

_____ B. Loss of Earth's original hydrogen (H_2) atmosphere

_____ C. Formation of Earth

_____ D. Melting of Earth, followed by formation of core and crust

_____ E. Formation of oldest known fossils

_____ F. The "Big Bang"—formation of the universe

_____ G. O_2 from photosynthetic prokaryotes produces aerobic atmosphere

_____ H. Origin of eukaryotes

_____ I. Volcanoes belch out atmosphere of H_2O, H_2S, CO_2, N_2, CH_4, NH_3

_____ J. Origin of life—first prokaryotes

_____ K. Appearance of animals

Exercise 2   (Modules 16.2 – 16.6)

Web/CD Thinking as a Scientist   *How Might Conditions on Early Earth Have Created Life?*

This flowchart summarizes experiment and theory concerning the origin of life. Fill in the boxes by choosing from the list of components. Fill in the ovals by choosing from the list of processes. Some are done for you.

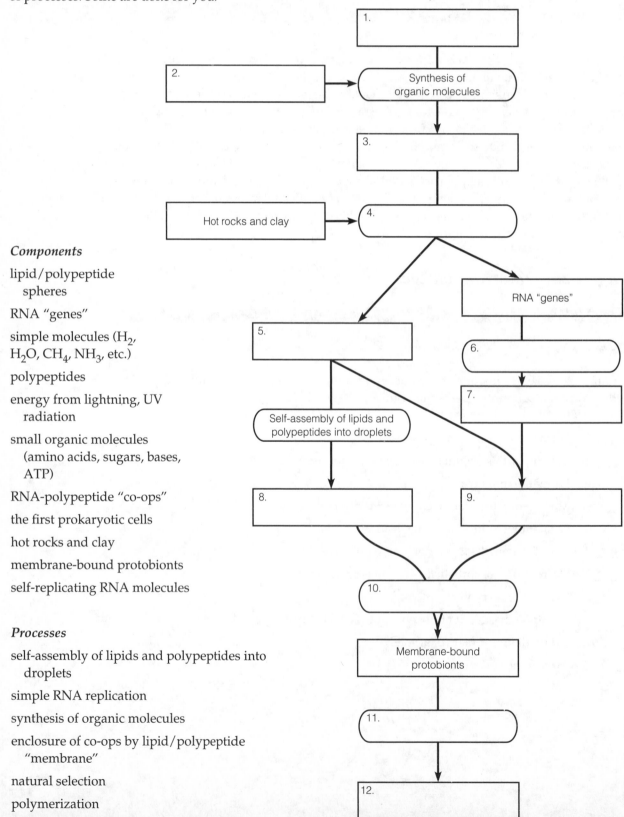

Components

lipid/polypeptide
  spheres

RNA "genes"

simple molecules (H_2,
  H_2O, CH_4, NH_3, etc.)

polypeptides

energy from lightning, UV
  radiation

small organic molecules
  (amino acids, sugars, bases,
  ATP)

RNA-polypeptide "co-ops"

the first prokaryotic cells

hot rocks and clay

membrane-bound protobionts

self-replicating RNA molecules

Processes

self-assembly of lipids and polypeptides into
  droplets

simple RNA replication

synthesis of organic molecules

enclosure of co-ops by lipid/polypeptide
  "membrane"

natural selection

polymerization

Exercise 3   (Modules 16.7 – 16.8)

There are two fundamentally different kinds of prokaryotes. State whether each of the following describes bacteria (B) or archaea (A).

_____ 1. genes lack introns

_____ 2. many rRNAs match eukaryote rRNAs

_____ 3. cell walls contain peptidoglycan

_____ 4. uninhibited by streptomycin

_____ 5. some membrane lipids have branched chains

_____ 6. some genes contain introns

_____ 7. complex RNA polymerase

_____ 8. membrane lipids have unbranched chains

_____ 9. simple RNA polymerase

_____ 10. inhibited by streptomycin

_____ 11. cell walls lack peptidoglycan

_____ 12. rRNAs different from eukaryotic rRNAs

_____ 13. more like eukaryotes

Exercise 4   (Module 16.11)

Web/CD Thinking as a Scientist   *What Are the Modes of Nutrition in Prokaryotes?*

Prokaryotes can be categorized according to their mode of nutrition. Identify whether each of the following prokaryotes is a photoautotroph (as are cyanobacteria), a chemoautotroph, a photoheterotroph, or a chemoheterotroph (as are most prokaryotes).

_____ 1. Energy source—sun; carbon source—organic compounds

_____ 2. Energy source—inorganic chemicals; carbon source—CO_2

_____ 3. Energy source—organic compounds; carbon source—organic compounds

_____ 4. Energy source—sun; carbon source—CO_2

Web/CD Activity 16B *Prokaryotic Cell Structure and Function*
Web/CD Activity 16C *Diversity of Prokaryotes*

These modules review some of the shapes, characteristics, and types of prokaryotes. Review them by completing this crossword puzzle.

Across

4. ____ are blue-green photosynthetic bacteria.

5. Prokaryotes ____ quickly in a favorable environment.

7. Archaea called ____ inhabit your digestive tract.

9. Prokaryotic ____ are smaller than those of eukaryotes, and differ in RNA and protein content.

10. Certain ____ block prokaryotic ribosomes, but not those of eukaryotes.

11. Archaea produce and consume vast amounts of ____, a greenhouse gas.

12. The ____ diverged from archaea early in the history of life.

14. Gram- ____ bacteria often cause disease.

16. A ____ of cyanobacteria may indicate polluted water.

17. Spherical prokaryotes are called ____.

18. ____ pili link prokaryotes during conjugation.

20. Certain proteobacteria ____ nitrogen.

21. Many archaea live in extreme environments such as hot ____.

22. Many bacteria produce dormant, resistant cells called ____.

23. Autotrophic prokaryotes can get energy from ____.

24. We can identify bacteria by using a ____ stain.

25. You might find an extreme ____ in very salty water.

27. A prokaryote may be able to swim using a propellerlike prokaryotic ____.

28. The earliest life form may have been a ____ that obtained energy from inorganic compounds.

29. Vibrios, spirilla, and spirochetes are curved or ____-shaped.

30. Archaea and bacteria are both ____.

Down

1. Cocci that occur in chains are called ____.

2. The ____ of a prokaryote has only about 1/1000 the DNA of a eukaryotes.

3. Some prokaryotes have folded ____ that perform metabolic function.

6. Hairlike appendages called ____ enable bacteria to stick to surfaces and each other.

8. ____ obtain carbon atoms from organic compounds.

13. The major kinds of prokaryotes are bacteria and ____.

15. ____ make their own organic compounds from inorganic sources.

16. Rod-shaped prokaryotes are called ____.

18. Cyanobacteria formed layered ____ in the ancient seas.

19. Gram- ____ bacteria include actinomycetes and mycoplasmas.

26. Chemoheterotrophs obtain ____ and carbon from organic compounds.

Exercise 6   (Modules 16.14 – 16.16)

Web/CD Activity 16C   *Diversity of Prokaryotes*

These modules outline some of the important roles of prokaryotes. To review them, fill in the blanks in the story below.

"I wish I could get rid of this cold," Matt said, as he wiped his red and swollen nose. "Bacteria are pests that evolved purely to make my life miserable. I wish they'd all disappear."

Alex looked up from his psychology textbook. "Colds are caused by viruses, not bacteria," he said. "And if bacteria disappeared, we'd be in deep trouble. We couldn't live on this planet if it weren't for bacteria."

"Really?"

"The first living things on Earth were [1]_____, the simple kinds of cells that we call bacteria and archaea. Prokaryotes eventually gave rise to creatures with more complex cells—[2]_____. That includes everything from algae to us. Incidentally, they changed the entire Earth. For example, they produced the [3]_____ in the atmosphere."

Matt sniffled. "Maybe they don't cause colds, but they do make people sick."

"Yes, there are many [4]_____, or disease-causing, bacteria," Alex replied. "About [5]_____% of all human diseases are caused by bacteria."

"So how do they make you sick?"

"Most bacteria cause illness by producing [6]_____. *Staphylococcus aureus*, for example, is a common skin bacterium, but it secretes substances called [7]_____ that can cause food poisoning or toxic shock in the body. A strain of *E. coli* bacteria from cattle produces an exotoxin that can cause bloody [8]_____. The deadly [9]_____ bacillus, used as a weapon by bioterrorists, also kills with exotoxins. Some bacteria have poisons called [10]_____ in their cell walls. A species called [11]_____ causes food poisoning that way."

Matt sneezed. "This doesn't feel like a simple cold. It could be pneumonia. Or maybe I inhaled anthrax [12]_____!"

"I doubt it's anthrax. If it is pneumonia, *then* we're talking *bacteria*. But healthy adults in developed countries don't get many bacterial infections these days. Sanitation measures, like [13]_____ and [14]_____ systems, keep many bacterial diseases from spreading, and—"

"You don't get pneumonia from dirty water, Mr. Know-It-All," Matt interjected.

"Right, but most forms of pneumonia, like most bacterial diseases, can be treated with [15]_____."

Matt sneezed. "I need to get to the emergency room. I think all bacteria should be wiped off the face of the Earth."

Alex interrupted, "Not so fast. We couldn't live without them. Many prokaryotes are involved in the cycling and recycling of various ⁱ⁶_____ between living things and the nonliving environment. Like the prokaryotes that live in the roots of ¹⁷_____—plants like beans and peas—that take ¹⁸_____ from the air and convert it into forms that plants can use."

"How does that help me?"

"It only helps you if you eat plants or animals that eat plants. I think you do, along with just about every other animal on this planet. By the way, ¹⁹_____ that live in soil and water also trap nitrogen, along with providing some of the oxygen we breathe."

Matt brightened. "I think I understand. There is another kind of recycling that prokaryotes do. They decompose ²⁰_____ matter, like ²¹_____ and ²²_____, and return chemicals to the environment. Then living things can use them again. If it weren't for prokaryotes, in no time we'd be up to our necks in . . . "

"Sewage. That's why prokaryotes are important in sewage-²³_____ plants. ²⁴_____ prokaryotes work on solid matter, called ²⁵_____ and ²⁶_____ bacteria break down liquid wastes. Some bacteria can even break down oil, and some are used to extract metals, like ²⁷_____, from low-grade ores, and might help us clean up toxic wastes from old mining sites. Using bacteria to clean up the environment is called ²⁸_____".

Matt sneezed. "I could use some to clean up my toxic wastes. This is no ordinary cold. Maybe it's the flu. Maybe it's Lemon disease."

Alex sighed. "You mean Lyme disease? Been bitten by a ²⁹_____ lately?"

Exercise 7   (Module 16.17)

This module explains how the complex organelles of eukaryotic cells may have arisen through (1) infolding of an ancestral prokaryote's plasma membrane and/or (2) engulfing of smaller prokaryotes by an ancestral prokaryotic cell (endosymbiosis). Review the module by answering the following questions on a separate sheet of paper.

1. Describe the endosymbiotic model.
2. Name two organelles that probably arose via endosymbiosis.
3. What are the two closest bacterial relatives of these organelles?
4. What is the evidence for endosymbiosis?
5. How might infolding have occurred?
6. Name two cell structures that probably originated via infolding of the plasma membrane of an ancestral prokaryotic cell.
7. What other organelle might have evolved from an archaeal endosymbiont?

Exercise 8   (Modules 16.18 – 16.24)

Web/CD Thinking as a Scientist   *What Kinds of Protists Are Found in Various Habitats?*

This phylogenetic tree tentatively outlines the evolution of various groups of eukaryotes. Most are informally called protists. Protists are a diverse collection of eukaryotes, simple to complex, with varying lifestyles. Read the modules about protists, and then review by matching each question with the correct branch(es) of the phylogenetic tree. Some answers are used more than once, and some questions have multiple answers.

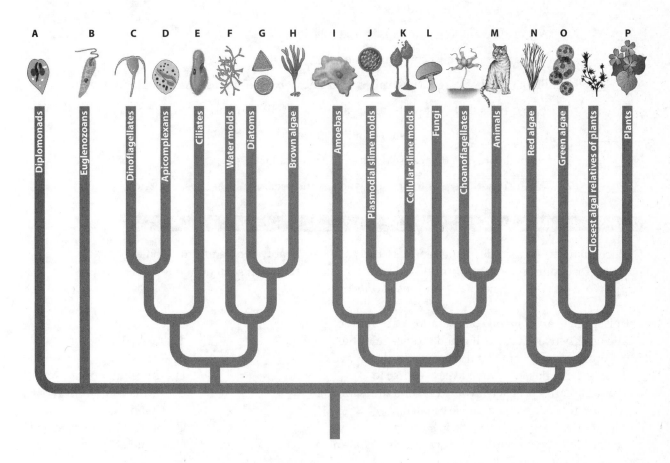

_____ 1. Three groups of organisms that are *not* protists

_____ 2. This group gets its name from the green flagellate *Euglena*

_____ 3. These three groups of organisms make up the stramenopiles

_____ 4. Three kinds of protists that decompose organic matter

_____ 5. Can be green single cells, balls or strands of cells, or complex multicellular organisms

_____ 6. The names of these three groups suggest that they are related to fungi, but they are not

_____ 7. The most ancient surviving lineage of eukaryotes

_____ 8. Fungus-like protists that decompose dead plants and animals in the water

_____ 9. Mostly free-living protists that move by means of cilia

_____ 10. Kelps

_____ 11. These three groups move via lobe-shaped extensions called pseudopodia

_____ 12. Algae characteristic of warm coastal tropical waters

_____ 13. You may see *Paramecium*, a representative of this group, in lab

_____ 14. Five groups of protists containing organisms informally known as "algae"

_____ 15. Close relatives of plants

_____ 16. These three groups are called alveolates, named for sacs under their membranes

_____ 17. Algae with complex, glassy, silica cell walls

_____ 18. Seaweeds are in these three groups

_____ 19. A flagellate in this group causes sleeping sickness

_____ 20. The largest and most complex algae

_____ 21. Blooms of these organisms, called "red tides," can poison fish and people

_____ 22. Two groups that exist part-time as individual cells and part-time as large "blobs"

_____ 23. *Giardia*, an intestinal parasite, is in this group

_____ 24. Animal parasites that cause diseases such as malaria

_____ 25. Some are serious plant parasites, such as the organism that causes potato blight

Exercise 9 (Module 16.25)

Multicellular life—seaweeds, plants, animals, and fungi—probably evolved from colonial protists. Test your understanding of this process by completing each sentence on the left with a word on the right. Each answer is used only once.

1. Cell specialization involved separation of somatic cells and _____.
2. The links between unicellular and multicellular life were _____.
3. Organisms moved onto land when green algae gave rise to _____.
4. At least three lineages may have given rise to _____.
5. Multicellularity probably evolved several times among _____.
6. All life functions happen in one cell in all _____.
7. Different cells do different jobs in _____.
8. The oldest known multicellular organisms include algae and _____ such as corals and worms.

A. Seaweeds
B. Animals
C. Protists
D. Unicellular organisms
E. Plants
F. Colonies
G. Multicellular organisms
H. Gametes

Testing Your Knowledge

Multiple Choice

1. Which of the following was probably *not* present in large amounts in the atmosphere at the time life is thought to have originated?
   a. water (H_2O)
   b. nitrogen (N_2)
   c. methane (CH_4)
   d. oxygen (O_2)
   e. carbon dioxide (CO_2)

2. Prokaryotes called _____ are similar in many ways to eukaryotic organisms.
   a. archaea
   b. bacteria
   c. protozoa
   d. cyanobacteria
   e. dinoflagellates

3. Biologists are interested in the role of clay in the origin of life. They think clay might have
   a. supplied the raw materials for organic compounds.
   b. catalyzed the formation of organic polymers such as proteins and RNA.
   c. formed primitive cell membranes that could grow and divide.
   d. catalyzed the formation of monomers such as amino acids and sugars.
   e. supplied the energy for metabolism in the first simple cells.

4. Which of the following is thought to have been the first step in the origin of life?
   a. cooperation among molecules
   b. formation of polypeptide spheres
   c. formation of organic monomers
   d. replication of primitive "genes"
   e. formation of organic polymers

5. *E. coli* bacteria, which live in human intestines, are shaped like tiny, straight sausages. They are
   a. bacilli.
   b. vibrios.
   c. spirochetes.
   d. cocci.
   e. spirilla.

6. Which of the following is a difference between bacteria and archaea?
   a. Archaea are unicellular, and bacteria are colonial.
   b. The genes of bacteria have introns, while archaea lack introns.
   c. They have different chemicals in their cell membranes and cell walls.
   d. Bacteria are autotrophic and archaea are heterotrophic.
   e. They look very different under a microscope.

7. Most prokaryotes
   a. obtain energy from sunlight and carbon from organic compounds.
   b. obtain both energy and carbon from inorganic compounds.
   c. obtain energy from inorganic compounds and carbon from CO_2.
   d. obtain energy from sunlight and carbon from CO_2.
   e. obtain both energy and carbon from organic compounds.

8. Which of the following is true of cyanobacteria?
   a. They are pathogenic.
   b. They are chemoheterotrophs.
   c. They are archaea.
   d. They are protists.
   e. They are photoautotrophs.

9. Archea called _____ live in salty environments, such as salt lakes.
   a. methanogens
   b. actinomycetes
   c. extreme halophiles
   d. apicomplexans
   e. extreme thermophiles

10. All algae are
    a. unicellular.
    b. green.
    c. autotrophic.
    d. prokaryotic.
    e. multicellular.

11. Different groups of seaweeds can generally be distinguished on the basis of
    a. color.
    b. size.
    c. whether they are multicellular or unicellular.
    d. whether or not they have true leaves, stems, and roots.
    e. whether they are autotrophic or heterotrophic.

12. You would *not* expect which of the following to cause disease?
    a. apicomplexan
    b. diatom
    c. euglenozoan
    d. gram-negative bacterium
    e. sprochete

13. Which of the following is *incorrectly* paired with its mode of nutrition?
    a. *Streptococcus*—chemoheterotroph
    b. red alga—photoautotroph
    c. animal—chemoheterotroph
    d. cyanobacterium—chemoheterotroph
    e. plant—photoautotroph

14. In the life cycle of a seaweed such as *Ulva*, a haploid spore develops into a(n)
    a. sporophyte.
    b. sperm.
    c. egg.
    d. gametophyte.
    e. zygote.

15. Until about 500 million years ago, all living things were
    a. asexual.
    b. autotrophic.
    c. aquatic.
    d. prokaryotic.
    e. unicellular.

Essay

1. If life could develop from nonliving chemicals 3 to 4 billion years ago, why can't the same thing happen at present?

2. Describe Stanley Miller's experiment that simulated conditions on the ancient Earth. How was the experiment carried out? What was its purpose? Its result?

3. Describe how molecules in the ancient seas may have cooperated to form the first "cells." What kinds of molecules may have been involved? How did they cooperate?

4. Name four diseases caused by bacteria. How do bacteria cause disease?

5. In what ways are prokaryotes useful and even vital to our well-being?

6. Primitive prokaryotes radically changed the Earth for all life that followed. Their effect on the environment is described as a "revolution." What did these prokaryotes do?

7. Explain the endosymbiosis model for the origin of eukaryotes.

8. How did multicellular eukaryotes—plants, animals, and fungi—probably arise?

Applying Your Knowledge

Multiple Choice

1. Which of the following discoveries would force scientists to revise their present theories regarding the origin of life on Earth?
    a. The Earth is found to be 6 billion years old, rather than 4.6 billion.
    b. Polypeptides can catalyze replication of small RNA molecules.
    c. There was a lot of oxygen gas in the atmosphere 4 billion years ago.
    d. Minerals in lava catalyze formation of polypeptides from amino acids.
    e. Lipids spontaneously form selectively permeable membranes.

2. The following are some major events in the early history of life.
    P. first heterotrophic prokaryotes
    Q. first genes
    R. first eukaryotes
    S. first autotrophic prokaryotes
    T. first animals
    Which answer below places these events in the correct order?
    a. PQSRT
    b. QSPTR
    c. QPSRT
    d. QSPRT
    e. SPQRT

3. Bacteria that live around deep-sea hot-water vents obtain energy by oxidizing inorganic hydrogen sulfide belched out by the vents. They use this energy to build organic molecules from carbon obtained from the carbon dioxide in the seawater. These bacteria might be described as
   a. photoheterotrophs.
   b. chemoautotrophs.
   c. photoautotrophs.
   d. chemoheterotrophs.
   e. none of the above.

4. The bacterium *Bacillus thuringensis* can withstand heating, dryness, and toxic chemicals that would kill most other bacteria. This indicates that it is probably able to form
   a. pseudopodia.
   b. endotoxins.
   c. endospores.
   d. pili.
   e. peptidoglycans.

5. In an experiment, a microbiologist put equal numbers of each of the following organisms into a flask of sterile broth consisting mostly of sugar and a few amino acids. She then placed the flask in the dark. Which of the organisms would be best able to survive and reproduce in this environment?
   a. chemoheterotrophic bacteria
   b. cyanobacteria
   c. diatoms
   d. extreme halophiles
   e. green algae

6. Which of the following is *not* evidence for the role of endosymbiosis in the origin of eukaryotes?
   a. Chloroplasts have their own DNA.
   b. The inner membrane of a chloroplast is similar to prokaryotic membranes.
   c. Mitochondria and chloroplasts are surrounded by two membranes.
   d. Mitochondria reproduce by binary fission.
   e. The DNA in the eukaryotic nucleus codes for some enzymes in mitochondria.

7. As she peered through the microscope, Paige said, "I know that this thing is supposed to be either a ciliate, a flagellate, or an amoeba, but I can't figure out which." Michelle replied, "That's easy . . . "
   a. "Watch how it moves."
   b. "How big is it?"
   c. "All you have to do is see whether it has a nucleus or not."
   d. "Watch and see what it eats."
   e. "Look at its chloroplasts."

8. Which of the following was probably *not* a direct evolutionary ancestor of a maple tree?
   a. a heterotrophic prokaryote
   b. a green alga
   c. an amoeba
   d. a colonial protist
   e. *All* of the above probably *were* ancestors of a maple tree.

9. Some biologists regard seaweeds as protists, even though most other protists are microscopic unicellular organisms. Other biologists think that at least some seaweeds should be considered plants, not protists. Which of the following would support the latter position?
   a. Certain seaweeds have been found to be heterotrophic.
   b. Certain seaweeds contain several kinds of specialized cells.
   c. Certain seaweeds undergo sexual and asexual reproduction.
   d. Certain seaweeds are found to be prokaryotic.
   e. Certain seaweeds have very complex cells.

10. Protozoans called choanoflagellates live in small clusters. They look very much like choanocytes, special feeding cells found in sponges, which are simple animals. Why might biologists find choanoflagellates of great evolutionary interest?
    a. They show how the very first organisms might have lived.
    b. They might show how the first heterotrophs lived.
    c. They might offer clues about the origin of multicellular organisms.
    d. They suggest what the first eukaryotes might have been like.
    e. They might offer clues about the first organisms to live in the sea.

Essay

1. No single simulation of conditions on the early Earth has produced all the fundamental building blocks of life. Under certain conditions, certain amino acids form. With different amounts of atmospheric chemicals, other amino acids are produced. Still other experiments generate nucleic acid bases, or sugars, or ATP. Do you think the fact that no one experiment has produced all these molecules is a problem for theories about the origin of life? Explain.

2. Scientists have found traces of amino acids on meteorites from space that have landed on the Earth. How would you suppose they interpret this finding? What might be a source of error? Why do you suppose they think amino acids from space are significant?

3. Ultraviolet light breaks chemical bonds in large organic molecules. (That is why it is dangerous to unprotected skin.) Does this fact lend support to hypotheses about the possible role of UV radiation in the origin of life, or does it present problems for these hypotheses? Explain.

4. The diphtheria bacterium *Corynebacterium diphtheriae* grows into a mass at the back of the throat and can kill its victim by suffocation. But diphtheria victims also suffer from nervous tremors, paralysis, and heart failure. How might a bacterium that grows in the throat cause symptoms in other parts of the body?

5. Some bacteria can reproduce as often as every 20 minutes. If a hundred of these bacteria were placed in a large flask of culture medium, how many would there be after 6 hours?

Extending Your Knowledge

1. Diseases such as the Black Plague, typhus, malaria, and African sleeping sickness have shaped history, politics, and geography. Can you think of other diseases that have helped to shape the modern world? There are many excellent books on this subject. If you are interested in this, you might want to explore the library.

2. When the average person hears the word "bacteria," do the beneficial effects of prokaryotes in ecology and human life come to mind, or does the person tend to think of harmful disease-causing organisms? You might want to conduct an informal poll.

3. The recent anthrax scare has renewed fear of biological warfare and bioterrorism. What do you think is the best way to protect people from biological weapons?

Plants, Fungi, and the Colonization of Land 17

Many people tend to take plants for granted. It is fun to run barefoot across a cool lawn or to kick through colorful leaves on a chilly autumn day. It is enjoyable to eat a crisp salad or a juicy red apple. In each of these instances, we appreciate plants for their textures, colors, and flavors. We seldom think of them from a biological point of view—growing, reproducing, adapting, evolving. But if plants are ignored, most of the time fungi are absolutely reviled. Fungi are all around us, but most of the time we fail to notice them, unless a piece of bread tastes a little strange, or an orange turns blue, or athlete's foot makes our toes itch. Aside from an occasional pretty mushroom, most people regard fungi with a sort of wary disgust. But fungi have their own way of living and reproducing, and they are important to us in several ways. This chapter examines plants and fungi in detail.

Organizing Your Knowledge

Exercise 1   (Modules 17.1 – 17.2)

Web/CD Activity 17A   *Terrestrial Adaptations of Plants*

Plants probably evolved from multicellular green algae called charophyceans, so the two kinds of organisms share many characteristics. But plants have unique adaptations for life on land that make them different from green algae. Complete the chart below by listing the differences between plants and green algae. Consider such things as support, nutrition, transport, and reproduction. An example is done for you. On the next page, list the similarities between plants and green algae.

Differences Between Green Algae and Plants

Green Algae	*Plants*
1. Supported by water	1. Leaves and stems usually contain rigid supporting elements containing lignin
2.	2.
3.	3.
4.	4.
5.	5.
6.	6.
7.	7.
8.	8.
9.	9.
10.	10.

Similarities Between Green Algae and Plants

1. _____

2. _____

3. _____

4. _____

Exercise 2   (Module 17.3)

Web/CD Activity 17B   *Highlights of Plant Evolution*

This module summarizes the four major steps in plant evolution that gave rise to four
major modern groups of plants. Each of the statements below describes one of those steps.
Number the steps in order, and fill in the names of the plant groups.

_____ A. The first plants that produced pollen and seeds arose. The modern seed plants are
         _____ and _____.

_____ B. Plants evolved from ancestral green algae ; one lineage gave rise to the _____,
         a group that included the mosses, liverworts, and hornworts.

_____ C. The first flowering plants, or _____, appeared.

_____ D. Vascular plants evolved that had roots and strong stems supported by lignin-hardened vascular
         tissues, unlike the _____. Their modern-day representatives are the
         _____ and the seed plants.

Exercise 3   (Modules 17.4 – 17.5)

Web/CD Activity 17C   *Moss Life Cycle*

Haploid and diploid generations alternate in plant life cycles. Study the diagrams in these
modules to review alternation of generations. Then fill in the blanks below to complete
the description of the moss life cycle.

        The moss life cycle, like that of all plants, is characterized by alternation of gener-
ations. Diploid individuals called [1]_____ produce [2]_____
plants called gametophytes, which in turn produce [3]_____ sporophytes.
Since it's a cycle, we could start at any point, but let's start with a spore. A haploid moss
spore grows into the haploid [4]_____ plant, the green, cushiony growth we
see on rocks or logs in a forest or bog. [5]_____ (eggs and sperm) develop in
the protection of special organs called [6]_____ that are part of the gametophytes. Moss [7]_____ have [8]_____ that enable them to swim
to the eggs, given a film of moisture produced by dew or raindrops. [9]_____,
the fusion of egg and sperm, produces a diploid [10]_____, which remains

protected in the female gametophyte. The zygote divides by mitosis and develops into the sporophyte, which consists of a [11]_____ attached to the gametophyte by a slender stalk. Within the sporangium, haploid [12]_____ are produced by the process of [13]_____. When these spores are mature, the sporangium opens and they scatter in the wind, to begin the cycle anew.

Exercise 4   (Module 17.6 – 17.7)

Web/CD Activity 17D   *Fern Life Cycle*
Web/CD Thinking as a Scientist   *What Are the Different Stages of the Fern Life Cycle?*

Ferns and lycophytes have vascular tissues, but do not produce seeds; they are often called "seedless vascular plants."

Identify the stages of the fern life cycle by labeling this diagram. Include the following: **haploid phase, diploid phase, sporophyte, gametophyte, zygote, sporangia, spores, sperm, egg, meiosis, fertilization,** and **mitosis and development.** One answer is used twice. Color the haploid part of the life cycle yellow and the diploid part of the life cycle gray.

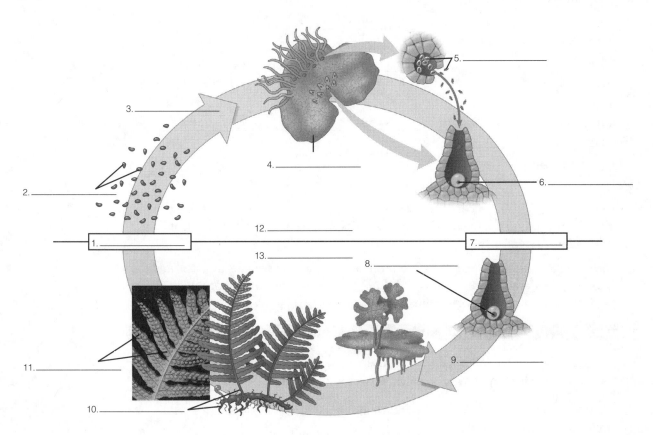

Exercise 5   (Module 17.8)

Web/CD Activity 17E   *Pine Life Cycle*

Two significant plant adaptations seen in gymnosperms such as pines are pollen and seeds. Complete the following sentences with a structure or stage in the pine life cycle. Select your answers from this list: **seed, pine tree, ovule(s), sperm, egg(s), pollen grain(s), male cone(s), female cone(s), embryo, seed coat,** and **zygote.** Some answers are used more than once.

1. A _____ is the diploid sporophyte generation of the pine life cycle.

2. The haploid gametophyte generation develops within the _____ and _____.

3. The small, soft _____ contain many sporangia. Meiosis occurs in the sporangia, producing many spores that develop into _____. These are male gametophytes.

4. Each scale in the larger, woody _____ bears two _____. Each of these develops as a sporangium covered by a tough integument.

5. The wind carries pollen grains to the female cones. Pollination occurs when a pollen grain lands on and enters an _____.

6. After pollination, meiosis occurs in the _____, producing a haploid spore that develops into the female gametophyte.

7. Over a period of months _____ are produced by the female gametophyte in the ovule. At the same time, the male gametophyte (the pollen grain) produces _____.

8. A tiny tube grows out of the _____, releasing a _____ to fertilize the _____.

9. A diploid fertilized egg, or _____, develops into a sporophyte _____. The whole ovule becomes a _____.

10. The _____ consists of the _____ and a food supply made from the remains of the female gametophyte, covered by a seed coat made from the ovule's integument.

11. The seed falls on the ground. When conditions are right, the seed germinates, and the embryo, over decades, grows into a _____, the adult sporophyte. It then produces cones, and the cycle begins again.

Exercise 6   (Module 17.9)

Flowers are responsible for the diversity and success of angiosperms—the flowering plants. Review the flower by matching each flower part with its function.

A. Petal

B. Style

C. Sepal

D. Ovary

E. Stamen

F. Carpel

G. Stigma

H. Filament

I. Anther

_____ 1. Eggs develop in this chamber

_____ 2. Consists of filament and anther

_____ 3. Produces pollen

_____ 4. Attracts pollinators

_____ 5. Female structure with ovary at its base

_____ 6. Protects the flower before it opens

_____ 7. Sticky tip that traps pollen

_____ 8. Stalk that supports anther

_____ 9. Between stigma and ovary

Exercise 7 (Module 17.10)

Web/CD Activity 17F   *Angiosperm Life Cycle*

Identify the stages in the life cycle of an angiosperm—a flowering plant—by labeling the diagram below. Include **haploid phase, diploid phase, sporophyte, anther, meiosis, ovary, ovule, female gametophyte, pollen grain (male gametophyte), egg, stigma, pollen tube, sperm, zygote, seed, seed coat, fertilization, embryo, fruit, food supply,** and **pollination.** Then color the haploid part of the life cycle yellow and the diploid part gray.

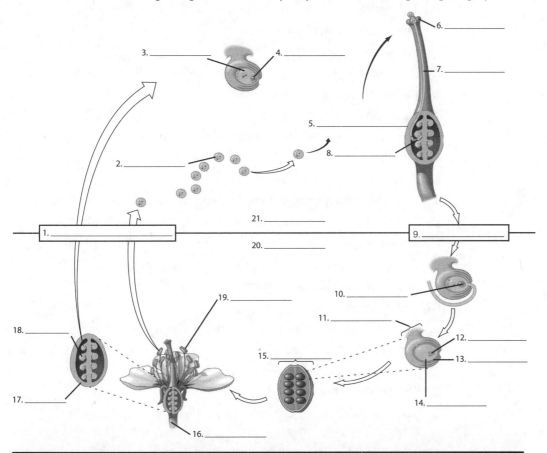

Exercise 8 (Modules 17.8 – 17.11)

Web/CD Activity 17E   *Pine Life Cycle*
Web/CD Activity 17F   *Angiosperm Life Cycle*

The life cycles of gymnosperms and angiosperms are similar, but angiosperms have added their own tricks of pollination and seed dispersal. Compare gymnosperms (conifers) and angiosperms in the chart below.

Characteristic	Gymnosperms (Conifers)	Angiosperms
Characteristic reproductive structures	1.	2.
Mode of pollination (wind, insects, etc.)	3.	4.
Interval between pollination and fertilization	5.	6.
Time required to produce seeds (pollination to seed dispersal)	7.	8.
Seed protection and dispersal	9.	10.

Exercise 9   (Modules 17.3 – 17.14)

Review your knowledge of the structure, life cycles, evolution, and uses of the major plant groups. Match each statement with a group (or groups) of plants. Some statements require more than one answer.

M. Mosses
F. Ferns
G. Gymnosperms
A. Angiosperms

_____ 1. Flowering plants

_____ 2. Two types of plants that produce seeds

_____ 3. These plants and their relatives formed coal deposits

_____ 4. The simplest vascular plants

_____ 5. Pines, firs, spruces, and cedars

_____ 6. A type of plant in which the gametophyte stage is dominant

_____ 7. Plants that produce fruits

_____ 8. Nonvascular plants

_____ 9. Two types of plants with flagellated swimming sperm

_____ 10. Conifers

_____ 11. Roses, apples, maples, and daisies

_____ 12. Plants with horizontal stems and leaves bearing sporangia

_____ 13. Plants with the shortest gametophyte and longest sporophyte stages

_____ 14. Two types of plants whose spores develop into pollen and ovules

_____ 15. The group that first developed good roots and rigid stems

_____ 16. Source of most lumber and paper

_____ 17. The simplest plants

_____ 18. Plants that produce seeds but not fruits

_____ 19. The majority of modern plants

_____ 20. Two types of plants without seeds

_____ 21. Source of most of our food

_____ 22. Many of these plants depend on animals for pollination and seed dispersal

Exercise 10   (Module 17.14)

Web/CD Activity 17G   *Connection: Madagascar and the Biodiversity Crisis*
Web/CD Thinking as a Scientist   *How Are Trees Identified by Their Leaves?*

This module presents some sobering statistics concerning the human threat to world plant diversity. Write in the number that completes each statement.

1. About _____ million acres of forest land are cleared every year.

2. About _____ percent of the world's forests are found in the tropics.

3. About _____ percent of tropical forests were destroyed in the last third of the twentieth century.

4. The forests of North America have shrunk by almost _____ percent over the last two centuries.

5. More than _____ percent of prescription drugs are extracted from plants.

6. Fewer than _____ of the worlds' _____ known plant species have been investigated as sources of medicines.

7. The All Species Foundation wants to catalogue every species within the next _____ years.

Exercise 11   (Module 17.15 – 17.20)

Web/CD Activity 17H   *Fungal Reproduction and Nutrition*
Web/CD Activity 17I   *Fungal Life Cycles*

Fungi are strange beasts, but as you get to know them, they will grow on you! These modules introduce some of the structures, roles, and types of fungi. We will review the fungus life cycle in the next exercise, but for now, review fungus vocabulary by matching each fungal phrase with the correct term.

_____ 1. Disperse fungi over great distances

_____ 2. Fungal infection

_____ 3. Hard-to-classify fungi with no known sexual stages

_____ 4. Threadlike fungal filament

_____ 5. Fungi named for their clublike spore-producing structures

_____ 6. A branching feeding network of fungal hyphae

_____ 7. The material of fungal cell walls

_____ 8. A mutually helpful association between plant roots and fungus

_____ 9. Black bread mold is one of these zygote fungi

_____ 10. Single-celled fungi that usually live in liquid or moist habitats

_____ 11. Phase of fungus life cycle where cells have two nuclei

_____ 12. Mutualistic relationship between fungus and photosynthetic organisms

_____ 13. Ants (and most other animals) can't break this down, but fungi can

_____ 14. The scientific study of fungi

_____ 15. About 90% of plants have partnerships with these fungi

_____ 16. The closest relatives of fungi

_____ 17. A fast-growing furry mycelium on bread or fruit

_____ 18. How fungi get their food

_____ 19. Hyphae of these fuse to start sexual reproduction

_____ 20. Aboveground reproductive structure of a sac or club fungus

_____ 21. A group of fungi that produce spores in saclike structures

_____ 22. The earliest line of fungi; they have flagellated spores

_____ 23. Fungus that attacks wheat crops

A. Chytrids

B. Fruiting body

C. Mycelium

D. Cellulose

E. Mycosis

F. Lichen

G. Absorption

H. Zygomycetes

I. Mating types

J. Spores

K. Animals

L. Mycology

M. Basidiomycetes

N. Rust

O. Mold

P. Imperfect fungi

Q. Hypha

R. Ascomycetes

S. Heterokaryotic

T. Chitin

U. Yeast

V. Glomerulomycetes

W. Mycorrhiza

Exercise 12   (Modules 17.15 – 17.17)

Web/CD Thinking as a Scientist   *How Does the Fungus* Pilobolus *Succeed as a Decomposer?*

Whether a fungus is a club fungus (basidiomycete), sac fungus (ascomycete), or zygote fungus (zygomycete), its sexual life cycle includes haploid, diploid, and heterokaryotic phases. Fill in the blanks to complete this paragraph about the life cycle of a club fungus.

The life cycle of a club fungus—the familiar "mushroom"—consists of
[1]_____ distinct phases. Under suitable conditions, haploid
[2]_____ germinate and produce long filaments called [3]_____.
These strands make up a network called a [4]_____. The haploid mycelium
grows through the substrate, secreting enzymes, digesting organic matter, and absorbing
nutrient molecules. There are different kinds of mycelia, called [5]_____.

If compatible mating types come into contact, their hyphae fuse, and the hybrid hyphae form their own mycelium. The [6]_____ within these hybrids do not fuse, however. This begins the [7]_____ phase of the fungus life cycle, which may last for weeks—or years! Each cell in this mycelium contains two genetically distinct nuclei. Eventually, the dikaryotic mycelium forms a [8]_____—the mushroom that pops up out of the ground. Under the mushroom's cap, haploid nuclei finally fuse, forming [9]_____ cells. Without undergoing mitosis, each of theses cells undergoes [10]_____, forming haploid spores. These spores, produced by the billions, are dispersed by wind, water, and animals to places where they can germinate and grow into new hyphae.

Exercise 13   (Modules 17.15 – 17.22)

Web/CD Activity 17H   *Fungal Reproduction and Nutrition*
Web/CD Activity 17I   *Fungal Life Cycles*
Web/CD Thinking as a Scientist   *How Does the* Pilobolus *Succeed as a Decomposer?*

The last few modules in this chapter discuss types of fungi, their ecological roles, and their importance to humans. After reading the modules, review your knowledge of fungi by completing this crossword puzzle.

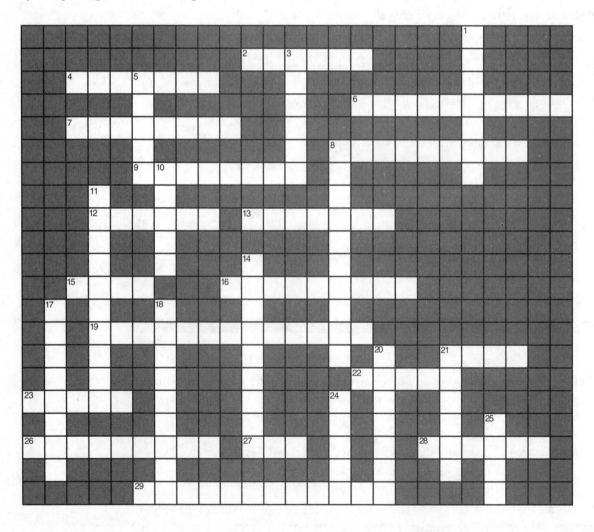

Across

2. A fungus consists of filaments called ____.
4. Fungi are responsible for the flavors of many ____.
6. A ____ is a biologist who studies fungi.
7. ____ is caused by a human fungal parasite.
8. Lichens are very sensitive to air ____.
9. ____ are the most highly prized edible fungi.
12. ____ are unicellular fungi.
13. Yeasts are used in baking and ____.
15. Fungi cannot move, but they ____ rapidly.
16. The word ____ describes the relationship between fungus and algae in a lichen.
19. Unlike plants, which are autotrophic, fungi are ____ eukaryotes.
21. Common mushrooms are ____ fungi, or basidiomycetes.
22. A parasitic fungus causes ____ elm disease.
23. A lichen consists of a fungus and green ___ or cyanobacteria.
26. Molds and ____ are typical fungi.
27. Yeasts, cup fungi, and molds are ____ fungi, or ascomycetes.
28. Club fungi can break down tough ____ in wood.
29. A mushroom is a fungal ____ structure.

Down

1. Athlete's foot is a human fungal disease, or ____.
3. Mycorrhizal fungi help many ____ absorb nutrients.
5. People have been poisoned by ____-infested grain.
8. Fungi produce ____ and other antibiotics.
10. Parasitic fungi called smuts and ____ attack crops.
11. A root and fungus can form a ____, a mutually beneficial partnership.
14. A mycorrhizal fungus helps a plant obtain ____.
17. A network of fungal hyphae make up a ____.
18. Some fungi are parasites, some are mutualists, and some ____ organic matter.
20. Fungi digest their food ___ their bodies.
21. Fungal cell walls are made of ____.
24. Fungi are classified in their own kingdom—Kingdom ____.
25. Leaf-cutting ____ grow fungi for food.

Testing Your Knowledge

Multiple Choice

1. Which of the following is *not* a difference between algae and plants?
   a. Plant cells have rigid cellulose walls, and algae cells do not.
   b. Plant zygotes and embryos are protected in moist chambers, and those of algae are not.
   c. Algae lack discrete organs—leaves, stems, roots—characteristic of plants.
   d. Plants have xylem and phloem, and algae do not.
   e. Plants have a waxy, waterproof cuticle, and algae do not.

2. Bryophytes (such as mosses) differ from all other plants in that most bryophytes
   a. do not produce flowers.
   b. have cones but no seeds.
   c. have flagellated sperm.
   d. lack vascular tissues.
   e. produce spores.

3. The gametophyte stage of the plant life cycle is most conspicuous in
   a. ferns.
   b. mosses.
   c. angiosperms.
   d. gymnosperms.
   e. seed plants.

4. Ferns and mosses are mostly limited to moist environments because
   a. their pollen is carried by water.
   b. they lack a cuticle and stomata.
   c. they lack vascular tissues.
   d. they have swimming sperm.
   e. their seeds do not store much water.

5. The diploid generation of the plant life cycle always
   a. produces spores.
   b. is called the gametophyte.
   c. is larger and more conspicuous than the haploid stage.
   d. develops from a spore.
   e. produces eggs and sperm.

6. During the Carboniferous period, forests consisting mainly of ____ produced vast quantities of organic matter, which was buried and later turned into coal.
   a. early angiosperms
   b. seedless plants
   c. giant mosses
   d. gymnosperms
   e. gymnosperms and early angiosperms

7. Which of the following best describes how fertilization occurs in a conifer?
   a. A sperm cell swims through a film of moisture to fertilize the egg.
   b. A pollen grain carried by wind fertilizes the egg.
   c. A pollen grain carried by wind produces a sperm that fertilizes the egg.
   d. A sperm cell carried by wind fertilizes the egg.
   e. A pollen grain swims through a film of moisture to fertilize the egg.

8. Most species of plants are
   a. non-seed-bearing plants.
   b. angiosperms.
   c. gymnosperms.
   d. plants other than angiosperms.
   e. nonvascular plants.

9. When you look at a pine or maple tree, the plant you see is
   a. a haploid sporophyte.
   b. a diploid sporophyte.
   c. a haploid gametophyte.
   d. a diploid gametophyte.
   e. none of the above.

10. In a flowering plant, meiosis occurs in the _____, producing a spore that develops into a female gametophyte.
    a. fruit
    b. seed
    c. stamen
    d. anther
    e. ovary

11. A fruit is a ripened
    a. seed.
    b. pollen grain.
    c. bud.
    d. ovary.
    e. anther.

12. How do fungi "find" things to eat?
    a. They produce huge numbers of tiny spores.
    b. They grow rapidly.
    c. They make their own food.
    d. They do all of the above.
    e. They do a and b only.

13. The body of a fungus consists of threadlike _____, which form a network called a _____.
    a. mycelia . . . sporangium
    b. hyphae . . . gametophyte
    c. mycelia . . . hypha

d. hyphae . . . mycelium
e. sporangia . . . fruiting body

14. Where and when does fertilization occur in the mushroom life cycle?
    a. underground, as a mycelium begins to spread
    b. on the surface of the ground, when a spore germinates
    c. in a mushroom, when nuclei of a heterokaryotic cell fuse
    d. underground, when hyphae of different mating types fuse
    e. in a mushroom, when eggs and sperm meet

15. Which of the following kinds of fungi would be considered the least useful or beneficial?
    a. mycorrhizal fungus
    b. yeast
    c. rust
    d. truffle
    e. decomposer

Essay

1. Compare a plant with a multicellular green alga, paying particular attention to plant adaptations to life on land.

2. Most nonbiologists consider seaweeds and fungi to be plants. Why? Why are seaweeds, fungi, and plants placed in separate kingdoms?

3. What are the seed plants? What adaptations have made them so successful?

4. What kinds of products do we obtain from forests? What kinds of trees supply most of our forest product needs? How can we obtain the forest products we need and still preserve our forests?

5. Sketch a flower, name its major parts, and describe their functions.

6. How do animals assist in angiosperm reproduction? How have the structures of angiosperms adapted to reflect this relationship with animals?

7. How do fungi obtain their food?

8. What two components make up a lichen? What are their roles?

Applying Your Knowledge

Multiple Choice

1. An explorer found a plant that had roots, stems, and leaves. It had no flowers but produced seeds. This plant sounds like a(n)
   a. fern.
   b. bryophyte.
   c. angiosperm.
   d. moss.
   e. gymnosperm.

2. Which of the following stages in the life cycle of a maple tree corresponds to the leafy, spongy plant in the moss life cycle?
   a. egg and sperm
   b. adult tree
   c. flower
   d. pollen grain and ovule
   e. zygote

3. Deep in the tropical rain forest, a botanist discovered an unusual plant with vascular tissues, stomata, a cuticle, flagellated sperm, conelike reproductive structures bearing seeds, and an alternation-of-generations life cycle. He was very excited about this discovery because it would be rather unusual for a plant to have both
   a. a cuticle and flagellated sperm.
   b. vascular tissues and alternation of generations.
   c. seeds and flagellated sperm.
   d. alternation of generations and seeds.
   e. cones and vascular tissues.

4. The pinyon pine lives in near-desert areas in western North America. This habitat is a bit unusual for gymnosperms because they
   a. have a long life cycle for such harsh growing conditions.
   b. possess flagellated sperm that must swim to the egg.
   c. produce extremely small quantities of pollen.
   d. lack vascular tissues and are unable to transport much water.
   e. produce cones rather than drought-resistant seeds.

5. Unlike most angiosperms, grasses are pollinated by wind. As a consequence, some "unnecessary" parts of grass flowers have almost disappeared. Which of the following parts would you expect to be most reduced in a grass flower?
   a. ovaries
   b. petals
   c. anthers
   d. carpels
   e. stamens

6. Fuchsia flowers are generally reddish, they hang downward, and their nectar is located deep in floral tubes. Fuchsias are typically pollinated by
   a. bees.
   b. flies.
   c. bats.
   d. butterflies.
   e. birds.

7. Some scum was found growing near the edge of a pond. Under a microscope, each of its cells were found to contain two nuclei. This means the scum must be
   a. some kind of alga.
   b. a fungus.
   c. a plant gametophyte.
   d. a liverwort.
   e. a plant sporophyte.

8. Strolling through the woods, you would be least likely to notice which of the following?
   a. a moss gametophyte
   b. a fern gametophyte
   c. an angiosperm sporophyte
   d. a fern sporophyte
   e. the heterokaryotic stage of a fungus

9. The diploid phase of the life cycle is shortest and smallest in which of the following?
   a. moss
   b. angiosperm
   c. fungus
   d. fern
   e. gymnosperm

10. Which of the following is a difference between plants and fungi?
    a. Plants have diploid and haploid phases.
    b. Fungi have cell walls.
    c. Fungi are autotrophic.
    d. In fungi, zygotes undergo meiosis to produce spores.
    e. Plants undergo sexual reproduction.

Essay

1. What characteristics does a fern share with the seed plants that evolved later? In what ways are ferns similar to mosses?

2. What kinds of plants are dispersed to new habitats by spores? By seeds? What are the advantages and disadvantages of each means of dispersal?

3. When his orchard was attacked by parasitic fungi, a farmer sprayed the trees with a powerful fungicide. The next season, most of the trees were free of the parasite, but they grew poorly and produced even less fruit than they had when they were infected. What might account for this change?

4. If you have ever had athlete's foot, you are probably aware that fungal infections are rather difficult to get rid of. What do you know about the structure and lifestyle of fungi that might make them particularly persistent pests?

5. Biologists give each species a Latin name, such as *Homo sapiens* for humans and *Acer macrophyllum* for the big-leaf maple. Why do you suppose this might be difficult for lichens?

Extending Your Knowledge

1. You might enjoy taking a walk through a nearby garden, park, campus, forested area, or even a backyard, and looking for examples of some of the plants and fungi, and their structures, mentioned in this chapter. For example, look for mosses (gametophytes and sporophytes), mushrooms (fruiting bodies, spores, and mycelia), ferns (sporophytes, sporangia, and spores), conifers (male and female cones, pollen, and seeds), and angiosperms (flowers and their parts, pollen, seeds, and fruits).

2. Are there any rare or endangered species of plants in your area? The Nature Conservancy buys property to set aside as preserves for endangered species. Your state probably has a native plant society concerned with preserving native plants and their habitats. If you would like to learn more about plant conservation or see some of these plants, you can contact local or state offices of these organizations. Find the Nature Conservancy on the Web at www.nature.org.

The Evolution of Animal Diversity

18

Human beings are animals, and our species evolved in an environment filled with animal life. Even now, when many humans live in cities, other species of animals affect our lives in many ways. We eat some animals, such as clams, chickens, and cattle, and make useful products from their shells, hides, fur, and bones. Bees make honey, and their cousins the mosquitoes make us miserable. Animals still work for humans in many parts of the world, tilling fields and carrying heavy loads. Worms enrich the soil, and coral animals build reefs that protect harbors. We choose some animals to live with us as pets. We marvel at the beauty of tropical fish and rain forest butterflies and at the size and strength of whales and crocodiles. Only in the last century have humans begun to appreciate the evolutionary relationships among all these fascinating animals and our own place in this story. The evolution of animal life is the subject of this chapter.

Organizing Your Knowledge

Exercise 1   (Module 18.1)

This module describes how we "draw the line" between animals and other organisms. Five major groups of living things are listed below. Use colored pens or pencils to draw lines separating the groups on the basis of the following characteristics.

1. Draw a black line between organisms that have simple cells and those that have complex cells (with organelles). Write "simple cells" on one side of the line and "complex cells" on the other.

2. Draw a blue line between organisms that are primarily unicellular and those that are multicellular. Label these groups.

3. Separate heterotrophic and autotrophic multicellular organisms with a green line, and label each group.

4. Draw a red line between those heterotrophs that ingest their food and then digest it, and those that digest food outside their bodies and then absorb the nutrients. Label each. On the animal side of the line, also note that animals have unique extracellular proteins and intercellular junctions, lack cell walls, are diploid (except for eggs and sperm), have unique embryonic stages, and most have muscle and nerve cells.

Plants   Fungi   Animals

Protists

Prokaryotes

Exercise 2 (Module 18.2)

Using Module 18.2 as a guide, sketch and briefly describe how animals may have evolved from colonial protists.

Exercise 3 (Module 18.3)

Symmetry—radial or bilateral—is an important feature of an animal's body plan. Compare animals having bilateral symmetry with those having radial symmetry by completing the following table.

Animals with Radial Symmetry	Animals with Bilateral Symmetry
Examples: sea anemone, jelly	1.
2.	Right and left sides
3.	Dorsal and ventral surfaces
No anterior or posterior ends	4.
No distinct head	5.
6.	Move actively through environment
Encounters environment equally from all directions	7.

Exercise 4 (Module 18.3)

Most animals have a body cavity, a space between the digestive tract and the body wall. (There are many advantages to having a body cavity; these are reviewed in the "Testing Your Knowledge" section.) There are two kinds of body cavities: A pseudocoelom is in direct contact with the digestive tract, and a middle layer derived from mesoderm lines the body covering. A coelom is a space within the mesoderm-derived layer, which covers the digestive tract and lines the body wall. The diagrams below show the body covering and the digestive tract of three animals. Complete the diagrams by sketching in the middle (mesoderm-derived) tissue layer. Label the layers and color the *body covering* blue, the *digestive tract* yellow, and the *middle tissue layer* red.

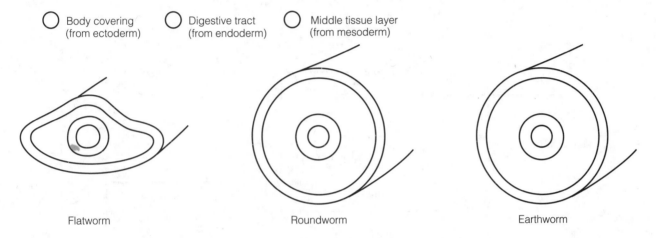

○ Body covering (from ectoderm)　　○ Digestive tract (from endoderm)　　○ Middle tissue layer (from mesoderm)

Flatworm　　　　　　　　　Roundworm　　　　　　　　　Earthworm

Exercise 5   (Module 18.4)

Web/CD Activity 18C   *Animal Phylogenetic Tree*

Module 18.4 introduces an animal phylogenetic tree based on body plan (tissues, symmetry, body cavity, and so on.) Suppose you found an animal and wanted to know to which phylum it belonged. You could use what biologists call a "key," a series of questions that leads you to the animal's identity. Such a key is given below, in the form of a flowchart. Before you can use this key, you will need to complete it. Some questions and phylum names are given; others are missing. Using information from this module, fill in the blanks to complete the key. You can get some answers by looking at the pictures in Module 18.4. (You might want to try out the key by "keying out" a real animal.)

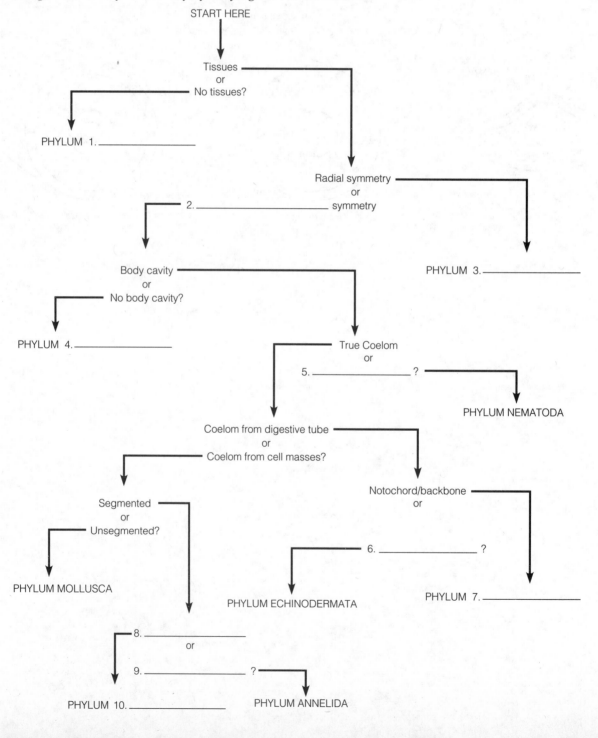

START HERE

Tissues
or
No tissues?

PHYLUM 1._____

Radial symmetry
or
2._____ symmetry

PHYLUM 3._____

Body cavity
or
No body cavity?

PHYLUM 4._____

True Coelom
or
5._____?

PHYLUM NEMATODA

Coelom from digestive tube
or
Coelom from cell masses?

Notochord/backbone
or

Segmented
or
Unsegmented?

6._____?

PHYLUM MOLLUSCA

PHYLUM ECHINODERMATA

PHYLUM 7._____

8._____
or

9._____?

PHYLUM 10._____

PHYLUM ANNELIDA

Exercise 5   (Modules 18.5 – 18.6)

Sponges and cnidarians are the simplest animals. Complete the following description of these two phyla by filling in the blanks.

Sponges, phylum [1]_____, and cnidarians, phylum [2]_____, are both simple animals. Most sponges and cnidarians live in the [3]_____, but some are found in fresh water. Sponges are usually irregular in shape, but all cnidarians are characterized by [4]_____ symmetry. This means their body parts are arranged in a circle around a central axis.

Sponges are by far the simpler of the two animals. A sponge is a simple tube perforated by tiny [5]_____. The body wall consists of [6]_____ layers of cells. The outer layer functions to protect the sponge. A gelatinous middle layer contains wandering amoebocytes and a skeleton made of flexible spongin or more rigid mineral-containing particles. The sponge's inner layer consists of cells called choanocytes bearing [7]_____, which move to create a current of water that [8]_____ the sponge through the small pores and [9]_____ through a large central opening. The choanocytes trap [10]_____ from the water and then engulf them by phagocytosis. The amoebocytes pick up food from the choanocytes and distribute it to other cells. They also make the [11]_____ fibers.

Unlike other animals, sponges lack both [12]_____ and muscles. In fact, their cells are relatively unspecialized, so the cell layers are not considered true [13]_____. It is likely that sponges are early offshoots of ancient colonial [14]_____ called choanoflagellates.

Cnidarians—animals such as [15]_____, sea anemones, and corals— are a bit more complicated. They have a [16]_____ cavity, muscles, and a [17]_____ system that enables them to respond to stimuli and coordinate muscle action. Unlike sponges, their cells are organized into [18]_____, groups of cells adapted to perform specific functions. But unlike more complex animals, they have only [19]_____ tissue layers, and most of their activities are carried out at the tissue level, not by the organs and [20]_____ of more complex creatures.

Cnidarians are radially symmetrical and come in two shapes. A [21]_____ is a tube with tentacles radiating from one end. It is usually fixed in place. A [22]_____ is a disk with a fringe of tentacles on the edge. [23]_____ are medusas and are able to move about in the water. Some cnidarians, such as the freshwater form called a [24]_____, illustrated in the text, exist only in the polyp form; some cnidarians exist only as medusae. Others have both medusa and polyp stages in their life cycles.

A cnidarian captures small prey and pushes it into its mouth with its [25]_____. Special cells called [26]_____ on the tentacles (characteristic only of cnidarians) sting and entangle the prey. The mouth of a polyp is on top of the body, in the center of the tentacles. A jelly's mouth is [27]_____, in the center of the umbrella. The mouth leads to a digestive sac called the [28]_____ cavity. Food is digested here, and fluid in the cavity circulates food particles around the body. The fluid in the cavity also keeps the flimsy body "inflated" and gives the cnidarian its shape. Because the gastrovascular cavity has only one opening, [29]_____ are expelled through the mouth.

Exercise 6 **(Modules 18.6 – 18.8)**

Review and compare the structures and lifestyles of cnidarians, flatworms, and round-worms by completing this chart.

	Cnidarians	Flatworms	Roundworms
1. Phylum name			
2. Examples			
3. Type of body symmetry			
4. Number of tissue layers			
5. Body shape(s)			
6. Body cavity			
7. Digestive tract			
8. Where they live			
9. Importance to humans			

Exercise 7 **(Module 18.9)**

This module discusses several of the structural and functional characteristics of mollusks. Match each of the statements on the left with a body structure on the right.

_____ 1. Modified to form a lung in land snails

_____ 2. Lacking in slugs

_____ 3. Used by a clam to capture food

_____ 4. Divided into hinged halves in bivalves

_____ 5. Functions in locomotion in most mollusks

_____ 6. Extracts oxygen from the water

_____ 7. Rasping organ used to scrape up food

_____ 8. Distributes nutrients, water, and oxygen around the body

_____ 9. Missing or internal in squids and octopuses

_____ 10. Outgrowth of the body surface that drapes over the animal

_____ 11. Modified to form tentacles in cephalopods

_____ 12. Body cavity around heart, kidney, and reproductive organs

_____ 13. Long projections on the back of a sea slug

_____ 14. "Crawling" movements of this structure propel gastropods

_____ 15. Used by a clam for digging and anchoring in mud or sand

(list continues)

A. Coelom
B. Radula
C. Gill
D. Foot
E. Mantle
F. Circulatory system
G. Shell

____ 16. Shoots out a jet of water to propel a squid

____ 17. Eyes of a scallop are along the edge of this structure

____ 18. A one-piece coiled structure in snails

____ 19. Lacking in terrestrial snails and slugs

____ 20. Secretes the shell

Exercise 8   (Module 18.10)

This module discusses annelids, the segmented worms, and the importance of segmentation. Review by filling in the blanks.

The next time you dig up an earthworm, or see one wriggling on the sidewalk, pause to appreciate its beauty and complexity. Earthworms are segmented worms of the phylum [1]_____. The name, which means "ringed," refers to the repeating ringlike [2]_____ that make up the worm's body. There are three main groups of annelids. Most live in the [3]_____, but many species live in [4]_____ and moist soil.

The most distinctive external characteristic of annelids is segmentation. Internally, each segment is separated from adjacent ones by membranous [5]_____. The [6]_____ system consists of a simple brain, a ventral nerve cord, and clusters of nerve cells in each segment. There are blood vessels serving each segment, and [7]_____ structures, which dispose of fluid wastes, are also repeated. A dorsal heart (actually an enlarged blood vessel) pumps blood via a [8]_____ circulatory system. The main blood vessels and the [9]_____ system are unsegmented.

What are the advantages of a segmented body? It probably is an adaptation to facilitate [10]_____. It gives the body greater [11]_____ and [12]_____. Longitudinal and [13]_____ muscles in each segment enable an earthworm to burrow, obtaining nutrients from the soil that passes through its digestive tract. Earthworms stir up the soil, and their [14]_____ improve its texture.

The largest group of annelids are the [15]_____. Most of these worms live in the [16]_____, where they wriggle along the bottom, burrow in the mud, or construct protective [17]_____. The mobile polychaetes move by means of segmental [18]_____. In tube-dwellers, these appendages are modified for [19]_____.

The third group of annelids are the [20]_____. Some suck [21]_____, but most are free-living [22]_____ that eat small animals. Most leeches live in [23]_____, but some are found in the sea or on land. Leeches have sharp [24]_____, and they secrete an anesthetic that enables them to slice painlessly through the skin and an anticoagulant that keeps blood flowing freely. The latter substance may be useful for dissolving [25]_____.

Annelids are not the only segmented animals. [26]_____ are segmented; this is seen clearly in the abdomen and in the thorax of an insect, where wings and legs are repeated. Animals with backbones are also segmented. In humans, segmentation is most clearly seen in the backbone and in the abdominal muscles.

208 *Chapter 18*

Exercise 9   (Modules 18.11 – 18.12)

Web/CD Thinking as a Scientist   *How Are Insect Species Identified?*

These modules concern phylum Arthropoda, a large and important group of invertebrates. Review your knowledge of arthropods by completing the crossword puzzle.

Across

1. Crabs and lobsters are ____.
6. The arthropod exoskeleton is made of ____ and protein.
9. ____ make up the largest order of animals.
10. An insect has ____ pairs of legs.
12. The ____ crab is a "living fossil" related to spiders.
13. ____ are sensory appendages on the head.
14. ____ are marine filter-feeding crustaceans.
16. Insects are the only invertebrates with ____.
18. Arthropods have ____ appendages.
20. A spider might hunt insects or catch them in a ____.
22. The ____ is an arachnid with pincers and a sting at the end of its tail.
23. ____ is shedding the old exoskeleton and growing a larger one.
24. Much of insect success can be attributed to their ability to ____.
25. A lobster uses its ____ for defense.
26. Arthropods have an ____ circulatory system.
27. The ____ are the most diverse group of arthropods.
28. ____ are multilegged carnivores.
29. many insects undergo ____ during their development.

Down

2. Crabs, grasshoppers, and tarantulas are all representatives of phylum ____.
3. ____, bees, and wasps are classified in order Hymenoptera.
4. Scorpions, spiders, ____, and mites are all arachnids.
5. The study of insects is called ____.
7. Every arthropod has a hard external skeleton called an ____.
8. An insect's body consists of head, thorax, and ____.
11. Many people are ____ to dust mites.
15. Unlike ____, which have similar segments the length of the body, most arthropods are divided into distinct groups of segments.
17. The arthropod body consists of groups of ____.
19. An insect's wings and legs are attached to its ____.
21. ____ are wormlike plant-eaters with many short legs.

Exercise 10   (Module 18.13)

Web/CD Activity 18A   *Characteristics of Invertebrates*

Echinoderms are a unique animal phylum. Circle the statements below that relate to echinoderms, and cross out statements that are not relevant to echinoderms.

1. Adults have radial symmetry

2. Live in salt water and fresh water

3. Larvae have radial symmetry

4. Have spines embedded under the skin

5. Most closely related to cnidarians

6. Lack segmentation

7. Have an exoskeleton

8. Good at regeneration

9. Have a water vascular system

10. Move and feed with tube feet

11. Adults have bilateral symmetry

12. Examples: sea urchins and sea stars

13. Live in salt water

14. Larvae have bilateral symmetry

15. Have an endoskeleton

16. Most closely related to chordates

17. Are segmented

18. Examples: clams and snails

Exercise 11   (Module 18.14)

Like all vertebrates, we are chordates, but so are some very simple animals that are not vertebrates—the lancelets and tunicates. All chordates, from tunicates to truck drivers, share four key chordate characteristics. Label these four characteristics on the drawing of a lancelet below.

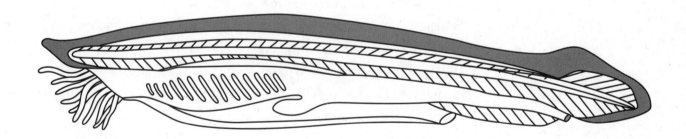

Exercise 12 (Modules 18.14 – 18.21)

Web/CD Activity 18B *Characteristics of Chordates*

Vertebrates are chordates with a skull and backbone made of bone or cartilage. Most have appendages with bony supports. These modules discuss the major clades of chordates, including the vertebrates. Match each statement below with the proper group(s) of vertebrates. Several questions have more than one answer.

A. Tunicates
B. Lancelets
C. Hagfish
D. Lampreys
E. Cartilaginous fishes
F. Ray-finned fishes
G. Coelocanths and lungfishes
H. Amphibians
I. Reptiles
J. Mammals

____ 1. The first tetrapods

____ 2. Endothermic vertebrates are found in these two clades

____ 3. First vertebrates with amniotic eggs

____ 4. Most live on land, but have aquatic eggs and larvae

____ 5. Two invertebrate chordates

____ 6. Have hair

____ 7. Two vertebrates that lack jaws

____ 8. Two groups with an operculum that helps pump water over their gills

____ 9. Lizards and snakes

____ 10. Proliferated after the dinosaurs died out

____ 11. Trout, bass, and perch

____ 12. Gave rise to amphibians

____ 13. Feed their young milk

____ 14. Evolved from reptiles

____ 15. Scales help them live on dry land

____ 16. Three groups of tetrapods

____ 17. Adults have pharyngeal slits, but not the other three chordate traits

____ 18. Have a swim bladder that provides buoyancy

____ 19. The young of most in this group develop inside their mothers

____ 20. Some in this clade have feathers

____ 21. Three groups with paired fins

____ 22. Invertebrate chordates that burrow in the sea bottom

____ 23. Placentals, marsupials, and monotremes

____ 24. Lobe-fins

____ 25. Human beings are in this group

____ 26. Populations are dramatically declining around the world

____ 27. The largest land animals ever were in this group

____ 28. Have a backbone, but no jaws

____ 29. Chondrichthyans

____ 30. Frogs and salamanders

____ 31. Give rise to reptiles

____ 32. Birds are in this clade

____ 33. "Sea squints"

____ 34. Skates, sharks, and rays

Exercise 13   (Modules 18.15 – 18.21)

Take another look at the chordates by labelling this phylogenetic tree. First, identify the major steps in chordate evolution by labelling the derived characters of each clade along the right side of the tree (1–9). Choose from:

A. Legs
B. Head
C. Lobed fins

D. Milk
E. Lungs or derivatives
F. Vertebral column

G. Brain
H. Jaws
I. Amniotic egg

Next identify the broad groupings of chordates by labelling the "bars" across the top (10–15). Choose from:

J. Tetrapods
K. Chordates

L. Amniotes
M. Craniates

N. Vertebrates
O. Jawed vertebrates

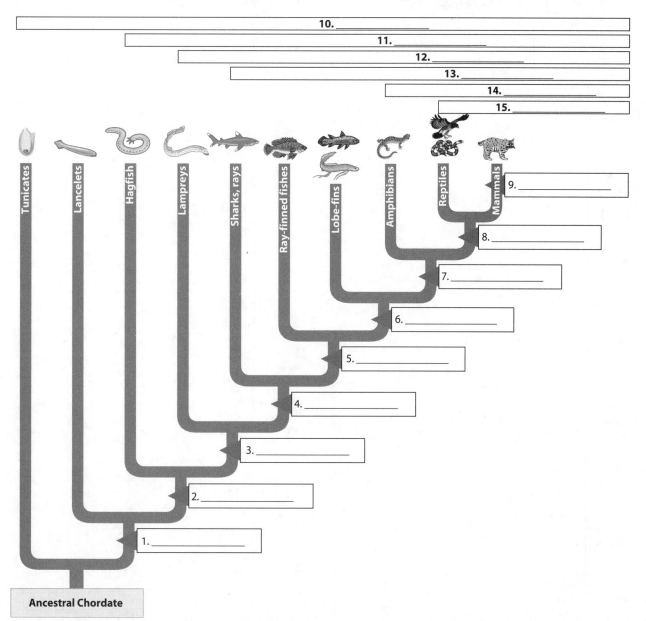

Exercise 14 (Module 18.22)

Web/CD Activity 15C   *Animal Phylogenetic Tree*
Web/CD Thinking as a Scientist   *How Do Molecular Data Fit Traditional Phylogenies?*

Molecular data have shaken the animal phylogenetic tree a bit. In the revised tree, characteristics such as body cavity type (coelom, and so on) are not as important as other more obscure characteristics such as feeding method and embryological development. Study the revised tree in Module 18.22, and then see if you can identify the structural characteristics that characterize each of the following branches of the tree.

____1. Sponges

____2. Everything but sponges

____3. Cnidaria

____4. All protostomes and deuterostomes

____5. Protostomes

____6. Deuterostomes

____7. Lophotrochozoa

____8. Ecdysozoa

A. Embryological development

B. Presence of true tissues

C. Feeding structure and larva

D. Radial symmetry

E. Bilateral symmetry

F. Molting of exoskeleton

G. Lack of true tissues

Testing Your Knowledge

Multiple Choice

1. Which of the following is *not* a characteristic of all animals?
   a. They are multicellular.
   b. They have tissues, organs, and organ systems.
   c. They are eukaryotes.
   d. They ingest their food.
   e. They are heterotrophic.

2. Animals probably evolved from colonial protists. How do animals differ from these protist ancestors?
   a. The protists were prokaryotic.
   b. Animals have more specialized cells.
   c. The protists were heterotrophic.
   d. The protists were autotrophic.
   e. Animals are able to reproduce.

3. A ____ is the simplest animal discussed in this chapter to have ____ .
   a. sponge . . . bilateral symmetry
   b. flatworm . . . a body cavity
   c. roundworm . . . a complete digestive tract
   d. jelly . . . a complete digestive tract
   e. snail . . . a body cavity

4. Which of the following animals does *not* have a body cavity?
   a. flatworm
   b. ant
   c. mouse
   d. clam
   e. earthworm

5. Which of the following phyla include numerous parasites and pests?
   a. roundworms and flatworms
   b. mollusks and roundworms
   c. annelids and flatworms
   d. annelids and roundworms
   e. mollusks and flatworms

6. Which of the following animals is *not* segmented?
   a. leech
   b. snail
   c. human being
   d. lobster
   e. salmon

7. Phylum ____ includes the largest number of species.
   a. Mollusca
   b. Arthropoda
   c. Annelida
   d. Chordata
   e. Echinodermata

8. The water vascular system of a sea star functions in
   a. movement of the tube feet.
   b. circulation of nutrients around the body.
   c. pumping water for swimming movements.
   d. waste disposal.
   e. keeping all parts of the body moist at low tide.

9. A ____ is a chordate but not a vertebrate.
   a. lamprey
   b. shark
   c. lancelet
   d. beetle
   e. frog

10. How do lampreys differ from other vertebrates?
   a. They have a skeleton made of flexible cartilage.
   b. They do not have jaws.
   c. They do not have paired appendages (fins or legs).
   d. all of the above
   e. b and c only

11. The first vertebrates to live on land were
   a. chelicerates
   b. reptiles.
   c. amphibians.
   d. cartilaginous fishes.
   e. mammals.

12. There are three major groups of mammals, categorized on the basis of their
   a. size.
   b. habitat.
   c. method of locomotion.
   d. teeth and digestive system.
   e. method of reproduction.

13. Zoologists have traditionally placed chordates and echinoderms on one major branch of the animal phylogenetic tree, and mollusks, annelids, and arthropods on another major branch. Which of the following is the basis for this separation into two branches?
    a. whether or not the animals have a skeleton
    b. type of symmetry
    c. whether or not the animals have a body cavity
    d. how the body cavity is formed
    e. whether or not the animals are segmented

14. Which of the following are most numerous and successful on land?
    a. mollusks and chordates
    b. annelids and arthropods
    c. arthropods and chordates
    d. annelids and chordates
    e. mollusks and arthropods

15. Which of the following is *not* a hypothesis suggested to explain the Cambrian explosion of animal diversity?
    a. increase in atmospheric oxygen levels
    b. development of more complex predator-prey relationships
    c. evolution of new regulatory/developmental genes
    d. evolution of vertebrates
    e. All of the above are hypothesis suggested to explain the Cambrian explosion

Essay

1. Describe the characteristics that separate animals from the other groups of living things.

2. Describe some of the characteristics that biologists consider important when deciding the phylum into which an animal should be classified.

3. What kinds of animals have a body cavity? What kinds lack a body cavity? Describe some of the advantages of having a body cavity.

4. Describe how the mantle, mantle cavity, and shells of snails, clams, and squids are modified for their different ways of life.

5. In terms of numbers of individuals and numbers of species, it could be argued that insects are the most successful creatures on Earth. What are some characteristics that have made them so successful?

6. The spadefoot toad of the southwestern United States is an unusual amphibian; it is capable of surviving in the desert. Few amphibians can tolerate dry desert conditions, but many reptiles—horned toads, rattlesnakes, and desert tortoises—thrive in hot, arid regions. In what ways are reptiles better adapted to life in the desert than amphibians?

7. Describe adaptations of birds for flight.

Applying Your Knowledge

Multiple Choice

1. Compare the two phylogenetic trees in Modules 18.4 and 18.22. The tree based on molecular data most drastically revises which of the following relationships?
   a. cnidaria and all other phyla
   b. annelids and mollusks
   c. sponges and all other phyla
   d. annelids and arthropods
   e. chordates and echinoderms

2. Which of the following includes the largest number of species?
   a. animals that are segmented
   b. animals with radial symmetry
   c. animals with a body cavity
   d. animals that are unsegmented
   e. animals with a backbone

3. Which of the following is radially symmetrical?
   a. a doughnut
   b. an automobile
   c. a spoon
   d. a peanut butter sandwich
   e. a wristwatch

4. A marine biologist dredged up a small animal from the bottom of the ocean. It was uniformly segmented, with short, stiff appendages and soft, flexible skin. It had a complete digestive system and a circulatory system but no skeleton. Based on this description, this animal sounds most like
   a. a lancelet.
   b. a crustacean.
   c. a mollusk.
   d. a roundworm.
   e. an annelid.

5. "Pill bugs" or "sow bugs," often found under rocks and logs in moist places, are perhaps most noticed for their ability to roll up into a ball when disturbed. Sow bugs are really crustaceans, not insects. Therefore, a sow bug does *not* have
   a. an exoskeleton.
   b. gills.
   c. three pairs of legs.
   d. antennae.
   e. jointed appendages.

6. Which of the following is thought to be most closely related to you?
   a. sea star
   b. snail
   c. earthworm
   d. jelly
   e. ant

7. There are only a few species of cartilaginous fishes, compared with the bony fishes. Cartilaginous fishes are mostly limited to a lifestyle of swimming fast in open water. Ray-finned fishes have adapted to many different lifestyles—clinging to seaweed, hiding in crevices, even burrowing in the bottom. This could probably be attributed to the fact that ray-finned fishes
   a. have more rigid skeletons.
   b. are smaller than cartilaginous fishes.
   c. have operculums and swim bladders.
   d. have lateral line systems and paired fins.
   e. are endothermic.

8. Which of the following is *not* thought to be in the lineage that led to human beings?
   a. an amphibian
   b. a dinosaur
   c. a jawless vertebrate
   d. a colonial protist
   e. a lobefin

9. Which of the following is *not* shared by birds and other reptiles?
   a. endothermic metabolism
   b. amniotic eggs
   c. backbone of vertebrae
   d. scales made of keratin
   e. gill structures in embryo

10. Imagine that you are a paleontologist (a scientist who studies fossils of ancient life forms). In a recent dig, you unearthed bones of all of the following. Which could you have found in the oldest sediments?
    a. amphibians
    b. placental mammals
    c. dinosaurs
    d. birds
    e. marsupials

11. You would expect to find the greatest number of phyla of animals _____ and the greatest number of species of animals _____.
    a. on land . . . in the sea
    b. in fresh water . . . in the sea
    c. in the sea . . . on land
    d. in the sea . . . in fresh water
    e. on land . . . in fresh water

Essay

1. Sponges have no muscles and cannot move. They have no nerve cells and cannot sense the environment around them. Why are they considered animals?

2. A flattened creature called *Trichoplax*, in phylum Parazoa (a small phylum not discussed in this chapter), is the simplest known animal. Its body consists of a simple ciliated outer layer over a core of unspecialized cells. It has no digestive tract, but it crawls over food and hunches its "back" to form a temporary hollow that serves as a digestive sac. What does this animal suggest about the early evolution of animals?

3. Imagine that you are a Peace Corps volunteer assigned to a small African village where many people are infected with pork tapeworms, which are spread from pig to pig and from pigs to people by eating infected meat. Resources are scarce; the poor villagers cannot afford expensive medicines. If these worms have life cycles like other tapeworms, suggest three ways the villagers could interrupt the worm's life cycle and prevent themselves from becoming infected.

4. Name what you consider to be a successful phylum of animals. What are your criteria for choosing these animals? What makes them successful?

5. Zoologists have found that certain marine snails and polychaete worms have similar ciliated swimming larvae. What does this evidence suggest about the evolution of annelids and mollusks? Is this reflected in the animal phylogenetic trees given in this chapter? Explain.

6. How do you know that a dog is a chordate? A vertebrate? Are all chordate and vertebrate characteristics seen in the adult dog? If not all are seen in the adult, what makes the dog a chordate?

7. Nearly all the land vertebrates in the Arctic and Antarctic are birds and mammals—polar bears, walruses, and penguins, for example. Why do you think there are so many birds and mammals, but virtually no reptiles or amphibians, in these regions?

Extending Your Knowledge

1. Learning more about the evolution of animals makes a visit to a zoo or museum much more rewarding. Many large cities have zoos, nature centers, aquariums, and natural history museums. You might enjoy examining the invertebrates of a coral reef up close in an aquarium exhibit, comparing placentals and marsupials at the zoo, and reliving the age of dinosaurs at a natural history museum. The American museum of Natural History in New York recently rearranged its Hall of Vertebrates to resemble a cladogram! If you want a real close-up experience, many museums, zoos, and aquariums train volunteers to act as guides and "explainers."

2. If you are interested in learning about, observing, and protecting wildlife, there are several organizations you can join. Three of the largest are National Audubon Society, 950 Third Avenue, New York, NY 10022; National Wildlife Federation, 1400 16th St. NW, Washington, DC 20036; and World Wildlife Fund, 1250 24th St. NW, Washington, DC 20037. Find them on the Web at www.audubon.org, www.nwf.org, and www.wwf.org.

Human Evolution

19

Their footprints, preserved in 3.5-million-year-old volcanic ash, tell a story: Two upright-walking early humans walked across the dusty African plain. Their tracks cross, and are crossed by, the tracks of other animals. Some of the animals may have been their prey, and some perhaps preyed on them. At one point the walkers stopped and perhaps looked back. Did they hear something? What were they thinking? More than any of the animals around them, these small creatures lived by their wits. Did they have a concept of who they were, where they came from, where their path had started, or where it might lead? These early humans were our ancestors. Perhaps they did not wonder about their origins, but we do. The origin and evolution of human beings is the subject of this chapter.

Organizing Your Knowledge

Exercise 1 (Module 19.1)

You are a primate. Review the characteristics of primates, especially anthropoids, and then explain which of these characteristics enables you to do each of the following.

1. Reach in and find a quarter among a pocketful of coins.

2. Catch a ball that is thrown to you.

3. Throw the ball back.

4. Pick up a postage stamp from your desktop.

5. Do a handspring.

6. Thread a needle.

7. Shoot a basketball.

Web/CD Activity 19A   *Primate Diversity*

Humans are primates. Review how we fit into this specialized order of mammals by matching the statements below with the major modern groups of primates (A–I) shown on the phylogenetic tree. (Some questions will require more than one answer.)

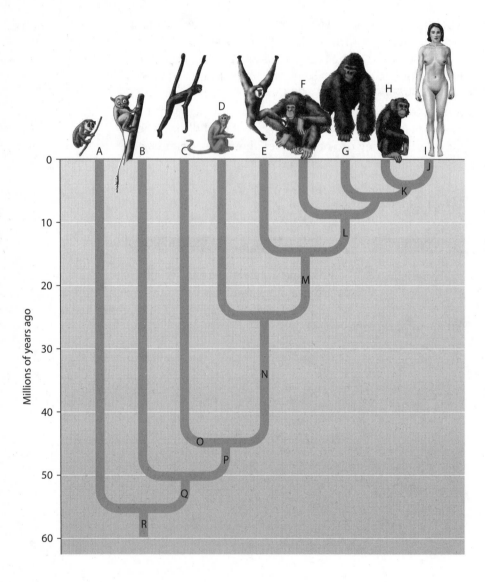

_____ 1. Hominids

_____ 2. These groups make up the hominoids

_____ 3. Apes

_____ 4. These are all anthropoids

_____ 5. New World monkeys

_____ 6. Old World monkeys

_____ 7. Two groups of prosimians

_____ 8. Tarsiers

_____ 9. Lorises, potoos, and lemurs

_____ 10. Chimps

_____ 11. The sifaka, from Madagascar

_____ 12. The golden lion tamarin

_____ 13. Baboon

_____ 14. Gibbons

_____ 15. Orangutan

_____ 16. Gorillas

_____ 17. Spider monkey, from South America

_____ 18. Rhesus macaque, from Africa

_____ 19. Grouped together because they have lack tails and have relatively large brains, long arms, and short legs

_____ 20. South American monkeys

_____ 21. African and Asian monkeys

_____ 22. Ape group that is most distantly related to other apes

_____ 23. Most primitive group of living primates

_____ 24. All the groups with a fully opposable thumb

_____ 25. Monkeys with flat faces, widely spaced nostrils, and prehensile tails

_____ 26. Monkeys with downward-pointing nostrils, lacking prehensile tails

_____ 27. Closest living relatives of humans

Now match the statements below with important steps in primate evolution (J–R). (One answer each.)

_____ 28. The long-sought "missing link" between humans and other apes.

_____ 29. The earliest anthropoid

_____ 30. The ancestral primate

_____ 31. This primate seems to have rafted to South America

_____ 32. Cro-Magnons

_____ 33. Ancestor of all primates except lemurs, potoos, and lorises

_____ 34. Late in 2004, fossils of this ancestor of all apes, named *Pierolapithecus catalaunicus*, were found in Spain

Exercise 3 (Modules 19.3 – 19.8)

Web/CD Activity 19B *Human Evolution*

These modules discuss the evolution of humans from primate ancestors, and derived characters that set us apart from other primates. Those unique human characteristics fall into three groups: Physical characteristics, psychological/behavioral characterisics, and characteristics related to reproduction and care of the young.

List three physical characteristics (changes in the skeleton and other structures) that distinguish humans from other primates:

1.

2.

3.

List three unique psychological or behavioral characteristics of humans (some are related to physical changes, such as a genetic difference):

4.

5.

6.

Now list three human characteristics related to mating and care of young (at least one of these is a physical change):

7.

8.

9.

Exercise 4   (Introduction, Modules 19.3 – 19.8)

Web/CD Activity 19B   *Human Evolution*

Only one hominid survives today—*Homo sapiens*. But paleoanthropologists have unearthed fossils of many earlier species. The hominid family tree is a tangled bush with
many branches. Some of these earlier hominids were our direct ancestors. Others, like the
Neanderthals, apparently were not. Review some of the important milestones of human
evolution by matching the correct hominid with each of the statements below. (To make
things easier, the hominids are listed in rough chronological order, but remember that
many species coexisted.)

_____ 1. The first hominid associated with sharp stone tools

_____ 2. A group of hominids that lived between 4.5 and 2.4 million years ago

_____ 3. The earliest known hominid

_____ 4. Earliest hominids who were fully bipedal

_____ 5. The first hominids to leave Africa

_____ 6. Represents a quantum leap in brain size, toolmaking, and reduced sexual
dimorphism

_____ 7. "Lucy"

_____ 8. Descended from *Homo heidelbergensis*

_____ 9. "Handy man"

_____10. First evidence of this species 160,000 years ago in Africa

_____11. Lived alongside our species, but went extinct 30–40,000 years ago

_____12. Early Europeans known as Cro-Magnons are early examples of this species

_____13. "Wise man"

_____14. Skeletons suggest these hominids formed stronger pair-bonds than their
ancestors

_____15. Capable of more complex art and symbolic thought than other species

_____16. "Upright man"

_____17. We found their tracks in African volcanic ash

_____18. A hominid with staying power who lived from about 2.8 million to about
200,000 years ago

_____19. Spread out of Africa to all continents in the last 50,000 years or so

A. *Sahelanthropus
   tchadensis*
B. Australopiths
   (*A. afarensis*, etc.)
C. *Homo habilis*
D. *Homo ergaster*
E. *Homo erectus*
F. *Homo nean-
   derthalensis*
G. *Homo sapiens*

Exercise 5 (Introduction, Modules 19.3 – 19.8)

Web/CD Activity 19B   *Human Evolution*

Over the last few decades, paleoanthropologists have made great strides in fleshing out the story of human evolution, shattering some of our old ideas. The quotes below reflect various lingering misconceptions and fallacies about human evolution. Counter each of these false statements with a more up-to-date correction.

1. "Scientists have very little evidence of our origins. They are still looking for the 'missing link.'"

2. "Human beings are descended from chimps."

3. "The key change that set us apart from other primates was the development of a larger brain. Walking upright came later."

4. "All hominid fossils form a single line of descent, from the ancestors of *Australopithecus*, through *Homo erectus*, to modern humans."

5. "Neanderthals, who lived in Europe 200,000 years ago, were the immediate ancestors of modern humans."

6. "Hominids first appeared in Africa, then spread around the globe. Modern *Homo sapiens* evolved in Europe."

Exercise 6   (Modules 19.9 – 19.12)

Our culture has enabled humans to alter nature to suit our own purposes. Which of the three stages in cultural evolution is reflected in each of the phrases below?

_____ 1. Permanent settlements first arose

_____ 2. The current stage in cultural evolution

_____ 3. Still practiced by the !Kung people of Africa

_____ 4. Probably caused the extinction of saber-toothed cats and other large animals

_____ 5. Began 10,000 to 15,000 years ago

_____ 6. Led to significant acceleration in population growth

_____ 7. Turned areas of the Middle East and North Africa into desert

_____ 8. People migrated to cities in search of factory jobs

_____ 9. The way of life of the earliest hominids

_____ 10. First allowed people to specialize in different occupations

_____ 11. The norm for humans over most of the last 100,000 years

Testing Your Knowledge

Multiple Choice

1. The lemurs of Madagascar are examples of
   a. Old World monkeys.
   b. anthropoids.
   c. prosimians.
   d. New World monkeys.
   e. apes.

2. Which of the following hominids apparently did *not* first appear in Africa?
   a. *Homo sapiens*
   b. *Homo erectus*
   c. *Australopithecus afarensis*
   d. *Homo neanderthalensis*
   e. *Homo heidelbergensis*

3. Which of the following is regarded as evidence of a major change that may have led to human speech?
   a. Fossils that show that bipedal locomotion freed the hands for use of sign language
   b. Discovery of a gene that shapes the brain for use of language
   c. Evidence that the tongue of *Homo sapiens* differed from that of *Homo erectus*
   d. DNA studies of Neanderthals and other fossil humans
   e. Changes in the nasal passages and voice box of early *Homo sapiens*

4. Biochemical evidence indicates that _____ are more closely related to humans than to other apes.
   a. the gorilla and the orangutan
   b. the chimpanzee and the baboon
   c. the orangutan and the chimpanzee
   d. the gorilla and the chimpanzee
   e. the orangutan and the gibbon

5. The earliest hominid fossils so far discovered are about _____ years old.
   a. 40,000
   b. 100,000
   c. 1 million
   d. 3 million
   e. 7 million

6. Fossil evidence suggests that humans evolved in
   a. Europe.
   b. South America.
   c. Asia.
   d. the Middle East.
   e. Africa.

7. Which of the following steps in human evolution appears to have occurred first?
   a. development of language
   b. large brain
   c. bipedalism
   d. development of culture
   e. use of tools

8. Which of the following appear to be the best explanations so far for differences in human skin color?
   a. Darker skin shields the skin from UV; lighter skin allows synthesis of vitamin D.
   b. Darker skin cools the body; lighter skin allows synthesis of folate.
   c. Darker skin warms the body; lighter skin prevents breakdown of folate.
   d. Darker skin prevents breakdown of folate; lighter skin allows synthesis of vitamin D.
   e. Darker skin warms the body; lighter skin cools the body.

9. Which of the following was the first hominid to live in Europe and Asia?
   a. *Homo sapiens*
   b. *Homo erectus*
   c. *Homo neanderthanlensis*
   d. *Australopithecus*
   e. *Homo habilis*

10. Which is thought to be the first hominid to use sharpened stone tools?
    a. *Homo erectus*
    b. *Homo habilis*
    c. *Australopithecus africanus*
    d. the Neanderthals
    e. *Homo sapiens*

Essay

1. List the specialized characteristics of primates.

2. Which primate is the closest living relative of humans? What is the evidence for this close relationship?

3. Describe the characteristics that separate each of the paired groups:
   a. anthropoids—prosimians
   b. Old World monkeys—New World monkeys
   c. monkeys—apes
   d. other apes—humans

4. What were three main milestones in the evolution of *Homo sapiens*?

5. The industrial revolution of the last three centuries has had major impacts on the biosphere. Did the earlier scavenger-gatherer-hunter and agricultural stages of cultural evolution cause any widespread, lasting environmental changes?

6. Compare the rates of human biological and cultural evolution over the last 100,000 years. How has the evolution of culture contributed to our success? How has it caused problems?

7. Describe genetic studies that point to when and where *Homo sapiens* originated. Do fossils support these data?

Applying Your Knowledge

Multiple Choice

1. Which of the following is least closely related to the others?
   a. human
   b. lemur
   c. gorilla
   d. squirrel monkey
   e. baboon

2. At the zoo, Tom saw a species of primate he had never even heard of before. He said, "It was called a white-faced saki. It had long, dark, spiky fur, a long fluffy tail, forward-facing brown eyes, a white forehead, and yellow cheeks." Based on this information, the white-faced saki could *not* be
   a. an ape.
   b. a prosimian.
   c. a New World monkey.
   d. an Old World monkey.
   e. an anthropoid.

3. Fossils show that *Australopithecus afarensis* males were substantially taller and heavier than females. By contrast, the males and females of *Homo erectus* were similar in size, as are modern humans. What does this suggest about these two hominids?
   a. Sex roles were more clearly defined in *Australopithecus* than in *Homo erectus*.
   b. *Australopithecus* females had to do most of the job of rearing the young.
   c. *Australopithecus* was more of a hunter-gatherer than *Homo erectus*.
   d. *Homo erectus* males and females formed more long-lasting pair-bonds.
   e. *Australopithecus* was an herbivore, while *Homo erectus* was more of a carnivore.

4. Trina saw an old Tarzan movie on television. The movie supposedly took place in Africa, but Trina easily spotted that it was not really filmed there. Which of the following could have tipped her off?
   a. Chimps like Cheetah do not live in Africa.
   b. The monkeys in the jungle all had prehensile tails.
   c. There were no prosimians shown in the forest.
   d. Tarzan wrestled with a gorilla.
   e. Only Old World monkeys were shown.

5. Which of the following categories includes all of the others?
   a. apes
   b. Old World monkeys
   c. anthropoids
   d. hominids
   e. New World monkeys

6. Which of the following probably coexisted for a time with *Homo sapiens*?
   a. *Australopithecus robustus*
   b. *Homo erectus*
   c. *Australopithecus afarensis*
   d. *Homo habilis*
   e. *Australopithecus africanus*

7. Most known species of hominids
   a. were the ancestors of modern humans.
   b. evolved during the last million years or so.
   c. had large brains.
   d. lived primarily in trees.
   e. were not our ancestors.

8. Which of the following is a hominoid, but not a hominid?
   a. *Homo neanderthalensis*
   b. lemur
   c. *Australopithecus africanus*
   d. gorilla
   e. squirrel monkey

9. The bonobo, an ape closely related to the chimpanzee, displays behaviors very similar to those of humans, and some anthropologists have suggested that the bonobo is the living primate most closely related to humans. Which of the following would be the easiest way to try to substantiate this idea?
   a. Look for fossils of bonobos, chimps, and humans.
   b. Study the DNA of bonobos and chimps.
   c. Determine which of the species are anthropoids.

d. Compare the DNA of bonobos and humans.

e. Compare the DNA of bonobos, chimps, and humans.

10. Paleontologists have found fossils of several kinds of australopiths. What is their place in human evolution?

a. They are all thought to be ancestors of modern humans.

b. They are all extinct side branches of the human family tree.

c. Some evolved into humans, others into apes.

d. They are the ancestors of various modern apes.

e. Some may have been our ancestors, others offshoots of our family tree.

Essay

1. Unlike many other mammals, such as horses or deer, humans are born at a quite "undeveloped," helpless stage in development. Wouldn't it be better for humans to spend a little longer period *in utero* and enter the world a little more self-sufficient? What evolutionary compromise seems to prevent this?

2. Language, use of tools, and self-awareness are often cited as human characteristics—traits that set us apart from other members of the animal kingdom. Do these characteristics really appear to be unique to humans? Explain.

3. Finding which of the following would require anthropologists to change their current ideas about human evolution the most? Why?

a. A large-brained, quadrepedal (walking on all fours) hominid 2.5 million years old

b. A small-brained, bipedal hominid 5.0 million years old

4. What are some of the current questions about human evolution that paleoanthropologists are trying to answer?

Extending Your Knowledge

1. Scavenging-gathering-hunting was the first stage in human culture, and it was the dominant mode of existence for all humans until very recently. Some modern humans—the !Kung of Africa, for example—still practice this way of life. But in most parts of the world, the scavenging-gathering-hunting lifestyle has been displaced by modern, machine-age industrial/urban culture. People such as the aborigines of Australia, who until recently lived off the land, have all but disappeared. European immigrants quickly overran the continent of North America and displaced most of the Native American scavenger-gatherer-hunters, in little more than a hundred years. Were there Native Americans in your vicinity when the continent was settled by Europeans? Who were they? Are their descendants still present, and do they retain elements of their traditional way of life? How could you find out more about them?

2. Some experts suggest that we are unable to deal effectively with environmental problems because our Stone Age minds are unable to see and cope with the dangers created by our accelerating technology. Neurobiologist Robert Ornstein and ecologist Paul Ehrlich present this idea in their book *New World, New Mind*. They think that evolution has shaped our brains to be better at identifying obvious, immediate dangers like saber-toothed tigers than subtle changes in the environment like global warming. Does anthropological evidence support this view? Can you think of other ways in which our behavior lags behind our technology? How might we get around this limitation?

3. Late in 2004, paleoanthropologists announced that they had discovered fossils of a previously unknown species of humans on the island of Flores in Indonesia. These tiny people, dubbed *Homo florensis,* stood only three feet tall and may have descended from Homo erectus. They used stone tools to hunt dwarf elephants. Amazingly, they held out on their home island until about 13,000 years ago, apparently coexisting with *Homo sapiens* nearby. It is unlikely that we would find any such hominids still alive—the world is pretty well-explored by now. But what do you think would have happened if "Flores man" had lasted just a little longer and been discovered by modern sailors?

Unifying Concepts of Animal Structure and Function

20

Take a moment to look at the skin on the back of your hand and appreciate its role in the structure and function of your body. The skin is an organ; along with hair and nails, it makes up the integumentary system. The skin's structure is well suited to its functions. Its outer layer consists of many layers of cells, dead and impermeable at the surface, living and reproducing below. The skin is tough, forming a barrier that keeps water in the body and bacteria out. It is also flexible, allowing the body to bend and move. It protects the body from changes in the external environment and at the same time actively responds to those changes. Nerve endings in the skin sense environmental variations, and sweat glands, blood vessels, and pigment cells respond to them. Other body systems interact with the skin: The digestive and respiratory systems supply it with food and oxygen, the excretory system disposes of its wastes, and the immune system protects it from infection. In exchange, the skin covers and protects these systems, playing a vital role in homeostasis—maintaining a constant internal environment suitable for continuing body functions. The skin illustrates the important connection between how the body is built and how it works. This chapter introduces the important concepts uniting animal structure and function.

Organizing Your Knowledge

Exercise 1   (Introduction and Module 20.1)

Web/CD Activity 20A   *Correlating Structure and Function of Cells*

These modules discuss how a gecko's feet enable it to climb walls and a bird's wings enable it to fly. You are probably familiar with many other examples of the correlation between animal structure and function. In a sentence or two, state how each of the following illustrates this correspondence of structure and function. The first one is done for you.

1. A whale's tail. *The tail is flattened into broad flukes, which propel the whale through the water.*

2. A hummingbird's beak.

3. Your hand.

4. A frog's legs.

5. A mosquito's sharp, tubular mouthparts.

6. A cow's multichambered stomach.

Exercise 2   (Modules 20.2 – 20.3)

Web/CD Activity 20B   *The Levels of Life Card Game*
Web/CD Activity 20C   *Overview of Animal Tissues*

Review the hierarchy of structural levels in an animal by filling in the blanks in the following paragraphs.

The body of an animal—a cat, for example—is organized on several hierarchical levels. The smallest parts of the cat that are alive are individual [1]_____, such as the muscle fibers in the wall of the stomach. Their function is to contract and move the contents of the stomach, mixing cat food with digestive juices. Many muscle cells cooperate to form a [2]_____, the second level of body structure and function. Besides muscle, there are three other kinds of tissues that make up the cat's stomach: [3]_____ tissue, [4]_____ tissue, and [5]_____ tissue. The stomach itself, formed of these four tissues, is an [6]_____, which performs the functions of storing and digesting food. The stomach, esophagus, intestines, and digestive glands make up the digestive system, which exemplifies the [7]_____ level of structure and function. It is one of a dozen or so systems that cooperate to form the cat—an [8]_____. This is the whole animal, the highest level of the hierarchy.

To review, starting from the top down: A cat is an [9]_____. It is composed of a number of [10]_____, each of which performs specific functions such as digestion or circulation. Each system is composed of [11]_____ such as the heart or stomach, which are built from four kinds of [12]_____. At the most fundamental level of the hierarchy, a tissue is composed of individual [13]_____.

Exercise 3 (Module 20.4)

Web/CD Activity 20D *Epithelial Tissue*

Epithelial tissue covers the body and lines body organs. Complete the following chart comparing four kinds of epithelium. Refer to the illustrations in Module 20.4 to aid you with the descriptions.

Tissue Type	*Description*	*Body Locations*	*Functions*
1.	2.	3.	4.
5.	6.	7.	Absorbs and secretes fluid in kidneys
8.	Single layer of thin, flattened cells	9.	10.
11.	12.	Lines the intestine	13.

Exercise 4   (Module 20.5 and 20.8)

Web/CD Activity 20E   *Connective Tissue*
Web/CD Activity 20F   *Muscle Tissue*
Web/CD Activity 20G   *Nervous Tissue*

Complete this crossword puzzle to review the structures, functions, and locations of connective tissues and the medical uses of artificial tissues.

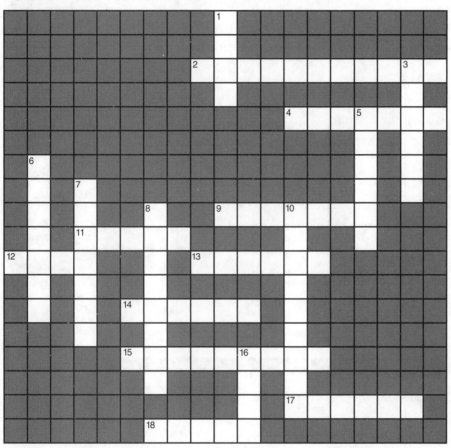

Across

2. Artificial skin is grown from cells called human____.
4. ____ tissue contains fat.
9. Bone matrix contains fibers embedded in ____ salts.
11. ____ is a connective tissue with a liquid matrix.
12. Connective tissue is one of ____ major categories of tissues.
13. The matrix is a web of ____ in a liquid, jelly, or solid.
14. Connective tissue cells are scattered in a non-living ____.
15. Cartilage supports the nose and ears, and forms discs between the ____.
17. Adipose tissue pads and insulates the body and stores "fuel" to give the body ____.
18. ____ connective tissue holds other tissues and organs in place.

Down

1. Artificial ____ has been successfully used to cover burns.
3. A ____ is a group of cells with a common structure and function.
5. Blood matrix is called ____.
6. Fibrous connective tissue forms ____ and ligaments.
7. ____ connective tissue has densely packed bundles of collagen fibers.
8. Some kinds of connective tissues contain rope-like ____ fibers.
10. ____ is a strong but flexible skeletal material, but it does not regenerate easily.
16. ____ is the most rigid connective tissue.

Exercise 5   (Modules 20.6 – 20.7)

Web/CD Activity 20F   *Muscle Tissue*
Web/CD Activity 20G   *Nervous Tissue*

Review nervous tissue and the three types of muscle tissue by matching each phrase on the left with a tissue type on the right.

_____ 1. Contractile tissue of the heart

_____ 2. Forms body communications system

_____ 3. Attached to bones by tendons

_____ 4. Branching, interconnected muscle cells

_____ 5. Cells characterized by axons and dendrites

_____ 6. Carries out voluntary body movements

_____ 7. Composed of neurons and other supporting cells

_____ 8. Cells are striated, or striped

_____ 9. Muscle cells that lack striations

_____ 10. Found in walls of digestive tract, bladder, arteries, and so on

_____ 11. Involuntary muscle of internal organs other than the heart

_____ 12. Figure 1 at the right

_____ 13. Figure 2 at the right

_____ 14. Figure 3 at the right

_____ 15. Figure 4 at the right

A. Smooth muscle
B. Cardiac muscle
C. Skeletal muscle
D. Nervous tissue

1.
2.
3.
4.

Exercise 6   (Modules 20.9 – 20.10)

Review the functions of human organs and organ systems by filling in the blanks in the following paragraphs.

An animal—be it a gecko or a human being—consists of a number of cooperating organ systems, each of which performs specific life functions. Each system consists of organs, which in turn carry out particular jobs. As you read these words, your brain receives nerve impulses from your eyes. It evaluates the information received and sends out responses via the spinal cord and nerves. These parts—sense organs, brain, spinal cord, and nerves—make up the [1]_____ system, one of two systems that control and coordinate body activities.

As you read, your eyes scan the page, and your hand moves to write answers in the blanks. These responses are carried out by muscles, which make up the [2]_____ system, the system responsible for body movements. To move the body, muscles pull against bones, which make up the [3]_____ system. This system also supports the body and protects delicate internal organs such as the brain.

Muscles, like all organs, require food and oxygen to function. Food is digested and absorbed by the [4]_____ system, whose parts include the stomach, intestines, and digestive glands such as the liver and pancreas. As you inhale and exhale, oxygen enters the body via the lungs, key organs of the [5]_____ system. The [6]_____ system—the heart, blood, and blood vessels—functions to transport food and oxygen to your muscles and other organs. Some of the fluid delivered to body tissues leaves the blood and is picked up by vessels of the [7]_____ system. The fluid passes through lymph nodes, where special cells called lymphocytes attack foreign substances and microbes. The skin, or [8]_____ system, is normally able to keep most disease-causing organisms out. When they get through, they are attacked by cells of the [9]_____ system—lymphocytes, which are produced and stored in the thymus, bone marrow, and spleen, as well as lymph nodes.

As the blood delivers its cargo of food and oxygen, it also picks up waste products that must be expelled by the body. The kidneys—the key organs of the [10]_____ system—remove metabolic waste products from the blood and dispose of them via the ureters, bladder, and urethra. The kidneys also regulate the osmotic balance of blood. This activity, and many others in the body, are controlled by chemical signals called hormones. Hormones are sent out by the glands of the [11]_____ system, which acts in concert with the nervous system to coordinate the activities of all the other body systems. Endocrine glands—the pituitary, thyroid, adrenals, and others—regulate such activities as digestion, growth, metabolism, and water balance. They even help to control the process of reproduction, via their effects on the testes and ovaries, the major organs of the male and female [12]_____ systems.

Exercise 7   (Module 20.11)

Compare body imaging methods by summarizing each method on the chart.

Method	Medium Used	Used For
Conventional X-ray	1.	2.
Computerized tomography (CT)	3.	4.
Magnetic resonance imaging (MRI)	5.	6.
Magnetic resonance microscopy (MRM)	7.	8.
Positron-emission tomography (PET)	9.	10.
Functional MRI	11.	12.

Exercise 8 (Module 20.12)

Animals must exchange materials with their environment. The cells of saclike or flattened animals (such as *Hydra* and flatworms) can exchange substances directly with the surrounding water. Larger animals, whose outer coverings are small compared with their volume, have specialized surfaces for exchanging materials with the environment. Below is a simplified diagram of an animal, similar to Figure 20.12A in the text. The circulatory system is shown. Add your own simple sketches and label the **digestive system, respiratory system, excretory system, interstitial fluid,** and **cells.** Draw and label arrows to show exchange of **food, nutrients, O_2, CO_2, metabolic wastes,** and **feces.**

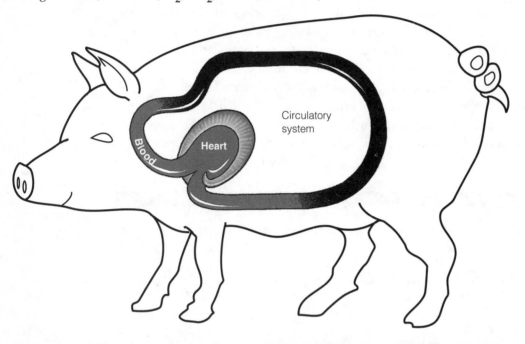

Exercise 9 (Modules 20.13 – 20.14)

Web/CD Activity 20H   *Regulation: Negative and Positive Feedback*

To review the concepts of homeostasis and negative feedback, the most important principles of animal function, read the following paragraph and then fill in the chart that follows.

Despite changes in the external environment, an animal can keep its internal environment remarkably constant. This maintenance of a constant internal environment is called homeostasis. The text states that animals such as the ptarmigan maintain relatively constant salt and water balance and body temperature. Conditions within the human body are also more or less constant, fluctuating within narrow limits. For example, an organ called the pancreas monitors and regulates the amount of sugar in the blood. This maintains a constant supply of fuel for body cells, even though the body's intake of food varies widely during the day. After a meal, when blood sugar rises, the pancreas sends out a chemical signal, a hormone called insulin. Insulin causes body cells to take up and store sugar, which lowers blood sugar to the optimum range—70 to 110 mg of sugar per 100 mL of blood. This illustrates negative feedback: An increase in blood sugar triggers a response that counteracts the increase. Between meals, as cells consume sugar, the concentration of sugar in the blood starts to decrease. The pancreas responds by reducing its output of insulin and stepping up its secretion of a second hormone, glucagon. Glucagon signals certain cells to release sugar from storage, raising blood sugar to the optimum level. Thus, despite changes in the external environment (timing and content of meals), blood sugar usually fluctuates within the narrow range that is best for cells.

Compare the regulation of blood sugar described in the preceding paragraph with the control of body temperature outlined in Modules 20.13 and 20.14. Fill in the chart by identifying the components of each homeostatic control system.

	Body Temperature	*Blood Sugar*
Type of change in external environment	1.	2.
Control center	3.	4.
Stimulus	5.	6.
Kind of signal sent by control center to effector	7.	8.
Effector	9.	10.
Response	11.	12.
Set point	13.	14.

Testing Your Knowledge

Multiple Choice

1. The four major categories of tissues are
   a. bone, muscle, blood, and adipose.
   b. nervous, epithelial, connective, and muscle.
   c. muscle, epithelial, bone, and cartilage.
   d. blood, nervous, connective, and muscle.
   e. simple squamous, simple cuboidal, simple columnar, and stratified squamous.

2. Which of the following levels of structure encompasses all the others?
   a. tissue
   b. cell
   c. organ
   d. organism
   e. system

3. How many organ systems make up your body?
   a. four
   b. hundreds
   c. twelve
   d. millions
   e. It depends on the size of the person.

4. Which of the following tissues produces voluntary body movements?
   a. smooth muscle
   b. simple cuboidal epithelium
   c. cardiac muscle
   d. skeletal muscle
   e. fibrous connective tissue

5. Neurons are specialized cells characteristic of
   a. muscle tissue.
   b. nervous tissue.
   c. connective tissue.
   d. epithelial tissue.
   e. all of the above.

6. All but one of the following systems are correctly paired with one of their parts. Which pair is *incorrect?*
   a. circulatory system—heart
   b. respiratory system—lung
   c. endocrine system—thyroid gland
   d. integumentary system—hair
   e. excretory system—intestine

7. Which of the following do the excretory, digestive, and respiratory systems have in common?
   a. They are present only in animals with backbones.
   b. They contain specialized surfaces for exchange with the environment.
   c. They work independently, without any control by the nervous system.
   d. They enable the animal to absorb needed materials from its environment.
   e. They are isolated from the animal's internal environment.

8. A change in the body often triggers a response that counteracts the change. This kind of response is known as
   a. negative feedback.
   b. empowerment.
   c. cause and effect.
   d. positive feedback.
   e. adaptation.

9. Homeostasis is
   a. exchange of materials with the surrounding environment.
   b. the idea that all vertebrates are built in a similar way.
   c. the correlation of structure and function.
   d. maintaining a relatively constant internal environment.
   e. cooperation of body parts to form tissues, organs, and systems.

10. An animal's "internal environment" is
    a. the blood.
    b. the interior of compartments like the heart and stomach.
    c. anyplace beneath the skin.
    d. any fluid inside the body.
    e. the interstitial fluid that surrounds the cells.

11. Which of the following are listed in the correct hierarchical order?
    a. system-tissue-organ
    b. cell-tissue-organ
    c. organ-tissue-system
    d. tissue-cell-organ
    e. organism-organ-system

Essay

1. Describe how the structure of a bird wing relates to its function.

2. Name the four major types of body tissues, and briefly describe the functions of each type.

3. Name the twelve organ systems of a vertebrate. Describe the function of each system in one sentence each.

4. What are the advantages of computerized tomography (CT) over conventional X-rays?

5. Compare a thermostat controlling room temperature with your brain controlling body temperature. Describe the following for each system: stimulus, control center, effector, and response.

6. Some animals can "breathe" through their skins, without the aid of lungs or gills, or absorb food and expel wastes through the surfaces of their bodies, without specialized digestive tracts or kidneys. Why are all these animals rather small?

Applying Your Knowledge

Multiple Choice

1. Bone does *not* show which of the following correlations between structure and function?
   a. It is rigid.
   b. Its cells are packed tightly together.
   c. It contains reinforcing fibers.
   d. It can grow with the animal.
   e. It contains canals for blood vessels and nerves that keep it alive.

2. When you sprain your ankle, the "straps" of tissue that hold the bones together are stretched and torn. What kind of tissue do you think is damaged in a sprain?
   a. stratified squamous epithelium
   b. visceral muscle
   c. fibrous connective tissue
   d. adipose tissue
   e. cartilage

3. Which of the following forms a thick protective barrier that keeps bacteria out of the body?
   a. skeletal muscle
   b. fibrous connective tissue
   c. stratified squamous epithelium
   d. cartilage
   e. simple columnar epithelium

4. An organ such as the heart or liver contains
   a. muscle tissue.
   b. nervous tissue.
   c. connective tissue.
   d. epithelial tissue.
   e. all of the above.

5. A new drug has been developed that impairs the movement of smooth muscle. It would affect the muscle
   a. that moves the arms and legs.
   b. of the heart.
   c. in the wall of the intestine.
   d. all of the above
   e. b and c only

6. Which of the following is *not* an organ?
   a. the stomach
   b. a blood vessel
   c. a neuron
   d. the heart
   e. a lung

7. A researcher wants to study the metabolic activity of various parts of exercising heart muscle. This might be accomplished by
   a. doing a CT scan.
   b. taking some X-rays.
   c. doing an ultrasound scan.
   d. using MRI.
   e. doing a PET scan.

8. Which of the following best illustrates homeostasis?
   a. All the cells in the body have much the same chemical composition.
   b. Cells of the skin are constantly worn off and replaced.
   c. When blood CO_2 increases, you breathe faster and get rid of CO_2.
   d. All organs are composed of the same four kinds of tissues.
   e. The lung has a large surface for exchange of gases.

Essay

1. You have probably read books or seen nature programs on television that describe adaptations of animals to their environments. Choose an animal and briefly describe how its body shows the correlation of structure and function.

2. What might interest an anatomist about each of the following: how a fish swims, how a penguin keeps warm, how an insect defends itself from its enemies? What might interest a physiologist about each of them?

3. Read the descriptions and look at the illustrations in the text of the following tissue types, and then explain how their structure correlates with their function: bone, simple squamous epithelium, and blood.

4. Briefly describe which organ systems might cooperate in delivering food and oxygen to your brain cells.

5. The parathyroid glands regulate the amount of calcium in the blood. They send out hormone signals that control how much calcium the intestine absorbs from food and how much calcium the kidneys excrete in the urine. What do you think the parathyroids cause to happen when blood calcium gets too high? What happens when blood calcium gets too low? How does this illustrate negative feedback?

Extending Your Knowledge

1. Many physicians specialize in treating particular body systems. Neurologists, for example, deal primarily with the nervous system. What about dermatologists? Can you think of others? (Physicians are listed by specialty in the telephone Yellow Pages.)

2. Many everyday devices use negative feedback to maintain some constant condition. For example, there is a valve that maintains a constant water level in a toilet tank. When the toilet is flushed, the water level drops. A float drops, opening a valve, which raises the water level and the float. When the water level reaches its "set point" (when the tank is full) the rising float shuts off the water. How does this system illustrate negative feedback? Can you think of other devices that work this way?

3. Have you or someone you know ever had a CT or MRI scan? How have these devices changed health care in recent years?

Nutrition and Digestion

21

What was the last thing you ate? Or might you be snacking right now? What makes you feel hungry? Why do you need to eat? What is in food that your body needs? Why are some kinds of food better for you than others? What happens to food after you eat it? How is the way you eat and process food different from the eating and digestion processes of other animals? What are calories? Vitamins and minerals? Fiber? And what are the latest recommendations regarding sodium, saturated fat, cholesterol, and dieting? These and many other questions about nutrition and digestion are considered in this chapter.

Organizing Your Knowledge

Exercise 1   (Module 21.1)

Test your knowledge of animal diets and feeding methods by filling in this chart. Most examples are from the module; others are animals with which you are probably familiar. For diet, you may choose from omnivore, herbivore, and carnivore. There are several feeding methods: Animals can be absorptive feeders, suspension feeders, substrate feeders, fluid feeders, or bulk feeders.

Animal	Diet	Feeding Method
1. Cow		
2. Earthworm		
3. Aphid		
4. Humpback whale		
5. Human being		
6. Tapeworm		
7. Shark		
8. Female mosquito		
9. Clam		
10. Grasshopper		

Some small, simple animals digest their food in multipurpose food vacuoles or larger gastrovascular cavities. But most animals have an alimentary canal, which processes food in four stages. Read about the four stages of food processing and kinds of digestive systems. Then connect the two by (1) labeling the drawings, and (2) coloring the drawings to show which digestive stages occur in each portion of these alimentary canals. On the drawings label **mouth, pharynx, esophagus, crop, gizzard, stomach, intestine,** and **anus.** Then color the alimentary canals to show where each stage in food processing occurs. Use yellow for **ingestion** (including swallowing and storage), red for **digestion,** green for **absorption,** and blue for **elimination.** (If two processes occur in the same area, mix the colors.)

Ingestion ◯   Digestion ◯   Absorption ◯   Elimination ◯

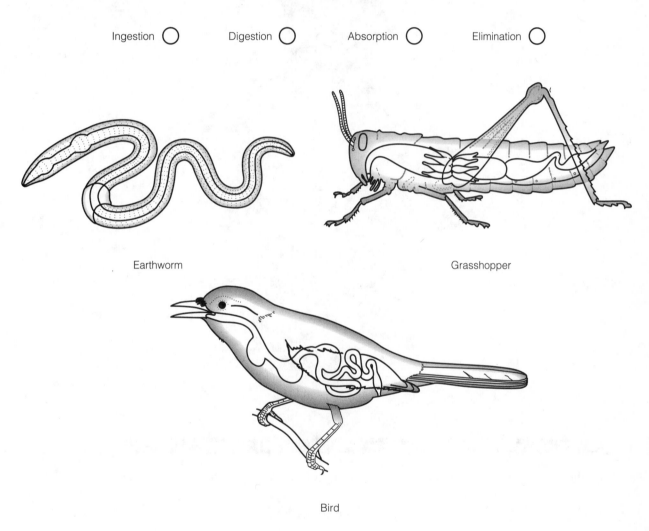

Earthworm   Grasshopper

Bird

Exercise 3   (Module 21.4)

Picture the parts of the human digestive system and their relationships to one another. Label and color these parts of the digestive system on the diagram below: **stomach, oral cavity, small intestine, esophagus, rectum, pancreas, gallbladder, mouth, large intestine, tongue, salivary glands, anus, liver, pyloric sphincter,** and **pharynx.**

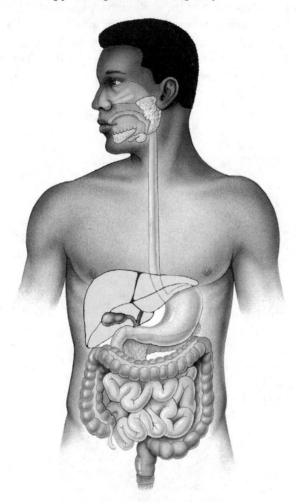

Exercise 4   (Modules 21.4 – 21.8)

Identify the part of the digestive tract described in each of the following statements. Some answers occur more than once.

_____   1. Kind of front tooth used for biting

_____   2. Largest digestive gland

_____   3. Muscles of esophagus that contract above a bolus

_____   4. The opening through which wastes are eliminated

_____   5. Where liver secretions are stored

_____   6. A ring of muscle that controls food leaving the stomach

_____   7. The throat

_____   8. The digestive gland above and to the right of the stomach

_____   9. Back teeth that grind and crush food

_____   10. Muscles of the esophagus that contract below swallowed food

_____   11. Where food goes after it leaves the stomach

_____   12. Digestive gland below the stomach

_____   13. Where taste buds are located

_____   14. The portion of the tract that wraps around the small intestine

_____   15. Pointed teeth that are overdeveloped in Count Dracula

_____   16. Where the tongue and teeth are located

_____   17. The portion of the canal between the esophagus and the small intestine

_____   18. The technical name for a swallowed ball of food

_____   19. Digestive glands that secrete into the oral cavity

_____   20. Where final steps of digestion and absorption of nutrients occur

_____   21. Where water is absorbed and feces are formed

_____   22. A flap of cartilage and connective tissue that keeps food out of the trachea

_____   23. The pyloric sphincter regulates its exit

_____   24. The muscular tube from the pharynx to the stomach

_____   25. The portion of the canal just before the anus

_____   26. The mouth opens into this portion of the canal

_____   27. Part of the digestive tract blocked when you choke

Exercise 5   (Modules 21.8 – 21.12)

Web/CD Activity 21A   *Digestive System Function*
Web/CD Thinking as a Scientist   *What Role Does Amylase Play in Digestion?*

The following story will help you to visualize the movement of food through the stomach and intestine and the processes that occur there. Write the proper word in each blank.

The experimental subject has been complaining of intestinal pain, but his doctors have been unable to pinpoint its cause. Your assignment is to inspect the linings of his stomach and small intestine. You step into the Microtron and are quickly reduced to microscopic size. You enter the subject via a drink of water, and the swallowing reflex sweeps you into the [1]_____, the tube to the stomach. A wave of muscle contraction, called [2]_____, propels you forward. Ahead, the opening broadens, and you tumble into the stomach.

The subject has been on a liquid diet, so the stomach is filled with clear fluid instead of [3]_____—the normal mixture of food particles and gastric juice. Huge folds in the stomach lining look like underwater ridges and valleys. Periodic waves of peristalsis sweep across this landscape like earthquakes. As you approach the stomach wall, you note that the lining is dotted with numerous pits—the openings of tubular [4]_____, which produce [5]_____, the stomach's digestive fluid.

You enter one of the pits and swim into the gastric gland. Your instruments indicate an increase in secretion of gastric juice. The presence of food—and you—in the stomach causes cells in the stomach wall to secrete a hormone called [6]_____, which (along with nerve signals) stimulates secretion of gastric juice.

Several kinds of cells line the walls of the gastric gland. Many of the cells are undergoing [7]_____, rapidly replacing cells that are damaged in this harsh environment. Deep in the gland, large parietal cells secrete [8]_____. Nearby chief cells secrete inactive pepsinogen, which quickly changes into the enzyme [9]_____

when it comes into contact with the acid. The function of pepsin is to begin the digestion of
[10]_____, preparing them for further digestion in the [11]_____.
Secreting the enzyme in inactive form [12]_____ the cells where it is produced. [13]_____ cells secrete thick [14]_____, which helps protect the stomach lining.

You exit the gland and continue your inspection of the stomach lining. The opening from the esophagus, where you entered the stomach, is sealed tightly. There is little
evidence of backflow of acid chyme into the esophagus, which might cause
[15]_____. Nowhere do you see any evidence of a [16]_____—an
open sore that develops in the stomach lining. It used to be thought that this happened
when there was too much pepsin or acid, or not enough [17]_____. Evidence
now points to [18]_____ by a prokaryote—*Helicobacter pylori*—which damages the stomach lining.

The stomach's exit is guarded by a tightly closed doughnut of muscle, the
[19]_____. As you approach, it opens to allow a squirt of liquid to leave the
stomach. Soon you enter the next portion of the digestive tract, the [20]_____.

This organ is much narrower than the cavernous stomach. It looks like a curving,
twisting tunnel. The first few inches is called the duodenum. Here you see a jagged sore,
tucked into a fold in the wall of the duodenum—a duodenal [21]_____. This
must be the cause of the problem. The medical team will have to deal with this.

Continuing your survey of the small intestine, you note an opening where fluid is
squirting into the duodenum. This is the duct through which [22]_____ from
the gallbladder and an alkaline, enzyme-rich solution from the [23]_____
enter the small intestine. The alkaline solution [24]_____ acid from the stomach. Bile, produced by the [25]_____ and stored in the gallbladder, contains
bile salts that break up [26]_____ into small droplets, a process called
[27]_____. This makes it easier for pancreatic enzymes to digest them. An enzyme called [28]_____ breaks fat molecules down into fatty acids and glycerol. Other pancreatic enzymes, trypsin and chymotrypsin, continue the digestion of
[29]_____ that began in the stomach. They break polypeptides down to
smaller polypeptides, and then peptidases break these smaller polypeptides down to
[30]_____. Pancreatic amylase hydrolyzes [31]_____ to form maltose, a disaccharide. Other enzymes digest it and other disaccharides to form
[32]_____. Nucleases digest DNA and RNA.

Contractions in the walls of the small intestine gently propel you along. The next
two portions of the small intestine are specialized for [33]_____ of nutrients.
The surface is highly folded, with numerous small projections called [34]_____.
You swim downward, and the villi surround you like the huge rubbery trunks of some in-
flatable forest. Their surfaces look and feel velvety, because the epithelial cells covering them
bear their own tiny projections called [35]_____. All these folds give the small
intestine a huge surface area—over 600 square meters! You press your light against the surface of one of the villi. Inside, you can see the ghostly outlines of [36]_____ vessels and a network of [37]_____. Nutrients are pumped or
[38]_____ through the intestinal epithelium and enter these vessels. The capillaries join to form a large blood vessel that carries blood directly to the liver, where nutrients
are converted to forms the body needs. For example, the liver stores excess glucose in the
form of a polysaccharide called [39]_____.

As you finish your inspection of the small intestine, you see the sphincter ahead that controls the movement of unabsorbed food material into the [40]_____, or colon. The colon is where [41]_____ is absorbed and undigested material is turned into [42]_____, which are stored in the rectum and expelled through the anus. That part of the digestive system is not on your itinerary! You work your way through the wall and into a blood vessel, and are soon back in the lab, munching on a sandwich.

Exercise 6   (Module 21.13)

Match each of the following animals with a description of its digestive tract.

_____ 1. Horse

_____ 2. Coyote

_____ 3. Frog

_____ 4. Cow

_____ 5. Rabbit

_____ 6. Deer

A. Digestive system gets simpler when it becomes an adult and changes to a carnivorous diet

B. Microbes in cecum digest cellulose, and some nutrients are absorbed in colon

C. Ruminant herbivore

D. Reingests fecal pellets to obtain nutrients

E. Carnivore with relatively simple digestive system

F. Rechews "cud," which is further digested by microbes and enzymes in four-chambered stomach

Exercise 7   (Module 21.14)

This module is short, but it is an important introduction to the subject of nutrition. See if you can summarize the module in a sentence of *exactly* 25 words.

Exercise 8   (Module 21.15)

Cellular metabolism breaks down the molecules in food and uses the energy released to build ATP, which in turn fuels body activities. The energy in food is measured in Calories (which are actually kilocalories). To get a handle on the energy contents of foods and the energy consumed by body activities, calculate how much you would need to exercise to use the calories in the food listed on the next page (similar to Table 21.15).

First, you need to calculate your own weight in kilograms. One kilogram equals 2.2 pounds, so:

$$\text{your weight (in kg)} = \frac{\text{your weight (in lb)}}{2.2} = \underline{\quad\quad}$$

Next you need to calculate how many kcal/min you use performing various activities:

kcal/min you use = kcal/kg/min (from table) × your weight (in kg)

Use the above formula to figure out how many kcal/min you would use jogging, swimming, and walking. Then write the figures in the spaces on the next page.

Finally, use the following formula to figure out how long you would have to exercise to use up the energy in a food item, and write the figures in the table:

$$\text{exercise time} = \frac{\text{energy in food}}{\text{kcal/min you use}}$$

Food-exercise energy equivalents:

	Jogging	*Swimming*	*Walking*
Speed	9 min/mile	30 min/mile	20 min/mile
kcal/kg/min	0.173	0.132	0.039
kcal/min for you:	_____	_____	_____

Number of minutes you would have to do the above exercises to consume the calories in:

Big Mac 560 kcal	_____	_____	_____
Cheese pizza—slice 450 kcal	_____	_____	_____
Coca-Cola—10 oz 144 kcal	_____	_____	_____
Whole-wheat bread—slice 86 kcal	_____	_____	_____

(You may want to look at the labels to get the energy contents of some of your favorite foods. You can record their exercise equivalents in the spaces below.)

_____	_____	_____	_____
_____	_____	_____	_____
_____	_____	_____	_____

Exercise 9   (Modules 21.16 – 21.22)

Vitamins, minerals, and certain fats and amino acids are all essential nutrients. They are substances an animal cannot make that must be obtained in food. These and other nutrients are listed on food labels. Paying attention to food labels can help us avoid obesity. There is a lot of detail in these modules, so try to focus on overall concepts. These concepts, the names of some nutrients, principles of food labeling, and the problem of obesity all relate to this crossword puzzle.

Across

2. The hormone leptin is produced by ____ (fat) cells.

4. The minerals sodium and ____ are important in acid-base and water balance.

8. Excess ____ -soluble vitamins are not stored, but excreted from the body.

9. A diet low in one or more essential amino acids can lead to ____ deficiency.

11. The vitamin debate is mainly about ____ .

14. The hormone _____ is made by fat cells and normally suppresses appetite.

15. ____ acid (vitamin-C) is important in collagen synthesis.

16. Most vitamins function as ____ .

17. There are ____ essential amino acids that must be obtained from the diet.

19. Humans naturally crave foods high in ____ .

21. We need the mineral ____ to make the hormone thyroxine.

22. A ____ is an essential organic nutrient needed in very small quantities.

27. All essential amino acids can be obtained from a combination of corn and ____ .

28. Packages list ____ and nutritional information.

29. ____ is a mineral important in bones, teeth, muscle, nerves, and blood clotting.

30. ____ -soluble vitamins (A, D, and so on) can be stored in the body.

31. An individual whose diet is deficient in one or more nutrients is said to be_____.

Down

1. Ingredients are listed on packages from greatest amount to ____ amount.

2. Vitamin D aids in the ____ and use of calcium and phosphorus.

3. The B ____ vitamins are important coenzymes in cellular metabolism.

5. The body can make fats by combining fatty _____ with molecules such as glycerol.

6. Calcium and ____ are important in building the skeleton.

7. Most victims of malnutrition are ____ .

10. Anorexia _____ is an eating disorder associated with compulsive aversion to body fat.

12. The amino acids that body cells cannot make are called ____ amino acids.

13. ____ is a component of hemoglobin and enzymes of energy metabolism.

18. The energy content of foods is expressed in ____ .

20. ____ are elements other than C, H, O, and N.

22. Vitamin A is important for normal ____ .

23. ____ contributes to a number of human health problems, such as diabetes and heart disease.

24. In earlier times, humans who stored more fat were able to survive _____ .

25. Essential amino acids are most easily obtained from ____ sources.

26. ____ are minimal standards of nutrients established by nutritionists for preventing deficiencies.

Exercise 10   (Modules 21.18 – 21.24)

It is important to read food labels because certain ingredients and nutrients are more healthful than others. But it is also important to not to obsess about food, and not to indulge in fad dieting. For a healthy lifestyle, one should try to maximize (↑) which of the following, minimize (↓) which, and seek moderate levels (→) of which others?

_____ 1. Blood cholesterol
_____ 2. Vegetable oils
_____ 3. Sodium
_____ 4. Fat intake
_____ 5. High-fiber foods
_____ 6. Exercise
_____ 7. Smoking
_____ 8. LDLs
_____ 9. Saturated fats
_____ 10. Carbohydrate intake

_____ 11. HDLs
_____ 12. Unsaturated fats
_____ 13. Foods rich in antioxidants
_____ 14. Alcohol intake
_____ 15. Fruits, vegetables, and other plant foods
_____ 16. Smoked and cured foods
_____ 17. Animal fats
_____ 18. Trans fats
_____ 19. Vitamin intake

Testing Your Knowledge

Multiple Choice

1. The box elder bug is an insect that sucks plant juices. It is a
   a. suspension-feeding omnivore.
   b. substrate-feeding herbivore.
   c. suspension-feeding herbivore.
   d. fluid-feeding herbivore.
   e. fluid-feeding carnivore.

2. Which of the following lists the four stages of food processing in order?
   a. ingestion, digestion, absorption, elimination
   b. digestion, ingestion, absorption, elimination
   c. ingestion, absorption, elimination, digestion
   d. ingestion, digestion, elimination, absorption
   e. absorption, digestion, ingestion, elimination

3. Which of the answers below would *not* work in the following sentence? "In order for the body to absorb and use _____ , they must be broken down by hydrolysis into _____ ."
   a. polysaccharides . . . monosaccharides
   b. amino acids . . . proteins
   c. fats . . . glycerol and fatty acids
   d. disaccharides . . . monosaccharides
   e. starch . . . monosaccharides

4. How does a gastrovascular cavity differ from an alimentary canal? The gastrovascular cavity
   a. stores food but does not digest it.
   b. is usually much larger.
   c. has only one opening.
   d. functions in digestion but not absorption.
   e. can use only small food particles.

5. In humans, most nutrient molecules are absorbed by the
   a. stomach.
   b. liver.
   c. small intestine.
   d. large intestine.
   e. pancreas.

6. The largest variety of digestive enzymes function in the
   a. large intestine.
   b. oral cavity.
   c. stomach.
   d. gallbladder.
   e. small intestine.

7. After nutrients are absorbed, the blood carries them first to the
   a. brain.
   b. pancreas.
   c. kidneys.
   d. liver.
   e. large intestine.

8. Digestion of proteins begins in the _____ , and digestion of polysaccharides begins in the ___.
   a. mouth . . . stomach
   b. stomach . . . small intestine
   c. stomach . . . mouth
   d. stomach . . . stomach
   e. small intestine . . . stomach

9. In humans, a major function of the large intestine is
    a. absorption of water.
    b. digestion of food molecules.
    c. breakdown of toxic substances.
    d. absorption of nutrients.
    e. storage of food before it is digested and absorbed.

10. The energy needed to fuel essential body processes is called
    a. essential nutrient level.
    b. metabolism.
    c. recommended daily allowance.
    d. optimum energy intake.
    e. basal metabolic rate.

11. Which of the following statements about fat is *incorrect?*
    a. Craving and storing fat may have been beneficial to early humans.
    b. It is best to strive to have as little body fat as possible.
    c. Too much fat in the diet can lead to cardiovascular disease.
    d. A moderate amount of fat contributes to a healthy immune system.
    e. In the U.S., the majority of people are overweight or obese.

12. Which of the following is *not* an essential nutrient?
    a. iron
    b. glucose, a monosaccharide
    c. methionine, an amino acid
    d. sodium
    e. pantothenic acid, a vitamin

13. Which of the following is an organic molecule needed by the body in small amounts?
    a. protein
    b. zinc
    c. vitamin C
    d. monosaccharide
    e. calcium

14. ____ are needed in the diet as components of teeth and bone, regulators of acid-base and water balance, and parts of certain enzymes.
    a. Amino acids
    b. Fats
    c. Minerals
    d. Nucleic acids
    e. Vitamins

15. To maintain health, it is best to try to
    a. decrease intake of unsaturated fats.
    b. decrease HDL levels in the blood.
    c. decrease intake of saturated fats.
    d. maintain a diet in which at least 35% of calories come from fat.
    e. increase LDL levels in the blood.

Essay

1. Why do the nutrient molecules contained in the food that you eat have to be digested before your body can use them?

2. In a few sentences, compare the digestive system of a grasshopper with that of a human.

3. Describe the substances in saliva and their roles in the function of the digestive tract.

4. Plants are difficult to digest. Describe how the digestive tracts and habits of three different herbivorous mammals are suited to a plant diet.

5. Briefly describe the potential problems with fad diets. What is the best approach to healthy, effective weight control?

6. Describe the kinds of information you can get from reading the labels on food. How might this information be useful?

Applying Your Knowledge

Multiple Choice

1. Which of the following do a wolf, a hummingbird, a termite, and an elephant have in common?
   a. All are omnivores.
   b. All are substrate feeders.
   c. All are ingestive feeders.
   d. All are carnivores.
   e. All are herbivores.

2. How would you expect the digestive system of a hawk, a carnivore, to compare with that of a finch, a seed-eater?
   a. The hawk would have a larger gastrovascular cavity.
   b. The finch digestive system would be longer (relative to body size).
   c. The hawk would have a gizzard, but the finch would not.
   d. The hawk digestive system would be longer.
   e. The hawk would lack both crop and gizzard.

3. Which of the following might make the most effective antiulcer medication? A chemical that
   a. stimulates parietal cells of the gastric glands.
   b. kills bacteria in the stomach.
   c. inhibits mucous cells of the gastric glands.
   d. stimulates secretion of bile.
   e. stimulates gastritis.

4. The lungs are folded into many small air sacs covered with blood vessels, which divide to form many small capillaries that increase the transfer of substances through their walls. The structures in the digestive system similar in function to these air sacs and capillaries are the
   a. villi.
   b. colon and rectum.
   c. gastric glands.
   d. high-density lipoproteins.
   e. sphincters.

5. Imagine that you have eaten a meal containing the following nutrients. Which would *not* have to be digested before being absorbed?
   a. protein
   b. polysaccharide
   c. disaccharide
   d. nucleic acid
   e. amino acid

6. Aunt Rose had her gallbladder removed and afterward
   a. could not eat foods containing large amounts of fat.
   b. had trouble digesting proteins.
   c. could eat monosaccharides and disaccharides but not polysaccharides.
   d. had to wash her food down with large quantities of water.
   e. needed to take an amino acid supplement.

7. Laxatives work in the large intestine to relieve constipation. Which of the following would probably *not* be an effective laxative? A substance that
   a. contains lots of fiber.
   b. promotes water absorption in the large intestine.
   c. speeds up movement of material through the large intestine.
   d. decreases water absorption in the large intestine.
   e. stimulates peristalsis.

8. It is important to get some vitamin B-1 every day, but it is all right if intake of vitamin A varies a bit. Why?
   a. Vitamin B-1 is an essential nutrient, and vitamin A is not.
   b. Vitamin A can be stored by the body, but vitamin B-1 cannot.
   c. The body needs much larger amounts of vitamin B-1 than of vitamin A.
   d. The body requires vitamin B-1, but vitamin A is just an "extra."
   e. Vitamin A is water-soluble, and vitamin B-1 is fat-soluble.

9. Why does your body need grams of carbohydrates each day but only milligrams of vitamins, thousands of times less?
   a. Carbohydrates are used up, but vitamins are reusable.
   b. The body needs carbohydrates to function, but not vitamins.
   c. Vitamins contain much more energy per gram.
   d. The body makes vitamins itself but needs to get carbohydrates from food.
   e. Carbohydrates are essential nutrients, but vitamins are not.

10. Joe is trying to decide whether to buy a candy bar or some gumdrops to reward himself for losing 10 pounds. Both are 300 calories. Most of the calories in the candy bar come from fat, and most calories in the gumdrops come from carbohydrate. If Joe wants to keep his weight under control, it would be best to buy the
    a. gumdrops, because the body tends to hoard carbohydrates.
    b. gumdrops, because the body tends to hoard fat.
    c. candy bar, because the body tends to hoard carbohydrates.
    d. candy bar, because the body tends to hoard fat.
    e. It doesn't matter; calories are calories, whatever the source.

Essay

1. James and Michael are both working out to build up their muscles. Muscle is mostly protein, so they both are on high-protein diets. James is eating twice as much meat as normal. Michael is a vegetarian, so he is consuming plant foods high in protein, such as beans and peanuts. What are the pros and cons of each diet?

2. It is sometimes necessary to remove a diseased portion of the digestive system, but usually a patient can get along on a modified diet. It would be most difficult to live without which of the following, and why? Stomach, large intestine, gallbladder, small intestine, salivary gland.

3. The pancreas does not actually secrete the protein-digesting enzymes trypsin and chymotrypsin. It secretes substances called trypsinogen and chymotrypsinogen, which are converted to trypsin and chymotrypsin by a substance secreted by the walls of the small intestine. What do you think is the reason for this two-step process?

4. NASA calculates that an astronaut uses an average of 2500 kcal of food energy per day. For every 4.83 kcal of food energy used, one liter of oxygen is consumed. How many liters of oxygen will the astronaut consume in a day?

5. Professor T. Herman Thimblebottom suspects that the element scandium is an essential mineral in the human diet. How might he go about demonstrating this? Why might it be difficult?

6. Look at some food labels. How many calories are contained in one serving of one of the foods? How far would you have to jog, swim, or walk to use the calories?

Extending Your Knowledge

1. Do you ever read the labels on food products? It can be very informative and entertaining. Look at some food labels. Which ingredients are present in the largest amounts? How much sugar does the product contain? (Remember, there are many words for sugar.) Which ingredients have value as nutrients, and which merely "improve" the taste, texture, appearance, or storage of the food? Is this food high in fat? Saturated or unsaturated fat? Calories? Sodium? How do more "natural" foods compare with more "processed" foods? How do "light" foods compare with regular ones?

2. Some advocates of a healthier diet suggest shopping only around the periphery of the supermarket, without venturing into the aisles. Do you think this would work? Why?

3. Are you eating a healthy diet and getting enough exercise? How would you evaluate your diet, given the information in this chapter? Outline changes in your diet and lifestyle that could decrease your chances of developing cancer or cardiovascular disease.

Gas Exchange

Most of the time we don't give gas exchange much thought. It just happens without thinking—a process "as natural as breathing." But have you ever been short of breath on a mountain hike or had "the wind knocked out of you" on an athletic field? Have you ever choked on food or been knocked down by a wave at the beach and come up coughing and sputtering? When breathing is interrupted, even for a few seconds, we realize how important every breath is. Breathing is part of the process of gas exchange. Not all animals breathe as we do, but they do need to obtain oxygen from their environment and dispose of carbon dioxide. This chapter is about the vital subject of gas exchange.

Organizing Your Knowledge

Exercise 1   (Module 22.1)

In most animals, there are three phases of gas exchange: **breathing, transport** of gases by the circulatory system, and **exchange** of gases with tissues. State which phase is interfered with in each of the following situations.

_____ 1. In the disease cystic fibrosis, thick mucus coats the inside of the lungs, blocking passage of gases.

_____ 2. A broken neck can paralyze the muscles of the chest.

_____ 3. Babies sometimes inhale small objects that can block the windpipe.

_____ 4. Anemia is a decrease in the oxygen-carrying protein hemoglobin.

_____ 5. During a heart attack, blockage of a blood vessel causes heart muscle cells to die from lack of oxygen.

_____ 6. An asthma attack narrows air passages into the lungs.

_____ 7. Bedridden patients sometimes get bedsores when blood vessels to the skin are pinched.

_____ 8. A mountain climber is breathing rapidly and his heart is beating strongly, but in the thin air there is not enough oxygen in his blood to diffuse into brain cells.

Exercise 2   (Module 22.2)

Match each of the following animals with a term (A–D) that describes it and a diagram (P–S) that shows its respiratory surface. Also color each respiratory surface yellow.

A. Lungs   B. Gills   C. Tracheae   D. Body surface

Animal	Term	Diagram
1. Beetle	____	____
2. Cat	____	____
3. Earthworm	____	____
4. Trout	____	____
5. Human	____	____
6. Chicken	____	____
7. Crayfish	____	____

P.   Q.   R.   S.

Exercise 3   (Modules 22.2 – 22.7)

Web/CD Activity 22A   *The Human Respiratory System*

Review gas-exchange mechanisms of different animals by filling in the blanks below.

Although some small animals, such as earthworms, use their [1]_____ as a gas-exchange organ, most animals have specialized organs that enable them to obtain [2]_____ and expel waste [3]_____. The part of an animal where gas exchange occurs is called the [4]_____ surface. Individual molecules of O_2 and CO_2 diffuse through a membrane only if they are dissolved in [5]_____, so the respiratory surface must be [6]_____. The surface must also be [7]_____ enough to take in sufficient oxygen for the body's needs and [8]_____ enough for gases to diffuse through it rapidly.

Most aquatic animals obtain dissolved oxygen from the surrounding water by means of [9]_____—outfoldings of the body surface. An advantage of exchanging gases in water is that the animal does not have to expend effort to keep the respiratory surface [10]_____. A disadvantage is that the concentration of available oxygen is much [11]_____ in water than in the air. The gills of a fish are efficient gas-exchange organs. Gill arches on each side of the fish's body bear numerous elongated gill filaments, and each of these bears numerous platelike [12]_____. The gills are red because the lamellae are filled with tiny [13]_____ covered by a thin layer of cells. The fish opens and closes its mouth and gill coverings to [14]_____ its gills, increasing contact between the water and the respiratory surface. Water flows past the lamellae in a direction [15]_____ to the flow of [16]_____ inside the lamellae. This countercurrent increases the efficiency of the gills. [17]_____ is transfer of a substance from a fluid moving in one direction to a fluid moving in the opposite direction. As water and blood flow past each other, [18]_____ diffuses from water to blood. As the blood picks up more and more oxygen, it comes into contact with water containing [19]_____ and [20]_____ available oxygen. Thus the countercurrent flow of water and blood creates a [21]_____ gradient that favors the diffusion of oxygen along the entire length of the lamella, greatly enhancing the efficiency of the gill.

Land animals obtain oxygen from the [22]_____. There are advantages to breathing air: It contains a [23]_____ concentration of oxygen than water, and it is [24]_____ to move than water. The biggest disadvantage of breathing air is that it tends to [25]_____ the respiratory surface. The [26]_____ system of an insect is an effective mechanism for gas exchange. [27]_____ tubes that branch throughout the body deliver gases directly to body cells without the help of the [28]_____ system. The tracheae branch and rebranch and end in tiny fluid-filled tubes that touch the surface of individual body [29]_____. [30]_____ dissolves in the fluid and diffuses into the cells, while [31]_____ diffuses out of the cells and is expelled from the insect's body.

Most land vertebrates have [32]_____ that function in gas exchange. The human lungs are in the chest cavity, above a thin sheet of muscle called the [33]_____ that functions in breathing. When you inhale, air enters through the [34]_____ and is warmed, humidified, and [35]_____ in the

^{36}_____ cavity. The air then passes into the ^{37}_____ (throat) and through the ^{38}_____, or voicebox. Here a pair of ^{39}_____ make the sounds that enable us to speak. When the vocal cords are ^{40}_____, high-pitched sounds are produced. When the cords are ^{41}_____, they make lower-pitched sounds. From the larynx, air passes through the ^{42}_____ into a pair of ^{43}_____, one leading to each lung. Inside the lungs, the bronchi branch into numerous narrow tubes called ^{44}_____. The surfaces of these respiratory passageways are covered by a film of ^{45}_____ that traps dust and other contaminants. The beating of numerous ^{46}_____ move the mucus and trapped particles out of the respiratory tract. The bronchioles end in clusters of tiny air sacs called ^{47}_____. There are ^{48}_____ of these tiny sacs in each lung, so their total surface area is enormous. The lining of each alveolus is a thin layer of epithelial cells that makes up the respiratory surface. Oxygen diffuses through a thin layer of moisture, through the epithelium, and into a network of ^{49}_____ that covers the surface of the alveolus. ^{50}_____ diffuses out of the blood and into the air within the alveolus. Thus the respiratory system works with the circulatory system in the process of gas exchange.

Exercise 4   (Module 22.5)

Web/CD Activity 22A   *The Human Respiratory System*

Use this diagram to review the parts of the human respiratory system. Label and color the parts in **bold** type. Color the **nasal cavity** purple, the **pharynx** blue, the **larynx** green, the **trachea** yellow, the **bronchi** orange, and the **bronchioles** red. Also color the surrounding **lung tissue** healthy pink (this woman is definitely a nonsmoker!) and the **diaphragm** brown.

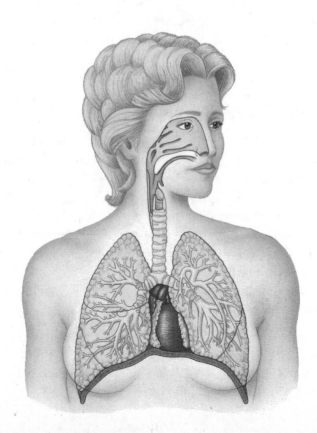

Exercise 5. (Module 22.6)

How do tobacco smoke and other air pollutants affect the respiratory system? Test your knowledge by completing this crossword puzzle.

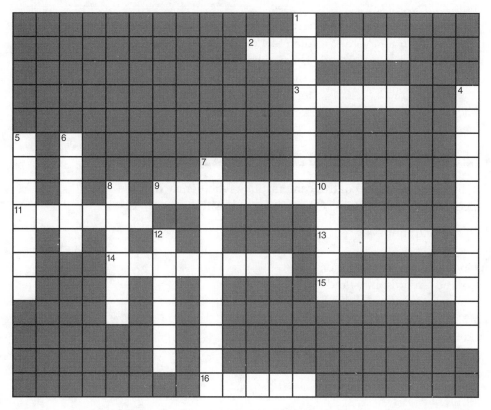

Across

2. Cigarette smoke irritates the lining of the ____, destroying their cilia and microphages.

3. ____ is an air pollutant associated with respiratory problems.

9. In ____, the alveoli lose their elasticity and the victim becomes short of breath.

11. Besides lung cancer, smoking also increases the risk of cancer of the pancreas, bladder, and ____.

13. Cigarette smoking destroys the ____ that normally sweep out pollutants.

14. When cilia are destroyed, only ____ can rid the lungs of pollutants.

15. Each year, ____ kills more than 440,000 Americans.

16. The ____ of a smoker are black, not pink.

Down

1. Carbon ____ is a harmful gas in polluted air.

4. ____ are defensive cells that engulf microorganisms and particles.

5. ____ years after quitting smoking, the risk of lung cancer and heart disease is similar to that of people who have never smoked.

6. On average, smokers die 13–14 ____ earlier than nonsmokers.

7. The ____ tissue lining the respiratory system is very delicate.

8. Lung ____ is the deadliest disease caused by smoking.

10. ____ and cilia protect the respiratory passages.

12. ____ dioxide is one of thousands of harmful pollutants in city air.

252 *Chapter 22*

Exercise 6 (Module 22.7)

State whether each of the following pertains to inhalation (I) or exhalation (E).

_____ 1. The diaphragm contracts.
_____ 2. The rib cage expands.
_____ 3. The diaphragm expands and arches upward.
_____ 4. Air enters the lungs.
_____ 5. The diaphragm moves downward.
_____ 6. Muscles between the ribs contract.
_____ 7. Muscles between the ribs relax.
_____ 8. The volume of the rib cage decreases.
_____ 9. Air pressure in the alveoli is less than that of the atmosphere.
_____10. The diaphragm relaxes.
_____11. Air pressure in the alveoli is greater than that of the atmosphere.
_____12. Air is forced out of the lungs.

Exercise 7 (Module 22.8)

Review control of breathing by the respiratory control centers: State whether each of the following changes would speed up (↑) or slow down (↓) your rate of breathing.

_____ 1. A rise in blood CO_2 concentration
_____ 2. Hyperventilating
_____ 3. A severe drop in blood oxygen concentration
_____ 4. An increase in pH of the cerebrospinal fluid
_____ 5. An increase in carbonic acid in the blood
_____ 6. A drop in blood pH
_____ 7. A decrease in blood CO_2 concentration
_____ 8. Holding your breath as long as you can, then releasing it
_____ 9. Vigorous exercise

Exercise 8 (Modules 22.8 – 22.10)

Web/CD Activity 22B   *Transport of Respiratory Gases*

The following story summarizes the cooperation of the respiratory and circulatory systems in gas exchange and transport. Fill in the blanks to complete the story.

Your respiratory system works together with your [1]_____ system in exchange and transport of gases. Imagine that you are riding a bicycle. Your leg muscles are working hard, consuming [2]_____ and producing [3]_____ as a waste product. Blood returning from the muscles therefore has a relatively [4]_____ concentration of oxygen and a [5]_____ concentration of carbon dioxide. One side of the heart pumps this oxygen-poor blood through the capillaries covering the [6]_____ in your lungs. You are breathing hard. A breathing control center in your [7]_____ responds to the [8]_____ in the pH of your blood caused by the increase in [9]_____. It speeds up the pace of nerve impulses sent to the [10]_____ and muscles between your [11]_____, and you breathe more rapidly. This helps expel the [12]_____ and meet the muscles' needs for more [13]_____.

The air in the lungs has a high partial pressure of [14]_____ and a low partial pressure of [15]_____ relative to the blood. [16]_____ diffuses out of the blood into the air inside the alveolus, moving from a region of [17]_____ partial pressure to a region of [18]_____ partial pressure. [19]_____ similarly diffuses down its pressure gradient from the air in the alveolus into the blood.

The oxygen that enters the blood is not very [20]_____ in water, so little oxygen is transported in dissolved form. Most oxygen is carried by a protein called [21]_____, contained within [22]_____ blood cells. A hemoglobin molecule consists of four polypeptide chains, each of which contains a heme group with an [23]_____ atom at its center. Each of these atoms can carry one [24]_____ molecule, so the millions of hemoglobin molecules in a red blood cell can carry millions of O_2 molecules.

The oxygen-rich blood that leaves the lungs returns to the heart, which pumps it out to the exercising [25]_____ of your legs. The blood passing through the capillaries in a muscle contains a [26]_____ partial pressure of O_2 than the muscle cells where it is being used up, so O_2 diffuses out of the blood and into the cells. The cells are making CO_2 at a fast pace, so the partial pressure of CO_2 is [27]_____ in the cells than in the blood. CO_2 diffuses [28]_____ the cells and [29]_____ the blood.

Some CO_2 is carried dissolved in blood plasma. Most of it enters [30]_____ blood cells, but most does not combine with [31]_____. Instead, enzymes in the blood cells cause most of it to react with [32]_____ molecules, forming carbonic acid (H_2CO_3). Each carbonic acid molecule then breaks apart, forming a hydrogen ion (H^+) and a [33]_____ ion (HCO_3^-). Hemoglobin picks up most of the H^+ ions, so it does not acidify the blood much. The bicarbonate ions diffuse out into the blood [34]_____, where they are part of the blood-[35]_____ system that stabilizes the pH of the blood. If blood pH drops, the bicarbonate ions combine with H^+ ions and remove them from the plasma. If pH [36]_____, the bicarbonate releases these H^+ ions back into solution.

When the blood from the muscles returns (via the heart) to the lungs, the events that formed bicarbonate ions are reversed. Bicarbonate and H^+ form carbonic acid, which breaks up to form water and [37]_____, which in turn diffuses out of the blood into the air of the alveoli. Thus the respiratory and circulatory systems continue to work in close cooperation as you continue your bicycle ride.

Exercise 9   (Modules 22.9 – 22.11)

Web/CD Activity 22B   *Transport of Respiratory Gases*

Number the following in order from first to last to show the path an O_2 molecule must follow through a mother to a cell in her fetus.

_____ A. The mother's heart pumps the oxygen-rich blood to her uterus.

_____ B. Oxygen diffuses out through the walls of capillaries in the uterus.

_____ C. The mother takes a deep breath of fresh air.

_____ D. Oxygen leaves the blood of the fetus and diffuses into a growing cell in the fetus's brain.

_____ E. Oxygen diffuses across the thin wall of an alveolus in the mother's lung and into a capillary.

_____ F. The mother's blood, now loaded with oxygen, returns from her lungs to the heart.

_____ G. Oxygen-rich fetal blood flows into the fetus through a vein in the umbilical cord.

_____ H. Oxygen diffuses through the wall of a capillary in the placenta and into the blood of the fetus.

_____ I. Oxygen attaches to hemoglobin in the mother's blood.

_____ J. Oxygen attaches to hemoglobin in fetal blood.

_____ K. The fetus's heart pumps the oxygen-rich blood out to its tissues.

Testing Your Knowledge

Multiple Choice

1. Which of the following has no specialized respiratory structures?
   a. crab
   b. earthworm
   c. salmon
   d. ant
   e. snake

2. The respiratory control centers are located in the
   a. heart.
   b. lungs.
   c. diaphragm and rib muscles.
   d. brain.
   e. large arteries.

3. When you exhale, the diaphragm
   a. relaxes and arches.
   b. relaxes and flattens.
   c. contracts and arches.
   d. contracts and flattens.
   e. contracts and arches, but only when you are exercising vigorously.

4. Why are bird lungs more efficient than human lungs?
   a. They use countercurrent exchange.
   b. They have more surface area than human lungs.
   c. They are able to concentrate the oxygen to much higher levels.
   d. Their alveoli are much larger.
   e. They use a one-way rather than an in-out air flow system.

5. Inhaled air passes through which of the following last?
   a. bronchiole
   b. larynx
   c. pharynx
   d. trachea
   e. bronchus

6. An advantage of gas exchange in water, compared with gas exchange in air, is that
   a. water usually contains a higher concentration of O_2 than air.
   b. water is easier to move over the respiratory surface.
   c. the respiratory surface does not dry out in water.
   d. ventilation requires less energy in water.
   e. the respiratory surface does not have to be as extensive in water.

7. In the blood, bicarbonate ions
   a. help transport oxygen.
   b. act as buffers to guard against pH changes.
   c. are transported by hemoglobin.
   d. attach to numerous CO_2 molecules, keeping them from solution.
   e. are poisonous and must constantly be removed.

8. Smoking destroys the cilia in the respiratory passageways. This
   a. makes it harder to move air in and out of the lungs.
   b. decreases the surface area for respiration.
   c. slows blood flow through lung blood vessels.
   d. makes it harder to keep the lungs clean.
   e. interferes with diffusion across the respiratory surface.

9. Most oxygen is carried by the blood ____. Most carbon dioxide is carried by the blood ____.
   a. attached to hemoglobin . . . in the form of bicarbonate ions
   b. dissolved in the plasma . . . dissolved in the plasma
   c. in the form of H^+ ions . . . in the form of bicarbonate ions
   d. attached to hemoglobin . . . attached to hemoglobin
   e. attached to hemoglobin . . . dissolved in the plasma

10. A disease called emphysema decreases the springiness of the lungs. This decreases ____ and makes it exhausting to breathe.
    a. the volume of air the lungs can take in
    b. respiratory rate
    c. blood flow through the lungs
    d. countercurrent exchange
    e. mucus production

11. The ____ is a structure specialized for diffusion of gases and nutrients between the blood of the mother and the fetus.
    a. uterus
    b. placenta
    c. lamella
    d. alveolus
    e. umbilicus

Essay

1. Compare the advantages and disadvantages of obtaining oxygen from water and obtaining it from the air.

2. How are gills and lungs similar? How are they different?

3. Describe how the structure, number, and arrangement of alveoli are well suited to their function in gas exchange.

4. Where are your vocal cords? How do they work? How do they produce high-pitched and low-pitched sounds?

5. When you inhale, does air flow into the lungs, causing them to expand? Or do the lungs expand, causing air to flow in? Explain.

6. Explain how countercurrent exchange in a fish gill enhances absorption of oxygen from water.

Applying Your Knowledge

Multiple Choice

1. Which of the following normally contains the highest concentration of oxygen?
   a. body cells
   b. inhaled air
   c. air in the alveoli
   d. blood entering the lungs
   e. blood leaving the lungs

2. Which of the following in a human is most similar in function to the gill lamellae of a fish?
   a. vocal cords
   b. bronchioles
   c. alveoli
   d. tracheae
   e. diaphragm

3. In which of the following does oxygen pass directly from the air, through a moist surface, to individual cells, without being carried by the blood?
   a. mouse
   b. ant
   c. shark
   d. earthworm
   e. frog

4. A fish opens and closes its mouth and gill covers. A dog pants. A marine worm waves long, filmy gills in the water. All of these movements
   a. are examples of ventilation.
   b. show how circulation aids respiration.
   c. are examples of breathing.
   d. slow diffusion of CO_2.
   e. enhance countercurrent exchange.

5. ____ in CO_2 in your blood, which causes ____ in pH, would cause your breathing to speed up.
   a. An increase . . . a rise
   b. An increase . . . a drop
   c. A decrease . . . a rise
   d. A decrease . . . a drop
   e. Actually, it is rise and fall of O_2, not CO_2, that controls breathing.

6. Which of the following would have the same O_2 content?
   a. blood entering the lungs—blood leaving the lungs
   b. blood entering the right side of the heart—blood entering the left side of the heart
   c. blood entering the tissue capillaries—blood leaving the tissue capillaries
   d. blood entering the right side of the heart—blood leaving the right side of the heart
   e. blood leaving the tissue capillaries—blood leaving the lungs

7. Patients with chronic lung disease and difficulty breathing often adapt to the high concentration of CO_2 in their blood. The breathing centers stop responding to CO_2 level. If such a patient has difficulty breathing, medical personnel are reluctant to give the patient pure oxygen. Based on what you know about control of breathing, why do you think this is the case?
   a. The patient's body would use the oxygen to make even more CO_2.
   b. The oxygen would increase concentration of bicarbonate, altering pH.
   c. Increased oxygen in the blood might slow or stop breathing.
   d. The body is not used to the oxygen, and the patient would overdose.
   e. The patient would breathe too fast and become tired out.

8. In an old science fiction movie, the hero tries to drown a giant ant by holding its head under water. Would this work? Why?
   a. Yes. Ants use lungs to breathe much as we do.
   b. Yes. The skin surface, covered with water, could not get O_2 from the air.
   c. No. Ants use gills for respiration, like crabs do.
   d. No. Ants breathe through holes in the sides of their bodies.
   e. No. The ant could get oxygen by diffusion from the water.

9. A zoologist compared the respiratory efficiency and swimming speed of different fish. He found that less efficient fish tended to have
   a. greater ventilation.
   b. a thicker respiratory surface.
   c. more hemoglobin.
   d. a faster heart rate.
   e. a more extensive respiratory surface.

10. A biochemist mixed 10 drops of acid with 100 mL of water, and the pH dropped from 7.4 to 5.0. She then mixed 10 drops of acid with 100 mL of blood. The pH dropped from 7.4 to 7.2. What is the reason for this difference?
    a. Blood is thicker than water.
    b. Blood is already very acidic, so the acid has less effect.
    c. Blood is saturated with oxygen; there is little room for acid.
    d. Blood contains buffers that reduce pH change.
    e. Water is already more acidic than blood; there is little room for more.

Essay

1. Trace the path of an oxygen molecule from the air to one of your brain cells, naming all the places and structures it passes through on its way.

2. You are on the team to design a robot that will patrol the devastated terrain around the Chernobyl nuclear power plant. The robot will function like a living organism, gathering organic debris for "food" and obtaining oxygen from the surrounding air. What features would you want to include in your design of its respiratory surface?

3. Carbon monoxide molecules in cigarette smoke and automobile exhaust attach to hemoglobin molecules where oxygen normally attaches, and they hold on more strongly than oxygen. What effect would this have on the body?

4. A man smokes a pack of cigarettes (20) a day for 40 years. If each cigarette shortens his life, on average, by 5 minutes, how much "before his time" will he die? (*Note:* On average, each cigarette *does*, in fact, shorten life by five minutes!)

5. In a submarine, the oxygen supply was accidentally interrupted, causing the oxygen content of the air to drop. A machine that removes carbon dioxide continued to function, so there was no corresponding buildup of CO_2. None of the sailors felt short of breath or noticed anything wrong until several individuals fainted. Why do you think they did not feel short of breath?

Extending Your Knowledge

1. Do you smoke, or are you close to a person who smokes? Quitting the smoking habit is one of the most important and effective changes you can make to improve your health and your quality of life. Quitting is not easy; the nicotine in tobacco is highly addictive. But there is help available. College health centers and wellness programs and state and county health departments usually have information, classes, and programs to help people stop smoking. The American Lung Association is a good place to start: www.lungusa.org. Thousands have kicked the smoking habit—and you can too. Good luck!

Circulation

If you are of average size, your heart pumps about 5 liters of blood every minute. This is about equal to the total volume of blood in your body. The blood rushes under pressure through thick-walled arteries to smaller arterioles and finally through a network of millions of tiny capillaries. Here the blood slows down and materials are exchanged between it and the extracellular fluid that bathes all body cells. A muscle cell, brain cell, or bone cell may be far from the external environment, but it is never more than a fraction of a millimeter from the blood in a capillary. Via the capillary net, every cell obtains food and oxygen and disposes of carbon dioxide and other wastes. Chapter 23 tells the story of blood, the vessels that carry it, and the heart that pumps it.

Organizing Your Knowledge

Exercise 1 (Modules 23.1 – 23.2)

These modules give an overview of circulatory systems—their basic functions and parts. Review the information by matching the following words and phrases. Each answer is used only once.

A. Ventricle	_____ 1. Network of small blood vessels
B. Open system	_____ 2. Circulatory system of vertebrates
C. Capillary	_____ 3. Inadequate for transport over a long distance
D. Atrium	_____ 4. Vessel that carries blood away from the heart
E. Blood	_____ 5. Solution in spaces between cells
F. Closed system	_____ 6. Heart chamber that pumps blood out via arteries
G. Capillary bed	_____ 7. Carries out circulatory functions in a jelly
H. Diffusion	_____ 8. Circulatory system of insects, spiders, and so on
I. Interstitial fluid	_____ 9. Heart chamber that receives blood from veins
J. Artery	_____10. Vessel that conveys blood from arteries to veins
K. Gastrovascular cavity	_____11. Vessel that returns blood to the heart
L. Vein	_____12. The circulatory fluid

Exercise 2 (Module 23.3)

The fish heart pumps blood through two sets of capillaries. As the blood of a fish passes through the tiny gill capillaries, it loses pressure. Therefore, once it has picked up oxygen, it delivers this oxygen to the capillaries in body tissues rather half-heartedly. The mammal heart is like two fish hearts side by side. Each side pumps blood only through one set of capillaries. The right heart pumps blood only to the lungs. The left heart then raises the pressure and sends this blood on its way to body tissues. On the next page, compare fish and mammal circulatory systems. Trace the flow of blood in each animal by numbering the parts blood passes through in order.

A. Fish

__1__ a. Ventricle

_____ b. Systemic capillaries

_____ c. Atrium

_____ d. Gill capillaries

B. Mammal

__1__ a. Right ventricle

_____ b. Left atrium

_____ c. Lung capillaries

_____ d. Right atrium

_____ e. Systemic capillaries

_____ f. Left ventricle

Exercise 3 (Module 23.4)

Web/CD Activity 23A   *Mammalian Cardiovascular System Structure*
Web/CD Activity 23B   *Path of Blood Flow in Mammals*

To review the mammalian cardiovascular system, start by labeling the parts indicated in this diagram. Then color the vessels that carry oxygen-rich blood red and those that carry oxygen-poor blood blue. Finally, trace the path of blood flow by numbering the circles (1–9).

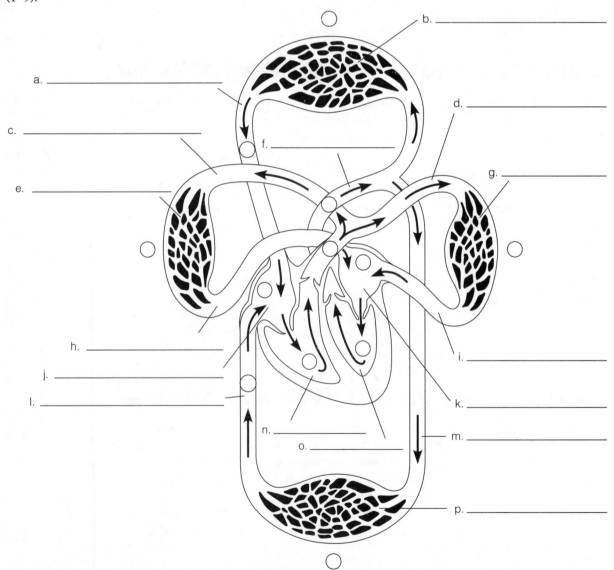

Exercise 4   (Module 23.4)

Web/CD Activity 23A   *Mammalian Cardiovascular System Structure*
Web/CD Activity 23B   *Path of Blood Flow in Mammals*

One way to learn about circulation and the parts of the heart is to trace the circulation of a drop of blood from one location in the body to another, naming all the structures that the drop of blood passes on its way. Use Figures 23.4A and 23.4B in the text. (A tip: Remember that blood always has to circulate from an artery to capillaries to a vein. There aren't any shortcuts. For example, to get from the capillaries of the big toe to capillaries of the little toe, blood must go back to the heart, to the lungs, back to the heart, and then to the little toe.)

Imagine a drop of blood starting in a brain capillary, circulating to the foot, and then circulating to the hand. The blood flows from the capillary into a vein that runs down the neck, and empties into the [1]_____, the large vein that serves the head and arms. From there, the blood enters the [2]_____ of the heart. This chamber pushes the blood into the [3]_____, which pumps it out through the [4]_____ to the lungs. In the capillaries of the lungs, the blood picks up oxygen. Then it returns to the heart via the [5]_____. It enters the [6]_____, which pumps the blood to the [7]_____. This chamber pumps blood out through the [8]_____, the largest blood vessel in the body. This vessel branches and rebranches, and finally the blood is delivered to [9]_____ in the foot, where nutrients and oxygen are dropped off at the tissues. The drop of blood travels back to the heart via leg veins, which join the [10]_____, which empties into the heart. The [11]_____ atrium and ventricle again pump blood through the [12]_____ circuit to the lungs. The [13]_____ atrium and ventricle then pump the blood out through the aorta into the [14]_____ circuit. This time the drop of blood flows down a large [15]_____ to the hand, where the blood again passes through a network of capillaries.

Try going on, tracing the flow of blood from the hand, to the intestine to pick up food molecules, or to the kidneys for filtration, and then back to the brain. Once you have done this a couple of times, try it without the diagram, or make your own sketch.

Exercise 5   (Modules 23.4 – 23.5)

Web/CD Activity 23A   *Mammalian Cardiovascular System Structure*
Web/CD Activity 23B   *Path of Blood Flow in Mammals*

Complete the following chart, comparing the structure of capillaries, arteries, and veins.

	Capillaries	*Arteries*	*Veins*
Carry blood from	1.	2.	3.
Carry blood to	4.	5.	6.
Thickness of walls (thick, thin, or in-between)	7.	8.	9.
Layers in walls (names)	10.	11.	12.
Valves? (yes or no)	13.	14.	15.

Exercise 6   (Module 23.6)

The following exercise relates to the function of heart valves and cardiac output. Indicate which of the following statements refer to the atrioventricular (AV) valves and which refer to the semilunar (SL) valves.

_____   1. Between atria and ventricles

_____   2. Open during systole

_____   3. At exits from ventricles

_____   4. Prevent backflow from ventricles to atria

_____   5. Prevent backflow from aorta and pulmonary artery to ventricles

_____   6. Close during systole

_____   7. "Lub" of "lub-dupp" heart sounds

_____   8. "Dupp" of "lub-dupp" heart sounds

Exercise 7   (Module 23.6)

Cardiac output is the amount of blood that the left ventricle pumps into the aorta per minute. It is equal to the amount of blood pumped per beat times the number of beats per minute.

1. What is your cardiac output right now? An average heart pumps about 75 mL of blood per beat. To calculate cardiac output, take your pulse, then multiply 75 mL times the number of beats per minute.

   Cardiac output = 75 mL × _____ beats per minute = _____ mL of blood per minute

2. Now run in place for a minute, take your pulse, and make the same calculation.

   Cardiac output = 75 mL × _____ beats per minute = _____ mL of blood per minute

3. How much did output change with exercise? _____ mL of blood per minute

Exercise 8   (Modules 23.7 – 23.8)

Web/CD Activity 23B   *Path of Blood Flow in Mammals*

Find a word or phrase in Module 23.7 or 23.8 that goes with each of the following.

_____ 1. Lowering these dietary lipids can reduce heart attack risk

_____ 2. Region that sets heart rate

_____ 3. Death of cardiac muscle cells due to heart vessel blockage

_____ 4. Relays signal to contract from atria to ventricles

_____ 5. Inserting a catheter to compress plaques in arteries

_____ 6. Buildup of plaques on the inside of blood vessels

_____ 7. Surgery that routes blood around blocked arteries

_____ 8. Location (outside heart) of centers that control heart rate

_____ 9. "Fight or flight" hormone that can speed up the heart

_____ 10. A device that "shocks" the heart to restore its rhythm

_____ 11. Imaging techniques used to "see" the heart

_____ 12. Vessel blocked in a heart attack

_____ 13. A wire mesh tube that opens an artery

_____ 14. An implanted device that triggers normal heartbeat

Exercise 9   (Modules 23.9 – 23.10)

Web/CD Thinking as a Scientist   *Connection: How Is Cardiovascular Fitness Measured?*

Pay special attention to Figure 23.9A; it contains a lot of information. Included are sizes, arrangement, and names of the blood vessels and the changes that occur in pressure and velocity as blood passes through them. After reading the modules and studying the figures, match each of the following statements with one of the blood vessels listed on the left.

A. Aorta

B. Arteries

C. Arterioles

D. Capillaries

E. Venules

F. Veins

G. Vena cava

_____ 1. Pressure is lowest here.

_____ 2. Pressure and speed drop the most in these vessels.

_____ 3. Pressure is usually measured in this kind of vessel.

_____ 4. Blood moves fastest here.

_____ 5. These vessels have the strongest pulse.

_____ 6. Blood moves most slowly here.

_____ 7. These vessels are the narrowest.

_____ 8. Diastolic pressure here might be 80 mm Hg.

_____ 9. Pressure here might be 20 mm Hg.

_____ 10. Pressure is highest here.

_____ 11. Systolic pressure here might be 120 mm Hg.

_____ 12. Velocity of blood increases sharply as blood flows through these vessels.

_____ 13. Hypertension may cause damage to the walls of these vessels, aggravating atherosclerosis.

_____ 14. Blood flows rapidly here, but there is no pulse.

_____ 15. Muscles and breathing help propel blood through these vessels.

Exercise 10   (Module 23.11)

The two mechanisms that control blood distribution, discussed in Module 23.11, sound very much alike. Sometimes, smooth muscle in arterioles *leading to* a capillary bed relax or contract to allow blood into the capillaries or divert it away. The second mechanism, illustrated in Figure 23.11B, involves muscle at the beginnings of the *capillaries themselves* that control blood flow. Nerve impulses and hormones can control these mechanisms. Sometimes changes in the tissues surrounding capillaries influence increase or decrease in vessel diameter. Can you imagine two physical or chemical changes that might occur in exercising leg muscles that might trigger vessel dilation and cause an increase in blood flow through the capillaries in the muscles? Write those changes below.

Exercise 11   (Module 23.12)

Through the thin walls of capillaries, materials are exchanged between the blood and the interstitial fluid that surrounds body cells. Complete this diagram showing the movement of blood through a capillary and exchange of materials with the interstitial fluid. Draw a red arrow to show blood moving from an arteriole into the capillary. Draw a blue arrow to show blood moving out of the capillary into a venule. (Why use two different colors?) Two forces are responsible for exchange of fluid between the blood and the interstitial fluid: Blood pressure tends to push fluid out of a capillary, and osmotic pressure tends to push water in. Draw large and small arrows at the arterial end of the capillary to show the relative strengths of blood pressure and osmotic pressure. Do the same at the venous end. Now draw arrows to show the net movement of fluid at the venous and arterial ends. (Do your arrows show net movement of fluid into or out of the capillary at the arterial end? Why? Is there net movement of fluid into or out of the capillary at the venous end? Why?)

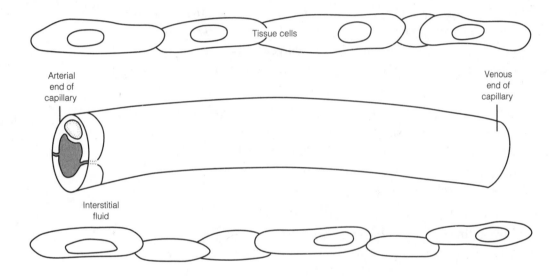

Exercise 12   (Modules 23.13 – 23.15)

Most of the information on blood is summarized in the figure in Module 23.13. Study the composition of blood, and then compare the three blood cell types by filling in the blanks in the following table.

Cell Type	Relative Size	Relative Numbers	Function
Platelets	Smallest	1.	2.
3.	4.	5.	Oxygen transport
Leukocytes	6.	Least numerous	7.

All the components necessary for blood clotting are present in blood all the time. Tissue damage activates them so that clotting occurs, in a sort of "chain reaction." Try to visualize blood flow and blood clotting by filling in the missing words in the following story.

You step into the Microtron, and you are quickly reduced to a size slightly smaller than a red blood cell. The support team injects you into a small artery in the arm. Blood pressure is fairly [1]_____ here. You feel a boom of pressure on your eardrums about once per second; this is simply the [2]_____, and it will gradually disappear as the blood in the artery flows into the narrower [3]_____ that lead to the capillary beds. Bright lights on the subject's arm enable you to see what is around you: Most of the cells around you are [4]_____, flexible disks carrying [5]_____ to the body's cells. There are also a few larger, irregular [6]_____, important in body defense. They slowly crawl along the blood vessel walls. Some even move against the current. It is best to avoid them, because some are [7]_____, capable of eating bacteria and debris. All around you are tiny "blobs." These must be [8]_____, which are involved in maintaining osmotic balance, defense, and blood clotting. There are also swarms of small fragments of cell cytoplasm, called [9]_____, that assist in the clotting process.

As you enter a [10]_____, blood slows almost to a stop, and the scene brightens. The walls of these vessels are a single layer of [11]_____, only one cell thick. You can even see gaps between cells, where fluid in the capillary is exchanged with the [12]_____ fluid. There isn't much room here, though. The capillaries are so narrow that the red blood cells have to line up single file in some places.

You are just under the skin of the fingertip. The team pricks the subject's skin with a pin. You are moving directly toward the wound, so you use your gripper to hang onto the vessel wall. The clotting process is already under way. The damaged lining of the vessel exposes [13]_____ to the blood. [14]_____ stick to the exposed tissue and release a cloud of chemicals. These chemicals cause even more platelets to adhere. But in this case, the damage is too serious for a platelet plug to stop the leak. Chemicals released from the platelets and damaged cells in the vessel wall activate blood enzyme. The enzyme then causes small blobs of [15]_____ floating in the blood to change shape and form sticky strands of [16]_____. These strands stretch like a tangle of cords across the hole in the vessel, trapping red blood cells. The blood strains against the fibrin clot, but finally it holds and leakage stops.

A phagocyte has caught your leg! You break free, but in the process you loosen a big chunk of the clot. It could travel to the heart, lodge in one of the [17]_____, and cause a [18]_____! You let the flow of blood carry you along. As blood leaves the capillary bed and enters a [19]_____, blood flow speeds up. The vessel walls thicken, and it gets darker again. You enter an even larger [20]_____, and the blood slows down even more. Ahead you can dimly make out the flaps of a [21]_____, which keeps the blood moving toward the heart. Fortunately, the clot is briefly caught in an eddy downstream from the valve. You use your laser to break it into fragments small enough to pose no threat to the subject. This is a good time to make your exit, and you are soon back in the lab discussing your adventure.

Exercise 14 (Module 23.16)

Stem cell research promises to cure several blood diseases. Review the potential of stem cells by matching each phrase on the left with a term from the list on the right.

_____ 1. Cancer of the white blood cells

_____ 2. Spongy bone tissue where blood cells develop

_____ 3. Unspecialized cell that differentiates to become blood cell

_____ 4. A cell that divides uncontrollably

_____ 5. A standard cancer treatment

_____ 6. Location where bone marrow is obtained for transplantation

_____ 7. The newest method of gathering stem cells gets them from here

_____ 8. A white blood cell

_____ 9. Where stem cells first form

_____ 10. A device used to gather stem cells from donor blood

A. Umbilical cord
B. Leukocyte
C. Cancer cell
D. Centrifuge
E. Leukemia
F. Embryo
G. Red marrow
H. Stem cell
I. Radiation
J. Pelvic bone

Exercise 15 (Summary)

This chapter introduces a lot of vocabulary connected with circulation. Circle the term that does not fit with the others in each of these groups, and briefly explain what the other terms in each group have in common.

1. venule   artery   atrium   capillary

2. fibrin   leukocyte   platelet   red blood cell

3. sphygmomanometer   systolic pressure   fibrinogen   hypertension

4. pulmonary veins   right ventricle   lungs   aorta

5. fibrin   leukocyte   platelet   hemophilia

6. semilunar valve   AV node   SA node   pacemaker

7. diastole   systole   pulse   leukocyte

8. oxygen   red blood cell   hemoglobin   epithelium

9. pulmonary artery   systemic circuit   aorta   left ventricle

Testing Your Knowledge

Multiple Choice

1. Rhythmic stretching of the arteries caused by heart contractions is called
   a. hypertension.
   b. heart murmur.
   c. hemophilia.
   d. pulse.
   e. diastole.

2. Which of the following animals has an open circulatory system?
   a. fish
   b. human
   c. frog
   d. jelly
   e. spider

3. Which of the following *cannot* move freely in and out of a capillary?
   a. sugar
   b. oxygen
   c. carbon dioxide
   d. water
   e. plasma protein

4. Heart valves function to
   a. keep blood moving forward through the heart.
   b. mix blood thoroughly as it passes through the heart.
   c. control the amount of blood pumped by the heart.
   d. slow blood down as it passes through the heart.
   e. propel blood as it passes through the heart.

5. Which of the following correctly traces the electrical impulses that trigger each heartbeat?
   a. ventricles, pacemaker, AV node, atria
   b. pacemaker, AV node, atria, ventricles
   c. atria, pacemaker, AV node, ventricles
   d. pacemaker, atria, AV node, ventricles
   e. pacemaker, AV node, atria, ventricles

6. A heart attack occurs when
   a. a heart valve malfunctions.
   b. a coronary artery is blocked.
   c. the heart is weakened by overwork.
   d. the aorta is blocked.
   e. the pulmonary artery is blocked.

7. The cells responsible for defense against infections are
   a. red blood cells.
   b. white blood cells.
   c. epithelial cells.
   d. platelets.
   e. pacemaker cells.

8. If cells are not receiving enough oxygen, a hormone signals the bone marrow to produce more
   a. leukocytes.
   b. fibrin.
   c. plasma.
   d. platelets.
   e. erythrocytes.

9. The primary sealants that plug leaks in blood vessels are
   a. platelets and fibrin.
   b. red blood cells and albumin.
   c. fibrin and white blood cells.
   d. white blood cells and platelets.
   e. hemoglobin and platelets.

10. Which of the following best describes an artery?
    a. carries blood away from the heart
    b. carries oxygenated blood
    c. contains valves
    d. has thin walls
    e. carries blood away from capillaries

11. Blood moves most slowly in
    a. capillaries.
    b. the aorta.
    c. veins.
    d. arterioles.
    e. venules.

12. In a fish, blood circulates through ____, while in a mammal, it circulates through ____ .
    a. two circuits . . . four circuits
    b. one circuit . . . two circuits
    c. four circuits . . . two circuits
    d. one circuit . . . four circuits
    e. two circuits . . . one circuit

13. The amount of blood flowing to skeletal muscles is greatly increased during exercise. This redirection of blood into muscles is accomplished by
    a. contraction of muscle in the walls of arteries.
    b. relaxation of muscle in walls of arterioles.
    c. opening of valves in veins.
    d. opening of valves in arteries.
    e. relaxation of muscle in the walls of veins.

Essay

1. Explain how the circulatory system changed to accommodate lung breathing and greater activity as land vertebrates evolved.

2. Cardiac output is the amount of blood the left ventricle of the heart pumps per minute. Cardiac output can be as much as four times greater when you are exercising than when you are at rest. What two things could the heart increase to increase cardiac output when you are exercising?

3. How does high blood pressure contribute to cardiovascular disease?

4. Briefly describe and explain changes in blood pressure as blood flows from arteries to capillaries to veins.

5. Describe the structural characteristics of capillaries that make them well suited for exchange of materials between blood and tissues.

Applying Your Knowledge

Multiple Choice

1. Just after blood leaves the left ventricle of the human heart, it passes through the
   a. pulmonary artery.
   b. left atrium.
   c. aorta.
   d. superior vena cava.
   e. right ventricle.

2. In which of the following animals are blood and interstitial fluid the same?
   a. grasshopper
   b. jelly
   c. fish
   d. dog
   e. sparrow

3. The heart specialist listened to Paul's heart through a stethoscope. Instead of the normal "lub-dupp, lub-dupp" heart sounds, he heard "siss-dupp, siss-dupp." The doctor said, "Hmm . . . I'm not sure it is anything to worry about, but I think there is something wrong with
   a. one of your coronary arteries."
   b. the pacemaker."
   c. an atrioventricular valve."
   d. your aorta."
   e. a semilunar valve."

4. Which of the following terms would be *least* useful in describing the circulatory system of a fish?
   a. capillary bed
   b. pulmonary artery
   c. ventricle
   d. atrium
   e. cardiovascular system

5. In circulating by the shortest route from the lungs to the foot, how many times would a drop of blood pass through the left ventricle?
   a. 0
   b. 1
   c. 2
   d. 3
   e. 4

6. A recording of the electrical activity of a patient's heart shows that the atria are contracting regularly and normally, but every few beats the ventricles fail to contract. Which of the following is probably not functioning properly?
   a. AV node
   b. semilunar valve
   c. coronary artery
   d. pacemaker
   e. AV valve

7. Which of the following functions most like a valve in a vein?
   a. a kitchen faucet
   b. a revolving door
   c. the volume control on a radio
   d. a subway turnstile
   e. a sliding patio door

8. In circulating from the brain to the arm, a drop of blood would *not* have to pass through which of the following?
   a. left atrium
   b. aorta
   c. superior vena cava
   d. pulmonary vein
   e. inferior vena cava

9. Emphysema damages the tissues of the lungs and slows pulmonary blood flow. This causes blood to back up, stretching and weakening the walls of the heart and blood vessels. Which of the following do you think would be most affected by this backup of blood from the lungs?
   a. aorta
   b. right atrium
   c. left atrium
   d. right ventricle
   e. left ventricle

10. The hormone erythropoetin (EPO) could be used to treat
   a. anemia.
   b. hypertension.
   c. leukemia.
   d. hemophilia.
   e. artherosclerosis.

Essay

1. A runner's heart rate is 160 beats per minute, and 90 mL of blood is pumped by the left ventricle with each beat. What is the runner's cardiac output?

2. The figures for blood pressure in an artery are usually given like this: 130/80 mm Hg. But it takes only one figure to specify blood pressure in a vein: 5 mm Hg. Why the difference?

3. Recall the forces that cause fluid to leave and reenter a capillary. How do you think high blood pressure would affect this balance of forces? How does this help explain that one of the symptoms of high blood pressure is swelling of the tissues with fluid?

4. Sometimes a baby is born with its large blood vessels reversed: The right ventricle pumps blood out through the aorta, and the left ventricle is connected to the pulmonary artery. The system is otherwise normal. How would this alter blood flow? Why would this be disastrous if not quickly corrected by surgery?

Extending Your Knowledge

1. What kinds of changes in the lifestyles of many Americans have led to the recent downturn in the incidence of cardiovascular disease? Have you made any changes in your health habits to avoid future heart or circulatory problems? Are there any changes you would like to make?

2. Have you had your blood pressure checked lately? Do you remember what your blood pressure was? How high does pressure have to be to indicate high blood pressure? Even young people can have high blood pressure. If you haven't had yours checked lately, it might be a good idea. Many pharmacies have simple devices you can use to check your blood pressure yourself.

The Immune System

If it were not for your immune system, you would have died long ago from measles, chicken pox, pneumonia, influenza, or any of hundreds of other diseases. Many times during your life, dangerous bacteria or viruses broke through the skin or other innate, non-specific defenses. Each time, your immune system protected you, and each time it was 100% successful. Microorganisms can overwhelm the body so quickly that 99% effectiveness won't do. The cells and antibodies of the immune system must specifically identify and destroy every invader and at the same time spare the body's own cells. Your immune system's remarkable ability to recognize and protect your "self" from "nonself" is the topic of this chapter.

Organizing Your Knowledge

Exercise 1 (Modules 24.1 – 24.3)

Innate immunity is the body's first line of defense. It is nonspecific, resisting any invader. Match each of the following components of innate immunity with a phrase from the right.

_____ 1. Natural killer cell

_____ 2. Lymph node

_____ 3. Macrophage

_____ 4. Interferons

_____ 5. Stomach acid

_____ 6. Complement system

_____ 7. Lysozyme

_____ 8. Hair

_____ 9. Neutrophil

_____ 10. Clotting proteins

_____ 11. Inflammatory response

_____ 12. Skin

_____ 13. Lymph

A. Proteins that burst bacteria

B. Seal off infected region

C. Protective enzyme in sweat and tears

D. Barrier that is first line of defense

E. Large phagocytic cell of interstitial fluid

F. Digests most microorganisms in food

G. White blood cell that engulfs bacteria

H. Attacks cancer cells and cells infected by viruses

I. Help cells resist viruses

J. Triggered by histamine, it disinfects and cleans injured tissues

K. Carries microbes to lymphatic organs

L. Filters inhaled air

M. May become swollen due to infection-fighting activity

Exercise 2   (Module 24.2)

Study this module on the inflammatory response. Then, from memory, fill in the missing words in this diagram.

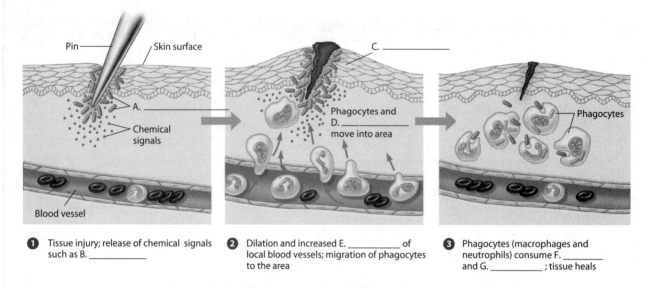

1 Tissue injury; release of chemical signals such as B. _____

2 Dilation and increased E. _____ of local blood vessels; migration of phagocytes to the area

3 Phagocytes (macrophages and neutrophils) consume F. _____ and G. _____ ; tissue heals

Exercise 3   (Modules 24.3 – 24.5)

These modules introduce some basic concepts and terminology related to the immune and lymphatic systems. To test your knowledge of this material, state the difference between the terms in each of the following pairs.

1. antigen—antibody

2. T cell—B cell

3. cell-mediated immunity—humoral immunity

4. lymphatic system—circulatory system

5. passive immunity—active immunity

6. blood capillary—lymphatic capillary

7. innate immunity—acquired immunity

Exercise 4 (Modules 24.4 – 24.8)

What is immunity? Why do you sometimes have to get sick to become immune to a disease? To answer these questions, fill in the blanks in the following story, using words and phrases from Modules 24.4 – 24.8.

Brian stopped by one afternoon to visit his friends Tom and Alicia and their 2-year-old daughter Samantha. Little did any of them suspect that Samantha was coming down with the measles. She had not yet been immunized, and she had caught the virus from a little boy in her day-care group.

Brian had never had the measles either. Once he had been exposed, and his 1_____ defenses had been breached, it was too late to do anything. It was up to the specific defenses of the 2_____ system to fight the invading viruses.

Both B and T lymphocytes were involved in the battle, but we will concentrate on the B cells. The viruses' protein coats contained protein molecules foreign to Brian's body; these 3_____ are what triggered his immune response. Throughout his body were many different types of B lymphocytes, each capable of responding to a different antigen. On each of these B cells were 4_____ that acted as receptors for various potential foreign antigens. Only a specific B-cell type possessed antibodies whose 5_____ were complementary to the shape of antigenic determinants on the measles virus antigens. Eventually the measles viruses encountered some of these B cells in a 6_____ node. Only these cells were "chosen" to be activated to fight the invading viruses. This process is called clonal 7_____.

The stimulated B cells began to multiply, forming a 8_____, a population of genetically identical 9_____ cells. These cells secreted 10_____ capable of locking on to the viral antigens and inactivating the measles viruses.

The initial phase of immunity described so far is called the 11_____ immune response. Unfortunately, Brian had to suffer through the measles, because this response is too slow to wipe out the invaders before they cause harm. It usually takes several 12_____ for lymphocytes to become activated and form effector cell clones, and by that time Brian was already sick.

After he recovered, however, Brian was 13_____ to measles. When his college roommate came down with measles just before final exams, Brian was safe. A second exposure to the same antigen triggers the 14_____ immune response, which is much quicker and stronger than the primary response and also lasts longer. The explanation for this is that the first exposure actually triggers the formation of two cell clones, the effector cells that fought the original infection and also a clone of 15_____ cells, which are held in reserve. Whereas effector cells may live only a few days, these other cells may last for 16_____. They are capable of mounting a quick and powerful secondary response.

Although Brian was protected from the measles, he still caught a cold the day before his math exam. But that is another story entirely.

Review the role of B cells in immunity by labeling and coloring the following diagram. First identify and label the **primary immune response** and the **secondary immune response.** Label **antigens,** and color them red. Label **B cells,** and color them light blue. Draw green antigen receptors on the B-cell types, and label and color green **antibodies** produced by effector cells. Label **effector cells,** and color them dark blue. Label **memory cells,** and color them purple. When you are done, explain the diagram.

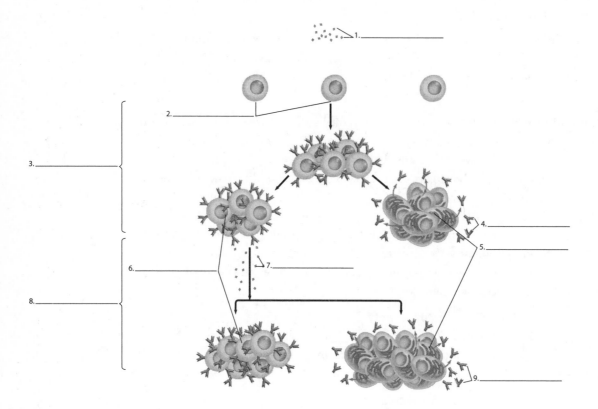

Exercise 6   (Modules 24.8 – 24.9)

The variable regions at the tips of an antibody's "arms" recognize antigens, and the constant region in the base of the Y helps destroy and eliminate antigens. As shown in Figure 24.9 in the text, there are a number of ways in which antibodies assist in destruction of antigens. These processes have been scrambled in the diagrams below. Identify each process, and label the diagrams accordingly. Choose from **phagocytosis, precipitation of dissolved antigens, neutralization, cell lysis, activation of complement,** and **agglutination of microbes.**

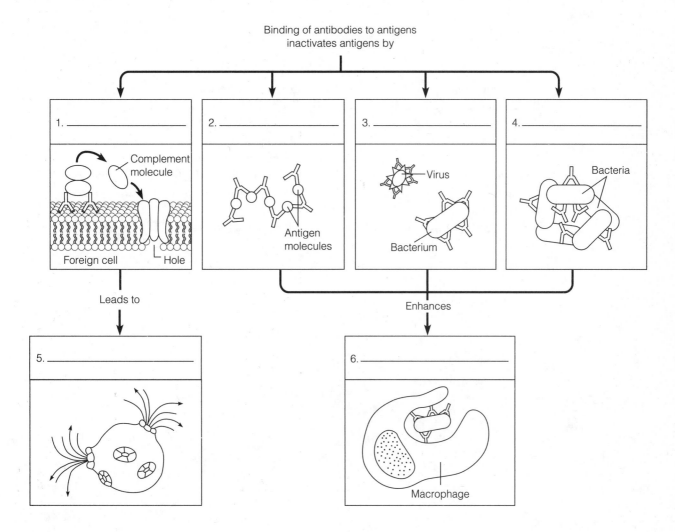

Exercise 7   (Module 24.10)

The usefulness of monoclonal antibodies stems from their ability to recognize and attach to specific molecules, which otherwise might be difficult to single out. See if you can think of several areas not mentioned in the text where this ability might be useful, and write your ideas on a separate piece of paper. Try not to limit your speculations to medicine. For example, how might monoclonal antibodies be useful in criminology, wildlife management, or archaeology?

Web/CD Thinking as a Scientist   *Connection: What Causes Infections in AIDS Patients?*
Web/CD Thinking as a Scientist   *Connection: Why Do AIDS Rates Differ Across the U.S.?*
Web/CD Activity 24A   *Immune Responses*

T cells are responsible for cell-mediated immunity. Contrast this arm of the immune system with humoral immunity, described in Module 24.9. Match each of the following components with its role in cell-mediated immunity (A–J) and then find each component on the diagram (P–Y).

Role
(A–I)

Diagram
(P–X)

_____ _____ 1. Interleukin-1
_____ _____ 2. Antigen-presenting cell
_____ _____ 3. Helper T cell
_____ _____ 4. T-cell receptor
_____ _____ 5. Self protein
_____ _____ 6. Perforin
_____ _____ 7. Interleukin-2
_____ _____ 8. Cytotoxic T cell
_____ _____ 9. Antigen
_____ _____ 10. B cell

A. Protein that recognizes antigens
B. Cell that secretes antibodies (humoral immunity)
C. Chemical signal that activates T and B cells
D. Macrophage that displays antigen
E. Evokes immune response
F. Cell that helps activate T and B cells
G. Body's own molecule that displays foreign antigens
H. Chemical signal from APC to T cell
I. Cell that attacks infected cells
J. Chemical that ruptures infected cells

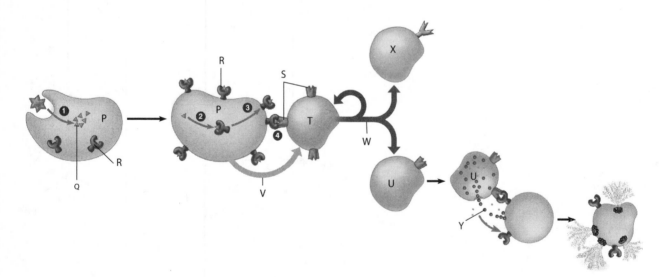

Exercise 9 (Module 24.12 and Introduction)

Web/CD Thinking as a Scientist   *Connection: What Causes Infections in AIDS Patients?*
Web/CD Thinking as a Scientist   *Why Do AIDS Rates Differ Across the U.S.?*

Understanding the immune system is important and useful. Understanding HIV and AIDS could save your life. Choose the correct italicized words or phrases to complete this story.

Lauren was worried and anxious because last week, her friend Robert told her that he had tested positive to antibodies against (1) *HIV, SCID,* the virus that causes AIDS. Now Lauren was waiting for the results of a blood test to determine whether she had been infected by Robert.

Lauren and Robert had had a sexual relationship that had ended about 6 months before. Because she was using birth control pills, they had never used (2) *an IUD, condoms, withdrawal,* which might have given them some protection. When he called, Robert told her that he was very sorry; he had no idea that he was HIV-positive. He said that he had probably been infected by a woman with whom he had had an earlier relationship. She had in fact experimented with drugs and had probably been infected by blood on a needle shared with a person infected with HIV. Robert was probably infected via contact with the woman's (3) *blood, skin, saliva* during intercourse. Lauren may have contracted HIV from Robert's blood or semen.

Until Robert called, Lauren had thought of HIV and AIDS mainly in terms of the poorer countries. She knew that in some (4) *Asian, African* countries, 40% of adults are infected with HIV. She also knew that HIV could be treated, but not cured, by drugs such as (5) *AZT, SCID.*

Now Lauren was forced to confront the harsh fact that American college students are at risk for HIV and AIDS. (In fact in the United States about a million people are living with HIV; each year more than 40,000 are newly infected with the virus.)

Lauren had learned about HIV and AIDS in her biology class. The virus mainly attacks the body's immune system, specifically (6) *B lymphocytes, helper T cells, cytotoxic T cells.* The virus contains RNA. When it enters a cell, this molecule directs the synthesis of DNA that becomes part of the cell's chromosomes. The viral instructions may remain dormant for as long as (7) *six months, ten years,* but eventually direct the cell to make more viruses. The viruses then destroy the host cell and leave to infect other cells. Because HIV attacks cells of the immune system, the body of the AIDS patient can be wracked by opportunistic infections. If untreated, death may result from disease or (8) *autoimmunity, cancer, allergies.* New combinations of drugs offer healthier, longer lives to individuals who are infected with HIV, but the drugs are expensive. It may take years to develop (9) *antibiotics, a vaccine* that could prevent the disease. At present, (10) *antibiotics, education* are/is the best weapon against AIDS.

Lauren had had a bad cold for the previous 2 weeks. At first she had thought nothing of it, but now she wondered if her immune system was beginning to falter. The telephone rang. She took a deep breath, sat down, and picked up the phone. The voice at the other end was professional. All Lauren could get out was, "Negative? The test was negative?"

She had never been so scared—or felt so relieved.

Exercise 10   (Modules 24.12 – 24.17)

Web/CD Activity 24B   *HIV Reproductive Cycle*

Sometimes the immune system fails to respond to invaders. Sometimes it goes overboard, responding to harmless substances in the environment, or even attacking the body's own tissues. Briefly describe how the normal function of the immune system goes awry in each of the following situations.

1. An immunodeficiency disease, like severe combined immunodeficiency

2. An allergy to cat hair

3. A cancerous brain tumor

4. Rejection of a heart transplant

5. Rheumatoid arthritis, an autoimmune disease

6. AIDS

7. Catching a cold after a stressful period, like final exams

Testing Your Knowledge

Multiple Choice

1. An antigen is
   a. a protein molecule that helps defend the body against disease.
   b. a type of white blood cell.
   c. an invading virus or bacterium.
   d. a foreign molecule that evokes an immune response.
   e. a body cell attacked by an invading micro-organism.

2. Your lymphatic system fights infection and
   a. delivers food and water to tissues.
   b. carries glandular secretions.
   c. maintains high blood pressure.
   d. allows red blood cells to approach cells more closely.
   e. drains fluid from tissues.

3. How do memory cells differ from effector cells?
   a. Memory cells are more numerous.
   b. Memory cells are responsible for the primary immune response.
   c. Memory cells attack invaders, and effector cells do not.
   d. Memory cells live longer.
   e. Memory cells are capable of producing antibodies.

4. Which of the following triggers tissue inflammation?
   a. accumulation of phagocytes in an injured area
   b. release of interferon by infected cells
   c. increased blood flow in an infected or injured area
   d. fever
   e. release of chemicals such as histamine by damaged cells

5. A cell capable of producing monoclonal antibodies is produced by fusing a lymphocyte with a
   a. tumor cell.
   b. red blood cell.
   c. bone marrow cell.
   d. T cell.
   e. macrophage.

6. A clone of lymphocytes
   a. produces different antibodies.
   b. lives in a specific area of the body.
   c. consists of immature cells, incapable of carrying out an immune response.
   d. makes antibodies against the same antigens.
   e. consists of both B cells and T cells.

7. Individuals infected with HIV
   a. can live for 15 to 20 years without symptoms.
   b. have little chance of developing AIDS.
   c. often die from autoimmune reactions.
   d. suffer from increased sensitivity to foreign antigens.
   e. can die from other infections.

8. Tissues are "typed" before an organ transplant to make sure that the _____ of donor and recipient match as closely as possible.
   a. T cells
   b. antibodies
   c. MHC markers
   d. histamines
   e. B cells

9. A vaccine contains
   a. white blood cells that fight infection.
   b. antibodies that recognize invading microbes.
   c. inactivated disease-causing microbes.
   d. a hormone that boosts immunity.
   e. lymphocyte antigens.

10. When you are immune to a disease
    a. antibodies against the disease are constantly circulating in your blood.
    b. certain lymphocytes are able to make the proper antibodies quickly.
    c. your innate defenses are strengthened.
    d. B cells are stimulated to quickly engulf invaders.
    e. antigens are altered so invaders can no longer attack your tissues.

11. An antibody is a
    a. protein that attaches to an antigen.
    b. foreign substance or organism.
    c. white blood cell that attacks invading bacteria or viruses.
    d. molecule on a body cell that identifies the cell as "self."
    e. large carbohydrate molecule that helps defend the body.

12. B lymphocytes
    a. attack cells that have been infected by viruses.
    b. engulf and destroy bacteria and viruses.
    c. multiply and make antibodies that circulate in blood and lymph.
    d. are responsible for cell-mediated immunity.
    e. do all of the above.

13. HIV infects mostly
    a. cells of the nervous system.
    b. B lymphocytes.
    c. cytotoxic T cells.
    d. macrophages.
    e. helper T cells.

14. The biggest difference between cell-mediated immunity and humoral immunity is
    a. how long their protection lasts.
    b. whether a subsequent secondary immune response can occur.
    c. whether clonal selection occurs.
    d. how they respond to and dispose of invaders.
    e. how fast they can respond to an invader.

15. Viruses and bacteria in body fluids are attacked by
    a. antibodies from B cells.
    b. cytotoxic T cells.
    c. complement proteins.
    d. helper T cells.
    e. antigens.

16. What do the antibodies secreted by plasma cells (the effector cells of humoral immunity) do to attack their targets?
    a. activate complement to punch holes in them
    b. clump cells together so that phagocytes can ingest them
    c. cause antigen molecules to settle out of solution
    d. attach to antigens and detoxify them
    e. all of the above

Essay

1. Explain how vaccination allows you to develop immunity to a disease without becoming ill. Include in your explanation the primary and secondary immune responses.

2. Compare humoral and cell-mediated immunity. In your comparison, discuss types of lymphocytes involved, roles of antibodies in the immune response, where invaders are attacked, and methods used to destroy invaders.

3. When biologists first started to work out the mechanisms of immunity, they found that the body produced antibodies that matched the shape of invading antigens. The researchers suspected that they would find that the immune system somehow analyzed the antigens and custom-built antibodies to fit. Did they find this to be the case? Explain.

4. Joan was telling her friend Jim why she had not been able to make it to biology class for several days: "My throat has been so sore I could hardly swallow," she croaked. "And the glands in my neck are really sore and swollen." Jim said, "We have been talking about this in class. They are not really 'glands,' you know, and the reason they are sore is . . ." Complete Jim's explanation.

5. Describe the inflammatory response and how it helps the body deal with injury or infection.

Applying Your Knowledge

Multiple Choice

1. Which of the following is not present until after the primary immune response occurs?
   a. memory cells
   b. macrophages
   c. helper T cells
   d. complement proteins
   e. antigens

2. The relationship between an antigen and an antibody is most like
   a. a battery and a flashlight.
   b. a hand and a glove.
   c. a hammer and a nail.
   d. a left foot and a right foot.
   e. a recipe and a cake.

3. A group of researchers have tested many chemicals and found several that have potential for use in treating malfunctions of the immune system. Which of the following would seem to have the most promise as a drug for inhibiting autoimmune diseases?
   a. Compound A13: acts like histamine
   b. Compound Q6: stimulates cytotoxic T cells
   c. Compound N98: a potent allergen
   d. Compound B55: suppresses specific cytotoxic T cells
   e. Compound M31: stimulates helper T cells

4. The body produces antibodies complementary to foreign antigens. The process by which the body comes up with the correct antibodies to a given disease is most like
   a. going to a tailor and having a suit made to fit you.
   b. ordering the lunch special at a restaurant without looking at the menu.
   c. going to a shoe store and trying on shoes until you find a pair that fits.
   d. picking out a video that you haven't seen yet.
   e. selecting a lottery prize winner by means of a random drawing.

5. Rhonda has been diagnosed as suffering from an immunodeficiency disease. Her doctor suspected Rhonda might have an immunodeficiency because
   a. Rhonda strongly rejected an organ transplant.
   b. Rhonda suffered from numerous allergies.
   c. Rhonda's blood showed high levels of numerous antibodies.
   d. Rhonda seemed to be immune to her own "self" molecules.
   e. Rhonda suffered from repeated, prolonged infections.

6. The idea behind vaccination is to induce _____ without the vaccinated individual having to get sick.
   a. passive immunity
   b. the primary immune response
   c. anaphylactic shock
   d. nonspecific defenses
   e. inflammation

7. Researchers found that when laboratory rats were already infected with a virus, they were better able to resist infection by a second completely different virus. The first infection apparently caused _____ , which protected the rats from the second infection.
   a. increased stress
   b. secretion of interferons
   c. production of antibodies
   d. passive immunity
   e. cell agglutination

8. Which of the following would be effective in eliminating bacteria but ineffective against viruses? (Hint: Viruses are not cells.)
   a. activation of complement proteins
   b. secretion of interferon by infected cells
   c. neutralization by antibodies
   d. agglutination by antibodies
   e. perforin secretion by cytotoxic T cells

9. An allergen acts like
   a. an antigen.
   b. histamine.
   c. interferon.
   d. an antibody.
   e. complement.

10. In a series of immune system experiments, the thymus glands were removed from baby mice. Which of the following would you predict as a likely result?
    a. The mice suffered from numerous allergies.
    b. The mice never developed cancerous tumors.
    c. The mice suffered from autoimmune diseases.
    d. The mice readily accepted tissue transplants.
    e. The mice were unable to produce an inflammatory response.

Essay

1. More people die each year from bee stings than from rattlesnake bites. The individual who is stung dies of anaphylactic shock, a massive, life-threatening allergic response. Anaphylactic shock usually occurs the second time an individual is stung by a bee. Why does this occur the second time and not the first time?

2. Researchers have not yet come up with a cure for the common cold, but they have made some interesting observations, among them: (1) There are more than 50 different known kinds of rhinoviruses, the viruses that cause colds; and (2) an average 2-year-old might catch three or four colds per year, but an 80-year-old catches a cold only once every 3 or 4 years. How do these findings relate to each other and the function of the immune system?

3. Before traveling to Africa, John got a "gamma-globulin" injection containing antibodies against hepatitis. Two years later, John planned another trip to Africa. His doctor recommended another gamma-globulin injection because John was no longer immune to hepatitis. What kind of immunity did John acquire from the first injection? Why didn't the immunity last?

4. After the flu season, blood samples were obtained from flu patients, and antibodies in the blood were analyzed. Researchers found that the antibodies produced by different patients in response to the same virus were often quite different. In some cases, the antigen-binding sites of the antibodies were completely different shapes. Explain how this could occur.

5. Antibodies are Y-shaped, with antigen-binding sites at the tip of each "arm" of the Y. How does the fact that an antibody has two antigen-binding sites help the antibody inactivate invaders? (Think what would happen if they had only one antigen-binding site.)

6. Organ transplant recipients must be very careful to avoid infections, especially viral infections. Explain why.

Extending Your Knowledge

1. Do you know anyone who has AIDS or is HIV-positive? Did you know anyone who died of AIDS? Is there anything you can do to help confront HIV and AIDS in your community and around the world?

2. Recently, several college campuses were disrupted by major measles epidemics. Are you immune to measles? Have you been immunized, or did you have measles when you were younger? How about other diseases? Do you know where to look or whom to ask to find out whether you are immune to specific diseases?

Control of the Internal Environment

A tiny sparrow hops across snow drifts on a subfreezing day, searching for food. How is it able to remain alive, even active, in such a hostile environment? Despite the cold, the sparrow's internal environment, the fluid that bathes its cells, remains remarkably constant. Numerous mechanisms keep it that way despite external fluctuations in temperature. Rapid cell metabolism allows the bird to generate its own body heat. Its feathers, circulatory system, and behavior control heat loss. It needs to keep eating to keep its temperature up, but metabolism of food alters body chemistry. To compensate, the sparrow's liver makes minute-to-minute adjustments in blood nutrient content and sweeps the blood for toxins. The kidneys actively sort the blood's chemical constituents, helping to maintain optimum solute content, adjust pH, get rid of waste products, and save water at the same time. This chapter discusses three processes that help maintain a constant internal environment—thermoregulation, osmoregulation, and excretion of wastes.

Organizing Your Knowledge

Exercise 1 (Introduction)

A black bear has many adaptations that help in thermoregulation—maintenance of its internal body temperature within narrow limits. Bears are endotherms, animals that derive most of their body heat from metabolism. Many other animals are ectotherms, which absorb heat from their surroundings. Which of the following are endotherms, and which are ectotherms?

_____ 1. Most fishes

_____ 2. A squirrel

_____ 3. Most invertebrates

_____ 4. A frog

_____ 5. Some insects

_____ 6. Some fishes

_____ 7. An eagle

_____ 8. A lizard

Exercise 2   (Modules 25.1 – 25.3)

Web/CD Thinking as a Scientist   *How Does Temperature Affect Metabolic Rate in* Daphnia?

Does the room temperature where you are right now feel comfortable? Do you feel a bit chilly? Or are you sweating because it is hot and humid? These modules discuss thermoregulation, and they provide numerous examples of methods animals use to regulate their internal temperatures. Animals regulate temperatures two ways: (1) by changing rate of heat production and (2) by adjusting rate of heat gain or loss. In the first column, state whether each of the following is:

A. A method of warming or cooling the body by regulating heat production
B. A method of warming the body by reducing heat loss
C. A method of warming the body by increasing heat gain
D. A method of cooling the body by increasing heat loss
E. A method of cooling the body by decreasing heat gain

Then, in the second column, identify which of the following adaptations is involved in each case:

P. Metabolic heat production
Q. Insulation
R. Circulatory adaptations
S. Evaporatic cooling
T. Behavioral responses

_____ _____   1. Moisture evaporates from a lizard's nostrils.
_____ _____   2. A robin fluffs up its feathers to trap more air near the skin.
_____ _____   3. A rabbit grows a thicker coat in the winter.
_____ _____   4. A lizard comes out of its burrow and turns broadside to the sun.
_____ _____   5. An elephant sprays itself with cold water.
_____ _____   6. Bees shiver.
_____ _____   7. Blood vessels dilate in a jackrabbit's ears.
_____ _____   8. You jump up and down and swing your arms to warm up on a cold day.
_____ _____   9. A kangaroo rat presses itself against the cool wall of its burrow.
_____ _____   10. A countercurrent heat exchanger cools the blood flowing to a duck's feet.
_____ _____   11. Hormones increase a mouse's metabolic rate.
_____ _____   12. A special network of vessels in a tuna retains warm blood, so muscles function better in cold water.
_____ _____   13. A cat licks itself, and saliva evaporates from its skin.
_____ _____   14. A bee seeks flowers that focus sunlight on its body.
_____ _____   15. A snake moves out of the sun and into the shade.
_____ _____   16. A chipmunk hibernates through the winter.

Exercise 3   (Modules 25.4 – 25.5)

To stay alive, animals must osmoregulate to maintain a correct balance of water and solutes in their body fluids. The concentration of seawater is "close enough" for many marine animals, so they are osmoconformers, simply matching the osmotic concentration of their environment. Other marine creatures, as well as freshwater and land animals, are osmoregulators. They actively move solutes and water in and out of their cells to maintain body fluid compositions different from their environments. Land animals have particular problems with loss of water and solutes. Use the information in the modules to complete this chart comparing osmoconformers and various osmoregulators.

	Marine Worm	*Freshwater Fish*	*Saltwater Fish*	*Human*
Osmoregulator or osmoconformer?	1.	2.	3.	4.
Tends to gain or lose water?	5.	6.	7.	8.
Tends to gain or lose solutes (ions)?	9.	10.	11.	12.
Method of compensating for gain or loss?	13.	14.	15.	16.

Exercise 4   (Module 25.6)

Breakdown of proteins and nucleic acids produces nitrogen-containing waste products. Different animals dispose of nitrogen in different ways. Summarize the kinds of animals that excrete each of the following nitrogenous wastes, and discuss the advantages and disadvantages of each.

	Ammonia	*Urea*	*Uric Acid*
Animals excreting this compound	1.	2.	3.
Advantages of excreting this compound	4.	5.	6.
Disadvantages of excreting this compound	7.	8.	9.

Exercise 5 (Module 25.7 – 25.8)

The liver is a complex organ that helps maintain body fluid homeostasis by breaking down and/or recycling nutrients, breaking down toxins, and converting nitrogenous wastes into forms suitable for excretion. Test your understanding of the liver by matching each of the statements on the left with a term from the list on the right.

_____ 1. Delivers blood from the intestine to the liver

_____ 2. Blood processed by the liver enters this vessel

_____ 3. The liver converts this toxic waste to a form that can be excreted

_____ 4. The nitrogenous waste synthesized by the liver for excretion

_____ 5. A nutrient the liver can break down for energy

_____ 6. Dangerous molecules broken down by the liver

_____ 7. A toxin that sometimes scars the liver

_____ 8. Liver damage caused by alcohol

_____ 9. These molecules made by the liver aid in blood clotting

_____ 10. These liver products carry fats and cholesterol to tissues

_____ 11. The liver stores glucose in this form for later use

_____ 12. A viral liver disease

A. Plasma proteins
B. Cirrhosis
C. Hepatic portal vessel
D. Glycogen
E. Hepatitis
F. Urea
G. Toxins
H. Inferior vena cava
I. Lipoproteins
J. Ammonia
K. Amino acid
L. Alcohol

Exercise 6 (Module 25.9)

Web/CD Activity 25A *Structure of the Human Excretory System*

The human excretory system performs important functions in fluid homeostasis. This exercise and the next will help you to become familiar with the kidneys and nephrons, their functional units. First, complete this diagram of the excretory system. Label a **renal artery,** and color it red. Label a **renal vein,** and color it blue. Label the **kidneys,** and color them brown. Then label and color the **ureters** (light brown), the **bladder** (yellow), and the **urethra** (green).

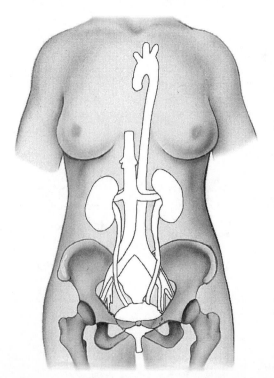

Exercise 7   (Module 25.9)

Web/CD Activity 25A   *Structure of the Human Excretory System*

Note in Figure 25.9 in the text how many tiny nephrons are arranged in each kidney. Label and color the following diagram of a nephron. Color arterioles and arterial capillaries red; color venous capillaries and the renal vein blue; and color the entire renal tubule, from Bowman's capsule through the collecting duct, yellow. Label **arteriole from renal artery, glomerulus, arteriole from glomerulus, capillaries, Bowman's capsule, proximal tubule, loop of Henle, distal tubule,** and **collecting duct.**

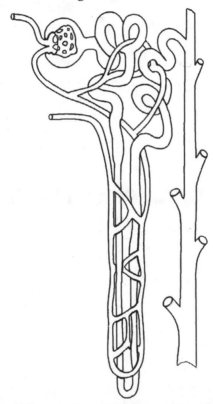

Exercise 8   (Modules 25.9 – 25.11)

Web/CD Activity 25A   *Structure of the Human Excretory System*
Web/CD Activity 25B   *Nephron Function*
Web/CD Activity 25C   *Control of Water Reabsorption*

Nephrons regulate the water and solute content of blood in a four-step process: filtration, reabsorption, secretion, and excretion. The next two exercises review nephron function, which is the heart of this chapter. After studying the modules, match each of the following parts of a nephron with its function.

A. Mainly functions in water reabsorption
B. "Refine" filtrate by reabsorption and secretion
C. Delivers blood to glomerulus
D. Carries urine to central "pelvis" of kidney; also functions in water reabsorption
E. Porous ball of capillaries where filtration occurs
F. Collects filtrate from glomerulus

_____ 1. Loop of Henle

_____ 2. Bowman's capsule

_____ 3. Proximal and distal tubules

_____ 4. Glomerulus

_____ 5. Collecting duct

_____ 6. Renal artery

Exercise 9   (Modules 25.10 – 25.11)

Web/CD Activity 25B   *Nephron Function*
Web/CD Activity 25C   *Control of Water Reabsorption*
Web/CD Thinking as a Scientist   *What Affects Urine Production?*

After studying kidney function, match each of the following components of blood with what happens to it in a nephron. Hint: Ask yourself, "Is this something the body wants to keep, or something it wants to get rid of?" (Answers may be used more than once.)

A. Not filtered
B. Filtered and mostly reabsorbed
C. Filtered, then mostly *not* reabsorbed, and
    finally excreted in urine
D. Filtered, mostly *not* reabsorbed, *also* secreted,
    and finally excreted in urine

_____ 1. Glucose
_____ 2. Water
_____ 3. Urea
_____ 4. H^+
_____ 5. Plasma protein
_____ 6. Amino acid
_____ 7. Red blood cell
_____ 8. Nicotine (drug)
_____ 9. Salt (NaCl)

Exercise 10   (Module 25.12)

Kidney dialysis has saved many lives, often as a stopgap until a kidney transplant can be arranged.

1. What are some advantages and disadvantages of dialysis, compared with a transplanted kidney?

2. What do you think would be some advantages and disadvantages of a kidney transplant, compared with dialysis?

Exercise 11   (Summary)

Review by filling in the blanks in the following scenario summarizing the roles of the kidneys and liver in regulating body fluid composition.

Tom finished his breakfast, downed his fourth cup of coffee (with milk and sugar), and dashed off to work. Digestion of milk proteins began immediately. The resulting amino acids were absorbed through the intestinal lining into Tom's blood. At the same time, enzymes split the sucrose (sugar) molecules, forming glucose and fructose, which also entered the blood. Caffeine and water molecules entered the blood immediately, without digestion.

Blood from Tom's intestine was carried via the [1]_____ to the liver. Liver cells absorbed some of the amino acids, sugars, fats, and caffeine. The fructose was immediately converted into glucose. The blood carried some of the glucose through the liver and into general circulation, where it was distributed to nourish body cells. The rest of the glucose was converted to [2]_____ and stored in the liver. It would later be turned back into [3]_____, as Tom's blood sugar dropped between breakfast and lunch. Meanwhile, the liver packaged some of the milk fats as lipoproteins and shipped them out to body tissues.

Some of the caffeine from the coffee made it through the liver and had the intended "wake-up" effect on Tom's brain as he was driving to work. Like other potentially harmful chemicals, the rest of the caffeine was [4]_____ by liver cells and the resulting less-toxic products were released back into the blood. All the caffeine would eventually be disposed of in this manner.

Some of the amino acids from the milk were carried by the blood to cells that could use them to make proteins. Other amino acids were broken down for energy or converted to other nutrients. Leftover amino groups were first changed into highly toxic [5]_____, but the liver cells quickly converted this into [6]_____, a much safer nitrogenous waste. This was also dumped into the blood exiting the liver.

The blood processed by the liver, with its cargo of glucose, amino acids, caffeine by-products, urea, and H^+ ions from the acid in the coffee, flowed through the hepatic vein to the [7]_____. This vessel carried the blood to the heart. The heart pumped the blood to the lungs and back, and then out through the aorta, whose branches distributed the blood to body tissues. About one-fifth of the blood pumped by each heartbeat flowed through the [8]_____ arteries to the kidneys.

In each of Tom's kidneys, branches of the renal artery delivered blood to a million [9]_____, the kidney's tiny working units. As a drop of blood passed through a [10]_____, a porous knot of capillaries, blood pressure filtered some of the blood through the pores into a funnel-like structure called [11]_____. Blood cells and large protein molecules were left behind in the blood, while smaller molecules were filtered. The fluid that collected in Bowman's capsule, called [12]_____, contained some of the water, glucose, H^+, and amino acids from Tom's morning coffee, plus urea and caffeine by-products from the liver. But filtration was just the beginning of the work of the nephron. The processes of [13]_____ and [14]_____ had to occur before the blood could leave the kidney and waste products could be [15]_____ from the body in urine.

As Tom exited from the freeway, the walls of the proximal tubules were beginning to reabsorb the glucose and amino acids in the filtrate via the process of [16]_____. Uptake of these substances, and others like salt, caused [17]_____ to be reabsorbed by osmosis, especially in the long [18]_____ of Henle and the collecting duct. These materials, valuable to the body, would reenter the blood and leave the kidneys through the [19]_____.

Filtration is not usually enough to get rid of all the acid the body needs to excrete, so Tom's kidneys actually removed some H^+ from the blood and added it to the filtrate, a process known as [20]_____. Although some of the urea was reabsorbed, most of it, along with the caffeine by-products, remained in the refined filtrate, which is called [21]_____ when it leaves the kidney.

Because the caffeine in Tom's coffee is a stimulant that raises blood pressure, Tom's kidneys were filtering his blood at a slightly higher than normal rate as he drove into the company parking lot. Urine left the kidneys through the [22]_____ and accumulated in the [23]_____. Tom had just enough time to stop at the restroom on his way to his office.

Testing Your Knowledge

Multiple Choice

1. Which of the following is an endotherm?
   a. mouse
   b. iguana
   c. frog
   d. trout
   e. all of the above except a

2. Which of the following describes the route of urine out of the body after it leaves the kidney?
   a. renal vein, bladder, urethra, ureter
   b. urethra, bladder, ureter
   c. renal vein, ureter, bladder, urethra
   d. ureter, bladder, urethra
   e. ureter, urethra, bladder

3. Blood flows through the hepatic portal vessel
   a. from the aorta to the kidney.
   b. from the intestine to the liver.
   c. from the kidney to the inferior vena cava.
   d. from the liver to the inferior vena cava.
   e. from the liver to the intestine.

4. As your kidneys regulate your body fluid composition, which of the following is the largest?
   a. the volume of filtrate formed by the nephrons
   b. the volume of urine excreted
   c. the volume of blood flowing through the nephrons
   d. the volume of solutes added to the filtrate by secretion
   e. the volume of filtrate reabsorbed

5. A countercurrent heat exchanger enables an animal to
   a. produce more heat when needed.
   b. reduce loss of heat to the environment.
   c. slow metabolism when food is not available.
   d. increase heat loss by evaporation.
   e. absorb heat from the environment.

6. Uric acid is the nitrogenous waste excreted by birds, insects, and many reptiles. An advantage of excreting uric acid is that it ____, but a disadvantage is that it _____ .
   a. saves water . . . costs energy
   b. saves energy . . . is highly toxic
   c. is not very toxic . . . wastes a lot of water
   d. is much more soluble in water than other wastes . . . costs energy
   e. saves water . . . is highly toxic

7. Which of the following is the most accurate and comprehensive description of the function of the kidneys?
   a. breaking down body wastes
   b. excreting wastes
   c. regulating fluid composition
   d. filtering the blood
   e. producing urine

8. Which of the following happens first as a nephron processes blood?
   a. excretion
   b. osmosis
   c. secretion
   d. reabsorption
   e. filtration

9. On a cold day, blood vessels in the skin
   a. dilate, allowing blood to keep the skin warm.
   b. constrict, forcing blood to flow through vessels to warm the skin.
   c. constrict, reducing heat loss from blood at the surface.
   d. dilate, causing blood to pass through the cold skin more quickly.
   e. dilate, preventing blood flow to the surface.

10. The animals in which of these pairs have similar problems regulating water balance?
    a. freshwater fish—saltwater fish
    b. land animal—freshwater fish
    c. osmoconformer—freshwater fish
    d. salmon in fresh water—salmon in salt water
    e. saltwater fish—land animal

11. The filtrate formed by the nephrons in the kidney is not the same as urine. The filtrate is first refined and concentrated by the processes of _____, forming the urine that leaves the body.
    a. filtration and secretion
    b. reabsorption and secretion
    c. reabsorption and excretion
    d. filtration and reabsorption
    e. secretion and excretion

12. Nitrogenous waste products are made from by-products of the breakdown of
    a. fats.
    b. starch.
    c. glucose.
    d. urea.
    e. proteins.

13. By definition, an ectotherm
    a. is "cold-blooded."
    b. is "warm-blooded."
    c. obtains most of its heat from its environment.
    d. derives most of its heat from its own metabolism.
    e. is none of the above.

14. Most aquatic animals excrete ammonia, while land animals excrete urea or uric acid. What is the most likely explanation for this difference?
    a. They have different diets.
    b. Land animals can get the energy needed to make urea or uric acid.
    c. Ammonia is very toxic, and it takes lots of water to dilute it.
    d. Land animals cannot afford the energy needed to make ammonia.
    e. Fish need to get rid of ammonia, but land animals need it to live.

15. A freshwater fish tends to _____ water by osmosis. As a consequence, its kidneys excrete _____ .
    a. gain . . . large amounts of dilute urine
    b. lose . . . small amounts of concentrated urine
    c. gain . . . large amounts of concentrated urine
    d. lose . . . large amounts of dilute urine
    e. gain . . . small amounts of concentrated urine

Essay

1. Describe how the skin can make adjustments that cool the body on a hot day and warm it on a cold day.

2. Explain how a goose can stand barefoot on ice without losing large amounts of body heat.

3. Compare the problems that freshwater and saltwater fish face in maintaining the water and solute balances of their body fluids. How does each kind of fish solve these problems?

4. Describe five behaviors that help different animals control their body temperatures.

5. The kidneys regulate body fluid composition by means of filtration, reabsorption, secretion, and excretion. Where does each of these processes occur? How does each contribute to the formation of filtrate and urine?

Applying Your Knowledge

Multiple Choice

1. Which of the following primarily involves heat transfer by convection?
    a. You roll down the car window to let the cool breeze blow through.
    b. The water in a lake is so cold that your legs become numb.
    c. You sweat profusely as you mow the lawn on a hot summer day.
    d. After sunset, you can feel heat from the warm pavement.
    e. As you lie on the sand, you can feel the sun's warm rays on your skin.

2. Which needs to drink the smallest amount of water to maintain its water balance?
    a. a sparrow
    b. a saltwater fish
    c. a freshwater fish
    d. a dog
    e. both a and b drink very small amounts

3. Humid weather makes you feel warmer because humid air
    a. interferes with heat loss by conduction.
    b. holds warm water vapor.
    c. interferes with heat loss by evaporation.
    d. prevents countercurrent heat exchange from occurring.
    e. increases metabolic heat production.

4. Pound for pound, a kidney uses as much energy as the heart. What do you think the energy is used for?
    a. to produce pressure for filtration
    b. for water reabsorption
    c. for the breakdown and detoxification of harmful substances
    d. to pump urine to the bladder
    e. for reabsorption and secretion of solutes by active transport

5. Look at the diagram of kidney dialysis, Figure 25.12, in the text. For kidney dialysis to work properly, the dialyzing solution should contain
    a. a higher solute concentration than blood.
    b. a higher concentration of urea than blood.
    c. a lower glucose concentration than blood.
    d. a lower concentration of urea than blood.
    e. a much smaller volume of fluid than the blood passing through it.

6. Which would have the toughest time surviving over the long term in the environment given?
   a. an osmoconformer in seawater
   b. an endotherm in a warm environment
   c. an ectotherm in a cold environment
   d. an osmoregulator in seawater
   e. an ectotherm in a warm environment

7. The loops of Henle in the kidneys of a desert kangaroo rat are much longer than those in a white laboratory rat because
   a. the kangaroo rat lives in an environment where water is scarce.
   b. the white rat's diet is much less varied than the kangaroo rat's.
   c. the kangaroo rat cannot always find food.
   d. the kangaroo rat produces more wastes.
   e. the kangaroo rat has less stress and lower blood pressure.

8. How does the filtrate produced by the filtration process of the glomerulus differ from urine? The filtrate
   a. contains very little water.
   b. contains a higher concentration of glucose.
   c. contains a much lower concentration of salt.
   d. contains a lower concentration of proteins.
   e. does all of the above.

9. The kidney's filtration process is nonselective, so
   a. many valuable substances are lost in urine.
   b. the proportions of substances in urine are the same as in blood.
   c. urine is much less concentrated than blood.
   d. it really has little control over body fluid composition.
   e. useful substances must be selectively reabsorbed.

10. Which of the following would be filtered from the blood but not normally found in urine?
    a. water
    b. red blood cell
    c. H^+ ions
    d. amino acid
    e. urea

Essay

1. In terms of heat loss, why is the windchill factor given in a weather report always a lower temperature than the air temperature?

2. An animal behaviorist found that a large part of a horned lizard's daily routine consisted of behavior related to thermoregulation. Very little of the lizard's time was occupied by searching for food. A pocket mouse in the same environment spent most of its time seeking food but very little time thermoregulating. Explain this difference in behavior.

3. Many kinds of poisons cause liver damage. Sniffing the solvents in glue can quickly injure the liver. Poisonous *Amanita* mushrooms also kill by destroying the liver. Over a longer period, alcohol causes liver damage. Why do you think the liver ends up being a "lightning rod" for all these different poisons?

4. You may have noticed that you never see pictures of Antarctic crocodiles lying in ambush for unwary penguins or seals. Why are birds and mammals more successful in the polar regions than crocodiles or frogs?

5. Imagine that you are working with a team of physiologists sampling and comparing the blood entering and leaving the liver. How might the blood entering and leaving the liver differ in each of the following circumstances? After the subject eats a large candy bar. After the subject consumes two shots of whiskey. After a 4-hour period during which the subject consumed no food.

Extending Your Knowledge

1. For a patient with kidney failure, a kidney transplant is the key to an improved, more normal life. For an individual with liver failure, an immediate transplant is essential for survival. Why is a liver transplant so important? If this is the case, why are liver transplants performed much less often than kidney transplants?

2. Suppose a friend or relative needed a kidney transplant. Would you volunteer to donate a kidney if your tissue types were compatible? What might be some reasons for and against donating one of your kidneys?

3. Side effects of many medications include liver or kidney damage. Why? Have any of the prescription or over-the-counter medications you use been implicated in causing liver or kidney impairment? Try looking at some of the labels or package instructions. How else might you find out about this?

Chemical Regulation

As you read this, your blood is carrying dozens of different hormones from endocrine glands to target cells all over your body. These hormones are chemical signals that regulate the activities of every cell. Hormones coordinate the activities of body organs, maintaining homeostasis. Their effects on their targets range from local adjustments in chemistry to control of large-scale body processes, such as growth and reproduction. Hormones are produced in minuscule amounts; cells are capable of responding to very subtle signals. Hormones carry specific messages, and each cell is equipped to respond only to certain ones. Different target cells may even respond differently to the same hormone. This subtle and intricate hormone communication and control system is the subject of Chapter 26.

Organizing Your Knowledge

Exercise 1 (Module 26.1)

This module describes the roles of chemical signals in control of body functions. Color and label the diagrams below, comparing the activities of endocrine cells, neurosecretory cells, and nerve cells. Note that the diagrams are not in the same order as those in the text. First, choose a color for each of the cells. Make hormone molecules from the endocrine cell green dots, hormone molecules from the neurosecretory cell blue dots, and neurotransmitter molecules red dots. Then label the following: **neurosecretory cell, hormone molecules, nerve signals, endocrine cell, neurotransmitter molecules, secretory vesicles, target cell, blood vessel,** and **nerve cell.**

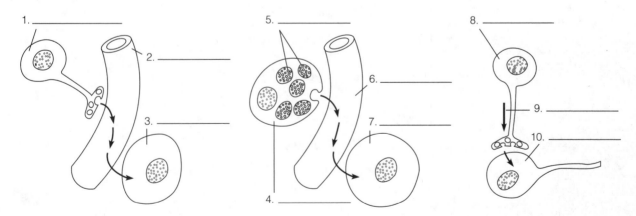

Exercise 2 (Module 26.1)

Module 26.1 contrasts the roles of the endocrine and nervous systems in controlling body activities. Check the statements that apply to each system. (Note: Some statements apply to both.)

	Endocrine	Nervous
1. A system of internal communication	_____	_____
2. Function involves electrical signals	_____	_____
3. Function involves chemical signals	_____	_____
4. Signals carried by neurons	_____	_____
5. Signals carried by body fluids	_____	_____
6. Signals carried by blood	_____	_____
7. Rapid messages	_____	_____
8. Slower messages	_____	_____
9. Split-second responses	_____	_____
10. Longer-lasting responses	_____	_____
11. Widespread effects	_____	_____
12. More pinpoint, localized effects	_____	_____

Exercise 3 (Module 26.2)

Web/CD Activity 26A   *Overview of Cell Signaling*
Web/CD Activity 26B   *Nonsteroid Hormone Action*
Web/CD Activity 26C   *Steroid Hormone Action*

Response of target cells to hormone, occurs in three steps: reception, transduction, and response. Just about everything you need to know about the mechanisms by which hormones affect their target cells is summarized in the two figures in Module 26.2. To review them, label the diagrams below. Choose from: **DNA, steroid hormone, enzyme, receptor protein, plasma membrane, target cell, new protein, signal transduction pathway, epinephrine, cellular response, nucleus, hormone-receptor complex,** and **relay molecules.**

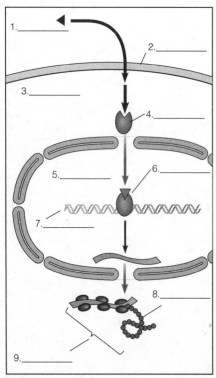

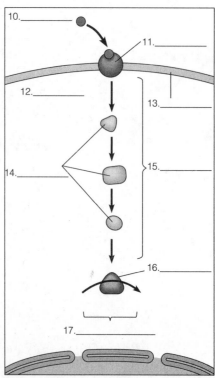

Web/CD Activity 26D   *Human Endocrine Glands and Hormones*

This module is an overview of the glands and hormones discussed in this chapter. You may want to refer back to the diagram and table as references as you read about glands and hormones in the modules that follow. Label the human endocrine glands on the diagram below, and color each one a different color. Choose from **thymus, testis, pineal gland, pituitary gland, adrenal glands, parathyroid glands, pancreas, thyroid gland, hypothalamus,** and **ovary.** Consulting Table 26.3 in the text, list under the name of each gland one of the hormones it produces.

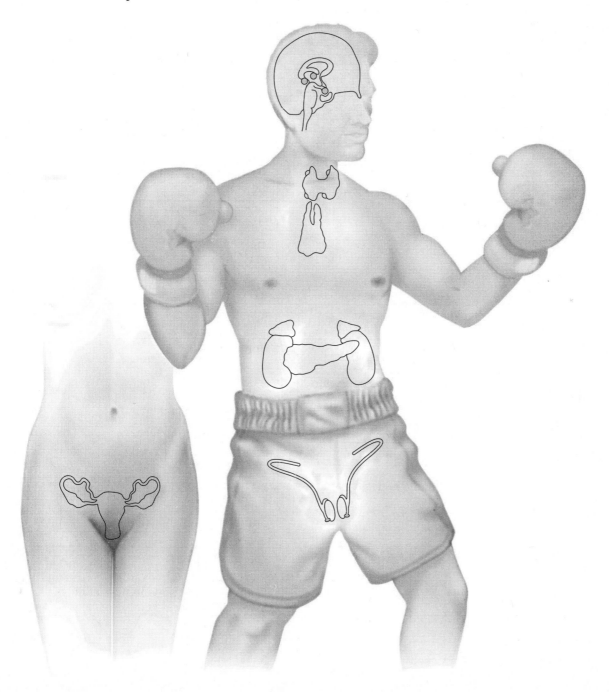

The table in Module 26.3 is a useful summary of glands, hormones, and hormone actions. Use it to complete the following chart by filling in the blanks. Note that the order of the glands here is different from the table in the text. (You may want to skip this for now and come back to it as a review after you have studied the glands and hormones in the following modules in more detail.)

Gland	Hormone(s)	Action
1.	2. & 3.	Control metabolic processes
	Calcitonin	4.
Ovaries	5.	6.
	Progesterone	7.
8.	9. & 10.	Trigger "fight or flight" response
11.	Glucocorticoids	12.
	13.	Regulate mineral gain and loss by kidneys
Pancreas	Glucagon	14.
	15.	16.
17.	18.	Raises blood calcium level
19.	Melatonin	20.
21.	22.	Stimulates growth
	Thyroid-stimulating hormone	23.
	Prolactin	24.
	25.	Stimulates adrenal cortex
	Luteinizing hormone	26.
	27.	Stimulates egg and sperm production
Posterior pituitary	Oxytocin	28.
	29.	Promotes water retention by kidneys
Thymus	30.	31.
32.	Androgens	33.
34.	Hormones that regulate anterior pituitary; hormones released by posterior pituitary	Affect pituitary

Exercise 6   (Module 26.4)

The hypothalamus and pituitary glands are closely related and tie the nervous and endocrine systems together. But the anterior and posterior lobes of the pituitary do different jobs and are controlled in different ways. State whether each of the following relates to the hypothalamus (H), anterior pituitary (A), or posterior pituitary (P).

_____ 1. Stores and secretes hormones actually made in the hypothalamus
_____ 2. Secretes releasing and inhibiting hormones that stimulate the anterior pituitary
_____ 3. Composed of nonnervous, glandular tissue
_____ 4. Part of the brain
_____ 5. Where the hormones oxytocin and ADH are released into the blood
_____ 6. Responds to releasing and inhibiting hormones from the hypothalamus
_____ 7. Secretes growth hormone, ACTH, and thyroid-stimulating hormone
_____ 8. Is the terminus of neurosecretory cells from the hypothalamus
_____ 9. Most hormones that it secretes influence other endocrine glands
_____ 10. Exerts master control over the endocrine system
_____ 11. Consists of nervous tissue and is actually an extension of the hypothalamus
_____ 12. Hypersecretion of one of this gland's hormones can cause gigantism
_____ 13. Negative feedback usually acts on this part of the system
_____ 14. Blood vessels carry releasing hormones from here to the anterior pituitary
_____ 15. Along with the brain, it secretes the body's natural painkillers
_____ 16. Secretes a hormone that affects the kidney's reabsorption of water

Exercise 7   (Module 26.5)

Web/CD Thinking as a Scientist   *How Do Thyroxine and TSH Affect Metabolism?*

Most people know at least a little about the thyroid gland. Once you understand that thyroid hormones stimulate metabolism, it is easy to remember the symptoms of over- and undersecretion and the effects of thyroid medication. Review the thyroid gland by matching each phrase on the right with a substance on the left.

A. Iodine
B. T_3
C. TSH
D. T_4
E. TRH

_____ 1. Secreted by hypothalamus; stimulates anterior pituitary
_____ 2. Secreted by anterior pituitary; stimulates thyroid
_____ 3. Hormone secreted by thyroid
_____ 4. Another hormone secreted by thyroid
_____ 5. Needed for the manufacture of thyroid hormones
_____ 6. T_3 and T_4 signal the hypothalamus to stop making this
_____ 7. Goiter occurs if there is not enough of this in the diet
_____ 8. Also called thyroxine
_____ 9. Sometimes called triiodothyronine

Two opposing, or antagonistic, hormones control the balance of calcium in the blood. To review calcium homeostasis, study Figure 26.7 in the text. Use it to help you choose the correct italicized words to complete the following paragraph.

Calcium is important for nerve impulse transmission, muscle contraction, transport of molecules through cell membranes, and (1) *digestion, protein synthesis, blood clotting.* Blood calcium concentration is held in a narrow range by the thyroid and parathyroid glands. If blood calcium drops, the (2) *thyroid gland, parathyroid glands* increase(s) secretion of (3) *parathyroid hormone, calcitonin.* This causes (4) *increased, decreased* reabsorption of calcium as the kidneys form urine, (5) *increased, decreased* absorption of calcium from food, and (6) *release of calcium from, deposition of calcium in* bone. If blood calcium concentration climbs too high, (7) *parathyroid hormone, calcitonin* is secreted by the (8) *thyroid gland, parathyroid glands.* This causes (9) *increased, decreased* reabsorption of calcium in the kidneys, and (10) *release of calcium from, deposition of calcium in* bone.

To understand how the pancreas controls blood sugar, you need to remember only two things: (1) the pancreas makes two hormones, insulin and glucagon, and (2) diabetics have high blood sugar, so many of them take insulin. Given these two facts, you can figure everything else out: If diabetics have to take insulin, insulin must make blood sugar go down. It must make cells take sugar out of the blood and use it or store it. That means glucagon must make blood sugar go up, by causing cells to get it out of storage and put it into the blood. Finally, an increase in blood sugar must trigger insulin secretion (to make sugar go down), so a drop in blood sugar must trigger glucagon secretion (to make blood sugar go up).

To review control of blood sugar level, choose either the word *increase(s)* or *decrease(s)* to complete each of the following statements.

1. Eating a meal rich in carbohydrates immediately causes blood glucose to _____.

2. When blood glucose _____, the pancreas secretes more insulin.

3. Insulin causes body cells to _____ their uptake and use of glucose.

4. Insulin also causes glycogen formation by the liver to _____.

5. Insulin therefore causes blood glucose to _____.

6. Between meals, blood glucose levels tend to _____.

7. When blood glucose _____, the pancreas secretes more glucagon.

8. Glucagon causes blood glucose to _____.

9. Glucagon _____ breakdown of glycogen in the liver and release of glucose to the blood.

10. Glucagon also triggers liver cells to _____ conversion of fats to glucose.

11. As blood glucose rises toward the set point, secretion of glucagon _____.

12. In type 1 diabetes, blood sugar _____ because the body is unable to produce insulin.

13. In type 2 diabetes, there is a _____ in cells' ability to respond to insulin.

14. Hypoglycemia results from an excess of insulin, which causes a sudden _____ in blood glucose.

Exercise 10 (Modules 26.9 – 26.10)

The adrenal gland is similar to the pituitary in that it is two glands in one. The parts have different jobs and send out different hormones. One of its parts (the medulla) is stimulated by nerve impulses to secrete its hormones, while the other part (the cortex) is stimulated by hormonal signals. The hormones from the adrenal medulla and cortex help the body deal with stress. After reading the modules and studying Figure 26.9 in the text, try to match each of the phrases on the right with adrenal stress hormones, E, M, or G.

E. Epinephrine and norepinephrine

M. Mineralocorticoids

G. Glucocorticoids

_____ 1. Increase breathing rate

_____ 2. Respond to short-term stress

_____ 3. Secreted by the adrenal cortex

_____ 4. Also secreted by the adrenal cortex

_____ 5. Increase blood volume and pressure in response to long-term stress

_____ 6. Triggered by nerve impulses from the hypothalamus

_____ 7. Secreted by the adrenal medulla

_____ 8. Cause proteins and fats to be broken down to make glucose

_____ 9. Triggered by ACTH from the pituitary

_____ 10. Also triggered by ACTH from the pituitary

_____ 11. Suppress the inflammatory response and immune system

_____ 12. Cause retention of sodium and water by kidneys

_____ 13. Dilate and/or constrict blood vessels, redirecting blood flow

_____ 14. Increase metabolic rate

_____ 15. Often prescribed to relieve pain from athletic injuries

Exercise 11 (Module 26.11)

Web/CD Activity 26D   *Human Endocrine Glands and Hormones*

Sex hormones are discussed in more detail in Chapter 27. After reading Module 26.11, try sketching on a separate sheet of paper a concept map for sex hormones. Refer to Chapter 3, Exercise 10, for a discussion of constructing a concept map. Include the following in your concept map: **testosterone, gonads, estrogens, hypothalamus, FSH and LH, progestins, testes, releasing factor, anterior pituitary, ovaries, sex hormones,** and **steroids.**

Exercise 12 (Summary)

On a separate sheet of paper, briefly describe how the items in each of the following groups are related to one another. They may affect one another, oppose one another, or have something in common.

1. luteinizing hormone   estrogens   androgens   follicle-stimulating hormone

2. oxytocin   antidiuretic hormone

3. glucagon   glucocorticoids   epinephrine

4. glucocorticoids   ACTH

5. antidiuretic hormone   mineralocorticoids   parathyroid hormone   calcitonin

6. thyroxine   calcitonin T_3

7. glucagon   insulin

8. glucocorticoids   epinephrine   mineralocorticoids   norepinephrine

9. ACTH   LH   growth hormone   FSH   TSH   prolactin

10. calcitonin   parathyroid hormone

11. oxytocin   prolactin

Testing Your Knowledge

Multiple Choice

1. Another system that works closely with the endocrine system to control body processes is the
   a. circulatory system.
   b. immune system.
   c. digestive system.
   d. nervous system.
   e. muscular system.

2. Every time you eat a cookie or candy bar, your blood sugar increases. This triggers an increase in the hormone
   a. thyroxine.
   b. epinephrine.
   c. adrenocorticotropin (ACTH).
   d. glucagon.
   e. insulin.

3. Every hormone
   a. is a protein.
   b. is produced in response to stress.
   c. is under the control of the pituitary gland.
   d. enters a cell and interacts with DNA.
   e. acts as a signal between cells.

4. Researchers have found increased levels of hormones from the _____ in the blood of students preparing for final exams. These hormones are produced in response to stress.
   a. thyroid gland
   b. pineal gland
   c. posterior pituitary
   d. adrenal glands
   e. parathyroid glands

5. Which of the following hormones have antagonistic (opposing) effects?
   a. thyroxine and calcitonin
   b. insulin and glucagon
   c. growth hormone and epinephrine
   d. ACTH and glucocorticoids
   e. epinephrine and norepinephrine

6. What is the role of a receptor in hormone action?
   a. It signals a cell to secrete a hormone.
   b. It informs a gland as to whether its hormones are having an effect.
   c. It enables a target cell to respond to a hormone.

d. It stops hormone action when it is no longer needed.
   e. It carries a hormone while it is in the blood.

7. When a boy goes through puberty, the steroid hormone testosterone "puts hair on his chest" by
   a. interacting with DNA in the nuclei of cells.
   b. causing cells to change shape.
   c. altering the permeability of plasma membranes.
   d. triggering nerve impulses in cells.
   e. turning enzymes on.

8. A hormone from the parathyroid glands works in opposition to a hormone from the _____ to regulate _____.
   a. posterior pituitary . . . metabolic rate
   b. thyroid gland . . . blood calcium
   c. pancreas . . . water reabsorption
   d. adrenal medulla . . . blood calcium
   e. thyroid gland . . . blood glucose

9. Some glands produce hormones that stimulate other endocrine glands. Which of the following hormones specifically acts to trigger secretion of hormones by another endocrine gland?
   a. thyroxine
   b. progesterone
   c. adrenocorticotropin (ACTH)
   d. antidiuretic hormone (ADH)
   e. melatonin

10. How is the level of thyroxine in the blood regulated?
    a. Thyroxine stimulates the pituitary to secrete thyroid-stimulating hormone (TSH).
    b. TSH inhibits secretion of thyroxine from the thyroid gland.
    c. TSH-releasing hormone (TRH) inhibits secretion of thyroxine by the thyroid gland.
    d. Thyroxine stimulates the hypothalamus to secrete TRH.
    e. Thyroxine and TSH inhibit secretion of TRH.

11. Steroid hormones are produced only by the
    a. adrenal medulla and pancreas.
    b. thyroid gland and pancreas.
    c. anterior and posterior pituitary.
    d. thyroid gland and gonads.
    e. gonads and adrenal cortex.

12. It usually takes much longer for sex hormones and other steroids to produce their effects than it takes for most nonsteroid hormones. Why?
    a. Steroids are bigger, slower molecules.
    b. Steroids usually must be carried longer distances by the blood.
    c. Steroids cause target cells to make new proteins, which takes time.
    d. Steroids must relay their message via a receptor.
    e. It takes longer for endocrine cells to make and secrete steroids.

13. The pituitary is actually two glands. The anterior pituitary secretes its hormones when stimulated by _____, and the posterior pituitary secretes its hormones when stimulated by

    _____ .
    a. hormones from the adrenal cortex . . . hormones from the thyroid
    b. hormones from the hypothalamus . . . nerve impulses from the hypothalamus
    c. hormones from the hypothalamus . . . hormones from the thyroid
    d. nerve impulses from the hypothalamus . . . hormones from the hypothalamus
    e. hormones from the pineal gland . . . hormones from the pancreas

14. Injections of a hormone are sometimes given to strengthen contractions of the uterus during childbirth. What hormone might this be?
    a. adrenocorticotropin (ACTH)
    b. thyroxine
    c. oxytocin
    d. insulin
    e. follicle-stimulating hormone (FSH)

15. Which of the following hormones has the broadest range of targets?
    a. ADH
    b. prolactin
    c. TSH
    d. epinephrine
    e. calcitonin

Essay

1. The pituitary is often called the master gland. In what way is this true? In what way is it misleading?

2. Compare how the adrenal cortex and adrenal medulla deal with stress.

3. How are hormones and neurotransmitters alike? How are they different?

4. Briefly describe an example of a hormonal disease or abnormality in which
    a. too much hormone is secreted.
    b. not enough hormone is secreted.
    c. hormone is secreted in a normal amount, but cells fail to respond.

5. Why are glucocorticoids useful in treating injuries? Why is prolonged use of glucocorticoids risky?

6. Describe two different situations in which a pair of hormones have opposite (antagonistic) effects on the body.

Applying Your Knowledge

Multiple Choice

1. Jet lag occurs when a person moves rapidly from one time zone to another, causing conflict between the body's biological rhythm and the new cycle of light and dark. Some scientists suspect that jet lag may result from disruption of a daily hormone cycle. Which of the following hormones do you think is the most likely suspect?
    a. epinephrine
    b. insulin
    c. melatonin
    d. estrogen
    e. prolactin

2. Which of the following hormones triggers secretion of the other two?
    a. thyroxine
    b. thyroid-stimulating hormone (TSH)
    c. TSH-releasing hormone (TRH)
    d. Any of the above can trigger secretion of the others.
    e. None of the above can trigger secretion of the others.

3. A tumor in an endocrine gland caused Jennifer to have weakened bones and unusually high levels of blood calcium. Which of the following was affected?
    a. anterior pituitary
    b. pancreas
    c. adrenal glands
    d. parathyroid glands
    e. thymus

4. Which of the following exerts control over all the others?
   a. adrenal cortex
   b. hypothalamus
   c. thyroid gland
   d. anterior pituitary
   e. testes

5. Because only the _____ gland uses iodine to make its hormones, radioactive iodine is sometimes used as a treatment for tumors of this gland.
   a. pituitary
   b. pancreatic
   c. thyroid
   d. adrenal
   e. testicular

6. Diabetes insipidus is an inherited endocrine malfunction (unrelated to the more common diabetes mellitus) in which the kidneys fail to reabsorb normal amounts of water. Victims of this disease produce gallons of urine each day, and their kidneys soon wear out. Treatment of this disease involves replacing a missing hormone. Which of the following do you think it is?
   a. glucagon
   b. epinephrine
   c. glucocorticoids
   d. antidiuretic hormone (ADH)
   e. thyroid-stimulating hormone (TSH)

7. In an experiment, researchers removed the _____ of young mice, and as a result, these mice were able to accept organ transplants without rejection.
   a. pineal glands
   b. thymus glands
   c. thyroid glands
   d. parathyroid glands
   e. adrenal glands

8. Tim once suffered a severe allergic reaction to a bee sting. The sting caused him to suffer a near-fatal drop in blood pressure called anaphylactic shock. Now he carries a kit containing a syringe of _____, which he can inject to speed up his heart if he reacts to a bee sting.
   a. insulin
   b. thyroxine
   c. testosterone
   d. calcitonin
   e. epinephrine

9. It has been found that certain salamanders fail to go through the normal transformation from tadpole to adult if there is a shortage of iodine in the pond water in which they live. This makes it impossible for the _____ to manufacture hormones necessary for normal development.
   a. thyroid gland
   b. posterior pituitary
   c. adrenal cortex
   d. pineal gland
   e. pancreas

10. As a young girl, Maria suffered a head injury that damaged her pituitary. An injury to the pituitary is particularly serious because of all the functions controlled by this gland. As Maria got older, she and her doctors found that all of the following except _____ were affected.
    a. metabolic rate
    b. growth
    c. her menstrual cycle
    d. milk production
    e. blood calcium level

Essay

1. Hypoglycemia is a condition in which blood sugar drops to abnormally low levels. What seems to be the cause of hypoglycemia? Why would it not be a good idea to try to correct this problem by eating more sugar?

2. One of the symptoms of severe diabetes mellitus is breath that has a sweetish, acetone smell. The smell comes from by-products of fat breakdown in the body. If another consequence of diabetes is excess sugar in the blood, why do the cells not just use the sugar instead of breaking down fat?

3. A tumor can cause enlargement of the thyroid gland. How could this result in abnormally high metabolic rate and body temperature? How could similar symptoms be produced by a tumor of the pituitary gland?

4. Some hormones, such as ADH, act on very specific targets (for ADH, the kidney). Other hormones, such as insulin, are able to affect every cell in the body. Based on what you have learned about how hormones exert their effects on target cells, speculate as to how some hormones can affect many targets, while other hormones affect only one.

5. A chemical called dioxin, or TCDD, is produced as a contaminant during some chemical manufacturing processes. Trace amounts of this substance were present in Agent Orange, a defoliant sprayed on vegetation during the Vietnam War. There has been a continuing controversy over its effects on Vietnam veterans exposed to it during the war. Animal tests have shown that dioxin can be lethal and can cause birth defects, cancer, liver and thymus damage, and immune system suppression. But its effects on humans are unclear, and even animal tests are uneven; a hamster is not affected by a dose that can kill a guinea pig. Researchers have discovered that dioxin exerts its effects like a steroid hormone. It enters a cell and binds to a receptor protein, which in turn attaches to the cell's DNA. How might this mechanism help explain the variety of effects of dioxin on different body systems and in different animals?

Extending Your Knowledge

1. Compare communication in the body via nerve impulses and hormones with everyday communication by telephone and letter. What are the characteristics of each? In what situations is each the most useful?

2. A survey indicates that over a quarter million U.S. high school students have used anabolic steroids, synthetic derivatives of testosterone. This number is in addition to the thousands of college and professional athletes who have used these hormones. Why do athletes use anabolic steroids? What are some of the dangers of steroid use? Do you know anyone who has used steroids in this way?

3. Do you know someone who has an endocrine disorder, such as diabetes? How does that person deal with his or her condition? Have there been recent discoveries or developments that have improved treatment of this condition?

Reproduction and Embryonic Development **27**

How old are you? When is your birthday? You celebrate your birthday as the beginning of your life, but birth is not really the beginning. Your birth was the culmination of a process that started months, even years, before. A couple of months before you were conceived, some of your father's cells divided to form sperm. One of these sperm contributed half your genes. Much earlier, when your mother was a baby, eggs began to form in her ovaries. A couple of weeks before fertilization, one of those eggs was brought to maturity. Meanwhile, hormones prepared your mother's uterus for your new life. A sperm penetrated the egg. It divided and redivided, creating a ball of cells that implanted itself in the wall of the uterus. Cells moved and specialized, shaping tissues and organs. Within weeks, fingers and toes appeared, and a tiny heart started beating. After 9 months of development and growth, you entered the world. These processes—sexual reproduction and embryonic development—are the focus of this chapter.

Organizing Your Knowledge

Exercise 1   (Introduction and Module 27.1)

These modules introduce some terms and concepts basic to an understanding of reproduction. Write the correct term in the space next to each phrase below.

_____ 1. Male and female reproductive systems in the same body

_____ 2. Small gamete usually propelled by a flagellum

_____ 3. When one parent produces offspring, without fusion of egg and sperm

_____ 4. Diploid cell formed by union of egg and sperm

_____ 5. General name for egg or sperm

_____ 6. Regrowth of lost body parts

_____ 7. Shedding of gametes into water

_____ 8. Large gamete that is not self-propelled

_____ 9. Sexual intercourse

_____ 10. Splitting of new individuals from existing ones

_____ 11. When egg and sperm unite inside the female's body

_____ 12. Breakup of the parent body into several pieces

Web/CD Activity 27B   *Reproductive System of the Human Male*
Web/CD Thinking as a Scientist   *What Might Obstruct the Male Urethra?*

These modules introduce the human reproductive systems. The diagrams below show a front view of the reproductive system of the human female and a side view of the reproductive system of the human male. See how many of the following parts you can label without looking at the modules, and then consult the modules for the remainder: **seminal vesicle, ovary, scrotum, urethra, epididymis, vagina, oviduct, vas deferens, prepuce, uterus, follicles, erectile tissue, testis, endometrium, cervix, glans, prostate gland, bulbourethral gland, corpus luteum, wall of uterus,** and **ejaculatory duct.**

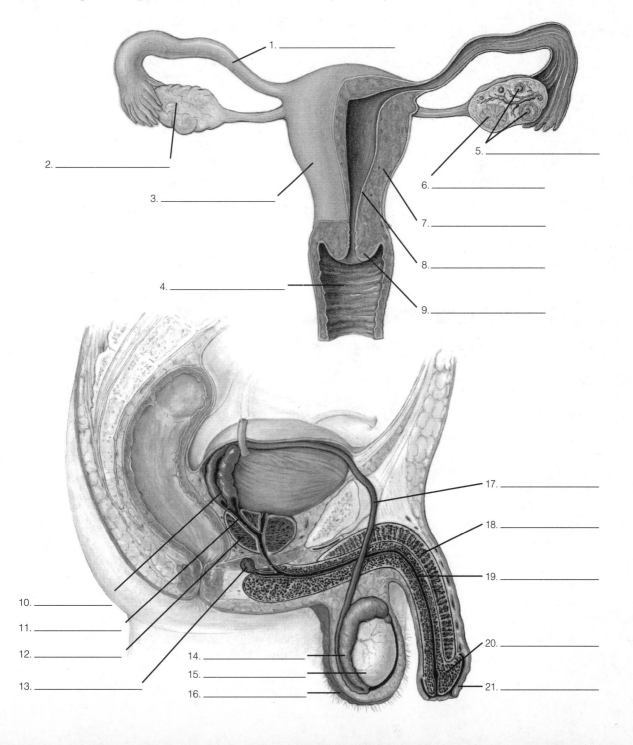

Exercise 3   (Module 27.4)

Both spermatogenesis and oogenesis produce haploid gametes, but there are several important differences in how they occur. Compare these processes in the chart below.

	Spermatogenesis	*Oogenesis*
Location of process	1.	2.
When primary cells form	3.	4.
Numbers and names of cells resulting from first meiotic division	5.	6.
Numbers and names of cells resulting from second meiotic division	7.	8.
Number of gametes produced from division of one primary cell	9.	10.
Total number of gametes produced per day or per month	11.	12.

Exercise 4   (Module 27.5)

Web/CD Activity 27A   *Reproductive System of the Human Female*

This module explains the monthly ovarian and menstrual cycles that coordinate the activities of the ovaries and uterus. You can use the table and diagrams in the text to get most of the words you need to complete the following story, which integrates the ovarian and menstrual cycles. (Figure 27.5 is particularly informative.)

Jennifer wasn't consciously aware of it, but a slight change was occurring in her brain. Her menstrual period had barely started, but already her [1]_____ was starting to secrete tiny amounts of a releasing hormone, which in turn signaled the pituitary gland to boost its output of [2]_____ and [3]_____, restarting her ovarian and menstrual cycles. FSH stimulated a [4]_____, dormant in her [5]_____ since before she was born, to begin to grow and secrete its own hormones. As the follicle grew, it started to secrete small amounts of [6]_____. This hormone in turn signaled the [7]_____, the lining of the [8]_____, to begin to grow and develop in preparation for possible implantation of a fertilized [9]_____. The estrogen from the [10]_____ also exerted [11]_____ feedback on the hypothalamus, keeping levels of [12]_____ and [13]_____ low during this [14]_____ phase of the ovarian cycle.

As the follicle grew in Jennifer's ovary, it secreted more and more estrogen. The higher level of estrogen caused the hypothalamus to signal the [15]_____ to release bursts of [16]_____ and [17]_____. The surge of LH caused the mature follicle to rupture, releasing an ovum in a process called [18]_____. This occurred on the [19]_____ day of Jennifer's 28-day cycle. The ovum entered the [20]_____, where fertilization could occur, and began to travel toward the uterus. The [21]_____ phase of the ovarian cycle had begun.

One of the effects of luteinizing hormone was to transform the ruptured follicle into a yellow, glandular structure, called the [22]_____, which continued to secrete estrogen, plus another hormone, called [23]_____. Both these hormones stimulated further growth and thickening of the uterine lining, further preparing it for the fertilized egg. These hormones also exerted strong negative feedback on the hypothalamus and pituitary, suppressing secretion of FSH and LH and so preventing additional [24]_____ from growing in the ovary. Eventually, the drop in LH caused the corpus luteum to degenerate and reduce its output of estrogen and [25]_____. (Because Jennifer's egg was not [26]_____, there was no hormonal signal sent from a developing embryo to maintain the corpus luteum.) Estrogen and progesterone were maintaining the lining of the uterus. As levels of these hormones dropped, the [27]_____ began to slough off. The monthly cycles were coming to an end, and [28]_____ had begun.

Exercise 5   (Module 27.6)

Human sexual responses occur in four phases (A–D) listed below on the left. Number the phases in order, and match each with a descriptive phrase (P–S) from the list on the right.

Phase	*Order*	*Description*	*Phrase*
A. Orgasm	_____	_____	P. Return to relaxed state
B. Resolution	_____	_____	Q. Increased breathing and heart rate
C. Excitement	_____	_____	R. Rhythmic contractions of reproductive systems
D. Plateau	_____	_____	S. Erection of penis and clitoris

Review sexually transmitted diseases by completing this crossword puzzle.

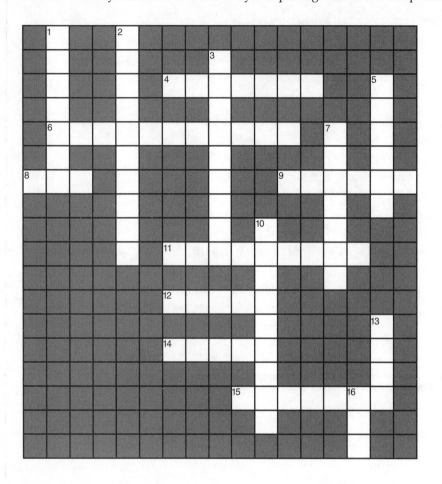

Across

4. Viral STDs are not ____, but bacterial STDs usually are.

6. Bacterial STDs are treated with ____.

8. ____ is short for "sexually transmitted disease."

9. A drug called valacyclovir is used to treat symptoms of genital ____.

11. Chlamydial infections, syphilis, and ____ are caused by bacteria.

12. Chlamydial infections and gonorrhea may cause no symptoms in ____.

14. A fungus called *Candida* causes ____ infections.

15. Latex ____ can usually prevent the spread of STDs.

Down

1. STDs are spread by sexual ____.

2. _____ is the best way to prevent STDs.

3. Trichomoniasis is caused by a ____.

5. Genital herpes has been linked to cervical ____.

7. AIDS and genital herpes are both caused by ____.

10. A genital discharge, sores, or painful ____ can be symptoms of STDs.

13. A virus called HIV is responsible for ____.

16. ____ usually have no symptoms of candidiasis or trichomoniasis.

Exercise 7   (Module 27.8)

Various methods of contraception can prevent pregnancy. Review them by matching each of the statements on the left with a contraceptive method listed on the right. Some answers are used more than once.

_____ 1. Refraining from intercourse around the time of ovulation

_____ 2. A barrier method used by women

_____ 3. Another barrier method used by women

_____ 4. High-dose birth control pills used after intercourse

_____ 5. Sterilization of the male

_____ 6. Removing the penis from the vagina before ejaculation

_____ 7. Long-term use may cause cardiovascular problems

_____ 8. A foam or jelly that kills sperm

_____ 9. Cutting the vas deferens to interrupt sperm path

_____ 10. Also called oral contraceptives

_____ 11. Time-release hormone inserted under the skin

_____ 12. Least effective method as typically used

_____ 13. Single-hormone tablet

_____ 14. A daily tablet containing synthetic estrogen and progesterone

_____ 15. Not really contraception; induces an abortion if pregnancy has already begun

_____ 16. Most effective method of contraception

_____ 17. Sterilization of the female

_____ 18. A barrier method used by men

A. Birth control pill

B. Vasectomy

C. Tubal ligation

D. Progestin minipill

E. Progestin implant

F. Rhythm method

G. Withdrawal

H. Male condom

I. Diaphragm

J. Cervical cap

K. Spermicide

L. Morning after pill (MAP)

M. Mifepristone

Exercise 8   (Module 27.9)

Embryonic development begins with fertilization. After reading the module and studying the illustrations, answer each of the following questions with one sentence.

1. How is the sperm able to get to the egg?

2. How does the sperm penetrate the egg?

3. In species whose eggs are fertilized externally, what keeps sperm of other species from fertilizing the egg?

4. Why is it important that only one sperm penetrate the egg?

5. What prevents more than one sperm from penetrating the egg?

6. In what ways does the egg change after fertilization?

Exercise 9 (Modules 27.10 – 27.12)

Modules 27.10–27.12 trace the development of an embryo from the zygote to development of rudimentary organs. The three stages in this process are cleavage, gastrulation, and organ formation. The illustrations below (combining development of a sea urchin and a frog) are in random order. Number them in the correct sequence (blanks A–K), and then label the following (blanks L–Z): **ectoderm, blastula, zygote, blastopore, digestive cavity (archenteron), blastocoel, endoderm, notochord, mesoderm, gastrula, neural tube, body cavity (coelom),** and **somite.** Some terms are used more than once.

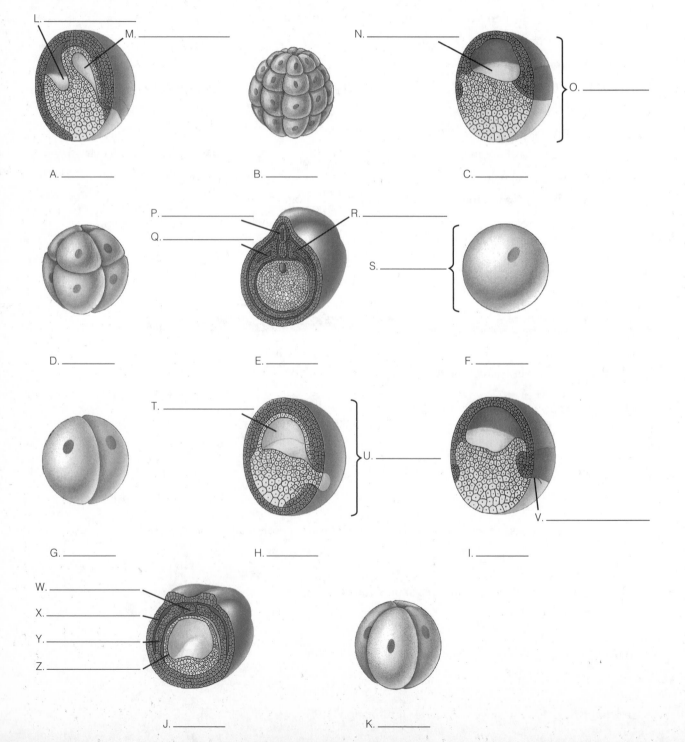

Exercise 10   (Module 27.12)

Web/CD Activity 27D   *Frog Development Video*

Gastrulation produces a three-layered embryo. Each of the three tissue layers gives rise to particular structures. After reading the module and studying the table, name the embryonic tissue layer from which each of the following structures develops—**ectoderm, mesoderm,** or **endoderm.** (You can figure out most of them just by remembering that ectoderm is on the outside of the embryo, endoderm is on the inside, and mesoderm is in between.)

_____ 1. Lining of digestive tract
_____ 2. Circulatory system
_____ 3. Pancreas and liver
_____ 4. Nervous system
_____ 5. Skeleton
_____ 6. Muscles
_____ 7. Lining of respiratory system
_____ 8. Skin epidermis

Exercise 11   (Modules 27.13 – 27.15)

Web/CD Thinking as a Scientist   *What Determines Cell Differentiation in the Sea Urchin?*

These three modules describe some of the processes that shape an embryo. Each of the following statements describes an experiment or observation that demonstrates one of the processes. Match the process with the experiment. Choose from **pattern formation, cell migration, programmed cell death, induction, differentiation,** and **changes in cell shape.**

1. Ectoderm from the side of a frog gastrula normally develops into skin. If a piece of this ectoderm is removed from the side and placed on the back of the embryo in contact with the notochord, it turns into neural tissue. This experiment illustrates _____.

2. A certain chemical is painted on a limb bud of a mouse. An additional set of toes develops at the site, with the little toe where the chemical is most concentrated. This experiment illustrates

   _____.

3. The reproductive system develops from mesoderm, except for the germ cells that will actually divide to form eggs and sperm. Embryologists have dyed these cells and followed them as they "crawl" from a distant site to the developing gonads, an example of _____.

4. When a tadpole changes into a frog, cells in its tail break down and the tail shrinks and eventually disappears, an example of _____.

5. If cells of a blastula are treated with chemicals that interfere with changes in the cells' internal cytoskeletons, the cells fail to fold inward properly to start the process of gastrulation. This experiment shows the importance of _____.

6. As a nerve cell matures and becomes specialized, it is no longer able to divide. Researchers are attempting to determine how genes are activated and deactivated in developing nerve cells. If turned-off genes could be turned back on, perhaps a nerve cell could be made to return to a dividing, embryonic state. This might be useful in repairing nervous system damage. These experimenters are attempting to reverse the process of _____.

Exercise 12   (Modules 27.16 – 27.18)

Many of the stages and processes seen in the development of other animals are also seen in human development and birth. Match the words and phrases below to review human development.

A. HCG

B. First trimester

C. Chorionic villi

D. Inner cell mass

E. Yolk sac

F. Fertilization

G. Prostaglandins

H. Blastocyst

I. Trophoblast

J. Fetus

K. Placenta

L. Labor

M. Gestation

N. Dilation

O. Oxytocin

P. Third trimester

Q. Allantois

R. Implantation

S. Amnion

T. Chorion

_____ 1. Outer layer that allows embryo to implant

_____ 2. Carrying developing young in reproductive tract

_____ 3. Name for embryo after several weeks' gestation

_____ 4. Help oxytocin induce labor

_____ 5. Hormone secreted by chorion; prevents spontaneous abortion

_____ 6. Structure that taps the mother's blood supply

_____ 7. Period when embryo grows rapidly and prepares for birth

_____ 8. Membrane that becomes the embryo's part of the placenta

_____ 9. Embedding of the blastocyst in the wall of the uterus

_____ 10. Forms fluid-filled sac around embryo

_____ 11. Trimester in which the biggest changes occur

_____ 12. Equivalent of frog blastula

_____ 13. The first stage of labor

_____ 14. In birds and reptiles it is important in waste disposal

_____ 15. Portion of the blastocyst that actually forms the baby

_____ 16. A pituitary hormone that initiates labor

_____ 17. Produces the embryo's first blood cells and germ cells

_____ 18. This process begins gestation

_____ 19. Outgrowths of the chorion that absorb nutrients and oxygen

_____ 20. Contractions of the uterus that bring about birth

Exercise 13   (Module 27.19)

In vitro fertilization, surrogate motherhood, and the possibility of sperm and embryo donation open a multitude of reproductive possibilities to infertile couples. A woman's eggs could be fertilized with her husband's sperm and implanted in her uterus. Alternatively, donated sperm or eggs could be used. Or the eggs could be implanted in another woman's uterus. How many combinations are possible? List as many as you can.

Testing Your Knowledge

Multiple Choice

1. Which of the following is a form of sexual re-production?
   a. budding
   b. fission
   c. fragmentation
   d. regeneration
   e. hermaphroditism

2. Which of the following correctly traces the path of sperm from their site of production out of a man's body?
   a. seminiferous tubule, vas deferens, epi-didymis, urethra
   b. epididymis, urethra, seminiferous tubule, vas deferens
   c. seminiferous tubule, epididymis, vas defer-ens, urethra
   d. epididymis, seminiferous tubule, vas defer-ens, urethra
   e. vas deferens, epididymis, urethra, seminif-erous tubule

3. A peak in _____ triggers ovulation on about the _____ day of the monthly cycle.
   a. progesterone . . . fourteenth
   b. LH . . . seventh
   c. FSH . . . second
   d. LH . . . fourteenth
   e. estrogen . . . twentieth

4. External fertilization occurs mostly in
   a. land animals.
   b. insects.
   c. aquatic animals.
   d. animals that reproduce asexually.
   e. mammals.

5. Human eggs are usually fertilized in the
   a. vagina.
   b. oviduct.
   c. ovary.
   d. cervix.
   e. uterus.

6. On its way to fertilize a human egg, a sperm cell does *not* have to pass through which of the following?
   a. oviduct
   b. vagina
   c. ovary
   d. vas deferens
   e. cervix

7. Which of the following hormones is the first to increase significantly every 28 days or so and initiates the ovarian cycle?
   a. progesterone
   b. follicle-stimulating hormone
   c. estrogen
   d. luteinizing hormone
   e. human chorionic gonadotropin

8. The fertilization envelope
   a. prevents an egg from being fertilized by more than one sperm.
   b. attracts sperm to an egg from some distance away.
   c. enables an egg to obtain nourishment from its environment.
   d. ensures that an egg is fertilized by sperm of the proper species.
   e. enables a fertilized egg to implant in the uterine wall.

9. After ovulation occurs, the empty follicle
   a. can be recycled to produce more eggs.
   b. changes into the corpus luteum and makes hormones.
   c. quickly degenerates.
   d. immediately initiates menstruation.
   e. becomes the site of implantation of a fertil-ized egg.

10. The first stage of embryonic development is _____ . This process produces _____ .
    a. gastrulation . . . a three-layered embryo
    b. gestation . . . a gastrula
    c. ovulation . . . a zygote
    d. cleavage . . . a hollow ball of cells
    e. parturition . . . a fetus

11. The recent increase in multiple births—twins and "super twins"—is attributed to
    a. couples having children earlier.
    b. better nutrition.
    c. use of contraceptives.
    d. fertility drugs.
    e. environmental chemicals.

Essay

1. Some animals, such as aphids, alternate be-tween sexual and asexual reproduction. Describe environmental conditions under which each form of reproduction might be most advantageous to the animal.

2. Name and describe the functions of the ex-traembryonic membranes that surround a human embryo.

3. Compare sperm formation in the testes with egg formation in the ovaries. When in life does each process begin? What triggers each to occur or be completed? How many cells result from division of a primary oocyte? A primary spermatocyte?

4. Draw a series of simple sketches that show how a fertilized egg develops into a three-layered gastrula.

Applying Your Knowledge

Multiple Choice

1. In an early stage of development of the frog nervous system, the bits of tissue that will form the backs of the eyes bulge from the growing brain. If one of these bits of tissue is removed and transplanted under the ectoderm in the area that will become the frog's belly, a lens develops in front of it. This experiment illustrates the importance of _____ in shaping an embryo.
   a. pattern formation
   b. cell aggregation
   c. induction
   d. cell migration
   e. cleavage

2. Birth control pills contain synthetic estrogen and progesterone. How might these hormones prevent pregnancy?
   a. They trigger premature ovulation, before an egg is mature.
   b. They cause the lining of the uterus to be sloughed off.
   c. They cause the corpus luteum to degenerate.
   d. They keep the pituitary from secreting FSH and LH, so ovulation does not occur.
   e. They prevent monthly development of the uterine lining.

3. Suppose you wanted to get started in the animal breeding business. You could start out with only one of most of the following animals, but you would have to start with at least two
   a. sea anemones.
   b. frogs.
   c. rotifers.
   d. sea anemones.
   e. sea stars.

4. Pregnancy tests detect a hormone in a woman's urine that is present only when an embryo is developing in her uterus. This hormone is secreted by
   a. the ovary.
   b. the chorion.
   c. a follicle.
   d. the fetus.
   e. the endometrium.

5. One difference between the blastula and gastrula stages of development is that
   a. blastula cells are more differentiated than gastrula cells.
   b. there are many more cells in a blastula.
   c. the blastula consists of more cell layers.
   d. the blastula is a solid ball of cells, but the gastrula is hollow.
   e. there is an opening from the cavity inside the gastrula to the outside.

6. Which of the following systems begin to take shape first in a frog?
   a. digestive and nervous systems
   b. nervous and muscular systems
   c. circulatory and digestive systems
   d. nervous and skeletal systems
   e. skeletal and circulatory systems

7. Doctors hoping to increase the chances of implantation of embryos fertilized in vitro might treat the recipient with
   a. progesterone.
   b. follicle-stimulating hormone.
   c. oxytocin.
   d. prolactin.
   e. all of the above.

8. Human embryonic development can be disrupted by diseases (such as rubella), drugs, alcohol, and radiation. These factors have the most severe effects on the baby if they are present during the first two months of gestation because at this time
   a. substances can most easily enter the embryo through the placenta.
   b. the embryo has not yet implanted in the wall of the uterus.
   c. the mother's hormone levels are highest.
   d. the most drastic and rapid changes are occurring in the embryo.
   e. the cells of the embryo are not yet fully activated.

9. If you wanted to examine the extraembryonic membranes surrounding a lizard or mouse embryo, you would have to cut through which of the following to see all the others?
   a. yolk sac
   b. amnion
   c. endoderm
   d. allantois
   e. chorion

10. The trick to developing a successful male contraceptic seems to be in figuring out a way to block sperm production without affecting androgen (testosterone) levels. (The latter tends to inhibit sexual desire.) Looking at Figure 27.3D, which of the following would seem to be the best approach for developing a male "pill"?
    a. block the hypothalamus
    b. block the anterior pituitary
    c. block releasing hormone
    d. block LH
    e. block FSH

Essay

1. Some populations of tropical fish called mollies consist of males and females that reproduce sexually, the way most fish do. Other populations consist only of females, whose eggs develop without fertilization. Which of these populations do you think would adapt most successfully to a change in its environment? Why?

2. A man who soaked in a hot tub for an extended period every day was found to be suffering from temporary sterility. It seems that his sperm production had been slowed down considerably. Explain why this might have occurred. (Note: This phenomenon cannot be relied on as a method of birth control!)

3. In general, land animals engage in more elaborate courtship behaviors before mating than do aquatic animals. Why might this be more important for land animals?

Extending Your Knowledge

1. Most mammals have an estrous cycle. Estrus is the period when ovulation occurs; the female is sexually receptive and mating occurs during this time. For some mammals, like mice, estrus occurs every few days. For dogs and cats, it might occur several times a year. Some wild animals, such as seals, go through estrus only once per year, during the mating season. Humans do not have estrus, and sexual intercourse can occur at any time during the ovarian and menstrual cycles. How would human life be different if women did have estrus? What do you think are possible reasons that estrus has been "lost" during human evolution?

2. How would our lives be different if we reproduced asexually, like some of the animals described at the beginning of this chapter? If we were hermaphroditic, like earthworms?

3. Few ethical or value issues evoke passionate differences of opinion like those involving human sexuality and reproduction. Do you and your friends discuss issues such as whether states should allow gay couples to marry or should restrict a woman's access to abortion? Do you and your friends differ on these issues? How do you discuss and try to resolve your differences?

Nervous Systems

As your eyes scan these words, the light and dark patterns printed on the page stimulate light-sensitive cells in your eyes. Changes occur in the membranes of nerve cells in the optic nerve, causing ions to pass in and out of the cells. The changes are transmitted like waves along the nerve cells all the way to visual areas in the back of the brain, in the cerebral cortex. In association areas of the cortex, these words are interpreted, compared with images and memories, and perhaps stored in memory themselves. At the end of each line of text, the brain's motor cortex sends nerve impulses to tiny muscles to move the eyes back to the beginning of the next line. At the same time, other areas of the brain monitor body temperature and blood pressure. They send impulses out by way of the spinal cord to your heart, blood vessels, and sweat glands to maintain homeostasis. In this way, your nervous system senses the environment, interprets it, and directs responses to your muscles and glands. Your nervous system will enable you to read and understand Chapter 28, which is all about nervous systems.

Organizing Your Knowledge

Exercise 1   (Module 28.2)

Web/CD Activity 28A   *Neuron Structure*

Review the structure of a neuron by labeling and coloring this diagram. Label the **cell body, axon, myelin sheath, dendrites, synaptic terminals,** and a **node of Ranvier.** Color the dendrites green, the cell body blue, the axon red, and the myelin sheath yellow.

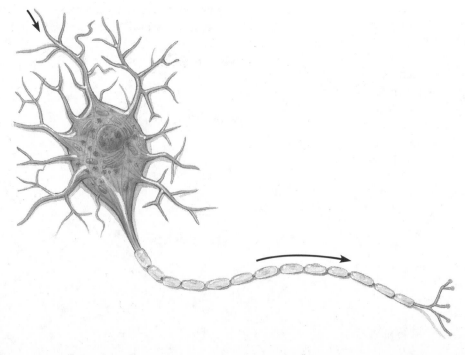

Exercise 2   (Modules 28.1 – 28.2)

Web/CD 28A   *Neuron Structure*

These modules describe the structures and functions of nervous systems and neurons. Review them by completing this crossword puzzle.

Across

2. Supporting cells, or _____, help neurons do their jobs.

4. A _____ is a cell specialized for carrying signals.

7. Motor _____ is conduction of signals from integration centers to effectors.

8. _____ is the interpretation of sensory signals and formulation of responses.

11. The _____ nervous system (PNS) consists of nerves outside the CNS.

13. _____ neurons carry information from sensory receptors in the CNS.

17. _____ receive messages and carry them to the neuron cell body.

18. The CNS consists of the _____ and spinal cord.

19. A Schwarm cell is one type of _____ cell.

20. An axon has many branches, each with a synaptic _____ at its end.

22. A _____ sheath insulates a neuron.

Down

1. _____ in the central nervous system integrate data.

3. Sensory _____ is conduction of signals from sensory receptors to the brain.

5. The _____ nervous system is abbreviated "CNS."

6. _____ neurons carry signals from the CNS to effectors.

9. A _____ is a bundle of neuron extensions wrapped in connective tissue.

10. The nucleus of a neuron is in the cell _____.

12. Interneurons are entirely within the _____.

13. The myelin sheath is destroyed in multiple _____.

14. Signals go faster when they jump along a neuron, between _____ of Ranvier.

15. The site of contact between two nerve cells is called a _____.

16. _____ are clusters of cell bodies belonging to neurons making up a nerve.

21. The _____ carries signals toward another neuron or an effector.

Exercise 3   (Modules 28.3 – 28.5)

Web/CD Activity 28B   *Nerve Signals: Action Potentials*
Web/CD Thinking as a Scientist   *What Triggers Nerve Impulses?*

How do neurons carry signals? What exactly is a nerve signal? These three modules explain how nerve cells transmit nerve signals, which are called action potentials. Number in sequence each of the following steps (A–E) in the generation and transmission of a nerve signal. Then match each of the steps with an explanation (P–T).

Sequence	*Explanation*	
_____	_____	A. Action potential occurs
_____	_____	B. Resting potential
_____	_____	C. Stimulus affects neuron
_____	_____	D. Action potential propagates along axon
_____	_____	E. Cell repolarizes

P. Na^+-K^+ pump moves Na^+ out of cell, K^+ into cell; K^+ leaks out but Na^+ can't get in. Membrane is + outside and – inside.

Q. Na^+ channels open, Na^+ leaks into cell, and cell depolarizes a bit; membrane becomes somewhat less + outside and less – inside.

R. Na^+ spreads out and causes adjacent area of membrane to depolarize.

S. Na^+ channels close, K^+ channels open, and K^+ leaves cell. Membrane becomes + outside and – inside.

T. Threshold reached and more Na^+ channels open, allowing Na^+ to rush into cell. Membrane becomes + inside and – outside.

Exercise 4   (Module 28.6)

Web/CD Activity 28C   *Neuron Communication*

Synapses are relay points where nerve impulses are transmitted from the synaptic terminal of a neuron to another cell. The following paragraph describes how this occurs. Complete the description by filling in the blanks.

Synapses are junctions between [1]_____ and receiving cells. Some are [2]_____ synapses, where action potentials pass from one neuron to the next. Most are chemical synapses, where chemical signals called [3]_____ pass between cells. A transmitting neuron and a receiving neuron are separated by a narrow gap, called the synaptic [4]_____. Neurotransmitter is contained in small vesicles in the synaptic terminals at the end of the [5]_____ of the transmitting cell. When an action potential arrives at the end of the transmitting cell's axon, chemical changes occur that cause the vesicles to fuse with the [6]_____. The neurotransmitter molecules are released into the synaptic cleft, rapidly diffuse across the cleft, and bind to [7]_____ molecules in the plasma membrane of the [8]_____. Commonly, this opens ion [9]_____ in the receiving cell's membrane, allowing ions to diffuse through the membrane and initiating new [10]_____ in the receiving cell. The neurotransmitter is quickly broken down by an [11]_____ or transported back into the signaling cell, and the ion channels close. This ensures that the transmission of a signal from cell to cell will be brief and precise.

Exercise 5 (Module 28.7)

Some neurotransmitters cause receiving cells to transmit signals, and other neurotransmitters inhibit the receiving cell's transmission of signals. Neuron X, diagrammed below, adds up all the excitatory and inhibitory impulses it receives. This particular neuron transmits its own impulses only when at least *two* neurons send it excitatory impulses at the same time. Each neuron sending it inhibitory impulses can cancel out the effect of one neuron sending excitatory signals. Given the inputs outlined below, state whether neuron X would or would not transmit nerve impulses in each case.

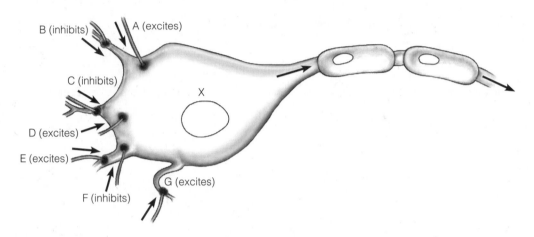

Neurons Transmitting

	A	B	C	D	E	F	G	X Transmits? (Y or N)
1.	Y	N	N	Y	Y	Y	N	_____
2.	N	Y	N	Y	N	N	Y	_____
3.	N	Y	N	Y	Y	N	Y	_____
4.	Y	N	N	N	Y	N	N	_____
5.	Y	Y	Y	Y	Y	Y	Y	_____
6.	N	Y	Y	N	N	Y	N	_____
7.	N	N	N	Y	N	N	N	_____

Exercise 6 (Modules 28.8 – 28.9)

These modules describe the functions of various neurotransmitters, and the effects of stimulants, tranquilizers, depressants, and antidepressants on the nervous system. Imagine a neuron in your brain that transmits impulses to other neurons that elevate mood. This "feel good" neuron transmits when it is stimulated by serotonin from nearby neurons. It is inhibited by GABA from its neighbors. With this in mind, explain how each of the following might affect your mood.

1. Hereditary shortage of serotonin

2. A drug, such as alcohol, that enhances the effects of GABA

3. A drug that stimulates serotonin receptors

4. A chemical that blocks the enzymes that normally break down GABA after it has docked with receptors in the receiving neuron

5. A drug that blocks receptors so that they cannot respond to GABA

6. A drug that blocks the enzymes that normally break down serotonin after it has docked with receptors in the receiving neuron

7. A drug that enhances the release of serotin by sending cells.

8. A drug that blocks the reuptake of serotonin from the synapse by sending cells.

Exercise 7   (Module 28.10)

The nervous systems of radially symmetrical animals are quite different from the nervous systems of animals with bilateral symmetry. Start by defining some terms used to describe nervous systems (question 1), then compare the nervous systems of two invertebrates (questions 2 and 3).

1. Match the descriptive terms on the right with their definitions.

_____ a. Concentration of the nervous system at the head end   A. Bilateral symmetry
_____ b. A headless, circular body   B. Nerve cords
_____ c. Presence of a central nervous system   C. Centralization
_____ d. A body with a head and tail and left and right sides   D. Radial symmetry
_____ e. A weblike network of nerve cells   E. Cephalization
_____ f. Lengthwise groupings of nerve cells   F. Nerve net

2. Which of the terms might describe a leech?

3. Which of the terms might describe a cnidarian (hydra)?

Exercise 8   (Modules 28.11 – 28.12)

Vertebrate nervous systems vary in complexity, but they all have certain features in common. Write the name of the part or division of the vertebrate nervous system that goes with each of the phrases below.

_____   1. Master control center

_____   2. Component of the peripheral nervous system (PNS) that transmits information to and from skeletal muscles

_____   3. Component of the motor division that regulates internal activities

_____   4. Fluid-filled spaces in the brain

_____   5. Layers of connective tissue around the brain and spinal cord

_____   6. Nerves that originate in the brain and terminate mostly in the head and upper body

_____   7. Myelinated axons and dendrites in the CNS

_____   8. Fluid that circulates around and through the CNS

_____   9. All nerves and ganglia outside the brain and spinal cord

_____   10. Nerve tissue of the brain and spinal cord consisting mainly of cell bodies and dendrites

_____   11. Brain and spinal cord

_____   12. Part of the CNS inside the vertebral column

_____   13. Nerves that originate in the spinal cord and extend to parts of the body other than the head

_____   14. Maintains a stable chemical environment for the brain

Exercise 9   (Module 28.13)

What is the overall effect of activity of the parasympathetic nervous system on the body? The sympathetic nervous system? One way to keep their actions straight is to remember that the sympathetic system triggers the "fight or flight" response to stress, gearing up the body for action. Effects of the parasympathetic system are best summarized in the phrase "rest and digest." It slows the body down, for the most part, but stimulates the digestive organs. With these effects in mind, see if you can state the effect of each system on the structures in the chart below.

Body Structure	Parasympathetic Effect	Sympathetic Effect
Stomach (via enteric nervous system)	1.	2.
Bronchi (lungs)	3.	4.
Genitals (male)	5.	6.
Heart	7.	8.
Salivary gland	9.	10.
Pupil	11.	12.

Exercise 10   (Modules 28.14 – 28.15)

Name each of the structures identified by numbers on this diagram of the human brain. Then match each structure or area with its function (A–G).

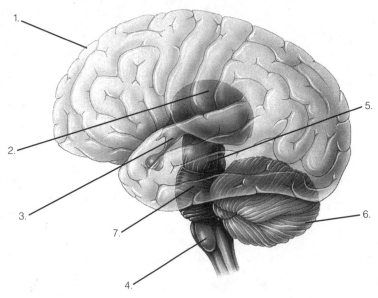

A. Coordination, balance, motor memory
B. Controls breathing
C. Visual reflexes, integrates auditory data
D. Controls breathing, circulation, swallowing, and digestion
E. Memory, learning, speech, emotions
F. Sorts and relays information to higher brain centers
G. Body temperature, hunger, thirst, sex, biological clock

Function

1. _____ _____
2. _____ _____
3. _____ _____
4. _____ _____
5. _____ _____
6. _____ _____
7. _____ _____

Exercise 11   (Modules 28.15 – 28.19)

These modules outline activities of various parts of the cerebrum. Review these activities and other aspects of the nervous system by filling in the blanks in the following story.

Liz appeared to be sleeping quietly, but she was dreaming, and her mind was active. Beneath her closed eyelids, her eyes darted back and forth rapidly, showing that she was in a state called [1]_____ sleep. Outside her window, a car squealed around a corner. An area in Liz's [2]_____ triggered arousal, and Liz woke up. She opened her eyes and looked at the clock—3:00 A.M. Then she remembered her dream. She had been riding her bike between two rows of trees as fluffy clouds passed overhead. "That's it! The idea for my art project," she thought. She fumbled for the light switch, grabbed a pencil and sheet of paper from the bookcase, and scribbled *clouds*. As she wrote, the movement of her hand was controlled by the [3]_____ cortex in the [4]_____ lobe of her cerebrum, which sent nerve impulses to the muscles of her arm via [5]_____ neurons. The [6]_____ helped the cerebrum by making the movements smooth and coordinated. Liz clicked off the light and soon went back to sleep.

The next morning, when she saw her scrawled *clouds* note, Liz felt happier and more energized than she had in days. She hadn't realized how the looming deadline for her term art project had been getting her down. (She also didn't realize that the 7_____ system, a group of centers in the forebrain, was actively involved in shaping these emotion, as well as in learning and memory.) Now she knew what she was going to paint for the project—a design based on clouds! She could picture the painting perfectly in her mind (actually, mostly in the 8_____ cerebral hemisphere, which specializes in spatial relations and pattern recognition). Liz started to make a mental list of the supplies she would need to do the painting.

In the art department of the campus bookstore, Liz gathered the materials she needed. She tested the texture of several watercolor papers with her fingertips. The nerve impulses from her fingers traveled along 9_____ neurons to her 10_____, which transmitted them up to her brain. In the brain, the 11_____ sorted the sensory impulses and relayed them to Liz's cerebral cortex. The 12_____ cortex in the 13_____ lobe of Liz's cerebrum interpreted the nerve impulses from her fingers, inferring from them the textures of the papers. The sensory information from her right hand was sent to the 14_____ side of the brain, but Liz could check it against her mental image because the nerve impulses were able to cross from one side of the brain to the other via the 15_____, a thick band of nerve fibers connecting the two cerebral hemispheres.

One of the shelves was empty, and Liz thought the missing paper might be just the one she needed. She read the stock number on the shelf, repeated it to herself a couple of times, and kept the number in her 16_____-term memory just long enough to ask a clerk whether the store had more in stock. No luck. Liz recalled the paper she had used for a couple of assignments the previous term. This information was stored in 17_____-term memory. The 18_____, part of the limbic system, had "labeled" the name of the paper for storage in memory and linked it with the emotion of happiness that Liz was feeling. Another limbic system center, the 19_____, assisted in this memory-formation process, as well as later recall.

Liz found the paper, paints, and a brush that she needed and headed for the checkout line, adding up the prices in her head (actually on the 20_____ side of her head, because 21_____ areas in the 22_____ cerebral hemisphere seem to be primarily responsible for mathematical calculations). The total was a bit more than she had anticipated, and a moment of anxiety about her budget caused Liz's heart and breathing to speed up a bit, under orders from the 23_____ and 24_____, two areas of the hindbrain.

With her purchases tucked under her arm, Liz set off for home on her bicycle. She crossed the park and turned onto a bike path flanked by tall maple trees. Fluffy clouds sailed overhead. Liz thought the scene seemed very familiar.

Review your knowledge of central nervous system disorders by matching each of the phrases on the left with one or more disorders from the list on the right. Answers may be used more than once, and three questions require more than one answer.

_____ 1. Sadness, loss of interest, sleeplessness, low energy, etc.

_____ 2. Two conditions occuring in about 1% of the population

_____ 3. Hearing voices, delusions, blunted emotions and speech

_____ 4. Two conditions whose risk increases with advancing age

_____ 5. Also called manic-depressive disorder

_____ 6. Often caused by a low level of the neurotransmitter serotonin

_____ 7. Two conditions associated with abnormal levels of the neurotransmitter dopamine

_____ 8. Characterized by neurofibrillary tangles and senile plaques

_____ 9. Symptoms result from death of neurons in the midbrain

_____ 10. Extreme mood swings, from high-to-low to low-to-high

_____ 11. Occurs in about 5% of the population

_____ 12. A motor disorder causing slowness in movement, rigidity, muscle tremors

_____ 13. Treated with SSRIs such Prozac, Paxil, and Zoloft

_____ 14. May eventually be treated by transplant of embryonic dopamine-secreting neurons

_____ 15. In a mild form, it is associated with the creativity of many famous artists, writers, and musicians

_____ 16. Progressive loss of memory, confusion, inability to care for self

A. Parkinson's disease
B. Schizophrenia
C. Bipolar disorder
D. Alzheimer's disease
E. Major depression

Testing Your Knowledge

Multiple Choice

1. Which of the following best describes an action potential?
   a. flow of electricity along a neuron
   b. passage of ions through the membrane of a neuron
   c. flow of neurotransmitter chemical along a neuron
   d. movement of tiny filaments of protein inside a neuron
   e. change in a neuron so that the inside becomes more negatively charged

2. A part of a neuron that carries signals toward the cell body is called
   a. a nerve.
   b. white matter.
   c. a neurotransmitter.
   d. a dendrite.
   e. an axon.

3. Which of the following maintains resting potential, the difference in electrical charge inside and outside a neuron membrane that enables the cell to transmit a signal?
   a. charges that pull sodium and potassium through the membrane
   b. opening of sodium and potassium channels in the membrane
   c. the myelin sheath, which prevents ions from entering or leaving
   d. transport and leakage of sodium and potassium into and out of the cell
   e. the mutual repulsion of sodium and potassium ions

4. The _____ contains association areas for speech, language, and calculation.
   a. right cerebral hemisphere
   b. medulla oblongata
   c. midbrain
   d. left cerebral hemisphere
   e. cerebellum

5. A stimulus triggers an action potential by
   a. causing sodium ions to leak into the neuron.
   b. triggering the release of neurotransmitter.
   c. causing potassium ions to leak out of the neuron.
   d. activating the sodium-potassium pump.
   e. causing sodium ions to leak out of the neuron.

6. The ____ of a primate, dolphin, or whale is much larger than this brain region in other mammals.
   a. brainstem
   b. hypothalamus
   c. cerebral cortex
   d. limbic sytem
   e. medulla

7. The autonomic nervous system
   a. is the part of the nervous system outside the brain and spinal cord.
   b. controls and coordinates voluntary movements.
   c. regulates the internal environment.
   d. integrates all sensory information from the environment.
   e. consists of the brain and spinal cord.

8. Which of the following correctly matches a part of the brain with its function?
   a. thalamus—responsible for learning and memory
   b. hypothalamus—relays sensory information to cerebrum
   c. cerebrum—controls breathing and circulation
   d. cerebellum—coordinates movements
   e. medulla oblongata—interprets visual information

9. The limbic system is involved in
   a. emotions, memory, and learning.
   b. speech and hearing.
   c. vision.
   d. sleep and wakefulness.
   e. control of heartbeat and respiration.

10. Which of the following is part of the central nervous system?
    a. cranial nerve
    b. spinal nerve
    c. spinal cord
    d. sympathetic nerve
    e. ganglion

11. Which of the following disorders does *not* seem to be caused by imbalance of neurotransmitters?
    a. Alzheimer's disease
    b. attention deficit hyperactivity disorder
    c. schizophrenia
    d. Parkinson's disease
    e. major depression

Essay

1. Sketch a chemical synapse and use your sketch to explain how a nerve impulse crosses a chemical synapse from the transmitting neuron to the receiving neuron.

2. Compare the functions of the parasympathetic and sympathetic divisions of the autonomic nervous system.

3. Briefly describe three major trends in the evolution of the vertebrate brain.

4. What are the three major functions of nervous systems? How are they illustrated as you answer this question?

5. A neuron can receive inputs from thousands of transmitting neurons, but it can respond in only one of two ways—it can either transmit action potentials or not transmit action potentials. How do transmitting cells signal the receiving cell, and what determines whether the receiving cell will or will not transmit action potentials itself?

Applying Your Knowledge

Multiple Choice

1. The axons of which of the following end in the central nervous system?
   a. motor neurons
   b. sensory neurons
   c. interneurons
   d. a and c
   e. b and c

2. A man was admitted to the hospital suffering from abnormally low body temperature, loss of appetite, and extreme thirst. A brain scan showed a tumor located in the
   a. hypothalamus.
   b. cerebellum.
   c. pons.
   d. right cerebral hemisphere.
   e. corpus callosum.

3. A drug that causes potassium to leak out of a neuron, increasing the positive charge on the outside, would
   a. make it easier to trigger action potentials in the neuron.
   b. cause the cell to release its neurotransmitter.
   c. speed up action potentials traveling the length of the cell.
   d. act as a stimulant.
   e. inhibit transmission of action potentials by the neuron.

4. What is the difference between a neuron and a nerve?
   a. One is sensory in function, the other motor.
   b. Nerves are found only in the central nervous system.
   c. They consist of different numbers of cells.
   d. Neurons are made of white matter, nerves of gray matter.
   e. Neurons are found only in vertebrates.

5. Which of the following animals is *least* cephalized?
   a. clam
   b. ant
   c. flatworm
   d. human being
   e. fish

6. After an accident, a young man began exhibiting bizarre psychological symptoms: He was convinced his parents had been replaced by lifelike robots because he experienced no feelings when he saw or remembered them. Apparently, the accident had damaged the man's
   a. medulla oblongata.
   b. cerebellum.
   c. basal ganglia.
   d. limbic system.
   e. reticular formation.

7. Alex became so dehydrated while playing tennis that his blood pressure started to drop. His _____ detected the pressure drop and sent signals via _____ to speed up the heart, compensating for the drop in pressure.
   a. hypothalamus . . . parasympathetic neurons
   b. medulla oblongata . . . sympathetic neurons
   c. cerebellum . . . sympathetic neurons
   d. medulla oblongata . . . parasympathetic neurons
   e. cerebellum . . . parasympathetic neurons

8. The gray matter of the cerebral cortex, where most higher-level "thinking" occurs, is composed mostly of
    a. interneuron cell bodies.
    b. myelinated axons of neurons.
    c. sensory neuron cell bodies.
    d. Schwann cells.
    e. motor neuron cell bodies.

9. Which of the following neurons would both receive nerve signals from other neurons and transmit signals to other neurons?
    a. a sensory neuron from the fingertip to the spinal cord
    b. a motor neuron from the spinal cord to the leg
    c. an interneuron in the brain
    d. all of the above
    e. both the sensory and motor neurons

10. Sensory receptors in the ear can detect a whisper or a shout and can transmit nerve signals about these sounds to the brain. How would sensory neurons relay information to the brain about the *loudness* of loud and soft sounds?
    a. The action potentials would travel at different speeds.
    b. The size of action potentials would vary.
    c. The number of action potentials would vary.
    d. The frequency of action potentials would vary.
    e. Action potentials would be routed to different parts of the brain.

Essay

1. Many nervous system drugs and poisons act at synapses. Explain how each of the following would alter the transmission of nerve impulses.
    a. Curare, a substance used on poison arrows by native South Americans, competes with acetylcholine for receptor sites on receiving neurons.
    b. Botulism toxin inhibits the release of acetylcholine.
    c. Diisopropyl fluorophosphate (DF) is used in warfare as a nerve gas. It blocks the enzyme that breaks down acetylcholine after it has crossed the synapse to the receiving neuron.

2. Why is transmission of a nerve signal along a neuron sometimes compared with the toppling of a row of dominoes? How does it differ from the toppling of dominoes?

3. Andrew's back was broken in an auto accident, and he was paralyzed below the shoulders. Explain why he is able to move his arms but not his legs. Like many people with such injuries, Andrew cannot feel his legs, but he does respond when the doctor tests his knee-jerk reflex. Explain how.

4. A victim of a severe head injury may live for years in a "persistent vegetative state"—unconscious but still alive and breathing. Explain how a person can continue to live even though the cerebrum, the largest part of the brain, ceases to function.

Extending Your Knowledge

1. Most adults require about 8 hours of sleep per night, teenagers more than 9 hours. How much do you get? Researchers are beginning to think that sleep deprivation is a major health problem in the United States. It may impair thinking, alertness, attention span, creativity, and judgment. Signs of sleep deprivation include needing an alarm clock to wake up, falling asleep within 5 minutes of going to bed (a well-rested person goes to sleep in 10 to 15 minutes), napping at will, and nodding off at work or in classes. Do you think you might be short of sleep? If you don't think you are getting enough sleep, consider adjusting or lightening your class or work schedule, taking naps, and giving sleep higher priority on your "things to do" list.

2. Have you taken a college psychology course? How do the facts and concepts in this chapter relate to what you learned in psychology? How was the coverage of the nervous system similar in biology and psychology? How did the coverage in the two courses differ? What are some reasons for possible differences?

The Senses

We take it for granted that humans experience the world through "the five senses"—sight, hearing, touch, taste, and smell. Sometimes we say that an acutely intuitive individual has a "sixth sense," but in reality we all have a sixth sense, and a seventh, and an eighth. The sense of touch is actually a complex of senses—touch, pressure, and temperature. We even have separate sensory receptors for heat and cold. Our inner ears respond not only to sound but also to changes in body position. Additional senses monitor the interior of the body, collecting information on blood pressure and the positions of muscles and joints. Pain warns us of injury or disease. Other animals possess sensory abilities that we can only imagine. A dog is aware of sounds and smells beyond our sensory range. Bees see ultraviolet light; rattlesnakes see infrared. Some creatures can even feel the Earth's magnetic field. This chapter is about the senses and how they reveal the world.

Organizing Your Knowledge

Exercise 1   (Modules 29.1 – 29.2)

As you read these words, several different processes occur in your eyes and brain before you understand what the words mean. Match each of these processes (A–F) with its description (L–Q), and number the processes in the order in which they might occur.

Order   Description

_____   _____   A. Transduction

_____   _____   B. Sensation

_____   _____   C. Transmission

_____   _____   D. Perception

_____   _____   E. Reception

_____   _____   F. Adaptation

L. Conscious understanding of sensory data

M. Conversion of a stimulus into electrical signals

N. Sending action potentials to the brain

O. Drop in sensitivity of receptors when stimulated repeatedly

P. Awareness of sensory stimuli by the brain

Q. Detection of stimuli by sensory cells

Exercise 2 (Module 29.3)

There are five general kinds of sensory receptors—pain receptors, thermoreceptors, mechanoreceptors, chemoreceptors, and electromagnetic receptors. To review them, complete this chart. Give the general category of receptors to which each example belongs, the kind of stimulus it responds to, and where you might find one of the receptors.

Example	General Receptor Type	Stimulus It Responds To	Locations
Infrared receptor	1.	Infrared radiation	2.
Touch receptor	3.	Contact and movement	4.
5.	6.	Nearby obstacles and animals in murky water	7.
8.	9.	10.	Eyes
Pain receptor	11.	Danger, injury, disease	Many body locations
12.	Mechanoreceptor	13.	Human ear
Heat receptor	14.	15.	16.
17.	18.	Body position	19.
Smell receptor	20.	21.	Nose
Magnetic orienting mechanism	22.	23.	Head of certain birds, fishes, mammals

Web/CD Activity 29A   *Structure and Function of the Eye*

Eyes can be simple or complex. The single-lens vertebrate eye is a complicated and sensitive sensory organ. To familiarize yourself with the major parts of the eye, label and color the diagram below. Include the **sclera, choroid, retina, pupil, iris, lens, cornea, optic disk** (blind spot), **vitreous humor, aqueous humor, artery and vein, fovea, optic nerve,** the **ciliary body** and the **suspensory ligament** attached to the lens.

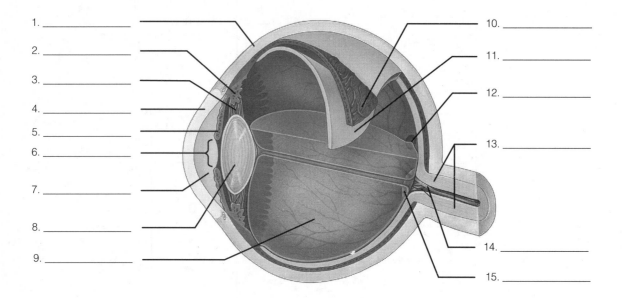

1. _____
2. _____
3. _____
4. _____
5. _____
6. _____
7. _____
8. _____
9. _____

10. _____
11. _____
12. _____
13. _____
14. _____
15. _____

Now match each of the following functions with a part of the eye by placing its letter next to the correct label on the diagram.

A. Carries nerve signals to the brain
B. Covers the front of the eye but lets light in
C. Supplies nutrients and oxygen to the lens, iris, and cornea
D. Photoreceptor cells here transmit action potentials to the brain
E. Regulates the size of the pupil to let more or less light into the eye
F. Bends light rays and focuses them on the retina
G. Muscles in this structure change the shape of the lens
H. Photoreceptors are highly concentrated at this center of focus
I. Works with the ciliary body in accommodation
J. Jellylike "filling" that maintains the shape of the eyeball
K. The fibrous "white" of the eye
L. Hole in the iris that admits light

Exercise 4   (Module 29.6)

Web/CD Activity 29A   *Structure and Function of the Eye*

The lenses in the eyes of a fish move back and forth to bring images into focus, but the lenses in human eyes focus by changing shape, a process known as accommodation. After studying Module 29.6, sketch the path of light rays and the shape of the lens when the eye focuses on a nearby object and a distant object.

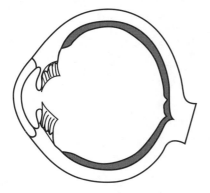
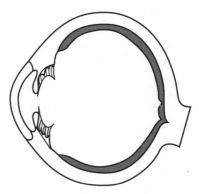

Nearby object   Distant object

Exercise 5   (Module 29.7)

There are a number of possible reasons for less-than-perfect vision. Some people are nearsighted, others farsighted. Some suffer from astigmatism. After reading about these vision problems, sort them out by completing the following chart.

Diagram			
Problem name	1.	2.	3.
Nearby or distant objects clear	4.	5.	6.
Nearby or distant objects unclear	7.	8.	9.
Cause(s)	10.	11.	12.
Lens used for correction	13.	14.	15.

Exercise 6   (Module 29.8)

This module discusses rods and cones, the sensory receptors of the retina. Indicate below the statements that relate to rods and those that relate to cones by placing check marks in the appropriate column. (Some statements relate to both.)

	Rods	Cones
1. Come in three types		
2. Stimulated by bright light		
3. Function in night vision		
4. Most numerous in the fovea		
5. Located in the back layer of the retina		
6. Contain a visual pigment called rhodopsin		
7. Can see different colors		
8. Stimulus transducers		
9. Absent from the fovea		
10. Most numerous at the outer edges of the retina		
11. Contain pigments called photopsins		
12. See only shades of gray		
13. Transmit signals to the visual cortex		
14. Deficient in color blindness		
15. Synapse with retinal neurons		
16. Most numerous receptors		
17. Most sensitive receptors		
18. Concentrated in fovea of a hawk		
19. More numerous in animals active at night		

Exercise 7   (Module 29.9)

The ear is a mechanical marvel, catching and amplifying sound waves and transforming them into action potentials. As you review the structure and function of the ear, label the following parts on the diagram at the top of the next page: **auditory nerve, eardrum, cochlea, oval window, anvil, semicircular canals, inner ear, auditory canal, stirrup, outer ear, pinna, hammer, Eustachian tube,** and **middle ear.** Color all structures of the outer ear red, the middle ear yellow, and the inner ear blue.

Now match each of the following functions with a part of the ear by placing its letter next to the correct label on the diagram.

A. Its hair cells transduce motion into action potentials
B. Equalizes pressure outside and inside eardrum
C. Other animals, such as dogs and horses, can turn it toward sounds
D. Transmits action potentials from ear to brain
E. Transmits vibrations to middle-ear bones
F. Collects and channels sound waves to eardrum
G. Transmits vibrations from eardrum to cochlea
H. Function in balance

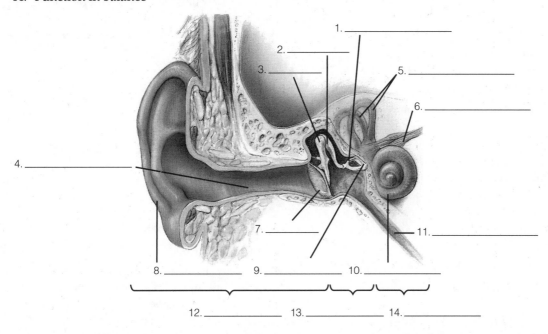

1. _____
2. _____
3. _____
4. _____
5. _____
6. _____
7. _____
8. _____
9. _____
10. _____
11. _____
12. _____
13. _____
14. _____

Exercise 8   (Modules 29.10 – 29.13 and Summary)

Modules 29.10 – 29.12 discuss the senses of balance, smell, and taste, and Module 29.13 reviews the role of the senses in the overall function of the nervous system. Review the senses (including vision and hearing) by matching each of the phrases on the right with a sensory structure from the list on the left. Some answers are used more than once.

A. Retina
B. Semicircular canal
C. Utricle and saccule
D. Hair cell
E. Olfactory receptor
F. Taste bud
G. Organ of Corti
H. Cone

____ 1. Receptors here detect changes in head movement
____ 2. Receptors here sense sour, salty, sweet, bitter, umami
____ 3. Odor molecules bind to proteins on its cilia
____ 4. Site of receptors for hearing
____ 5. Light receptor of the retina
____ 6. Receptors here sense position of head relative to gravity
____ 7. Receptor of hearing
____ 8. Site of photoreceptors
____ 9. Balance receptor
____ 10. Chemoreceptor of the nasal cavity
____ 11. This kind of receptor may lose sensitivity as we age

Testing Your Knowledge

Multiple Choice

1. Insect eyes are much better at seeing _____ than are our eyes.
   a. colors
   b. in dim light
   c. movement
   d. fine detail
   e. distant objects

2. A thermoreceptor in the skin converts heat energy into action potentials. This conversion process is called
   a. sensation.
   b. transduction.
   c. reception.
   d. integration.
   e. perception.

3. Which of the following correctly traces the energy of sound waves into the ear?
   a. auditory canal–eardrum–ear bones–cochlea
   b. eardrum–auditory canal–cochlea–ear bones
   c. auditory canal–ear bones–eardrum–cochlea
   d. eardrum–auditory canal–ear bones–cochlea
   e. eardrum–ear bones–auditory canal–cochlea

4. Which of the following are *not* correctly paired?
   a. mechanoreceptor—stretch receptor
   b. electromagnetic receptor—photoreceptor
   c. chemoreceptor—taste bud
   d. mechanoreceptor—touch receptor
   e. electromagnetic receptor—hair cell

5. When you focus your eyes on a nearby object, the lenses
   a. decrease in diameter.
   b. move forward.
   c. become more rounded.
   d. become flatter.
   e. move backward.

6. As light passes into the eye, it goes through which of the following first?
   a. lens
   b. pupil
   c. aqueous humor
   d. cornea
   e. vitreous humor

7. The brain determines the loudness of a sound from
   a. the part of the organ of Corti stimulated by the sound.
   b. the frequency of action potentials received.
   c. the size of air pressure changes in the middle ear.
   d. the part of the brain receiving nerve impulses from the ear.
   e. the size of the action potentials received.

8. Josh is color blind, so he has a lot of trouble picking out clothes. Which sensory structures below are affected in a color-blind person?
   a. organ of Corti
   b. cones
   c. hair cells
   d. rods
   e. utricle and saccule

9. Which of the following is *not* a mechanoreceptor?
   a. taste bud
   b. hair cell
   c. touch receptor
   d. pressure receptor
   e. stretch receptor

10. Eating carrots really is good for your eyes. Carrots contain vitamin A, which is used to make a substance called rhodopsin, which
    a. is a visual pigment that absorbs light.
    b. provides energy for the function of rods and cones.
    c. colors the iris of the eye.
    d. stimulates the neurons in the retina to form branches and connections.
    e. keeps the lens clear and transparent.

Essay

1. Bats can navigate in total darkness by listening to the echoes of their high-pitched clicks. In a few sentences, describe three other examples of animal sensory capabilities beyond the range of human senses.

2. Explain the difference between sensation and perception.

3. What is farsightedness? What eye defect causes it? How is farsightedness corrected?

4. Explain how a stimulus (light, chemical, temperature change, movement) triggers a nerve impulse in a sensory receptor.

5. How are the sense of smell and sense of taste similar? What are the differences between the senses of taste and smell?

Applying Your Knowledge

Multiple Choice

1. You look all over for your glasses and then find them on your forehead. You couldn't feel them because of
   a. perception.
   b. transduction.
   c. sensation.
   d. adaptation.
   e. accommodation.

2. Michael and Laura are both looking at an illustration in her psychology textbook. Michael says, "It is an old woman looking down toward the left." Laura says, "I see a young woman looking away from us." How can they look at the same picture and see different things?
   a. Perhaps one of them has astigmatism.
   b. Their sensations differ.
   c. They sense the same thing, but their perceptions differ.
   d. It may take some time before their eyes accommodate to the picture.
   e. They perceive the same thing, but their sensations differ.

3. Whales and dolphins are known to send out clicking sounds and listen to the echoes. This suggests that they might find their prey in the same way
   a. a rattlesnake finds a mouse in total darkness.
   b. a salmon locates its home stream.
   c. an owl locates a mouse in a dark barn.
   d. an electric eel finds its prey in a muddy river.
   e. bats navigate in dark caves.

4. As you read this sentence, how do the photoreceptors of your retina tell the brain whether an area is light (the paper) or dark (the print)?
   a. Either rods or cones send signals to the brain.
   b. Nerve signals are sent to different areas of the brain.
   c. Signals are sent to the brain at different frequencies.

   d. Accommodation adjusts for light and dark areas.
   e. Larger (white) or smaller (dark) action potentials are sent to the brain.

5. A fish detects vibrations in the water around it by means of its "lateral lines," rows of sensory receptors along each side of the body. Based on what you know about sensory receptors, the lateral line receptors are probably most similar to
   a. receptors in the organ of Corti.
   b. rods and cones.
   c. receptors in taste buds.
   d. receptors in the retina.
   e. olfactory receptors.

6. Damage to the nerve from the saccule and utricle to the brain could result in
   a. loss of sense of taste.
   b. blindness.
   c. dizziness.
   d. loss of sense of smell.
   e. deafness.

7. There may be as many as 50 different kinds of smell receptors, each activated by different molecules. We can differentiate among thousands of different smells because each substance stimulates a different combination of these receptors. In much the same way,
   a. our eyes can distinguish many different colors.
   b. our ears can differentiate between sounds of many different pitches.
   c. pain receptors can distinguish between mild and severe pain.
   d. our eyes can adjust to light of differing intensities.
   e. our ears can determine how loud a sound is.

8. It is said that if you are seasick, it is better to look out at the water than at the boat. Why?
   a. It fools your brain into thinking that you are not really moving.
   b. This stimulates the saccule and utricle, organs of equilibrium.
   c. Seeing that you are moving reduces conflict between vision and equilibrium.
   d. Keeping your head level reduces activity of the semicircular canals.
   e. Actually, looking at the water just makes seasickness worse.

9. The eyes of a nocturnal animal, such as an owl, could be expected to have a larger proportion of _____ than do our eyes.
   a. chemoreceptors
   b. cones
   c. rods
   d. hair cells
   e. mechanoreceptors

10. The carotid body is a structure in the wall of the main artery carrying blood to the head. In it are special receptors that monitor blood pressure. These receptors would belong to which of the following groups?
    a. chemoreceptors
    b. thermoreceptors
    c. photoreceptors
    d. mechanoreceptors
    e. electromagnetic receptors

Essay

1. Theresa is suffering from a mysterious disorder called Ménière's disease. Her symptoms are quite debilitating: loud roaring sounds and dizziness. Why do these symptoms occur together? What part of the body must be affected by Ménière's disease?

2. The eyes contain three kinds of cones—green, blue, and yellow. Which are stimulated when we see blue? Yellow? White? Black?

3. Some researchers believe that human beings might have an ability to sense direction that is distinct from the traditional "five senses." The human brain contains magnetite, a magnetic mineral, and it has been suggested that perhaps we have some ability to find our way by sensing the Earth's magnetic field. Describe an experiment that could test this suggestion.

4. Most sensory receptors are sensitive to a sudden change in a stimulus, but soon adapt—become less sensitive—to a continuing stimulus. For example, thermoreceptors quickly adapt, and a very hot bath soon feels comfortable. How might this be useful? Pain receptors are an exception; they do not readily adapt. Why do you think this is the case?

5. There are many different kinds and causes of deafness. Explain how each of the following would result in deafness: damage to hair cells from repeated loud sounds, brain injury, arthritis of the middle-ear bones, a torn eardrum, buildup of earwax in the auditory canal.

Extending Your Knowledge

1. Anosmia is loss of the sense of taste or smell. It sometimes results from a brain injury or as a side effect of medication. Do you think anosmia could be inconvenient or dangerous in any way? Do you think it would be worse to lose your sense of smell or your sense of taste? Why?

2. Have you ever felt your ears ringing after listening to loud music from a stereo or at a concert? Do you ever think about the possibility that listening to loud music could permanently damage your hearing? Are your friends aware of this danger to their hearing? What, if anything, should be done to warn people about this danger or protect them from it?

3. Most people and animals are attracted to sweet and fatty foods. They find many poisonous substances bitter and undesirable. How do you think such taste preferences might have evolved?

4. Neurologists and psychologists are fascinated by people affected by a condition called synesthesia ("senses together"). It has been shown that these individuals can "hear" colors, "feel" sounds, or "see" pain. What do you think might be the physical cause of synesthesia?

How Animals Move

30

Look at the palm of your hand. Primates are the only mammals that can turn their arms palm up. Our lower arm bones are able to pivot at the elbow and cross, so we are able to rotate our hands through almost a full circle. The arrangement of bones and muscles that allows us to do this is a heritage passed down from our tree-living ancestors. Although other mammals cannot do the palm-up trick, their skeletons and muscles show equally remarkable adaptations to other methods of movement. A horse runs on very long middle toes. A sea lion's legs have become flattened paddles, and its muscles can store extra oxygen for extended dives. Every hop of a kangaroo stores energy in elastic leg tendons; rebound of the tendons helps power the next hop. Running, swimming, and hopping are all movements produced by muscles pulling on bones. Skeletons, muscles, and movement are the subjects of this chapter.

Organizing Your Knowledge

Exercise 1   (Module 30.1)

This module introduces various forms of animal locomotion. Test your knowledge of locomotion by filling in the blanks.

Mammals range in size from tiny shrews to great whales. They have become adapted to a variety of lifestyles on land, in the water, and in the air.

Most mammals, from mice to musk oxen, walk or run on land. The surrounding air offers little resistance to movement, so [1]_____ is not much of a problem, even for the fastest runners. But a mammal that walks or runs must expend considerable effort supporting itself against the force of [2]_____. To maintain balance and stability when it walks, an animal employs the principle of the [3]_____, keeping three feet on the ground most of the time. The fastest land animal, the cheetah, is of course a mammal. When it runs, all of its feet may leave the ground at once, but its [4]_____ stabilizes its position.

Some mammals crawl or burrow—moles and gophers, for example. Unlike the earthworm, which moves by waves of muscular contractions called [5]_____, burrowing mammals use their legs for digging. The mole's coat is velvety smooth, reducing [6]_____ as it tunnels underground.

Millions of years ago, a group of shore-dwelling mammals took up life in the sea. Their descendants are the whales and dolphins. Because of their buoyancy, [7]_____ is no problem for these aquatic animals, and some have grown to enormous size. Water is much more dense than air, but a whale's tapered, [8]_____ shape enables it to slip through the water with little effort. The whale's tail forms a pair of broad, flat flukes, and the whale bends its body [9]_____ to push it through the water.

Bats are the only flying mammals. A bat's wings are modified arms, with skin stretched between elongated fingers. The bat's wings, like those of airplanes and birds, are [10]_____. As a bat beats its wings, air travels farther over the top surface than the bottom surface, making the pressure underneath the wings [11]_____ than the pressure on top. This generates [12]_____ and allows the bat to overcome the pull of [13]_____.

Exercise 2   (Module 30.2)

There are three main kinds of skeletons. Summarize your knowledge of skeletons by completing this chart.

	Hydrostatic Skeleton	Exoskeleton	Endoskeleton
Example animals	1.	2.	3.
Description of skeleton	4.	5.	6.
Materials skeleton is made of	7.	8.	9.
Functions of skeleton	10.	11.	12.
Drawbacks of skeleton	13.	14.	15.

Exercise 3   (Module 30.3)

Web/CD Activity 30A   *The Human Skeleton*

The human skeleton is unique in several ways, but it has the same overall pattern as the skeleton of a mouse or a duck. You may find it useful to know the names of some of the major bones of the skeleton. Label these bones on the diagram below: **humerus, sternum, femur, tarsals, tibia, carpals, radius, phalanges, ulna, vertebrae, skull, clavicle, metacarpals, ribs, scapula, fibula, pelvic girdle, metatarsals, shoulder girdle,** and **patella.** (Note: The letters relate to Exercise 4, below.) Before you leave this exercise, color the bones of the axial skeleton green and the appendicular skeleton yellow.

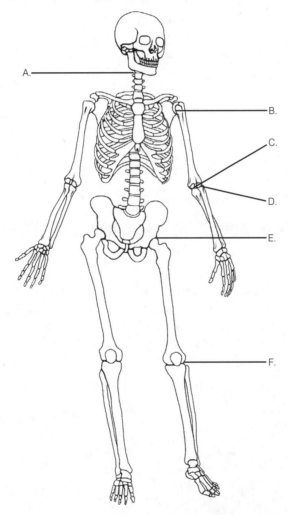

Exercise 4   (Module 30.3)

Web/CD Activity 30A   *The Human Skeleton*

Which of the letters in Exercise 3 identifies each of the following types of joints? (Three are from the module; you will have to figure out the other three yourself.)

1. Ball-and-socket joints: _____

2. Hinge joints: _____

3. Pivot joints: _____

Exercise 5   (Modules 30.4 – 30.6)

Review bone structure, growth, disease, and injury by matching each phrase on the left with a word or phrase from the right. Some answers are used more than once.

_____ 1. Fat stored in central cavity
_____ 2. Made of protein fibers, calcium, and phosphorus
_____ 3. Outer hard layer of bone
_____ 4. Cushions ends of bone at joints
_____ 5. Decrease after menopause may cause osteoporosis
_____ 6. Bone break that occurs from long-term repeated forces
_____ 7. Shafts of long bones made of this at first
_____ 8. Delivers food and hormones to bone
_____ 9. Produces blood cells
_____ 10. Site of red marrow
_____ 11. Bone grows until plates of this material are replaced by bone
_____ 12. Makes up the shaft of a long bone like the tibia
_____ 13. Connects a muscle to a bone
_____ 14. Covers outside of bone
_____ 15. Joint damage or disease
_____ 16. Weakening of bones that often occurs in older women
_____ 17. Protein fibers that make matrix flexible
_____ 18. Deficiency of this in the diet may contribute to osteoporosis

A. Spongy bone
B. Cartilage
C. Calcium
D. Red bone marrow
E. Blood vessel
F. Fibrous connective tissue
G. Yellow bone marrow
H. Compact bone
I. Osteoporosis
J. Collagen
K. Stress fracture
L. Matrix
M. Arthritis
N. Tendon
O. Estrogen

Exercise 6   (Modules 30.7 – 30.9)

Web/CD Activity 30B   *Skeletal Muscle Structure*

Many muscles work in opposing pairs to move the skeleton. Each muscle is composed of many muscle fibers. Each fiber contains smaller parts, which are responsible for muscle contraction. Review the contractile machinery of muscle by labeling these parts in the diagram below: **biceps muscle, triceps muscle, thin filament, muscle fiber, myofibril, tendon, thick filament, sarcomere,** and **Z line.**

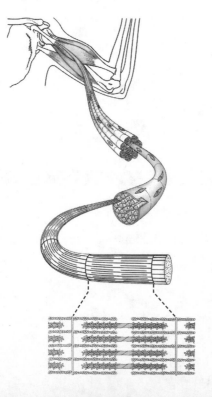

Exercise 7   (Modules 30.8 – 30.10)

Web/CD Activity 30B   *Skeletal Muscle Structure*
Web/CD Activity 30C   *Muscle Contraction*
Web/CD Thinking as a Scientist   *How Do Electrical Stimuli Affect Muscle Contraction?*

Try to visualize the inner workings of a muscle by filling in the blanks below.

Your new mission is to view contracting muscle cells close up. You step into the Microtron, and a moment later, you are working your way against the current in a small arm vein. Ahead are smaller, thin-walled [1]_____ that exit from the muscle and join to form the vein. You stay close to the wall and enter the smallest vessel.

Through the translucent wall of the capillary, you can make out a nearby surface that seems to be made of shiny white ropes. This is the [2]_____ that connects the end of the biceps to the lower arm bones. Also visible is a smoother, shinier cylindrical structure—one of the [3]_____ that controls contraction of the muscle. Ahead, you see the wrinkled blob of a white blood cell slipping through a gap in the vessel wall, and you follow it into the intercellular fluid. Close up, the nerve you saw a moment ago appears frayed—its neurons branch and rebranch to control every muscle cell. You follow one of the neurons; it branches to about 50 cells. The neuron and the cells it controls make up a [4]_____—a group of muscle fibers that work together.

Following one of the axons to its end, you finally arrive at a [5]_____, where the axon synapses with a muscle cell. Each cell is called a muscle [6]_____. You are deep inside the muscle now, so you radio for more light and ask for the subject to tense her arm.

Suddenly a cloud of particles sprays from the end of the axon, the membrane of the muscle fiber seems to shimmer, and you are tossed by a surge of pressure into a nearby tangle of connective tissue fibers. She's moving the arm just a little bit, but to you it feels like an earthquake! You contact your colleagues: "Hey, take it easy! Try it again, but gently!" This time you hang on tightly. Again there is a spray of particles—the release of [7]_____ from the end of the axon, signaling the muscle fiber to contract. When the neurotransmitter molecules contact the muscle fiber membrane, the shimmer you saw before is repeated—an [8]_____ spreading across the cell. Another earthquake, but this time you are ready.

You grip the membrane of the muscle fiber. It is pocked with numerous openings, where infoldings of the membrane form [9]_____ that carry action potentials deep into the cell. You swim inward, following the action potentials, as the cell continues to receive impulses and contract. You can press your powerful spotlight against the tubule wall and illuminate the inside of the fiber. Inside are numerous [10]_____, large bundles of parallel protein filaments. Transverse [11]_____ lines, made of protein, separate repeating [12]_____, the basic contractile units of the muscle fiber. The filaments themselves are of two types: The [13]_____ filaments look like twisted strings of beads, the beads themselves being globular [14]_____ molecules. The [15]_____ filaments are made of elongated [16]_____ molecules, each with a head that can reach out and pull against the thin filaments.

You watch the thick and thin filaments [17]_____ along each other with each volley of nerve impulses. As each action potential travels into the cell via the tubules, you see a cloud of [18]_____ ions released from storage in the endoplasmic reticulum. These ions quickly attach to the thin filaments, freeing up binding sites for the myosin [19]_____ on the adjacent thick filaments. Meanwhile, the myosin molecules of the thick filaments are writhing like a bundle of worms. Their heads use energy from [20]_____ molecules to detach from the thin filaments, straighten, attach, and bend, several times per second. The hundreds of myosin heads in one thick filament pull on all the surrounding thin filaments. This creates the pull that causes the thick and thin filaments to slide together, [21]_____ the sarcomere, and on a larger scale, the muscle fiber.

The nerve impulses cease. You can see [22]_____ ions being pumped out of the cytoplasm and back into the [23]_____. As the binding sites on the thin filaments close up, the myosin heads let go, and the muscle stops contracting.

Exercise 8   (Modules 30.11 – 30.12)

Muscles illustrate the relationship between structure and function. Aerobic and anaerobic exercise change the function of a muscle, and as a result they actually alter its structure. State whether each of the changes below relates primarily to aerobic or anaerobic exercise.

_____   1. Increases muscle endurance

_____   2. Pushes muscle fibers to the point where they are not getting enough oxygen

_____   3. Increases the size of muscle fibers and muscle "bulk"

_____   4. Causes muscle to become more efficient

_____   5. Increases muscle strength

_____   6. Chosen by sprinters, weightlifters, and other "power" athletes

_____   7. Increases resistance to muscle fatigue

_____   8. Chosen by distance runners and cross-country skiers

_____   9. Improves circulation and gas exchange

_____   10. Increases size and number of mitochondria in muscle cells

Testing Your Knowledge

Multiple Choice

1. _____ connects a muscle to a bone.
   a. Cartilage
   b. A neuromuscular junction
   c. A tendon
   d. A myofibril
   e. A motor unit

2. Inside a muscle fiber, _____ trigger(s) contraction and _____ provide(s) the energy.
   a. myosin . . . actin
   b. calcium ions . . . ATP
   c. actin . . . myosin
   d. calcium ions . . . myosin
   e. ATP . . . calcium ions

3. Which of the following is associated with osteoporosis?
   a. age
   b. calcium in the diet
   c. exercise
   d. estrogen
   e. all of the above

4. Muscles are arranged in pairs,
   a. so if one is injured, the other can take over.
   b. doubling their strength.
   c. because one pulls while the other pushes.
   d. enabling them to perform opposing movements.
   e. so they can take turns contracting and resting.

5. Which of the following is the last part of a long bone to harden?
   a. the central shaft
   b. the heads
   c. cartilage between the heads and the shaft
   d. the central cavity
   e. cartilage at the very ends

6. Which of the following correctly pairs an example of a joint with its general type?
   a. ball and socket—elbow
   b. pivot—shoulder
   c. hinge—hip
   d. hinge—elbow
   e. ball and socket—knee

7. Which of the following skeletons works poorly on land?
   a. endoskeleton
   b. hydrostatic skeleton
   c. exoskeleton
   d. a and b
   e. b and c

8. A stronger muscle contraction occurs when the brain
   a. activates muscle cells more quickly.
   b. sends stronger nerve impulses to the muscle.
   c. activates the motor units of the muscle one at a time.
   d. signals a larger number of motor units to contract.
   e. sends nerve impulses to the muscle one at a time instead of in bursts.

9. Which of the following animals has an endoskeleton?
   a. clam
   b. sea star
   c. sponge
   d. a and b
   e. b and c

10. Which of the following is part of the human appendicular skeleton?
    a. vertebral column
    b. scapula
    c. rib
    d. sternum
    e. skull

Essay

1. How do the shape of its wings and their movement through the air lift a bird off the ground?

2. What are the three major functions of a skeleton? Describe each, using parts of your own skeleton as illustrations.

3. What kinds of animals have jointed exoskeletons? What are the major advantages and disadvantages of exoskeletons, compared with endoskeletons?

4. What two forces must be overcome by a moving animal? Which is more of a problem on land? In the water?

5. Without looking it up, make a sketch showing the arrangement of thick and thin filaments in a muscle cell when the cell is relaxed and when the cell is contracting.

Applying Your Knowledge

Multiple Choice

1. Which of the following shortens when a muscle fiber contracts?
   a. thin filament
   b. myosin molecule
   c. sarcomere
   d. actin molecule
   e. thick filament

2. Which of the following ranks the parts in order, from largest to smallest?
   a. muscle, myofibril, filament, fiber
   b. fiber, muscle, myofibril, filament
   c. muscle, fiber, filament, myofibril
   d. myofibril, muscle, fiber, filament
   e. muscle, fiber, myofibril, filament

3. His doctor suspects that Jon might be suffering from leukemia. The doctor orders a sternal puncture, a test that samples red marrow. This test is performed to determine
   a. whether the marrow holds sufficient fat reserves.
   b. how fast Jon is growing.
   c. the rate of blood flow through the bone tissues.
   d. whether blood cell production is occurring normally.
   e. how fast cartilage is forming in the bones.

4. For every bone in the arm and hand, there is a corresponding bone in the leg and foot. Which of the following matches corresponding bones?
   a. humerus—femur
   b. metacarpals—tarsals
   c. radius—femur
   d. carpal—patella
   e. humerus—tibia

5. An animal that crawls or burrows faces problems similar to those faced by an animal that
   a. swims.
   b. runs.
   c. flies.
   d. walks.
   e. hops.

6. A scallop escapes from danger by clapping its shells together, which shoots out a stream of water and causes the scallop to hop backward. This is most like the movement of
   a. an earthworm.
   b. a water beetle.
   c. a squid.
   d. a bird (but under water instead of in the air).
   e. a whale.

7. Smooth muscle is a type of muscle found in the internal organs—the walls of blood vessels and the intestine, for example. It is called "smooth" because the cells of smooth muscle lack the alternating light and dark bands seen in the muscle fibers that move the skeleton. Which of the following do you think might best explain this difference? Smooth muscle
   a. does not require nervous stimulation to contract.
   b. does not contain the regular, repeating filaments of skeletal muscle.
   c. cells are much larger than the cells of skeletal muscle.
   d. motor units are much smaller than those of skeletal muscle.
   e. requires much less blood flow than skeletal muscle.

8. Which of the following drugs would cause muscle spasms or cramps (uncontrolled contractions)? A drug that
   a. blocks the release of calcium ions from endoplasmic reticulum.
   b. prevents the release of acetylcholine from motor neurons.
   c. blocks neurotransmitter receptors on muscle fiber membranes.
   d. blocks the enzyme that breaks down acetylcholine after contraction.
   e. prevents attachment of myosin heads to thin filaments.

9. If you were to cut through a muscle fiber and look at the cut end under a powerful microscope, what would the contractile parts of the cell look like?
   a. many crisscrossing lines
   b. thousands of tiny dots
   c. thousands of overlapping circles
   d. numerous irregular splotches
   e. many overlapping, parallel lines

10. A wildlife biologist came across the carcass of a deer. He suspected that the deer had died of starvation. To check this guess, he could
    a. examine the cartilage at the ends of the bones for signs of wear.
    b. crack open a bone and check the yellow bone marrow.
    c. have the deer's tissues tested for signs of rheumatoid arthritis.
    d. look for signs of spongy bone in the ends of long bones.
    e. evaluate the amount of fat stored in the red bone marrow.

Essay

1. Look at the skeleton of the frog in Module 30.2. In what ways is the frog's skeleton modified for its way of life?

2. Both roundworms and earthworms possess hydrostatic skeletons. The body of a roundworm is a single elongated sac, filled with fluid. The roundworm is able to move only by thrashing from side to side. Compare this with the structure and movement of an earthworm, and explain the difference.

3. The wings of birds come in several different shapes. Many small songbirds have short, stout wings. The wings of an eagle are long and broad, while those of a gull are long and narrow. How are these different wing shapes (structures) well suited to the activities and environments (functions) of these different birds?

4. A map of the human cerebral cortex shows that the area of the brain that controls muscle activity of the hands is equal in size to the area that controls all the muscles below the neck. The total size of the muscles moving the hands is much smaller than all the muscles of the trunk, legs, and so on. Explain in terms of the motor units that make up the muscles why so much nerve tissue is required to control the hands.

5. Early reconstructions of dinosaur skeletons showed them with their legs splayed out to the sides, like present-day alligators and lizards. Most dinosaur experts now think that the legs of dinosaurs were directly under the torso, like those of a dog or elephant. Why do you think they consider this important?

Extending Your Knowledge

1. There are actually two main kinds of muscle tissue: slow (red) muscle and fast (white) muscle. Slow muscle contracts slowly in response to nerve impulses. It is able to store oxygen and keep working with more endurance for extended periods. White muscle contracts quickly but has much less endurance. Most muscles contain a mixture of both types, in different proportions in different muscles. With this information in mind, speculate about the following: Which muscle type would you expect to predominate in the muscles that are moving your eyes as you read this? In your thigh muscles? Why is the breast muscle of a duck dark (red) meat but a chicken breast white meat? Which type do you think has been found in higher-than-average proportions in the leg muscles of world-class marathoners?

2. Millions of Americans regularly engage in bodybuilding exercise at health clubs, weight rooms, and gyms. Millions of others engage in aerobic activities like swimming, running, cycling, and dance aerobics. How does bodybuilding differ from aerobic exercise? Do you participate in any of these activities? What is their value, in your opinion? What effects does each have on muscle tissue, respiration, and the cardiovascular system? The experts say that 20 minutes of aerobic exercise at least three times a week is important for overall cardiovascular fitness, weight control, and stress reduction. How much exercise do you get each week? Is it enough? If you would like to get more exercise, how could you make it a higher priority?

Plant Structure, Reproduction, and Development

31

Tulips are natives of Turkey, not Holland. Tulip bulbs were first brought to the Netherlands in 1590. The Dutch were enchanted by the beauty of the flowers and by stories of their immense value to Turkish sultans. Demand quickly exceeded supply, and growing and selling bulbs became a big business. Greed and speculation drove prices up. Family heirlooms and land were sold to buy bulbs. Fortunes were made overnight. The frenzy was known as "tulipmania." At its peak, one bulb was traded for "two loads of wheat, four loads of rye, four fat oxen, eight fat pigs, 12 sheep, two barrels of butter, 1000 pounds of cheese, two hogsheads of wine, four barrels of beer, a silver beaker, a suit of clothes, and a bed." In 1637, the market collapsed, thousands of businesses were ruined, and much of Europe was thrown into a depression. Nowadays, it's tech stocks, not flowers that attract this kind of interest from investors. But flowering plants are important to us in many ways. Just how are plants important? What kind of flower is a tulip? What is the function of a flower? What is a bulb and what does it do for the plant? These and many other questions are answered in this chapter on plant structure, reproduction, and development.

Organizing Your Knowledge

Exercise 1 (Introduction and Modules 31.1 and 31.2)

Look around you and reflect a moment on all the products and "services" supplied by plants. In the space below, list four specific examples from your surroundings illustrating how we use plants. Name the plants if you can. Then list four things that plant researchers might investigate that might benefit us.

Exercise 2   (Module 31.2)

The figures below show differences between the two major groups of flowering plants—monocots and dicots—but the figures are scrambled. List the figures characteristic of monocots and dicots, and briefly describe the characteristic shown in each.

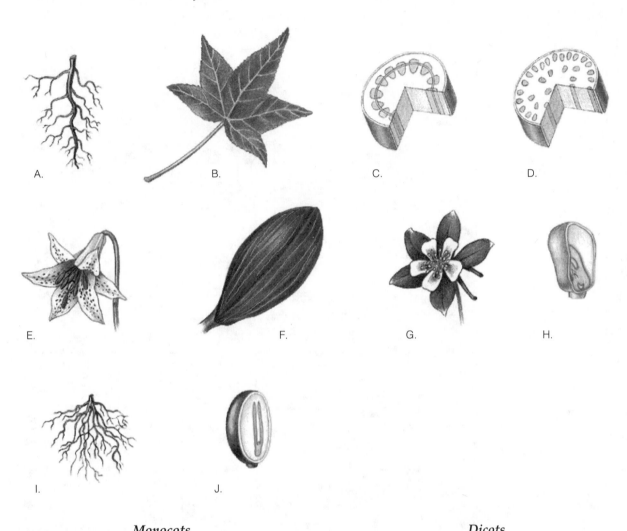

A.   B.   C.   D.

E.   F.   G.   H.

I.   J.

	Monocots		**Dicots**
Figure	*Description*	*Figure*	*Description*
_____	_____	_____	_____
_____	_____	_____	_____
_____	_____	_____	_____
_____	_____	_____	_____
_____	_____	_____	_____

Exercise 3   (Modules 31.3 – 31.6)

Web/CD Activity 31A   *Root, Stem, and Leaf Sections*
Web/CD Thinking as a Scientist   *What Are the Functions of Monocot Tissues?*

These modules introduce plant structure—cells, tissues, organs, and systems. Review the vocabulary of plant structure by completing this crossword puzzle. All answers are found in Modules 31.3 – 31.6.

Across

4. Dicots have one main _____, with many smaller secondary roots.

6. A root _____ is an extension of an epidermal cell that enhances absorption.

8. A _____ is an underground shoot with swollen leaves that store food.

9. The _____ system consists of stems, leaves, and reproductive structures.

12. Sieve-tube members are served by _____ cells.

13. A leaf "stalk" is called a _____ .

14. The _____ is ground tissue that makes up most of a root.

16. The middle _____ is a sticky layer between walls of adjacent cells.

18. _____ dominance is inhibition of axillary buds by the terminal bud.

19. A _____ is the point where a leaf is attached to a stem.

20. _____ is the parenchyma cells in a leaf specialized for photosynthesis.

25. Plant cells are surrounded by rigid _____.

26. _____ cells are the most abundant cells in most plants.

30. A _____ is a long, slender sclerenchyma cell.

32. A _____ is a vascular bundle in a leaf.

35. A _____ is an enlarged rhizome that stores food.

36. Hard _____ is the main chemical component of wood.

37. A _____ is an irregular cell that makes nutshells or seeds hard.

38. A waxy coating called the _____ helps the plant retain water.

39. _____ fills the central portion of a stem and may function in food storage.

40. A sharp cactus _____ is a modified leaf.

41. A _____ is a tiny pore in a leaf.

42. _____ cells regulate air flow through stomata.

43. The _____ system anchors the plant, and absorbs water and minerals.

Down

1. The _____ tissue system makes up much of a young plant and has many functions.
2. The cell structure that carries out photosynthesis is called a _____.
3. A plant cell contains a fluid-filled central _____.
5. A _____ is a horizontal stem.
7. A root, stem, or leaf is composed of three _____ systems.
10. The _____ bud is the growth point at the tip of a stem.
11. Plant cell walls are made of _____.
13. _____ carries sugar from the leaves to other parts of the plant.
15. _____ carries water up from the roots.
17. _____ buds form where leaves join a stem and are usually dormant.
21. A _____ plate is the perforated end of a food-conducting cell.
22. The _____ is the main site of photosynthesis in most plants.
23. _____ cells are the dead, hardened supporting cells that form the plant "skeleton."
24. The _____ covers and protects a plant.
27. _____ cells support the growing parts of a plant.
28. A _____ , or "runner," is a stem that grows along the surface of the ground.
29. The _____ forms a barrier that regulates flow into root vascular tissue.
31. Vascular tissues of stems and leaves are arranged in vascular _____.
33. The _____ tissue system is composed of xylem and phloem.
34. A _____ is a modified leaf that helps a plant cling and climb.

Exercise 4   (Module 31.6)

Web/CD Activity 31A   *Root, Stem, and Leaf Sections*
Web/CD Thinking as a Scientist   *What Are The Functions of Monocot Tissues?*

The cross-sectional diagrams below will help you to visualize the internal structures of roots, stems, and leaves. Label the following: **dicot stem, monocot stem, root, leaf, cortex, cuticle, endodermis, stoma, pith, epidermis, guard cell, mesophyll, vascular bundle, phloem, xylem,** and **vein.** Color the dermal tissue system brown, the vascular tissue system purple, and the ground tissue system yellow.

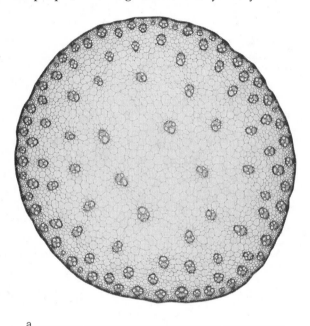

a. _____

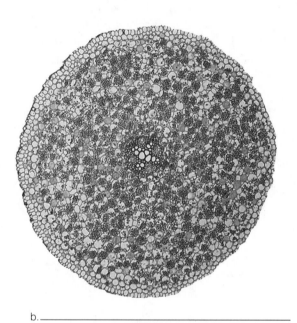

b. _____

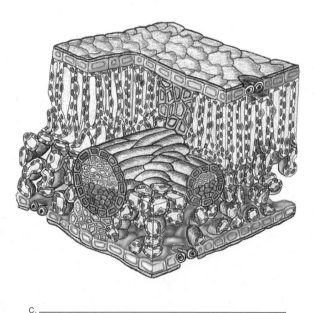

c. _____

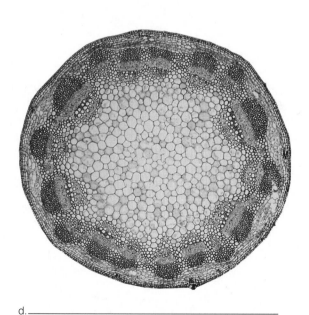

d. _____

Exercise 5   (Modules 31.7 – 31.8)

Web/CD Activity 31B   *Primary and Secondary Growth*

These modules discuss plant growth. There are different kinds of growth and different kinds of plant life cycles. After reading the modules, match each term on the left with a phrase on the right.

_____	1. Primary xylem	A. Lengthwise growth
_____	2. Biennial	B. Makes up the wood of a tree or shrub
_____	3. Secondary growth	C. Growth point at the tip of a shoot or root
_____	4. Indeterminate growth	D. Thick, waxy cells that protect woody stem
_____	5. Perennial	E. A plant that completes its life cycle in a year or less
_____	6. Primary phloem	F. Tissue just outside the vascular cambium of a tree trunk
_____	7. Sapwood	G. Ceasing to grow after reaching a certain size
_____	8. Primary growth	H. Meristematic tissue that forms cork
_____	9. Cork	I. A plant that takes 2 years to complete its life cycle
_____	10. Secondary phloem	J. A plant that can grow in girth
_____	11. Vascular cambium	K. Older layers of secondary xylem in a tree trunk
_____	12. Annual	L. Pushed outward by development of secondary phloem
_____	13. Determinate growth	M. A plant that continues to live for many years
_____	14. Heartwood	N. Young secondary xylem that still conducts water
_____	15. Secondary xylem	O. Rapidly dividing cells between xylem and phloem
_____	16. Cork cambium	P. Dividing cells along the length of a root or stem
_____	17. Apical meristem	Q. First water-conducting vascular tissue of a plant
_____	18. Woody plant	R. Growth in diameter
_____	19. Lateral meristem	S. Growing as long as a plant lives

Exercise 6   (Module 31.7)

Label the following on this diagram of a root tip: **root cap, root hair, vascular cylinder, cortex, epidermis,** and **apical meristem.** Color the root cap red, the zone of cell division orange, the zone of cell elongation yellow, and the zone of cell maturation blue.

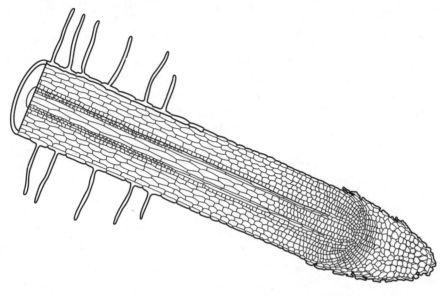

Exercise 7   (Module 31.7)

Web/CD Activity 31B   *Primary and Secondary Growth*

Try to picture the growth and structure of a maturing twig by identifying each of the layers in the stem. Choose from the following: A. cortex   B. secondary xylem   C. vascular cambium   D. primary phloem   E. epidermis   F. primary xylem   G. cork cambium   H. secondary phloem   I. pith   J. cork

Near the tip of the twig: 1. Name the layers, from outside to inside:_____
                                 2. Which layers are lateral meristems?_____

After a few years' growth: 3. Name the layers from outside to inside:_____
                                 4. Which layers are lateral meristems?_____
                                 5. Which layers make up the bark?_____
                                 6. Which layer makes up wood?_____
                                 7. Which layers present near the tip have now disappeared?_____

Exercise 8   (Module 31.8)

Web/CD Activity 31B   *Primary and Secondary Growth*

The diagram below will also help you visualize the secondary growth of a tree trunk. Label each of the following on the diagram: **vascular cambium, secondary phloem, sapwood, secondary xylem, heartwood, cork, cork cambium,** and **bark.** Color the vascular cambium green. Color the oldest xylem red, the newest xylem pink, the oldest phloem dark blue, and the newest phloem lighter blue.

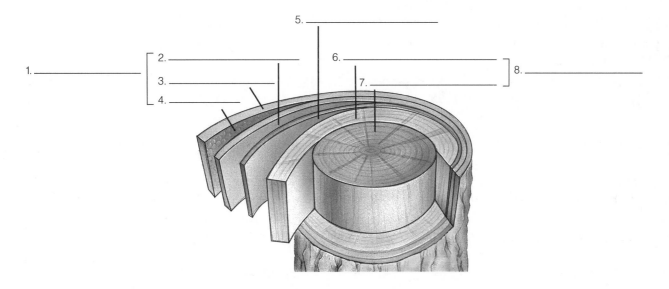

Exercise 9   (Module 31.9)

In the space below, sketch a flower, and label the following parts: **sepal, stigma, stamen, style, filament, carpel, anther, petal, ovary, ovule,** and **pollen grain.** At the risk of being a little sexist, color the female parts of the flower pink and the male parts blue.

Web/CD Activity 31C   *Angiosperm Life Cycle*
Web/CD Activity 31D   *Seed and Fruit Development*
Web/CD Thinking as a Scientist   *What Tells Desert Seeds When to Germinate?*

The following story will help you review plant reproduction and development. Complete the story by choosing the correct italicized word or phrase in each set.

The spring rains have begun. A blackberry seed has lain undisturbed on the ground near a rail fence since September. Now (1) *sunlight, warmth, water* causes the seed to swell. The process of (2) *pollination, germination, fertilization* is beginning. The protective (3) *endosperm, carpel, seed coat* splits open, a tiny root emerges, and the leaves of the tiny (4) *ovule, pollen grain, embryo* unfurl toward the sun. At first the plant's leaves are pale; its seed leaves, or (5) *cotyledons, anthers, sepals,* have absorbed the (6) *pollen, endosperm, sperm* stored within the seed to obtain food for the plant's first few days of growth.

As the days grow longer and warmer, apical meristems add new cells to the tips of the blackberry's roots and stems. By midsummer, the plant has produced several branches and dozens of leaves and has begun to climb up on the fence. Clusters of white flowers, resembling small wild roses, sprout at lateral buds. The blackberry plant represents the diploid (7) *gametophyte, sporophyte* phase of the blackberry's life cycle. The flowers are the reproductive structures where (8) *mitosis, germination, meiosis* produces haploid (9) *seeds, spores, fruits,* which undergo mitosis to become haploid (10) *gametophytes, sporophytes.*

Each flower begins as a bud, enclosed by green, protective (11) *petals, carpels, sepals.* As a flower opens, white (12) *petals, carpels, sepals* attract bees. The bees swarm over the flowers, collecting nutrient-rich pollen grains from the (13) *carpels, sepals, anthers* at the tips of the (14) *stamens, carpels, stigmas,* the male reproductive structures of the flowers. As they buzz from flower to flower, the bees deposit some of the pollen grains on the sticky (15) *stigmas, anthers, stamens* at the tips of the female reproductive structures, called (16) *anthers, carpels, sepals.* Each blackberry flower has numerous stamens and carpels, so the bees are busy.

Each pollen grain is actually a tiny (17) *male, female* gametophyte, containing a tube nucleus and two (18) *egg, seed, sperm* nuclei. (19) *Fertilization, Pollination, Meiosis* occurs as a bee rubs a pollen grain onto a stigma. The pollen grain germinates, and a pollen tube grows down through the carpel into the swollen (20) *anther, ovary, sepal* at its base. The two (21) *egg, sperm, tube* nuclei now travel down the pollen tube to (22) *a sepal, a carpel, an ovule,* which contains the (23) *stigma, stamen, embryo sac,* the tiny haploid female (24) *gametophyte, sporophyte.* The sperm nuclei enter the ovule, and one of them joins with the (25) *egg, anther, pollen grain* within, forming a diploid (26) *ovule, embryo sac, zygote.* This is fertilization. The other sperm nucleus combines with a large diploid cell, forming a triploid cell. This cell will divide to form the (27) *pollen, endosperm, ovule,* which will become enclosed in the seed and nourish the developing embryo. The joining of two sperm nuclei with an egg cell and another diploid cell is called double (28) *fertilization, germination, pollination* and is unique to flowering plants.

In the ovary at the base of each carpel, each ovule with its zygote now begins to develop into a (29) *seed, stamen, fruit.* The zygote divides, and one of the cells formed divides to become the (30) *flower, endosperm, embryo*. Because the blackberry is a dicot, the embryo sprouts two (31) *sepals, cotyledons, seed coats*, which will later help the embryo absorb nutrients stored in the seed. The outer coating of the ovule dries and thickens and becomes the seed coat.

Each blackberry flower contains multiple carpels. The flower petals shrivel and fall away, and the (32) *sepal, anther, ovary* at the base of each carpel grows and ripens. Eventually, the seeds end up in a cluster of ripened ovaries, which form (33) *a simple fruit, an aggregate fruit, a multiple fruit*, called a blackberry.

The blackberries are hard, green, and sour at first, but they become soft, dark purple, and sweet as (34) *starch, organic acid, sugar* accumulates in them. The fruits protect the seeds and aid in their dispersal. A blackbird perches on a swaying cane, twists off a fat berry, and gulps it down. The soft flesh of the blackberry is easily digested, but the seeds pass through the bird's digestive system unharmed, protected by their tough (35) *endosperm, seed coats, sepals*. A few hours after eating the berry, the bird perches on a fence and deposits the seeds in its droppings. Over the winter, the seeds will undergo a period of (36) *dormancy, differentiations, fertilization*. In the spring, the seeds will sprout, and the fence will make an ideal trellis for the next generation of blackberries.

Exercise 11   (Modules 31.14 – 31.15)

Match each phrase with a type or method of vegetative reproduction or propagation.

A. Cuttings
B. Root sprouts
C. Plant cloning
D. Fragmentation
E. Runners

_____ 1. Grass slowly spreads into a garden bed.
_____ 2. Pieces of a houseplant grow roots in water.
_____ 3. A clump of tulips expands each year.
_____ 4. Many small trees grow from a stump.
_____ 5. House plants grow from cells in a test tube.

Testing Your Knowledge

Multiple Choice

1. Which of the following is *not* a characteristic of dicots?
   a. two seed leaves
   b. parts of flowers in fours or fives
   c. taproot
   d. vascular bundles arranged in a ring
   e. veins in leaves usually parallel

2. Most of the photosynthesis in a plant is carried out by _____ in the leaves.
   a. collenchyma cells
   b. water-conducting cells
   c. parenchyma cells
   d. sclerenchyma cells
   e. food-conducting cells

3. How do cells in a meristem differ from other cells in a plant?
   a. They continue to divide.
   b. They photosynthesize at a faster rate.
   c. They are growing.
   d. They are differentiating.
   e. They store food.

4. "Angiosperm" is another name for
   a. plant.
   b. pollen.
   c. stored food in a seed.
   d. flowering plant.
   e. fruit.

5. In the process of pollination, pollen grains are transferred from the _____ to the _____ .
   a. ovary . . . anther
   b. stigma . . . ovary
   c. anther . . . sepal
   d. carpel . . . stigma
   e. anther . . . stigma

6. Which of the following is *not* a method of vegetative reproduction?
   a. runners
   b. germination
   c. stump sprouts
   d. cuttings
   e. fragmentation

7. What is endosperm?
   a. male reproductive cells in plants
   b. stored food in a seed
   c. cells that make up the bulk of a pollen grain
   d. the fleshy part of a fruit such as an apple or strawberry
   e. plant chromosomes

8. After fertilization, the _____ develops into a seed, and the _____ develops into a fruit.
   a. ovule . . . ovary
   b. pollen grain . . . ovule
   c. ovary . . . ovule
   d. egg . . . ovule
   e. egg . . . ovary

9. Lengthwise growth of a root into the soil results mainly from
   a. cell division in the apical meristem.
   b. elongation of cells.
   c. cell division in the vascular cambium.
   d. differentiation (specialization) of root cells.
   e. pulling by root hairs.

10. Which of the following is closest to the center of a daisy stem?
    a. cortex
    b. phloem
    c. pith
    d. epidermis
    e. xylem

11. Cell division would be slowest in which of these tissues?
    a. apical meristem of root
    b. cork cambium
    c. epidermis
    d. vascular cambium
    e. apical meristem of terminal bud

Essay

1. Briefly describe four functions of a plant's roots. Which of these is primarily a job of the epidermis? The vascular cylinder? The cortex?

2. Sketch a cross section of a young dicot stem, showing pith, xylem, phloem, cortex, and epidermis. What do these layers have to do with the three tissue systems in the stem? How does the arrangement of tissues in a dicot root differ from their arrangement in a stem?

3. What is the sporophyte stage of a flowering plant? What reproductive structures does it produce? What is the gametophyte stage? What reproductive cells do gametophytes produce? Which of the preceding are haploid and which are diploid?

4. What are meristems, and where would you find them in an apple tree? Which are responsible for primary growth? Secondary growth?

5. Briefly explain how the shape and structure of the following cells are well suited to their functions: sclerenchyma cells in a stem, water-conducting cells in a root, epidermal cells in a leaf.

6. Explain how the vascular cambium enables a tree to grow in diameter.

7. Starting with pollination, briefly explain how fertilization occurs in a flowering plant. What is the major difference between this process and fertilization in animals?

8. Briefly describe six different ways in which roots, stems, or leaves are modified for storage, reproduction, or other specialized functions.

Applying Your Knowledge

Multiple Choice

1. The vascular cambium in the trunk of a large, woody rhododendron shrub lies between
   a. secondary phloem and secondary xylem.
   b. secondary xylem and pith.
   c. primary xylem and primary phloem.
   d. secondary phloem and cortex.
   e. primary phloem and pith.

2. Which of the following is correctly matched with its tissue system?
   a. xylem—ground tissue system
   b. phloem—dermal tissue system
   c. cortex—ground tissue system
   d. pith—vascular tissue system
   e. All of the above are correctly matched.

3. The shoot system of a beavertail cactus consists of broad paddlelike structures covered with spines. The spines are modified _____ , so the flat green paddles must be modified _____ .
   a. buds . . . leaves
   b. buds . . . stems
   c. leaves . . . stems
   d. stems . . . roots
   e. stems . . . leaves

4. Cell division in the vascular cambium adds to the girth of a tree by adding new _____ on the inside of the cambium layer and _____ on the outside.
   a. phloem . . . xylem
   b. xylem and phloem . . . bark
   c. pith . . . xylem and phloem
   d. xylem . . . phloem
   e. xylem . . . cortex

5. A vandal killed a historic oak tree on the village green by "girdling" it with a chain saw. He cut through the bark and into the sapwood all the way around the tree. Why did the tree die?
   a. The leaves could not get water.
   b. Oxygen could not get to the roots.
   c. The roots could not get food.
   d. The leaves could not get food.
   e. The roots could not absorb water.

6. Artichoke hearts are tender and tasty. The leaves are tasty too, but most of an artichoke leaf is fibrous and impossible to chew. The leaves must contain lots of
   a. parenchyma cells.
   b. phloem.
   c. meristematic tissue.
   d. sclerenchyma cells.
   e. epidermal cells.

7. Based on your own experience with these plants, which do you think is correctly paired with the word describing its life cycle?
   a. apple tree—perennial
   b. rose bush—annual
   c. marigold—perennial
   d. oak tree—biennial
   e. tulip—annual

8. Plants growing in harsh environments such as deserts, sand dunes, and arctic tundra often reproduce vegetatively. This is because
   a. there are few animals available to pollinate them.
   b. they are members of plant families that only reproduce asexually.
   c. fruits would freeze or dry out in these environments.
   d. vegetative reproduction is not as risky or costly as making seeds.
   e. seeds would be eaten by hungry animals in these environments.

9. Which of the following would be least useful in figuring out whether a plant is a monocot or a dicot?
   a. life cycle—perennial, biennial, annual
   b. numbers of flower parts
   c. pattern of veins in leaves
   d. number of seed leaves
   e. arrangement of vascular tissue in stem

10. A cross section of part of a plant exposes epidermis, a thick cortex, and a central cylinder of xylem and phloem. This part is a
    a. fruit.
    b. seed.
    c. stem.
    d. root.
    e. bud.

Essay

1. Amy told Paul that a giant sequoia is the largest living thing, bigger than ten blue whales. Paul replied, "Yes, the sequoia is bigger, but it is an unfair comparison because *most of a tree is not alive*." Do the facts back up Paul's assertion? Explain.

2. Imagine pounding a large spike into a tree. Name (in order) all the tissue layers it would pass through as it penetrates into the heartwood.

3. What part of a plant are you eating when you consume each of the following: garlic, walnut, cabbage, carrot, cauliflower, artichoke, asparagus, cucumber, broccoli, rice, and potato?

4. When he was a boy, Grandpa nailed a horseshoe to a maple tree, about 5 feet from the ground. The tree has since grown much taller, and the diameter of its trunk has quadrupled, but the horseshoe is still about 5 feet up. Explain this in terms of how trees grow.

5. Palms are monocots. Do you think that you could kill a palm tree by girdling it (as in multiple choice question 5 in the "Applying Your Knowledge" section)? Why or why not?

6. Examine the diagram of a growing root, Figure 31.7B in the text. Why does it make sense that lateral roots must start growing from the inside of the main root rather than simply projecting from the outside?

7. Most food for humans is derived from the seeds of plants. Why does it make sense that seeds are a good source of nutrition?

Extending Your Knowledge

1. Do you like to grow houseplants? Are houseplants products of controlled breeding, or are they free-living somewhere in the wild? Have you ever wondered where they originally came from? What do you think their natural environment might be like? What are the functions of their varied shapes, colors, and textures? If you are interested in the answers to these questions, there are many gardening and horticulture books that discuss the origins of common houseplants.

2. Many individuals are concerned that wild plants be conserved and protected, especially in the tropics, where plant diversity is greatest and wild plants appear to be most endangered by human activities. They argue, "Humans use only a tiny number of plants for food and that there are undoubtedly unused and undiscovered species that could be better sources of food than any plants now in use." Others reply, "People are set in their ways. The fact that we use so few plants for food shows that people are unwilling to try new ones. While it might be important to preserve wild plants, it is unlikely that a new plant will become an important food source." What do you think? Are any new food plants becoming available? Have you tried anything new lately? Have a significant number of people anywhere in the world taken up a new staple crop in the recent past? Give an example if you know of one.

3. Keep a list of the plants and plant products you use in a single average day. Which species are represented? How many species? What categories of functions do the plants serve in your life? Do you think that the number of plants and plant products used in everyday life has increased or decreased in the last century? Is the number likely to increase or decrease in the future? Why or why not?

Plant Nutrition and Transport

Is there a houseplant where you live that you have had for a long time? Maybe a spider plant or a Boston fern or a Swedish ivy? Have you bought other plants that soon died? Why did one plant thrive and the other wither? Was it the temperature in the room or the amount of sunlight or the amount of water you provided? Did you underwater or overwater? Why can too much water kill a plant? Was the soil the right texture? What is the proper texture? Were the proper amounts of the right nutrients available in the soil? What nutrients do plants need? Which nutrients do they obtain from soil, from water, and from the air around them? Once a plant absorbs these nutrients, how are they transported to where they are needed? These are questions relevant to any plant, and they are the subject of this chapter.

Organizing Your Knowledge

Exercise 1   (Module 32.1)

Review the nutrients plants require by filling in the blanks in the paragraph below.

Plants need four main kinds of nutrients to survive and grow:
[1]_____, [2]_____, [3]_____, and
[4]_____. Most of a plant's mass is made of organic molecules built from
[5]_____, which is obtained from the [6]_____. Hydrogen for organic molecules comes from [7]_____, which is obtained from the
[8]_____ and split during the process of photosynthesis. Many substances manufactured by the plant require elements other than carbon, hydrogen, and oxygen. Proteins, for instance, contain [9]_____. Chlorophyll contains
[10]_____ and [11]_____, and ATP and nucleic acids contain
[12]_____ and [13]_____. These [14]_____ are absorbed from the soil by the plant's roots. Finally, the plant takes in
[15]_____ which is necessary for aerobic respiration from the air and soil.

Exercise 2   (Module 32.2)

On the diagram below, trace the uptake of water and minerals by a root. Show water moving from cell to cell (the intracellular route) with a blue arrow. Trace water moving along cell walls, between cells (the extracellular route), with a red arrow. The Casparian strip forces all water and solutes to pass through endodermal cells, allowing the endodermis to control movement of water and solutes into the vascular tissues. Color the endodermis and Casparian strip yellow. Label the following: **endodermis, root hair, epidermis, cortex, xylem vessels, Casparian strip, intracellular route, extracellular route,** and **plasmodesma.**

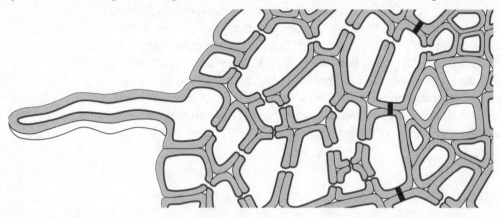

Exercise 3   (Modules 32.3 – 32.4)

Web/CD Activity 32A   *Transpiration*
Web/CD Thinking as a Scientist   *How Are Water and Solute Potentials Calculated?*
Web/CD Thinking as a Scientist   *How Is the Rate of Transpiration Calculated?*

Review the movement of water and ions up a plant stem by filling in the missing words in the phrases below, then numbering each of the phrases in order. Start with what happens in the roots and continue with what pulls water up from the top of the plant.

_____ A. Water molecules _____ from cells bordering the air spaces
        within the leaf.
_____ B. Inorganic ions are pumped into the _____ of the roots; this causes
        water to enter these cells by _____.
_____ C. Water molecules leaving leaf cells are stuck to nearby water molecules, a phe-
        nomenon known as _____. They also stick to the cellulose mole-
        cules in the walls of xylem cells, a phenomenon known as _____.
        As water molecules diffuse from leaf cells, they tug on adjacent water molecules.
_____ D. Water accumulating in the xylem of roots pushes xylem sap upward a few me-
        ters, a phenomenon known as _____.
_____ E. As water molecules in the leaf pull upward on adjacent water molecules,
        they in turn tug on water molecules in the xylem, all the way down to
        the roots. This occurs because water molecules stick to each other by
        _____ bonds.
_____ F. _____ cells in leaf epidermis take up K+ ions. Water enters by
        _____, and the cells swell. This causes the _____ to
        open, and water evaporates from the leaf, a process called _____.
_____ G. Water molecules are pulled up from the roots, bit by bit, as water molecules
        evaporate from the leaves. This explanation for the ascent of xylem sap is called
        the _____ mechanism.

Exercise 4 (Module 32.5)

Web/CD Activity 32B   *Transport in Phloem*

Phloem transports sugar from where it is produced to where it is used via a pressure flow mechanism. Review this mechanism, paying particular attention to Figure 32.5B in the text. Then test your understanding by stating whether each of the following refers to a sugar source or a sugar sink.

_____ 1. Where phloem sap has its highest sugar concentration

_____ 2. A beet root or potato tuber during the summer

_____ 3. Any photosynthesizing part of a plant

_____ 4. Phloem sap flows away from this area

_____ 5. Where sugar is pumped into the phloem by active transport

_____ 6. Where water enters phloem sap by osmosis

_____ 7. A bulb in early spring

_____ 8. Where water (hydrostatic) pressure is highest

_____ 9. Where sugar leaves the phloem by active transport

_____ 10. A leaf on a summer day

_____ 11. Where water (hydrostatic) pressure is lowest

_____ 12. Where phloem sap has its lowest sugar concentration

_____ 13. Phloem sap flows toward this area

_____ 14. A growing bud in early spring

Exercise 5 (Modules 32.6 – 32.7)

Seventeen essential elements are required by plants: iron (Fe), manganese (Mn), nitrogen (N), hydrogen (H), calcium (Ca), phosphorus (P), zinc (Zn), potassium (K), molybdenum (Mo), copper (Cu), carbon (C), oxygen (O), magnesium (Mg), boron (B), chlorine (Cl), sulfur (S), and nickel (Ni). Which of these nutrients are the answers to each question below?

_____ 1. Which nine nutrients are macronutrients, needed in large quantities?

_____ 2. Which six macronutrients are the major ingredients of organic compounds?

_____ 3. Which eight nutrients are micronutrients, needed in smaller amounts as cofactors that help enzymes?

_____ 4. Shortage of which nutrient is the most common deficiency of plants?

_____ 5. Which macronutrient combines with proteins to "glue" cells together?

_____ 6. Which macronutrient is the main solute for osmotic regulation, causing opening and closing of stomata by guard cells, for example?

_____ 7. Which micronutrient is a metallic component of the cytochromes involved in electron transport?

_____ 8. Deficiency of which nutrient is the second most common plant deficiency?

_____ 9. Which macronutrient is part of chlorophyll and a cofactor of several enzymes?

Exercise 6   (Module 32.8)

Web/CD Activity 32C   *Absorption of Nutrients from Soil*
Web/CD Thinking as a Scientist   *Connection: How Does Acid Precipitation Affect Mineral Deficiency?*

Match each of the phrases on the left with a component of soil from the list on the right.

_____   1. Fine clays and nutrients dissolved in water accumulate in this
                soil layer
_____   2. Roots exchange these for nutrient ions bound to clay particles
_____   3. Not bound tightly by soil particles
_____   4. Partially decomposed organic matter
_____   5. Their negative charges keep + ions from draining away
_____   6. Where most decomposing organic material and organisms are found
_____   7. Mainly composed of partly broken-down rock.
_____   8. Adhere to clay particles
_____   9. Diffuses into soil water from air spaces in the soil.
_____ 10. Held in tiny spaces between soil particles
_____ 11. Break down organic matter

A. + ions, like Ca^{2+}
B. B horizon
C. C horizon
D. Clay particles
E. Air
F. Water
G. Bacteria and
      fungi
H. H^+ ions
I. Topsoil
J. Humus
K. − ions, like NO_3^-

Exercise 7   (Modules 32.9 – 32.11)

Web/CD Activity 3D   *Connection: Genetic Engineering of Golden Rice*

State how each of the following would be harmful or helpful to the soil or the environment.

1. Applying an excess of commercially produced inorganic fertilizer to a field

2. Crop rotation; alternating different crops

3. Plowing and planting in rows up and down hills

4. Flooding fields with irrigation water and blocking flow to prevent runoff

5. Using fertilizers that contain organic material

6. Using perforated pipes to drip water onto soil

7. Planting "super" varieties of wheat, maize, and rice

Exercise 8 **(Modules 32.11 – 32.15)**

Web/CD Activity 32D   *Connection: Genetic Engineering of Golden Rice*

Review the adaptations of plants for obtaining nitrogen by filling in the blanks in the following story.

^{1}_____ is often the hardest to obtain of all plant nutrients but is essential to plants for making ^{2}_____, ^{3}_____, and other organic molecules. Ironically, the ^{4}_____ is 80% N_2 gas, but plants cannot use nitrogen in that form. Most plants must obtain nitrogen from the soil in the form of ^{5}_____ (NH^{4+}) or ^{6}_____ (NO^{3-}) ions. Certain soil bacteria, called ^{7}_____ bacteria, are able to convert N_2 to ammonium. Other bacteria, called ^{8}_____ bacteria, decompose organic matter in the soil and produce additional ammonium. Soil bacteria called ^{9}_____ bacteria then convert ammonium to nitrate, which is absorbed. Plant enzymes then convert nitrate back to ammonium, which is used to build ^{10}_____.

In some plants, roots may form associations with fungi, called ^{11}_____, which assist in absorption of nutrients. This is a ^{12}_____ beneficial arrangement—both plant and fungus gain. The fungus is good at absorbing water and inorganic ions, some of which are transferred to the plant. The fungus may secrete ^{13}_____, which stimulate roots to grow, and ^{14}_____ which help protect the plant from certain diseases. In return, some of the plant's ^{15}_____ nourishes the fungus.

Other specializations for obtaining nutrients have evolved in various groups of plants. Some plants, such as ^{16}_____ and ^{17}_____, are parasitic. They simply tap into the vascular systems of other plants and steal food and nutrients. ^{18}_____ plants, such as the sundew and ^{19}_____, often grow in bogs where the soil is poor in ^{20}_____ and other minerals. These plants obtain the minerals they need by trapping and digesting ^{21}_____.

The roots of plants in the ^{22}_____ family, such as peas, beans, and alfalfa, have swellings called ^{23}_____ that house ^{24}_____ bacteria. Again, the arrangement is mutually beneficial. The legume plant provides the bacteria with ^{25}_____, and the bacteria convert atmospheric ^{26}_____ into ammonium. Excess ammonium may actually enter the soil, and for this reason, farmers often ^{27}_____ crops, alternating between legumes and nonlegumes such as ^{28}_____.

Testing Your Knowledge

Multiple Choice

1. Nitrogen fixation is
   a. using nitrogen to build molecules such as proteins and nucleic acids.
   b. converting nitrogen in the air into a form usable by plants.
   c. recycling nitrogen from organic matter in the soil.
   d. absorbing N_2 from the soil.
   e. an unhealthy interest in nitrogen.

2. Guard cells
   a. control the rate of transpiration.
   b. push water upward in a plant stem.
   c. protect the plant's roots from infection.
   d. control water and solute intake by roots.
   e. protect nitrogen-fixing bacteria in root nodules.

3. Which of the following is a macronutrient?
   a. hydrogen   d. nitrogen
   b. calcium   e. all of the above
   c. phosphorus

4. Soil can easily become deficient in _____ , because these ions are negatively charged and do not stick to negatively charged clay particles.
   a. potassium
   b. calcium
   c. magnesium
   d. nitrate
   e. ammonium

5. Which of the following would trigger opening of stomata?
   a. extreme heat
   b. loss of potassium by guard cells
   c. nightfall
   d. swelling of guard cells due to osmosis
   e. all of the above

6. The sundew plant has to digest insects because
   a. it obtains nitrogen from their flesh that it cannot get from the soil.
   b. it has lost the ability to perform photosynthesis.
   c. it lives in a dry environment and needs the moisture in their bodies.
   d. it needs to get rid of insects that accidentally get stuck in its hairs.
   e. its flowers are fertilized by pollen in their digestive tracts.

7. Ammonifying bacteria in the soil
   a. convert ammonium to nitrate.
   b. fix nitrogen.
   c. convert nitrogen in organic molecules into ammonium.
   d. change nitrate into ammonium.
   e. use nitrate to make amino acids that plants can use.

8. Mycorrhizae are
   a. nutrients required by plants in relatively small amounts.
   b. plants such as mistletoe that parasitize other plants.
   c. medium-sized soil particles.
   d. cells that control evaporation of water from leaves.
   e. associations of roots with beneficial fungi.

9. Which of the following is a sugar source?
   a. a green leaf
   b. a developing fruit
   c. a growing root
   d. a growing shoot
   e. a tree trunk

10. What is the main source of energy that moves water upward in the trunk of a tree?
    a. musclelike contraction of xylem cells
    b. evaporation of water by the sun
    c. pressure exerted by root cells
    d. breakdown and release of energy of sugar molecules
    e. osmotic changes caused by alterations in salt content

11. All water and solute molecules must pass through _____ before they can enter the vascular system and move upward to the leaves.
    a. a stoma
    b. a root hair cell
    c. an endodermal cell
    d. an epidermal cell
    e. a cortex cell

12. A plant does *not* obtain which of the following substances from soil?
    a. magnesium
    b. nitrogen
    c. carbon
    d. oxygen
    e. phosphorus

Essay

1. Explain what causes water to enter the xylem of a tree root and move upward in the xylem of the trunk to the leaves.

2. Describe an experiment to find out whether the element molybdenum is an essential plant nutrient.

3. In a growing potato plant, what causes sugar to flow from a leaf to the potato? Early in the spring, what causes sugar to flow from the potato to the growing shoot?

4. Explain how root hairs obtain ions that cling to the surfaces of clay particles in the soil.

5. Organic fertilizers such as manure and compost and inorganic commercial fertilizers contain the same nutrient ions needed by plants. In addition to this, what does the humus in organic fertilizer contribute to the soil?

Applying Your Knowledge

Multiple Choice

1. The roots of many aquatic plants have special structures that project above the surface of the water. For example, cypress trees (which grow in swamps) have "knees" that extend upward above water level. Which of the following is the most logical function of these structures?
   a. obtaining carbon dioxide for photosynthesis
   b. nitrogen fixation
   c. obtaining oxygen for the roots
   d. transpiration
   e. absorbing trace minerals

2. Helen had a terrarium on her windowsill containing various houseplants. She wondered why the glass was often fogged with water droplets. Her friend Sara, who has had a biology class, tried to impress Helen by explaining, "The water evaporates from the leaves—it's a process called _____."
   a. root pressure
   b. adhesion
   c. photosynthesis
   d. pressure flow
   e. transpiration

3. In an apple tree, sugar might flow from _____ to _____ .
   a. a developing apple . . . a leaf
   b. the trunk . . . a leaf
   c. a growing root . . . a growing shoot tip
   d. a leaf . . . a developing apple
   e. a growing shoot tip . . . the trunk

4. Soil could be deficient in any of the following nutrients. If you had to supply one of them, which would be needed in the smallest amount?
   a. sulfur
   b. phosphorus
   c. nitrogen
   d. potassium
   e. iron

5. Which of the following is a difference between transport by xylem and transport by phloem?
   a. Active transport moves xylem sap but not phloem sap.
   b. Transpiration moves phloem sap but not xylem sap.
   c. Phloem carries water and minerals; xylem carries organic molecules.
   d. Xylem sap moves up; phloem sap moves up or down.
   e. Xylem sap moves from sugar source to sink, but phloem sap does not.

6. What keeps the force of gravity from pulling water molecules down from the leaves?
   a. upward pressure from the roots
   b. high water pressure in the leaves
   c. the Casparian strip blocks them from moving out
   d. movement of water toward a sugar sink
   e. cohesion and adhesion of water molecules

7. Jon is performing a chemical analysis of xylem sap. He should not expect to find much of which of the following?
   a. nitrogen
   b. sugar
   c. phosphorus
   d. water
   e. potassium

8. A botanist discovered a mutant plant that was unable to produce the material that forms the Casparian strip. This plant would be
   a. unable to fix nitrogen.
   b. unable to transport water or solutes to the leaves.

c. able to exert greater root pressure than normal plants.

d. unable to control the amounts of water and solutes it absorbs.

e. unable to use its roots as a sugar sink.

9. Normally when an aphid feeds by puncturing plant tissues, it does not have to suck the sap out. An inexperienced aphid, however, accidentally inserted its feeding tube in the wrong place and found the fluid in its gut being sucked out through the feeding tube. It had punctured
   a. the Casparian strip.
   b. a root nodule.
   c. a xylem cell.
   d. a phloem tube.
   e. a stoma.

10. Professor Timothy Schmidlap claims to have discovered a new macronutrient required for plant growth. Most of his colleagues are skeptical of his claim. Why might they consider it unlikely?
    a. All the nutrients required for plant growth have already been found.
    b. It is very difficult to prove that a plant needs a certain nutrient.
    c. Plants need thousands of nutrients; a new one is not significant.
    d. Any nutrient needed in large amounts has probably been found already.
    e. His colleagues are jealous and want to claim credit for the discovery.

Essay

1. Farmers in Lower Quasiland found swellings full of bacteria on the roots of their bean plants, so they sprayed the plants with antibiotics. The swellings disappeared, but the plants no longer grew well. Examining the plants more closely, the farmers concluded that the cause of the difficulty was fungi associated with the roots. They treated the plants with a fungicide, but then the plants did even worse. Explain what is happening.

2. Sometimes on very hot days, water evaporates so quickly from the leaves of a plant that empty "bubbles" occur in xylem tubes. This can be harmful to the plant. Explain why.

3. The jade plant, a familiar houseplant, has a peculiar adaptation to the hot, dry environment where it grows naturally. Its stomata open at night, carbon dioxide enters the leaves, and the carbon dioxide is stored in a chemical form. The stomata remain closed during the day, when the stored carbon dioxide is used in photosynthesis. How does this differ from most other plants? How might it be advantageous in the jade plant's environment?

4. A plant does not use potassium for a building material or energy source, but it stops doing photosynthesis and dies if it runs out of potassium. Why?

5. Denise and Roger dreamed of starting an organic homestead in southeast Alaska. They bought a plot of forested land, cut down the fir and spruce trees, built a cabin, and then had the soil tested. An accumulation of acid from decaying conifer needles, in combination with the heavy rainfall, had caused the soil to become quite deficient in a particular macronutrient. What nutrient was most likely to be in short supply? Why was it and not other nutrients washed from the soil ? How might they enrich the soil with this nutrient without using commercial fertilizers?

Extending Your Knowledge

1. More and more supermarkets are stocking organic produce. What does "organic" mean? Is there a legal definition of "organic" in your state? (It varies.) Do you buy organic fruits and vegetables when you have the chance? Why or why not? What are some reasons people buy organic produce? What are some reasons they don't buy more?

2. Do you have a "green thumb"? Planting a garden or growing houseplants can be an interesting and relaxing pastime. Some people save money by growing their own vegetables. Others like to surround themselves with fresh flowers. Some people take a more scientific interest in growing ferns or cacti. Wherever you live there is somewhere to grow plants—a sunny window, a balcony, pots along the sidewalk. You may even want to find out whether community garden plots are available in your city.

Control Systems in Plants

33

While most of us worry about depletion of the ozone layer that protects us from ultraviolet light, agricultural researchers are concerned that an excess of ozone in smog is killing crops. The U.S. government figures that ozone is responsible for 90% of crop losses from air pollution, at an estimated cost of $1–$5 billion per year. Biologists in England have discovered that plants exposed to a burst of ozone produce twice as much ethylene—a plant hormone—than normal plants. Interestingly, it is the ethylene, not the ozone, that seems to be responsible for most of the damage to the plants. This discovery may point the way to a method for preventing crop losses caused by air pollution. Ethylene is a plant hormone, just one of many phytochemicals, or "plant chemicals." This chapter is about how hormones and other mechanisms control plant responses to the environment.

Organizing Your Knowledge

Exercise 1   (Module 33.1)

A series of experiments demonstrated that phototropism is due to the effect on shoot elongation of a chemical (named auxin) formed in the shoot tip. The following figures (on the next page) illustrate some of those experiments, and others. Predict whether each of the numbered grass shoots will grow to the left, right, straight up, or show no growth in response to the experimental conditions shown. Assume that each seedling has germinated in the dark and has just been placed under the experimental conditions.

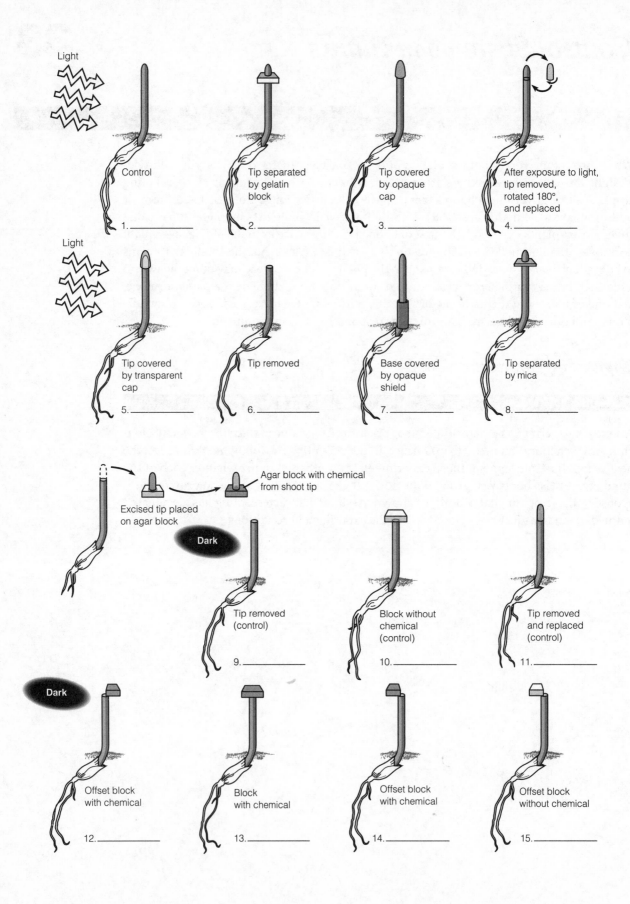

Exercise 2   (Modules 33.2 – 33.4)

Web/CD Thinking as a Scientist   *What Plant Hormones Affect Organ Formation?*

Module 33.2 introduces the major classes of plant hormones, while the next two modules focus on auxin and cytokinins, two important hormone classes. Review the effects of auxin and cytokinins by filling in the blanks in the following statements.

A. A number of different [1]_____, which promote elongation of stems, have been identified in plants. Others have been synthesized artificially.

B. [2]_____ promote cell division. Several are natural; some are produced in the lab.

C. The best-known natural auxin is [3]_____.

D. Most auxin synthesis occurs in the [4]_____ at the tip of a shoot. It moves down the stem and stimulates cells to elongate.

E. Cytokinins are produced in growing tissues, mainly [5]_____, [6]_____, and [7]_____. Those that originate in roots affect stems by traveling upward in [8]_____.

F. [9]_____ promotes stem elongation, perhaps by weakening [10]_____ and allowing cells to stretch.

G. The same concentration of auxin that causes shoots to elongate [11]_____ elongation of roots. A lower concentration [12]_____ root elongation.

H. Auxin transported down the stem from the terminal bud [13]_____ growth of axillary buds, causing the plant to grow straight up without branching.

I. If the terminal bud is pinched off, [14]_____ buds become active, forming lateral branches, and the plant becomes [15]_____. [16]_____ moving upward from the roots are thought to activate the axillary buds.

J. The ratio of [17]_____ and [18]_____ probably shapes the complex growth pattern of most plants. Auxin from the terminal buds [19]_____ branching, while cytokinins from the roots [20]_____ it. The [21]_____ axillary buds usually begin to grow before those closer to terminal buds, reflecting the higher ratio of [22]_____ to auxins in the lower parts of the plant.

K. The ratio of auxin and cytokinins also helps to coordinate root and shoot growth. A growing [23]_____ system produces cytokinins, which cause the [24]_____ system to branch.

Exercise 3   (Modules 33.5 – 33.6)

Gibberellins and abscisic acid each have a number of functions. State whether each of the following refers to gibberellin (G) or abscisic acid (ABA).

_____ 1. Slows down growth

_____ 2. Stimulates cell division and elongation in stems

_____ 3. Breakdown by winter cold may trigger spring germination

_____ 4. Responsible for maintaining seed dormancy

_____ 5. Produced by fungi, it causes "foolish seedling disease"

_____ 6. Causes dwarf plants to grow tall

_____ 7. Opposes the effects of gibberellins in seeds

_____ 8. Primary signal that enables plants to withstand drought

_____ 9. When released from an embryo, it signals a seed to germinate

_____ 10. If it is washed from seeds, they will germinate

_____ 11. Acts antagonistically to ABA in seeds
_____ 12. With auxin, it causes fruit to develop, even without fertilization
_____ 13. Causes leaf stomata to close
_____ 14. Causes grapes to grow larger and farther apart

Exercise 4   (Module 33.7)

Web/CD Activity 33A   *Leaf Abscission*

Review the maturation and ripening of fruit and leaf abscission by filling in the blanks in the following paragraphs.

Although auxin and gibberellins cause fruits to develop and mature, a hormone called [1]_____ triggers ripening. It is also involved in other processes related to [2]_____. Ethylene diffuses through the fruit in the air spaces between cells and brings on changes in the fruit such as [3]_____ and changes in color. As cells age, they produce more ethylene, speeding up the [4]_____ process. Ethylene can also diffuse through the air to a fruit. [5]_____ are often picked green and ripened by exposure to ethylene. Ripening of apples is slowed by flooding storage areas with [6]_____, which prevents buildup of [7]_____.

Ethylene probably triggers autumn color changes, drying, and fall of leaves from [8]_____ trees such as maples and oaks. Dropping its leaves allows a tree to save [9]_____. Fall colors are [10]_____ made in the fall and other substances revealed by the breakdown of green [11]_____.

The signals for leaf drop are the [12]_____, [13]_____ days of autumn. As a leaf ages, it produces less [14]_____. Cells in the [15]_____ layer, at the base of the leaf stalk, begin to produce ethylene. The ethylene causes enzymes to digest cell walls in the abscission layer. The wind and the weight of the leaf cause the stalk to break away, and the leaf falls.

Exercise 5   (Modules 33.4 – 33.8)

Review use of plant hormones in agriculture by writing the name of the hormone next to the summary of its uses.

_____ 1. Can make fruits develop without fertilization of seeds; with auxin, used for producing seedless fruits. Induce biennials to seed during the first year of growth.

_____ 2. Triggers ripening of stored fruit. May be used to thin fruit or loosen it for picking.

_____ 3. In low doses, prevent trees from dropping fruit. In larger amounts, used to cause fruit drop for thinning. A synthetic variant, 2,4-D, used as herbicide.

_____ 4. Hormone spray used to keep cut flowers fresh and make Christmas trees branch.

Exercise 6   (Modules 33.2 – 33.8)

Review the actions of the five classes of plant hormones by matching the hormones with their overall functions.

_____ 1. Abscisic acid

_____ 2. Cytokinins

_____ 3. Gibberellins

_____ 4. Ethylene

_____ 5. Auxin

A. Stimulate cell division and growth, affect root growth; oppose auxin by stimulating branching

B. Stimulates stem elongation; affects root growth, fruit development, tropisms

C. Inhibits growth, maintains dormancy, closes stomata during stress

D. Promote seed germination and bud development, stem elongation, leaf growth, flower and fruit development

E. Promotes fruit ripening, leaf drop, aging; opposes some auxin effects

Exercise 7   (Module 33.9)

Tropisms are directed growth responses that enable plants to adapt to environmental circumstances. Complete this chart comparing tropisms.

Name	Description	Mechanism
Phototropism	1.	2.
3.	Growth movement in response to gravity	4.
5.	6.	Contact triggers greater growth on opposite side, causing tendril to bend toward support

Exercise 8   (Module 33.10)

What is a circadian rhythm? A biological clock? See if you can explain the role of biological clocks in plant circadian rhythms in *exactly* 30 words.

Exercise 9   (Modules 33.11 – 33.13)

Web/CD Activity 33B   *Flowering Lab*

Research with *Arabidopsis* is helping plant scientists understand phenomena such as flowering. Short-day plants flower when the night exceeds a certain minimum length. They are inhibited from flowering when red light (or white light, which includes red) shortens a long night. (Far-red light following red or white light reverses this effect.) Long-day plants flower when night length is shorter than a certain maximum. They flower when red or white light shortens a long night. The graphs below are similar to those in the text. Test your understanding of flowering and day (night) length: Under each graph, write "yes" if the plant will flower under the conditions described. Write "no" if the plant will not flower.

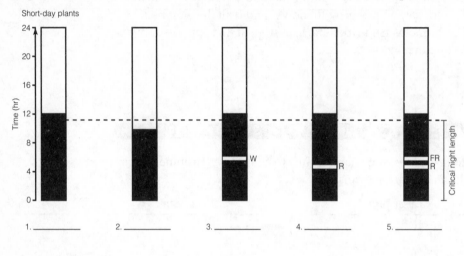

Short-day plants

1. _____   2. _____   3. _____   4. _____   5. _____

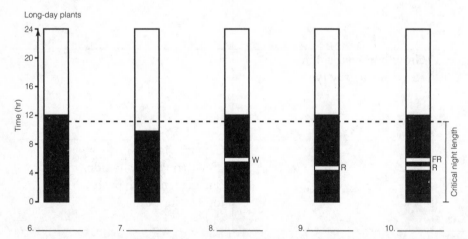

Long-day plants

6. _____   7. _____   8. _____   9. _____   10. _____

Exercise 10   (Modules 33.14 – 33.15)

Plants protect themselves with a variety of defenses. Match each plant part or process on the left with a description of its role in defense.

_____ 1. epidermis
_____ 2. chemical that attracts wasps
_____ 3. canavanine (an amino acid)
_____ 4. cell wall
_____ 5. alarm hormone
_____ 6. *R* gene
_____ 7. thorn
_____ 8. Avr gene

A. triggers defensive reaction distant from infection
B. physical barrier that keeps pathogens out of plant
C. physical defense that injures or repels herbivores
D. lures predators that kill harmful caterpillars
E. toughens to slow spread of pathogen in plant
F. codes for a "signal" molecule that warns a plant of infection
G. codes for receptor that matches and inactivates pathogen
H. alters insect proteins so insect dies

Testing Your Knowledge

Multiple Choice

1. Most plants flower when
   a. the soil reaches a certain temperature.
   b. they deplete soil nutrients.
   c. the days are the right length.
   d. a certain number of days have passed since they last flowered.
   e. the nights are the right length.

2. Plant tropisms appear to be controlled by
   a. auxin.
   b. tropogens.
   c. gibberellins.
   d. cytokinins.
   e. phytochromes.

3. The abscission layer
   a. causes a shoot to bend toward light.
   b. secretes cytokinin.
   c. is the location of the biological clock in a plant.
   d. detects light and measures photoperiod.
   e. is where a leaf separates from a stem.

4. _____ stimulate fruit development, so they are sprayed on crops to produce seedless fruits or vegetables.
   a. Ethylene and auxin
   b. Auxin and gibberellins
   c. Gibberellins and abscisic acid
   d. Ethylene and cytokinins
   e. Abscisic acid and auxin

5. Branching is inhibited by _____ from the tip of a growing shoot, but this effect is countered by _____ from the roots.
   a. cytokinins . . . auxin
   b. gibberellins . . . ethylene
   c. auxin . . . cytokinins
   d. gibberellins . . . abscisic acid
   e. auxin . . . abscisic acid

6. A synthetic auxin called 2,4-D is used
   a. to make fruit ripen.
   b. as a weed killer.
   c. to produce seedless fruits.
   d. to make plants produce larger numbers of seeds.
   e. to keep fruit from ripening.

7. A chemical change in a substance called phytochrome
   a. causes a plant to bend toward light.
   b. triggers fruit drop.
   c. enables a plant to measure day length.
   d. is responsible for gravitropism.
   e. allows a plant to deal with stresses like shortage of water.

8. A biological cycle with a period of about 24 hours is called
   a. thigmotropism.
   b. a circadian rhythm.
   c. a photoperiod.
   d. abscission.
   e. a biological clock.

9. Seeds of many desert plants will not germinate until a heavy rain washes away their
   a. phytochrome.
   b. abscisic acid.
   c. gibberellins.
   d. auxin.
   e. ethylene.

10. Plant chemicals called brassinosteroids and isoflavones have effects on humans similar to the effects of
    a. aspirin.
    b. vitamin C.
    c. cholesterol.
    d. growth hormone.
    e. estrogen.

11. Researchers have discovered that plant chemicals such as the salicylic acid of aspirin
    a. attract pollinators.
    b. stimulate plant growth.
    c. prevent or fight infections.
    d. can be used as stomach remedies.
    e. protect plants from grazers.

Essay

1. When a plant is tilted, it bends back toward a vertical position, even in the dark. How does a plant "know" which way is up, and what do hormones have to do with this?

2. How do we know that circadian rhythms are internal and not responses to environmental cycles? In what way are environmental cues important in circadian rhythms?

3. A farmer's almanac says, "To keep your stored apples crisp all winter, don't put bruised or wormy apples in the same barrels with the good ones." State the scientific reason why this is good advice.

4. Describe the changing balance of hormones that causes the leaves of deciduous trees to fall in the autumn. What does this change do to the structure of the leaf?

5. Explain the difference between long-day and short-day plants. How do plants "know" how long the day is?

Applying Your Knowledge

Multiple Choice

1. As leaf lettuce matures, the basal edible leaves suddenly send up a tall flowering shoot. After the plant "bolts" like this, it no longer produces broad, tasty leaves. Suppose you wanted to prevent bolting so that you could harvest lettuce longer. You might look for some way to interfere with the effects of
   a. abscisic acid.
   b. gibberellins.
   c. cytokinins.
   d. ethylene.
   e. gravitropism.

2. If the seeds in a cantaloupe are poorly fertilized and fail to develop, the cantaloupe itself will develop poorly and will probably drop off the vine. How does lack of seeds produce this effect?
   a. There is not enough ethylene in the cantaloupe for ripening to occur.
   b. The auxin concentration of the cantaloupe is too high for fruit growth.
   c. Abscisic acid causes the cantaloupe to shrivel and drop off.
   d. There is not enough auxin to stimulate development and stop fruit drop.
   e. Increased cytokinins cause the vine to break at the abscission layer.

3. According to the graph in Figure 33.3B in the text, an auxin concentration of 10^{-4} g/L would
   a. inhibit stem elongation and stimulate root elongation.
   b. stimulate stem elongation and inhibit root elongation.
   c. inhibit both stem and root elongation.
   d. stimulate both stem and root elongation.
   e. It is impossible to predict this from the graph.

4. If you wanted your plants to branch more, you might try spraying them with
   a. cytokinins.
   b. auxin.
   c. gibberellins.
   d. ethylene.
   e. abscisic acid.

5. When a plant structure such as a leaf is injured, it produces _____ , which may cause the part to age and drop off.
   a. cytokinins
   b. phytochrome
   c. auxin
   d. abscisic acid
   e. ethylene

6. An Alaskan trapper was worried about being attacked by grizzly bears, so he left the lights in his cabin on all the time. Plants near the cabin flowered a month early. Which of the following best explains this?
   a. It was due to phototropism.
   b. They must have been long-night plants.
   c. The lights must have emitted far-red light.
   d. They must have been long-day plants.
   e. They must have been short-day plants.

7. Once a flower is pollinated, changes occur that make it less attractive to insects. Its petals, for example, shrivel and fall off. Pollination must
   a. increase the output of cytokinins in the flower.
   b. block the flow of auxin from the roots.
   c. trigger the release of ethylene in the flower.
   d. increase the formation of phytochrome, which sets the biological clock.
   e. trigger the formation of gibberellins in the flower.

8. Which of the following grass seedlings will probably bend toward the light?
   a. tip covered with a cap made of black plastic
   b. tip separated from base by a gelatin block
   c. tip cut off
   d. tip separated from base by aluminum foil
   e. tip cut off; agar with auxin placed on cut surface on side toward light

9. A biologist growing plant cells in a laboratory dish wanted to cause them to _____ , so he treated them with cytokinins.
   a. enlarge
   b. become dormant
   c. grow roots
   d. produce auxin
   e. divide

10. A plant flowers only if days are shorter than 10 hours. Which of the following would cause it to flower?
    a. 8 hours light, 8 hours dark, flash of white light, 8 hours dark
    b. 12 hours light, 6 hours dark, flash of white light, 6 hours dark
    c. 8 hours light, brief dark period, 8 hours light, 8 hours dark
    d. 8 hours light, 8 hours dark, flash of white light, flash of far-red light, 8 hours dark
    e. 6 hours light, 6 hours dark, 6 hours light, 6 hours dark

Essay

1. During the 1800s, when natural gas was used for lighting, it was found that leaking gas from gas mains sometimes caused shade trees to drop their leaves. Based on what we now know about plant hormones, explain why this probably occurred.

2. A plant scientist hoped to make asparagus shoots grow faster by spraying them with auxin. She found in the lab that a 0.0001 M concentration of auxin increased elongation slightly. To increase its effect, she increased the auxin concentration to 0.01 M (100 times stronger) for spraying in the field. The results were not at all what she expected. The growth rate was less than normal! Explain why.

3. Predict the effects on a shoot of grafting it onto a much larger root system. What would be the role of hormones in these effects?

4. If you pinch off a leaf but leave the leaf stalk, the stalk soon falls from the stem. Pinching off the leaf must upset the balance between which two hormones? Which hormone, now in higher concentration in the stalk, causes the stalk to drop? Which hormone could you smear on the stalk to test whether this hypothesis is correct? What do you predict would happen?

5. In an experiment, soybean plants were planted at weekly intervals from the beginning of May through June. All the plants flowered at the same time—early September. Why?

6. Sometimes when grains are exposed to moisture they begin to germinate and are no longer usable. If you wanted to develop a hormone treatment to prevent this, what would you start with? Why?

7. Shoots exhibit positive phototropism and negative gravitropism. How could you find out which effect is stronger?

Extending Your Knowledge

1. Plans for a space colony include growing plants for food and oxygen. In low Earth orbit, there will be weightless conditions and a sunrise and sunset every 90 minutes. In what ways might these conditions create difficulties in raising plants? Can you suggest ways in which plant hormones might be used to overcome any of these problems?

2. How do nurseries and greenhouses use plant hormones to propagate and grow plants? How do they get plants to flower in the proper season or out of season? If there is a nursery in your area, you may want to arrange a visit to find out how the control systems of plants are manipulated to get the plants to do what a grower wants them to do.

3. Do you try to consume supplements or foods containing particular phytochemicals, such as nettles, echinacea, soy, isoflavones, or *Gingko biloba*? How can you find out whether these extracts are effective and, more importantly, safe?

The Biosphere: An Introduction to Earth's Diverse Environments

34

Much of biology, and much of your textbook, deals with what goes on inside individual organisms—systems, cells, genes, and biological molecules. But there are biological levels of organization larger and more inclusive than the individual organism—populations, communities, and ecosystems. These higher levels are the realm of ecology—the study of interactions between organisms and their environment. Some ecological questions are simple: How much rainfall does a maple tree need to survive? Others are complex: What processes affect the concentrations of oxygen and carbon dioxide in the atmosphere? This chapter introduces the ecological view of life.

Organizing Your Knowledge

Exercise 1   (Modules 34.1 – 34.3)

Web/CD Activity 34A   *Connection: DDT and the Environment*

These modules introduce the subject of ecology. Review the relationships between the various levels of study in ecology by completing the concept map on the next page. Place the following terms on the map: **abiotic components, biosphere, ecology, groups belonging to same species, populations, interactions, habitats, ecosystems, organisms, biotic components,** and **nutrients.**

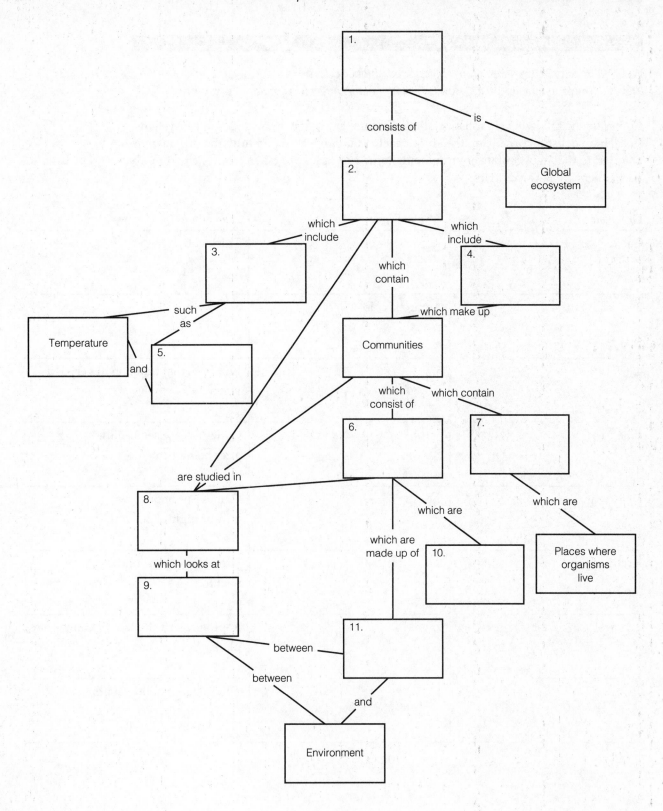

Exercise 2 (Modules 34.4 – 34.5)

Web/CD Activity 34B   *Adaptations to Biotic and Abiotic Factors*
Web/CD Thinking as a Scientist   *How Do Abiotic Factors Affect Distribution of Organisms?*

Complete the following chart listing abiotic factors that shape organisms and ecosystems. In each case, list an organism, the abiotic factor that affects it, and how the organism is adapted to that aspect of its environment. Some examples are taken from the text modules; others are new examples.

Organism	Abiotic Factor	Adaptation
Cactus	1.	Thick cuticle
2.	Antarctic cold	3.
4.	5.	Enzymes that function in very hot water
Pine tree	6.	Cones release seeds only after being heated
Desert rat	7.	Comes out of burrow only at night
Arctic poppy	8.	Grows low to ground, next to rocks
Tropical rain forest tree	9.	Rapid growth shades out competing plants
10.	11.	Thick coat of hollow hairs

Exercise 3 (Module 34.6)

Practice using concepts and terminology of climate by filling in the blanks in the following paragraphs.

The continent of North America stretches almost from the North Pole to the equator. Its life forms—from polar bears to iguanas—reflect this diversity in climate. The climate is warmest nearest the equator, where the sun's rays strike the Earth most [1]_____. Near the [2]_____, the sun's rays strike at more of an angle, so their energy is spread out over a larger area. The [3]_____ of the Earth in relation to the sun causes the seasons. In [4]_____, the Northern Hemisphere is tilted toward the sun. In [5]_____, North America tilts away from the sun.

In addition to temperature differences, wind and rain are important aspects of climate. The sun's heat [6]_____ air and evaporates [7]_____ near the equator. The warm air is less dense than cold air, so it [8]_____. The air cools as it ascends, and its moisture condenses and falls as [9]_____, watering the tropical forests of Guatemala, Costa Rica, and Panama. After losing their moisture, the high-altitude air masses spread away from the equator. Cool air is denser than warm air, so the air [10]_____ around [11]_____ degrees north and south of the equator. This dry air is responsible for the [12]_____ landscape of Mexico and the southwest United States. Some of the dry air spreads back toward the equator, and some picks up [13]_____ and spreads toward temperate latitudes. This air will cool and drop its moisture as it spreads farther from the equator, into areas such as the eastern United States.

The rise and fall of air masses and their resulting spread from place to place is deflected by the [14]_____ of the Earth to create the [15]_____ winds. Wind rushes back to replace rising air at the equator, creating the [16]_____, which dominate the tropics. The Earth's rotation causes the trade winds to blow from the [17]_____ in the Northern Hemisphere.

Ocean [18]_____ are created by the combined effects of unequal heating of surface waters, [19]_____, the Earth's [20]_____, and the locations of the [21]_____. Currents profoundly affect local climate. The cold Japanese current creates the foggy conditions that water the redwood forests of northern California. The [22]_____ warms the southeastern United States, as well as western Europe. Thus, [23]_____ moderate the climate of nearby land.

Landforms shape local climate. Altitude influences temperature and [24]_____. Moisture may be captured as air masses move over mountain ranges, creating dry [25]_____ downwind. The sagebrush landscape of Nevada lies behind the Sierra Nevadas, for example.

Exercise 4   (Modules 34.7 – 34.8)

Web/CD Activity 34C   *Aquatic Biomes*

All aquatic communities are shaped by similar environmental factors. Review words and concepts related to marine and freshwater environments by completing this crossword puzzle.

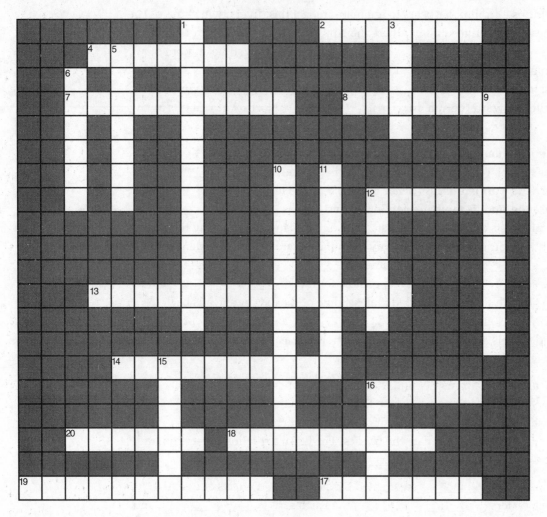

Across

2. A ____ is an ecosystem intermediate between aquatic and terrestrial ones.
4. The bottom of the sea, a lake, or a river is called the ____ zone.
7. The ____ zone is flooded by tides every 12 hours.
8. Light does not reach the ____ zone.
12. Motile animals like fishes and whales live in the ____ zone.
13. ____ is only possible in the photic zone.
14. Nitrogen and ____ may be limiting nutrients in lakes and ponds.
16. ____ reefs are diverse communities in warm shallow tropical waters.
17. Wetlands include swamps, bogs, and ____.
18. ____ of an estuary may vary from 1% to 3%.
19. ____ are drifting animals of the pelagic zone.
20. Many countries prohibit disposal of ____ and other wastes at sea.

Down

1. Drifting algae and cyanobacteria make up the ____.
3. The availability of ____ has a major effect on organisms in the sea.
5. An ____ is a coastal area where fresh and salt water meet.
6. Seasonal ____ in a lake brings nutrients to the surface and oxygen to the bottom.
9. The ____ shelves are the submerged edges of continents.
10. ____ vents are densely populated communities of the deep sea.
11. Near a river's mouth, the water may be murky with ____.
12. The ____ zone is the area near the surface where light can penetrate.
15. Decomposition on a lake bottom may deplete the water of ____.
16. Near a river's source, the water is usually cool and ____.

Exercise 5   (Modules 34.9 – 34.18)

As you read about the biomes, a good way to learn their characteristics is to make a chart that compares them. Using a large sheet of paper, list the *name* of each biome in the left column, and lay out columns for each of their characteristics across the top: *location, description, temperature, rainfall, special environmental conditions* (fire, etc.), *organisms*, and *human impact*. Once you have covered all the biomes, look at the CD-ROM and do the next few exercises, which compare biomes in various ways.

Exercise 6   (Modules 34.9 – 34.18)

Web/CD Activity 34D   *Terrestrial Biomes*

Temperature and precipitation are the most important factors shaping biomes. The figure below is a "climograph" outlining the temperature and rainfall limits within which each of the biomes occurs. Label the biomes on the graph. You do not need to know absolute temperatures and rainfall amounts; just ask yourself which biomes are hotter or colder, wetter or drier. Choose from **desert, tundra, temperate grassland, coniferous forest, temperate forest,** and **tropical forest.** Then color the areas on the graph to match the colors on the map in Module 34.9.

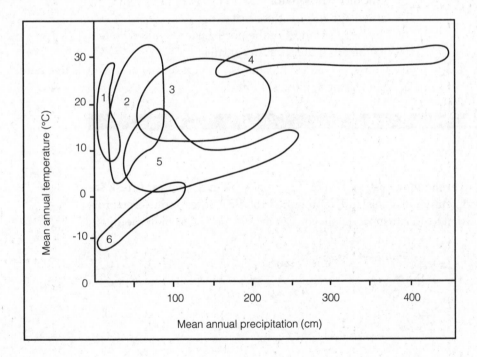

Exercise 7 **(Modules 34.9 – 34.18)**

Web/CD Activity 34D   *Terrestrial Biomes*

Match each of the phrases on the right with one of the biomes from the list on the left.

A. Tundra

B. Desert

C. Tropical forest

D. Temperate broadleaf forest

E. Chaparral

F. Coniferous forest

G. Temperate grassland

H. Savanna

_____ 1. Grows around the Mediterranean

_____ 2. Among the most complex and diverse of all biomes

_____ 3. Prairies, pampas, veldts, and steppes

_____ 4. The taiga of Canada, Alaska, Siberia

_____ 5. May be hot or cold but is always dry

_____ 6. Permafrost occurs here

_____ 7. May occur in rain shadow

_____ 8. Characterized by deciduous trees such as maples and oaks

_____ 9. Richest farmland in the United States

_____ 10. Antelope, zebras, lions, and cheetahs live here

_____ 11. Closest to the North Pole

_____ 12. Unlike other biomes, this one is growing in size

_____ 13. Characterized by mild, rainy winters and hot, dry summers with fires

_____ 14. Some characteristics of savannas but in colder areas than savannas

_____ 15. Straddles the equator

_____ 16. Grassland with scattered trees

_____ 17. The biome of most of the northeastern United States

_____ 18. Scrubland of dense, spiny shrubs

_____ 19. Moose, elk, snowshoe hares, beavers, bears, and wolves live here

_____ 20. The warmest, rainiest biome

Exercise 8 **(Modules 34.9 – 34.18)**

Web/CD Activity 34D   *Terrestrial Biomes*

Color and label this map to show the distribution of these major biomes: **tropical forest, savanna, desert, chaparral, temperate grassland, temperate broadleaf forest, coniferous forest,** and **tundra.** (If you want to go into more detail, you can include polar and high-mountain ice areas.)

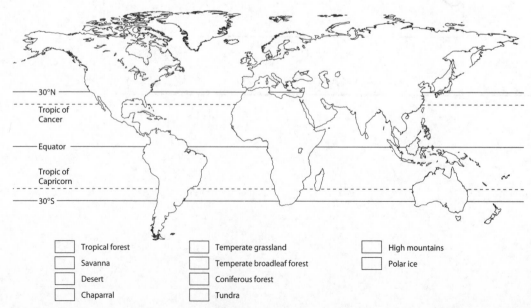

30°N

Tropic of Cancer

Equator

Tropic of Capricorn

30°S

☐ Tropical forest

☐ Savanna

☐ Desert

☐ Chaparral

☐ Temperate grassland

☐ Temperate broadleaf forest

☐ Coniferous forest

☐ Tundra

☐ High mountains

☐ Polar ice

Testing Your Knowledge

Multiple Choice

1. Which of the following is the most complex?
   a. community
   b. individual
   c. species
   d. ecosystem
   e. population

2. Change of seasons is caused by
   a. change in the distance of the Earth from the sun.
   b. the tilt of the Earth.
   c. variations in output of solar energy.
   d. changes in wind patterns.
   e. gravitational pull of the sun and moon.

3. The kind of tropical forest in a given area—thorn forest, deciduous forest, or rain forest—depends mostly on
   a. rainfall.
   b. average temperature.
   c. temperature difference between day and night.
   d. distance from the equator.
   e. prevalence of fire.

4. The tundra biome gets little precipitation, but during the short summer season it is very wet, with many marshy areas, ponds, and bogs. Why?
   a. Water cannot soak into the frozen ground.
   b. It does not rain much, but the tundra gets a lot of snow.
   c. The ground is so flat that the water cannot run downhill.
   d. It is flooded by melting snow from the surrounding mountains.
   e. The tundra does not look wet; it is a cold desert.

5. If you were to compare the source (beginning) of a river to its mouth (end), you would probably find that
   a. the water is colder near the mouth.
   b. there are more nutrients in the water near the source.
   c. there are more phytoplankton in the water near the mouth.
   d. the current is swifter near the mouth.
   e. silt accumulates near the source.

6. Which of the following occurs in areas a bit too warm for coniferous forest and a bit too dry for temperate forest?
   a. desert
   b. tropical thorn forest
   c. temperate grassland
   d. chaparral
   e. tundra

7. Which of the following is dominated by plants that drop their leaves in the winter to conserve water?
   a. desert
   b. coniferous forest
   c. tropical rain forest
   d. temperate broadleaf forest
   e. savanna

8. Winds tend to blow toward the equator because
   a. this is a very rainy area.
   b. equatorial air is heated by the sun and rises.
   c. air flows "downhill" from the North Pole.
   d. the Earth's rotational speed is fastest at the equator.
   e. the oceans are broadest at the equator.

9. Which of the following describes the climatic conditions characteristic of the coniferous forest?
   a. mild with occasional drought and fires
   b. cold and snowy
   c. hot or cold but always rainy
   d. cold and dry
   e. mild, rainy winters and hot, dry summers

10. The primary ecological factor determining the distribution of deserts is
    a. windiness.
    b. elevation.
    c. moisture.
    d. temperature.
    e. fertility of soil.

11. Which of the following is *not* an abiotic factor that shapes ecosystems?
    a. soil minerals
    b. predators
    c. fire
    d. rainfall
    e. storms

12. Which of the following would you expect to find in a rain shadow?
    a. a desert
    b. an ocean
    c. a rain forest
    d. a river system
    e. tundra

Essay

1. Ecology is the study of the interaction between organisms and their environment. We tend to think in terms of the environment affecting organisms. But interaction works both ways. Briefly describe two situations in which organisms (other than humans) change their environment.

2. In the post–World War II period (the 1940s and 1950s) there was a widespread belief in the United States that technology could provide a simple solution to many of our human problems. How and why has that attitude changed?

3. Explain why rain forests occur along the equator, while deserts are found 30 degrees north and south of the equator.

4. Which biomes in North America have been most affected by human activities? How have they been changed, and what kinds of activities have changed them?

5. Using two specific biomes as examples, explain the conditions that determine why a particular biome occurs in a specific area.

Applying Your Knowledge

Multiple Choice

1. Which of the following shows how biotic environmental factors can affect an organism?
   a. Mice have the highest reproductive rate of any common mammal.
   b. Maple trees will not grow in waterlogged soil.
   c. Some shrubs grow only where forest fires scorch their seeds.
   d. Trout will not live in shallow, warm water.
   e. Monarch butterflies live only where there are milkweed plants for food.

2. An ecologist studying how cattle grazing affects population dynamics of native animals like pronghorns and prairie dogs is focusing on
   a. populations.
   b. the biome.
   c. the ecosystem.
   d. the community.
   e. individuals.

3. Which of these biomes has been most altered by human activities in North America?
   a. temperate broadleaf forest
   b. coniferous forest
   c. desert
   d. tundra
   e. None of the above has been altered much in North America.

4. Which of the following might eat phytoplankton?
   a. great white shark
   b. beaver
   c. seal
   d. shrimp
   e. seagull

5. Which of the following lists only abiotic environmental factors?
   a. food, temperature, fire, wind
   b. soil minerals, oxygen level, light, predators
   c. food, parasites, predators, competitors
   d. wind, temperature, soil minerals, light
   e. light, rainfall, food, temperature

6. In the Sierra foothills of California, between the flat valley floors and mountain forests, there is a zone of grassland with scattered oak trees. Although it is too narrow to show on our biome map, this "oak woodland" would probably be classed as an example of
   a. coniferous forest.
   b. savanna.
   c. temperate deciduous forest.
   d. chaparral.
   e. tropical deciduous forest.

7. Which of the following countries is *not* paired with a biome that occurs there?
   a. Ecuador—tropical forest
   b. Australia—desert
   c. Argentina—temperate grassland
   d. Panama—coniferous forest
   e. France—temperate broadleaf forest

8. Which of these biomes generally lies between tundra and temperate broadleaf forest?
   a. savanna
   b. coniferous forest
   c. desert
   d. tropical forest
   e. chaparral

9. Most _____ must live in both the benthic and photic zones.
   a. sharks
   b. seaweeds
   c. clams
   d. whales
   e. phytoplankton

10. Bob was reading the newspaper. "More people burned out of their homes in Southern California," he remarked. His wife, Barb, looked up from her reading. "That's a shame," she said. "But I guess you have to expect a fire once in a while if you live in the _____ ."
    a. temperate grassland
    b. tundra
    c. chaparral
    d. coniferous forest
    e. savanna

Essay

1. Name five abiotic factors that might be important to the life of a tree in a mountain forest. Name one biotic factor that might affect the tree. How does the tree adapt to changes in these factors on a day-to-day basis? How has the tree species become adapted to these factors over many generations?

2. In general, temperature is an important abiotic factor shaping biomes on land, but it is less important in the ocean. In the ocean, light is an important abiotic factor, but it is less important on land. Explain why these factors differ in importance on land and in the sea.

3. Sailing ships going from Europe to North America generally sailed south, then west, and on the way back to Europe first sailed north, then east. From the information given in Module 34.6, can you explain why?

4. Climatologists are becoming more and more certain that human activities that produce carbon dioxide and other greenhouse gases are creating a global warming trend. Predicting the effects of global warming on the climate and organisms in a particular place stretches even the most complex computer models. That said, in very general terms, what effect will global warming probably have on the distribution of biomes? Which might shrink? Which might grow? In which directions will they probably expand or shrink?

5. A marine biologist was asked to speak on the topic of "Feeding the Growing Human Population by Harvesting the Sea." She said, "We can't do it. The sea is about as productive as a desert. The photic zone and the nutrients in the benthic zone are too far apart in most places." Explain what she meant.

Extending Your Knowledge

1. Large-scale ecological patterns become much more obvious from the air. Next time you fly, look for interactions between organisms and their environment. Transitions from one biome to another are readily apparent. Also look for effects of large bodies of water, elevation, rain shadows, rock and soil type, and human activities.

2. When was the last time you visited the zoo? Zoos and aquariums are getting away from the old "one-of-each-kind-of-animal-in-a-cage" format to a much more ecological approach, with large groups of animals and plants in natural settings. Exhibits are very educational, with lots of information on the roles and interactions of the animals in their ecosystems. The zoo may be the only place you will ever get to see the creatures of the tundra, or a South American rain forest, or the African savanna. At most zoos and aquariums, you can work as a volunteer guide or "explainer," or even take part in research projects.

Behavioral Adaptations to the Environment 35

Animal behavior is one of the most accessible areas of biology. To see it, you don't need to dive to a coral reef or travel to the Serengeti. You don't even need a microscope. There are examples around you every day: A spider spins its intricate web. How does it know what to do? A robin sings its song from a fence post. Why does it sing, and why only from certain perches? A moth circles a lightbulb, but a cockroach scurries away. Why do they react differently? A cat pounces on a ball, and a dog perks up its ears at the sound of its master's voice. What kinds of behaviors are they born with? What kinds of behaviors are learned? These everyday examples are reminders of the interconnectedness of animals and their adaptations to their environments. By studying the chapter, you will gain some tools you can use to understand the animal behavior you see every day.

Organizing Your Knowledge

Exercise 1   (Modules 35.1 – 35.3)

We can approach any animal behavior in terms of proximate (How?) questions—those related to environmental stimuli and the mechanisms of an animal's response. ("How does the bee know where to go?") We can also approach behavior in terms of ultimate (Why?) questions—those related to natural selection, the behavior's contribution to reproductive success. ("Why does the bee do that? How does it help the bee survive and reproduce?") In addition, we can also ask to what extent a behavior is genetically programmed (innate), and to what extent the behavior is modified by experience (learned). For each behavior described below, briefly describe what appear to you to be its proximate and ultimate explanations, and to what extent you think the behavior seems to be genetic (innate) or modified by experience (learned).

1. A frog will automatically attempt to catch any small, moving, insect-sized object. It will starve if given live insects that cannot move.

2. Gulls that nest close together in large colonies are able to remember the intricate patterns of spots on their own eggs and return only their own eggs to the nest if they roll out or are removed.

3. On its first flight each morning, a bee circles the hive. After foraging, it can find its way back. If the hive is moved even a short distance after the "orientation flight," the bee will not be able to locate the hive when it returns.

4. Male red-winged blackbirds defend territories (with food and nest sites) and respond aggressively—with threats and attacks—to the red wing patches of other males. If a male's wing patches are painted over, it can enter territories of other males and is ignored by them.

5. A monkey can soon figure out how to unbolt a door and escape from a cage.

6. After it has been stung by a bee, a toad will not try to catch other bees.

7. Monarch butterflies breed in the United States, but they are basically a tropical species, unaccustomed to much cold weather. Monarchs that hatch from eggs in the United States unerringly fly to a small wintering ground in the mountains of Mexico.

8. Newly hatched chicks will run for cover if a cardboard cutout of a hawk "flies" over the barnyard.

Exercise 2 (Module 35.3)

An innate (genetically programmed) behavior often takes the form of an automatic, stereotyped response—a fixed action pattern—to a specific stimulus—called a sign stimulus. The following chart lists some examples of innate behaviors from Modules 35.2 and 35.3, and from the previous exercise. Fill in the blanks for species, sign stimulus, and fixed action pattern.

Species	Sign Stimulus	Fixed Action Pattern
Kittiwake chick	1.	2.
3.	Flying shape	4.
5.	Open mouth of chick	6.
7.	Red wing patches	8.
9.	10.	Catches insect with tongue
11.	12.	Smile
13.	Egg near nest	14.
15.	16.	Forms strong pair-bond with female vole

Exercise 3   (Modules 35.4 – 35.11)

Web/CD Thinking as a Scientist   *How Can Pillbug Responses to the Environment Be Tested?*

There are several kinds of learning, from simple to complex. State whether each of the following describes an instance of habituation, imprinting, spatial learning using a cognitive map, association, social learning, or problem-solving. Some examples are from the modules; others are new.

_____ 1. A bird pecks at a caterpillar and gets a beakful of stinging hairs. After a couple of such experiences, the bird leaves this kind of caterpillar alone.

_____ 2. A salmon is found stuck in a pipe, trying to return to a hatchery pond.

_____ 3. During a sensitive period following hatching, gulls learn to identify their own young and keep them separate from the chicks in nearby nests.

_____ 4. A hermit crab will at first withdraw into its shell if you tap on the glass of its aquarium. After a few taps it no longer responds.

_____ 5. A bird called a gray jay buries food in hundreds of locations, then months later returns unerringly to dig it up and eat it.

_____ 6. A chimp learns to reach bananas by either hitting them with a stick or stacking up boxes. When the bananas are placed too high for either the stick or the stacked boxes, the chimp uses both the boxes and the stick.

_____ 7. If a bird does not hear its species' song shortly after hatching, it will not be able to sing it later.

_____ 8. A dolphin learns the commands "Throw ball" and "Retrieve ring." The first time the trainer says, "Retrieve ball," the dolphin does it correctly.

_____ 9. A monarch butterfly hatches and grows up in Ohio, then finds its way alone to wintering grounds in Mexico.

_____ 10. A cat learns to come running into the kitchen when it hears the electric can opener.

_____ 11. A monkey washes dirt from his food. Other monkeys observe this and start washing their food, too.

_____ 12. At first, young ground squirrels dash into their burrows at any unusual sight or sound. Later they learn to run only from real danger.

_____ 13. When their mother was killed, ducklings on a farm learned to follow a tractor.

_____ 14. Bees circle before going out foraging, and remember landmarks to find their way back to the hive.

_____ 15. Young vervet monkeys see and hear their parents give a warning call when they see a snake. Next time the young monkeys give the warning call, too.

Exercise 4   (Modules 35.2 – 35.11)

Web/CD Thinking as a Scientist   *How Can Pillbug Responses to the Environment Be Tested?*

To help you better understand innate and learned behaviors, sketch a concept map on a separate sheet of paper. Constructing a concept map is discussed in Chapter 3, Exercise 10.

Web/CD Thinking as a Scientist   *How Can Pillbug Responses to the Environment Be Tested?*

How do biologists figure out whether a worm displays negative phototaxis or a mouse has a bioclock? These modules describe some of the ways in which behaviors enable animals to survive in their environments—avoiding predators, locating food, finding their way, and attracting mates. The modules also describe some experiments and observations that have given us evidence about the causes, mechanisms, and reproductive value of various animals' behaviors. The following scenarios describe some observations and hypotheses about animal behavior. Briefly describe the kind of observations or experiments that could be done to test each hypothesis. Some examples are from the chapter, and others are new.

1. Turtles seen off the coast of Brazil are thought to lay their eggs on Ascension Island, in the middle of the Atlantic Ocean.

2. A kind of flatworm lives under rocks in streams and ponds and comes out at night to feed. It is probably negatively phototactic.

3. A crab is most active at high tide and becomes inactive at low tide, roughly 6 hours later. It is thought that this is not due to an internal "bioclock" but rather that the crab can somehow sense the tidal pull of the moon.

4. Homing pigeons travel even on cloudy nights. It is thought that they do not need to use vision to find their way to the vicinity of their home.

5. Birds become restless in the fall. It is thought that an environmental change other than a drop in temperature makes them ready to migrate.

6. The kangaroo rat is a nocturnal desert rodent. It may possess an internal bioclock that signals it when to emerge from its burrow.

7. Fruit flies don't live long enough to learn much. Their courtship "dance" seems to be innate.

8. Hummingbirds may find it "worth their while" to fight more fiercely over a patch of flowers when the flowers contain more nectar (food).

Exercise 6   (Modules 35.12 – 35.21)

Web/CD Activity 37A   *Honey Bee Waggle Dance*

Review foraging, mating, and social behaviors by filling in the blanks in the following story.

The gray wolf is the top predator of forests, grasslands, and tundra of North America. Wolves are social animals, cooperating in packs to hunt large mammals such as deer, moose, and caribou. Thus they are neither food [1]_____ nor "generalists" but somewhere in between. A typical wolf pack consists of 5 to 20 animals, led by a [2]_____ "alpha" male and his mate. Unlike lone hunters such as the cougar, wolves do not ambush their prey. The pack tracks an animal over long distances, finally exhausting it, attacking from all directions, and eventually dragging their victim down for the kill. The wolf pack is successful in about one chase in ten. Deer often outrun the pack, and moose often fight wolves off.

In the summer, when larger prey are harder to corner, wolves often change their [3]_____ and seek out smaller animals, such as beaver and hares. These smaller prey are easier to catch, and when wolves are alone, they must pursue smaller prey. Why then do wolves work in a pack to go after (usually unsuccessfully) prey sometimes much larger than they are, only to share their kill with other members of the pack? Apparently it is more [4]_____ to cooperate in a pack and tackle large prey than to catch smaller prey working alone. This is an example of [5]_____— feeding behavior that provides maximum energy gain for minimal energy expense.

The wolf pack constantly roams and defends a [6]_____ of from 100 to 1000 square kilometers. Like cheetahs and many other large territorial mammals, wolves use [7]_____ markers to stake their territorial claim. When on the move, they stop to urinate every few minutes. The territory assures a [8]_____ supply for the wolf pack and a safe place to raise their [9]_____.

The dominant feature of social behavior within the pack is a [10]_____ hierarchy. The alpha male leads the pack in chasing prey and in territorial defense. A dominance hierarchy may help to [11]_____ the wolf population. By allocating resources unequally, the hierarchy ensures that at least some of the wolves will receive an adequate food supply. It also minimizes the amount of energy spent in [12]_____ within the pack, energy that could more efficiently be used in hunting, defense, and reproduction.

An essential feature of social behavior is [13]_____, the exchange of information between individuals. Wolves use a varied repertoire of barks, yips, whines, and howls, similar to those of dogs. Body postures signal status in the pack hierarchy. The ears and tail of the alpha male are held erect. An ears-back, tail-down posture signifies submissiveness and lower rank. Thus wolves use three major channels of communication: [14]_____ signals to mark territory, and [15]_____ displays and [16]_____ within the pack.

The mating season, from late fall to late winter, is a time of turmoil for the wolf pack. In most cases, the dominant male and female are the only individuals in the pack who will mate, and only with each other; they are therefore [17]_____, maintaining a [18]_____ that may last for years. Keeping other individuals from mating increases certainty of the dominant male's [19]_____. During mating season, there is an increase in [20]_____ behavior, both threats and actual fighting, as wolves jockey for position in the hierarchy. Much of the behavior is in the form of [21]_____—mostly growls and posturing. Even if the dominant male and female are not challenged, lower-ranking individuals may test each other, seeking an opportunity to move up in rank.

Unlike many species that engage in a lengthy [22]_____ ritual, mating between the dominant male and female in the pack is performed with the minimum of fuss required to neutralize [23]_____ behavior. The pair may mate several times a day over a 2-week period. In the spring, the female gives birth to four to seven pups. They are given much food and attention by all members of the pack, not just their parents. This helping or self-sacrificing behavior is an example of [24]_____. Many of the individuals in the pack are aunts, uncles, even siblings of the pups. Their caring behavior could be explained in terms of [25]_____ selection. Even though most of the adults in the pack do not get to mate, their sharing and helping improve the chances that copies of their [26]_____ will be passed on. The pups nuzzle and lick the muzzles of adults, a signal that causes the adults to regurgitate food. (This behavior is also seen between adults, even those who are not related. This might be [27]_____ altruism, where the favor of food sharing may be repaid later by the beneficiary.) In 3 to 5 months, the cubs are able to travel with the pack. Later some will leave the pack, but most stay. Thus, social ties in the wolf pack are also family ties.

Exercise 7   (Module 35.20)

Web/CD Activity 37A   *Honey Bee Waggle Dance*

There are three main channels of animal communication: chemicals (scents), movements (visual displays or touch), and sounds (vocalizations, etc.). Communication in a number of species is described in several of the modules in this chapter. A list of species is given below. Most are from the chapter; you are probably familiar with most of the others. Place the name of each species on the chart to show how you think that species communicates. For example, grasshoppers communicate almost totally by means of sound; "grasshopper" goes in the "sounds" corner. House cats use sounds, movements, and scents; they are placed near the center of the triangle, roughly equally distant from the names of all three communication channels. Add the following to the chart: **honeybee, ring-tailed lemur, wolf, cricket, human, domestic dog, common loon, frog,** and **white-crowned sparrow.** It is not important in this exercise for you to be sure of the "right" answer. The important thing is to think about the variety of forms of animal communication.

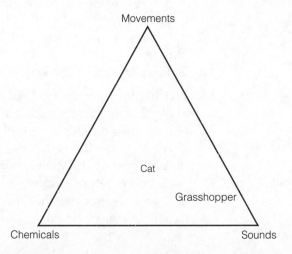

Exercise 8   (Modules 35.11 – 35.23)

Web/CD Activity 37A   *Honey Bee Waggle Dance*

Review some of the vocabulary of animal behavior by matching each of the phrases on the left with a term from the list on the right. Each answer is used once.

____ 1. Study of the evolutionary basis of social behavior
____ 2. Signaling or exchange of information
____ 3. A behavior that communicates, changing behavior of another animal
____ 4. May reinforce inherited social behavior in humans
____ 5. When self-sacrifice of relatives leads to altruism
____ 6. Helps an animal look for food
____ 7. Most efficient feeding behavior
____ 8. Help or sacrifice repaid later
____ 9. Minimizes agonistic behavior before mating
____ 10. Interaction between individuals, usually of same species
____ 11. Area defended against members of same species
____ 12. Study of the biological basis of human behavior
____ 13. May be promiscuous, monogamous, or polygamous
____ 14. The ability to store, process, and use sensory information
____ 15. Ranking by social interactions
____ 16. Self-sacrifice
____ 17. Threats or combat
____ 18. A signal that resolves conflict

A. Reciprocal altruism
B. Optimal foraging
C. Sociobiology
D. Kin selection
E. Courtship ritual
F. Social behavior
G. Evolutionary psychology
H. Territory
I. Agonistic behavior
J. Dominance hierarchy
K. Reconciliation behavior
L. Search image
M. Communication
N. Cognition
O. Culture
P. Altruism
Q. Signal
R. Mating system

Testing Your Knowledge

Multiple Choice

1. Which of the following *incorrectly* pairs an animal with a signal it might use to communicate?
    a. Ring-tailed lemur—scent
    b. Chimp—vocalization
    c. Common loon—scent
    d. Bee—movement
    e. Sparrow—vocalization

2. Scientists think that some birds may find their way during migration by
    a. imprinting on faint odors.
    b. listening to high-pitched sounds.
    c. observing the stars.
    d. seeing wavelengths of light that we cannot see.
    e. sensing the gravitational pull of the sun and moon.

3. Kineses and taxes are simple, automatic movements in response to stimuli. Movement of a moth toward a light is positive phototaxis. What is the difference between kineses and taxes?
    a. Taxes are directed toward a stimulus, kineses away.
    b. Kineses are innate, and taxes are learned.
    c. Kineses are more random, and taxes are more directed.
    d. Kineses are performed much faster than taxes.
    e. Taxes are innate, and kineses are learned.

4. Which of the following is a fixed action pattern?
    a. A goose rolls an egg back into its nest.
    b. A sparrow defends its territory.
    c. A wolf tracks its prey.
    d. A robin eats a distasteful bug, spits it out, and never eats one again.
    e. A sea gull returns to the same island breeding grounds each year.

5. Ducklings learn to follow their mothers. This is an example of
   a. habituation.
   b. problem solving.
   c. altruism.
   d. imprinting.
   e. kin selection.

6. Which of the following appears to be the primary method by which worker bees communicate the location of nectar to other members of the colony?
   a. They secrete a particular combination of chemical signals.
   b. They produce complex, high-pitched sounds.
   c. They fly to the nectar, and other bees follow.
   d. They perform a sequence of movements in the hive.
   e. They lay down a chemical trail to the source of food.

7. Birds sing, squirrels chatter, and wolves urinate on bushes to
   a. warn other members of the group of danger.
   b. mark their territories.
   c. attract mates.
   d. assert dominance.
   e. defend themselves from predators.

8. In _____, an animal learns to associate one of its behaviors with reward or punishment.
   a. rheotaxis
   b. trial-and-error learning
   c. fixed-action-pattern learning
   d. habituation
   e. imprinting

9. The ultimate question about any animal behavior is
   a. "How much does it cost?"
   b. "How does it increase reproductive success?"
   c. "How does an animal do it?"
   d. "What is the stimulus that triggers it?"
   e. "What's in it for me?"

10. When animals engage in _____, they often perform displays that make them look as large and dangerous as possible.
    a. courtship rituals
    b. altruism
    c. kin selection
    d. kineses
    e. agonistic behavior

11. The key idea of sociobiology is that
    a. human social behavior is the same as animal social behavior.
    b. social behavior has an evolutionary basis.
    c. little behavior is inherited; most is learned in social situations.
    d. almost all human and animal behavior is preprogrammed.
    e. animal behavior and human behavior are so different they cannot be compared.

12. Some animals carry out their mating rituals one-on-one, while others congregate to choose mates. What advantage might there be in animals carrying out such a *group* mating ritual?
    a. It reduces aggression.
    b. It prevents individuals from accidentally mating with relatives.
    c. It saves energy.
    d. It enables individuals to compare potential partners.
    e. A large group attracts more attention.

Essay

1. Describe the functions of imprinting in the lives of birds. How does it increase fitness?

2. What are the advantages of a dominance hierarchy to a high-ranking individual? What are the advantages of the hierarchy to the group?

3. Describe four functions of courtship rituals.

4. Explain why communication is essential to social behavior, and give five examples of animal communication.

5. Describe two methods birds use to find their way during long-distance migrations.

6. What are the advantages of fixed action patterns over learned behavior? In what situations do fixed action patterns seem to be especially important?

7. Name three species of animals that cooperate in social groups, and give three general benefits or advantages of social cooperation in these animals.

Applying Your Knowledge

Multiple Choice

1. A frog may at first be startled by tree branches swaying in the wind, but it soon gets used to these kinds of unimportant changes in its environment. This is an example of
   a. a fixed action pattern.
   b. imprinting.
   c. altruism.
   d. habituation.
   e. trial-and-error learning.

2. Bees can see colors we cannot see, and they can detect minute amounts of chemicals we cannot sense. But unlike many insects, bees cannot hear very well. A behavioral biologist would probably give which of the following as the ultimate explanation of their poor hearing?
   a. Bees are too small to have functional ears.
   b. Good hearing must not contribute much to a bee's reproductive success.
   c. If a bee could hear, its tiny brain would be swamped with information.
   d. This is an example of altruism.
   e. If bees could hear, the noise of the hive would drive them crazy.

3. Chimpanzees usually maintain their rank in the dominance hierarchy with charging displays. Sometimes actual combat occurs briefly, but usually aggression is ritualized. Why don't chimps fight more?
   a. Chimps are not strong enough to inflict much injury.
   b. Dominant females keep order and prevent fights from escalating.
   c. Natural selection favors avoiding injury and saving energy.
   d. Chimps spend most of their time mating; they don't have time to fight.
   e. Chimps are too hard-pressed finding food to spend much time fighting.

4. Western fence lizards are sometimes called bluebellies because the undersides of the males are bright blue. If one male invades another's territory, the territory owner protests vigorously, doing "push-ups" to display his belly and, if necessary, pushing the invader out. If an intruder's blue belly is first painted brown, he is completely ignored by the territory owner. The blue belly of the intruder apparently acts as a _____ , triggering _____ .
   a. search image . . . optimal foraging
   b. sign stimulus . . . altruism
   c. sign stimulus . . . agonistic behavior
   d. search image . . . altruism
   e. fixed action pattern . . . imprinting

5. An aquaculture facility hatched salmon eggs and released young fish into a river leading to the ocean. The fish fed and grew in the ocean, and in a few years they returned to the facility. Because the number of returning fish was low, a scientist suggested that the facility add a harmless chemical to the water draining from the fish ponds into the river. She hoped the chemical would increase the number of returning fish in a few years by
   a. decreasing their agonistic behavior.
   b. enabling them to imprint on the facility.
   c. stimulating habituation in the fish.
   d. improving the salmons' search image.
   e. promoting optimal foraging.

6. Every morning Brian turns on the light in the room and then feeds the fish in his aquarium. After a couple of weeks of this routine, he noticed that the fish came to the surface to feed as soon as he turned the light on, whether or not any food was present. This illustrates
   a. habituation.
   b. positive phototaxis.
   c. imprinting.
   d. association.
   e. spatial learning.

7. Wildlife biologists raising condors in captivity have found that it is important not to let condor chicks see their human "parents," so they feed the chicks with a puppet that looks like the head of an adult condor. This is to allow normal _____ in the chicks.
   a. imprinting
   b. habituation
   c. agonistic behavior
   d. social learning
   e. kin selection

8. Which of the following best illustrates optimal foraging?
   a. A robin will repeatedly attack any red object near its territory.
   b. Musk oxen will form a circle to fend off a wolf attack.
   c. Bats emerge to feed at about the same time each night.
   d. A blackbird will warn others in the flock if it senses danger.
   e. A sunbird will more fiercely defend flowers that produce more food.

9. Remembering facts that you read in a textbook would probably be considered
   a. optimal foraging.
   b. imprinting.
   c. spatial learning.
   d. habituation.
   e. social learning.

10. Which of the following sayings best summarizes the idea of reciprocal altruism?
    a. "A rolling stone gathers no moss."
    b. "You scratch my back, and I'll scratch yours."
    c. "A penny saved is a penny earned."
    d. "A bird in the hand is worth two in the bush."
    e. "Beauty is in the eye of the beholder."

Essay

1. Some animals, such as wolves and lions, hunt in groups, while others, like foxes and house cats, are loners. Every method of getting food involves tradeoffs; there is no one best way to do it. What do you think are the advantages of hunting in a group? The disadvantages?

2. Bats are the major predators of a species of moth. When one of these moths hears the high-pitched sound of a bat's sonar, it folds its wings and dives to the ground. How would a behavioral biologist describe the proximate cause of the moth's behavior? The ultimate cause?

3. European birds called wagtails eat insects. During the winter, when food is scarce, a wagtail may defend a territory, on which it captures an average of 20 insects per hour. Or the wagtail may join a flock that ranges widely over the countryside. A bird in a flock averages 25 insects per hour. If this is the case, why

don't the birds always travel in flocks? Discuss in terms of energy gain and expenditure.

4. Cancer patients often suffer nausea and loss of appetite as a side effect of the powerful drugs used in chemotherapy. They have a particularly tough time eating foods that they have in the past eaten just before drug treatment, even if the foods are subsequently given to them days or weeks after the effects of the drugs have worn off. Explain this "food aversion" in terms of your knowledge of learning in animals. How might this kind of learning be helpful to an animal?

5. Chickens are polygamous and sparrows are monogamous. Explain how these different mating systems might have been shaped by the needs of their young.

6. Many bird species use special alarm calls to signal the approach of danger. Biologists ask, "What possible advantage to a bird is there in warning other members of the flock? Isn't the bird that gives the warning call putting itself in danger? If so, how could this behavior possibly continue to exist? How could it improve a bird's fitness? Wouldn't the genes for warning other members of the flock be eliminated by natural selection?" The fact is, alarm calls and other forms of altruism—self-sacrifice—continue to exist. How might altruism improve the alarm-giver's fitness? List the two general kinds of explanations that biologists give for altruistic behavior.

7. Most people think babies are "cute." They like to hold them, play with them, and protect them. How might a sociobiologist explain this?

Extending Your Knowledge

1. Sometimes people ask, "Why are scientists wasting time and money studying animal behavior? Why aren't we putting our resources into research more relevant and useful to humans?" What do you think? Is animal behavior research relevant or useful? Can you cite specific examples?

2. Chimpanzees, ants, lions, and wolves sometimes attack other groups and kill individuals of their own species, but human beings are the

only animals that carry out "warfare" in an organized way on a mass scale. Does our knowledge of animal behavior offer any insight into human aggression and violence? Are humans more aggressive than other animals? Do other animals control their aggression in some way that we have left behind? What role does our technology play in the scale and ferocity of our wars? Or is knowledge of animal behavior irrelevant when it comes to the subject of war?

3. As mentioned in the introduction to this study guide chapter, animal behavior is one of the aspects of biology most easily observed in everyday life. Just for fun, keep a list of all the examples of animal behavior that you observe in a day. What kinds of behavior did you see? Which species were involved?

Population Dynamics

A marine biologist observes humpback whales in the Gulf of Alaska, zoologists band migrating shorebirds in Brazil, and a primatologist tracks orangutans in the forests of Borneo. Foresters map the spread of gypsy moths in a New England forest, a Peace Corps volunteer monitors crop pests in Kenya, wildlife biologists watch sea turtles lay their eggs on a Costa Rican beach, and United Nations public health workers compile statistics on the growth of the world's human population. The theme that unites all these activities is the study of population dynamics. We need to know where living things live; how numerous they are; the way they are distributed on land or in water; their birth rates, death rates, and life spans; the environmental conditions that affect them; and how they are adapted to these conditions. The many aspects of population dynamics are the subject of this chapter.

Organizing Your Knowledge

Exercise 1   (Modules 36.1 – 36.2)

Web/CD Activity 36A   *Techniques for Estimating Population Density and Size*

This exercise will allow you to work with the concepts of population density, dispersion pattern, and sampling. The map on the next page represents a meadow on the edge of the city of Mapleton. It is surrounded by developed and farmed land but has remained relatively undisturbed. Developers plan to build a subdivision that would cover the meadow. The Mapleton Open Space Alliance would like the meadow to remain as public open land. They note that the dwarf hawthorn, an uncommon shrub, is found in the meadow. It is considered a "sensitive species" by the state conservation department. The city council has asked for a construction delay until the status of the shrub is determined. You have been sent to determine the density of the hawthorn population in the meadow, as well as that of a deer mouse that may also be present. Use the map of hawthorn distribution on the next page for your survey, and answer the following questions.

      The area of the meadow is 16.8 hectares. (A hectare is a metric unit of area equal to about 2.2 acres, so the meadow totals about 37 acres.) This is too big an area to count every shrub, so you will have to look at sample plots. On the ground, this would be done with ropes and measuring tapes. You can choose random samples by merely dropping a penny on the map, drawing a circle around it, and counting the "shrubs" inside. On the scale of the map, the area covered by a U.S. penny equals 0.2 hectare.

1. Take ten samples. How many hectares does this total? _____

2. What is the total number of shrubs in the ten samples? _____

3. What is the density of hawthorns in shrubs per hectare? _____

4. What is the total number of hawthorns in the meadow? _____

5. How could you make your count more accurate? Why not do this?

6. Look at the map again. What is the pattern of dispersion of the shrubs? What might cause this pattern of dispersion?

You would also like to know the number of deer mice in the meadow. For this, it will probably work best to use the mark-recapture method. To learn about this method, you will have to look at Web/CD Activity 36A: Techniques for Estimating Population Density and Size.

7. Why does the mark-recapture method work better for mice than the method used to count the plants?

8. One night, you trap 40 mice, mark them, and let them go. Two nights later, you again trap 40 mice, and ten of them are marked. What is the total number of mice in the meadow? _____

9. What is the population density of mice in the meadow, in animals per hectare?

    _____

10. What do you have to assume about the mice and your method for your results to be valid? Could you be wrong? Why or why not?

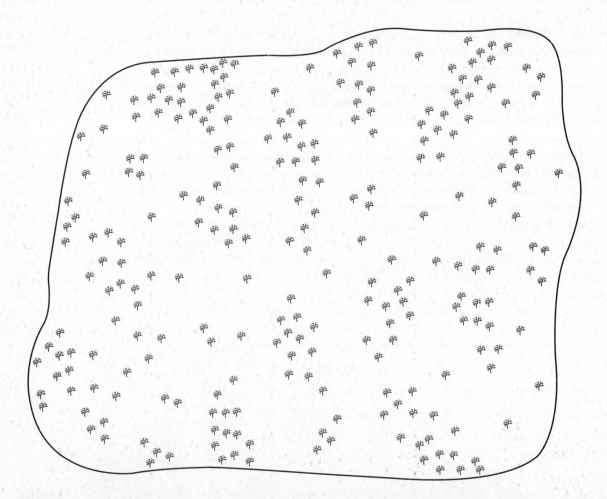

Exercise 2   (Module 36.3)

Web/CD Activity 36B   *Investigating Survivorship Curves*

Check your understanding of life tables and survivorship curves by matching each phrase on the right with a term on the left. Answers may be used more than once.

A. Life table
B. Survivorship curve
C. Type I survivorship
D. Type II survivorship
E. Type III survivorship

_____ 1. Graph of percent alive at the end of each age interval
_____ 2. Tabulation of deaths and chance of surviving
_____ 3. Most young die, but a few live to old age
_____ 4. Originally used to set life insurance rates
_____ 5. Characteristic of oysters
_____ 6. Death rate constant over life span
_____ 7. Characteristic of lizards and squirrels
_____ 8. Most offspring live a long life and die of old age
_____ 9. Characteristic of humans and many other large mammals

Exercise 3   (Module 36.4)

Models devised by ecologists describe two kinds of population growth. Exponential growth is described by this equation: $G = rN$. The rate of growth, G, depends on N, the size of the population, multiplied by r, the population's intrinsic rate of increase. Intrinsic rate of increase, r, is calculated by subtracting the death rate from the birth rate. Exponential growth is unregulated. The bigger the population, the faster it grows. This cannot be sustained for long in real populations, but it is interesting as a theoretical possibility. Populations of fast reproducers like bacteria and insects can grow at near-exponential rates for short periods.

Let's calculate and graph the exponential growth of a population of aphids for which $r = 40\%$ per week. Remember that $G = rN$. If there are 10 aphids to start with, the number of aphids added by the end of the first week (G) is equal to rN, or 0.4×10, which equals 4. So the total population (N) after one week is $10 + 4 = 14$.

1. Starting with the new total (N) of 14, how many aphids will be added (G) in the second week? _____ (Round off fractions.)

2. What will the total population (N) be at the end of the second week? _____

3. Aphids added in the third week? _____ Total after third week? _____

4. Aphids added in the fourth week? _____ Total? _____

5. Total at end of the fifth week? _____ sixth week? _____ seventh week? _____ eighth week? _____ ninth week? _____ tenth week? _____

6. Graph the size of the aphid population *(N)* versus time (in weeks) below.
   Population size was 10 at time = 0. Label the axes of the graph.

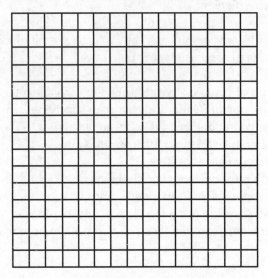

7. How would you describe the shape of this graph?

8. Could this kind of growth continue indefinitely? Why or why not?

Real environments will not support exponential growth. Populations are limited by space, food supply, or other factors that slow growth. The population may level off at a density the environment can maintain—carrying capacity. This so-called logistic growth is described by the equation $G = rN(K - N)/K$, where K represents carrying capacity.

The following data chart the growth of a population of deer on a small protected island off the coast of British Columbia, recorded over a 50-year period:

1945	92	1975	814
1950	151	1980	765
1955	295	1985	688
1960	603	1990	740
1965	861	1995	729
1970	920	2000	738

9. Graph the growth of the deer population below. Label the axes. How would you describe the overall shape of the graph?

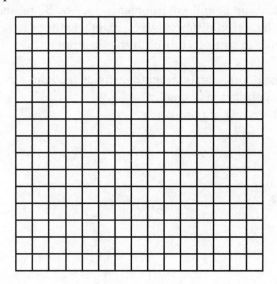

10. What happened to the population during the 1970s?

11. What may have caused the population density to level off?

12. What is your estimate of the carrying capacity of the island for deer?

13. What is the biological term for this kind of population growth?

Exercise 4   (Modules 36.5 – 36.6)

Population growth is limited by both biotic and abiotic environmental factors. Biotic and abiotic factors affect birth rates and death rates in different ways. State whether each of the following words or phrases relates more to biotic factors or to abiotic factors.

_____ 1. Have more effect when the population is larger (density dependent)
_____ 2. Have less effect when the population is smaller
_____ 3. Effect does not depend on density of population
_____ 4. Competition for food
_____ 5. Fire
_____ 6. Predation
_____ 7. Stress produced by crowding
_____ 8. Competition for nest sites
_____ 9. Storms
_____ 10. Drought
_____ 11. Disease
_____ 12. Heat and cold
_____ 13. Habitat disruption by humans
_____ 14. Cause populations to stabilize in size, presumably near carrying capacity
_____ 15. Cause rapid population growth followed by unpredictable crashes
_____ 16. Seem to cause boom and bust cycles among predators and prey
_____ 17. Limit the snowshoe hare population
_____ 18. Limit the lynx population
_____ 19. Effects of the nonliving environment
_____ 20. Effects of other organisms
_____ 21. Reduce clutch size as song sparrow population grows
_____ 22. Responsible for lemming boom and bust population cycles

Exercise 5   (Module 36.7)

Natural selection shapes different life history traits under different environmental conditions. Some populations exhibit *r*-selection, and others *K*-selection. Compare these contrasting life histories by completing this chart.

Characteristic	*r-Selection*	*K-Selection*
Life history emphasis	1.	Stability near carrying capacity
Relative body size	2.	3.
Number of offspring per reproduction	4.	5.
Age at first reproduction	Younger	6.
Emphasis on ____ of offspring	7.	Quality and care
Examples	8.	9.

Exercise 6   (Modules 36.8 – 36.10)

Web/CD Activity 36C   *Human Population Growth*
Web/CD Activity 36D   *Analyzing Age-Structure Diagrams*

Review the dynamics of human population growth by filling in the blanks in the following essay.

The world's human population now totals more than [1]_____ billion people. We are the most numerous large animal on Earth, and our huge numbers are a result of thousands of years of accelerating population growth, a phenomenon unprecedented in the history of life.

The human population grew relatively slowly and steadily until about 1650, when there were about [2]_____ people on earth. There was even a decline in the human population around 1450 when the [3]_____ wiped out about a fourth of the people of Europe. After that, population growth accelerated. The population doubled to 1 billion by 1850, again to 2 billion by 1930, and again to 4 billion by 1975. At current growth rates, world population could reach more than [4]_____billion by the year 2025.

Population growth rate depends on [5]_____ rate and [6]_____ rate. ZPG, or [7]_____, is when birth rate equals death rate. This can occur when a high birth rate is balanced by a high death rate. ZPG also occurs when a low birth rate is balanced by a low death rate. Such is the case in most [8]_____ countries, such as Sweden, where populations are stable or even declining.

Through most of human history, parents tended to have [9]_____ families, but few children survived to adulthood, so the population grew slowly. Improved nutrition, health care, and sanitation, brought about during the Industrial Revolution, caused a decrease in the death rate without a corresponding drop in birth rate, so overall human population growth accelerated. Whereas the developed countries have moved through a demographic [10]_____ to both low birth and death rates, most [11]_____ countries are still in that process. In these nations, such as Mexico and Afghanistan, the drop in [12]_____ has lagged behind the drop in [13]_____, so their populations continue to grow rapidly. Because most people live in the developing countries, this is the biggest factor in world population growth. In fact, of the next billion people added to the global human population, [14]_____% will be born in the developing countries!

The [15]_____ structure of a population—the proportion of individuals in different age groups—can help us predict growth trends and social changes. The age structure of a [16]_____ nation shows an even distribution among age groups. Each couple averages [17]_____ children (or less), just enough to replace themselves, and the population is stable. But even if the population is at ZPG, there can be problems. For example, in Italy and the United States, a relatively small number of younger workers will have to pay for [18]_____and [19]_____ for a large number of senior citizens. The situation is different in a developing country such as Afghanistan. There the age structure is [20]_____. Couples tend to have large families, and each generation of children outnumbers their parents. At a growth rate of 2.4%, the population of Afghanistan could double in 30 years! Such countries are pressed to increase [21]_____ production, expand [22]_____, and deal with many young [23]_____ workers.

Just to accommodate all the people expected to live on the Earth by 2025, we will have to ^{24}_____ food production. But we have already strained the Earth's resources and severely impacted our environment. Technology allows us to grow more food, but many people are starving and undernourished. Agricultural lands are depleted. Fresh ^{25}_____ for drinking and irrigation is becoming scarce. Pesticides pollute soil and water, and kill not only pests but their natural ^{26}_____. Livestock exceed the ^{27}_____ of grazing lands, and ocean ^{28}_____ are depleted.

Though populations in the developing world are growing much ^{29}_____ than those of the developed countries, the per capita environmental impact of developed countries is vastly greater. The ^{30}_____—the amount of land needed to support an individual's demands on the environment—is one way to measure and compare our ecological impact. An American has an ecological footprint of about 8.4 hectares, while a resident of India gets by on about 0.8 hectares. For many developed nations, and the world as a whole, the ecological footprint ^{31}_____ available resources. The world's richest countries, which account for only ^{32}_____ percent of global population, consume 86% of the world's resources. For example, the United States consumes more energy than Africa, Latin America, India, and China combined! A child born in the United States has an impact on the environment greater than 40 African children.

We can reduce our impact on the Earth through the practice of ^{33}_____ resource management—harvesting fish or trees without damaging the resource. We can minimize environmental and health risks by using natural ^{34}_____ controls to manage crop pests. A diet less reliant on ^{35}_____ would reduce impact on farmland. There is hope that social change may have a direct impact on explosive population growth in the developing world. As women's status and ^{36}_____ increase, they choose to practice ^{37}_____ control and limit family size. In the developed countries, people are becoming aware of the disproportionate impact they have on resources and the environment.

How many people can the Earth support? Some experts believe that we have already exceeded K, the earth's ^{38}_____. Others project that our planet could support 10 to 15 billion people, or even more. There is no question that the overall human birth rate will eventually come into line with the death rate, and that we will reach zero population growth. But the questions that remain are as follows: Will the balance come about by informed choice, or will it be imposed upon us? Will it occur through a decrease in the ^{39}_____ rate or an increase in the ^{40}_____ rate? Finally, what kind of life will there be in the future for all the people, and other species, who share the Earth?

Testing Your Knowledge

Multiple Choice

1. In wild populations, individuals most often show a _____ pattern of dispersion.
   a. random
   b. density-dependent
   c. equilibrial
   d. uniform
   e. clumped

2. A population would grow exponentially
   a. if it were limited only by biotic factors.
   b. until it reaches carrying capacity.
   c. if there were no limiting factors.
   d. if it were a *K*-selected population.
   e. if it showed logistic growth.

3. Which of the following would be most likely to have density-dependent effects on population growth?
   a. fire
   b. storms
   c. drought
   d. food supply
   e. cold

4. The effects of which of the following environmental factors would probably *not* change as a population grows?
   a. disease
   b. limited food supply
   c. competition for nesting sites
   d. weather
   e. predation

5. Human population growth was slow and gradual for a long period, but it turned sharply upward
   a. when death rates increased.
   b. when birth rates increased.
   c. when birth rates decreased.
   d. when death rates decreased.
   e. when the population reached carrying capacity.

6. A broad-based, pyramid-shaped age structure is characteristic of a population that is
   a. growing rapidly.
   b. at carrying capacity.
   c. stable.
   d. limited by biotic factors.
   e. shrinking.

7. A population grows rapidly at first and then levels off at carrying capacity if it is
   a. limited by biotic factors.
   b. limited by abiotic factors.
   c. an *r*-selected species.
   d. growing exponentially.
   e. characterized by uniform dispersion.

8. Which of the following is the most accurate comment on the Earth's carrying capacity (*K*) for people?
   a. *K* is smaller now than it was a thousand years ago.
   b. The human population is still a long way from *K*.
   c. Our technology allows us to increase *K*, but not indefinitely.
   d. When it comes to humans, the concept of *K* is irrelevant.
   e. The human population has already vastly exceeded *K*.

9. Which of the following would be true of a species whose life history is shaped by *r*-selection?
   a. Members of the species take a relatively long time to reach reproductive age.
   b. They are regulated mostly by density-dependent factors.
   c. They produce large numbers of offspring.
   d. The population usually stabilizes near carrying capacity.
   e. They give their young lots of care.

10. In the models that describe population growth, *r* stands for
    a. population density.
    b. a time interval.
    c. total number of individuals in the population.
    d. rate of increase.
    e. carrying capacity.

11. According to Figure 36.9B, the population of which of these countries is most in balance with its ecological capacities?
    a. USA
    b. Norway
    c. China
    d. India
    e. Canada

12. When needed resources are unevenly distributed, organisms often show a(n)_____ dispersion pattern.
    a. density-dependent d. random
    b. clumped e. uniform
    c. exponential

Essay

1. Describe the exponential and logistic population growth curves, explain their shapes, and explain why they are different.

2. Describe some factors that might affect carrying capacity.

3. Describe three biotic and three abiotic factors that might restrict population growth. Explain how the effects of these two kinds of factors on populations differ.

4. What has enabled the Earth's human population to continue to grow for a long period at near-exponential rates? How does this differ from other species?

5. What are population cycles? What kinds of hypotheses have been suggested to explain them?

6. Compare life histories shaped by *r*-selection and *K*-selection, and give examples of each.

Applying Your Knowledge

Multiple Choice

1. A particular species of tropical fish has only a few offspring and takes care of them for an extended period. We might also expect the fish population
   a. to be controlled mostly by abiotic factors.
   b. to show exponential growth.
   c. to live in a harsh environment.
   d. to start reproducing very young.
   e. to be relatively stable, near carrying capacity.

2. Gorillas have a relatively low birth rate. They take good care of their young, and most gorillas live a long life (if unmolested by humans!). The gorilla survivorship curve would look like
   a. a line that slopes gradually upward.
   b. a relatively flat line that drops steeply at the end.
   c. a line that drops steeply at first, then flattens out.
   d. a line that slopes gradually downward.
   e. a horizontal line.

3. Locust populations go through periods of sudden explosive growth, followed by a sudden decline in numbers. Their numbers are probably regulated by
   a. predation. d. random dispersion.
   b. biotic factors. e. abiotic factors.
   c. logistic growth.

4. Seagulls fiercely defend the areas around their nests in their cliff-top breeding colonies. Within the colony they would show a _____ dispersion pattern.
   a. uniform d. density-independent
   b. random e. clumped
   c. dense

5. A wildlife biologist is trying to predict what will happen to a bear population if bear hunting is banned. He had the equations all worked out but then realized that he had grossly underestimated the amount of food available to the bears. To make his predictions more accurate, he will have to go back to his equations and
   a. decrease *N*. d. increase *K*.
   b. increase *N*. e. decrease *r*.
   c. decrease *K*.

6. An ecologist would suspect a population is growing rapidly if it
   a. has a broad-based age structure.
   b. is near its carrying capacity.
   c. is limited only by density-dependent factors.
   d. shows a clumped pattern of dispersion.
   e. is far below its carrying capacity.

7. An oak tree produces thousands of acorns, but very few grow into mature oak trees. The oak tree exhibits a _____ survivorship curve.
   a. Type I
   b. Type II
   c. Type III
   d. Type I or II
   e. Type I or III

8. When birth rate equals death rate
   a. a population grows rapidly.
   b. the size of a population remains constant.
   c. biotic limiting factors do not affect the population.
   d. a population is in danger of extinction.
   e. a population goes through up-and-down cycles.

9. If you wanted to see what percentage of the population of Thailand is under 10 years old, you could look at
   a. a logistic curve for the population.
   b. the population's age structure.
   c. a life table for the population.
   d. a plot of population density.
   e. the population's survivorship curve.

10. To determine the density of a rabbit population, you would need to know the number of rabbits and
   a. the factors that limit population growth.
   b. their birth rate.
   c. the area in which they live.
   d. their population growth rate.
   e. their pattern of dispersion.

Essay

1. A fisheries biologist is interested in determining the population density of smallmouth bass in a lake. Using a net, she captures and tags 100 bass. A week later, she again catches 100 bass, and out of these fish, 5 are tagged. How many smallmouth bass are there in the lake?

2. Tim said, "It says here that Japan has a much greater population density than the United States." Carl replied, "That can't be right. There are a lot of people in Japan, but surely the population of the United States is much larger." What would you say to clear this up for them?

3. Biologists figure that the carrying capacity of a particular river is 100,000 salmon. They want to manage the catch of salmon so that the salmon population is at a size where replacement of the fish that are caught will happen at the fastest rate. Would you recommend that they catch fish until the salmon population stands at 80,000, or 50,000, or 10,000? Explain why. (Hint: They are assuming logistic growth of the salmon population.)

4. The temperature seldom dips below freezing along the Pacific coast, so robins can stay there all winter long. But during rare spells of subfreezing weather, the birds have a hard time finding food and shelter. Birds that find a warm place sheltered by evergreen branches have the best chance of making it through a cold snap, but there are only a limited number of such places. After a week of cold, only a fraction of the robin population may remain. Explain how both density-dependent and density-independent factors affect the robin population in this case.

5. Population explosions are often seen when animals are introduced into a new area. Examples include the spread of starlings across North America, a plague of rabbits in Australia, and the spread of cane toads in Florida. Why are these kinds of population explosions more likely in areas where the animals are not native?

6. The most intensive and rigorous studies of populations have been undertaken on islands and in lakes. Why are populations easier to study on islands or in lakes?

Extending Your Knowledge

1. If you bend a wire coat hanger into a rough square, the area enclosed by the square is roughly 0.05 square meter. You can use this "quadrat" to survey population densities of plants or slow-moving animals. You might toss it onto a lawn, for example, and count the plants inside to compare densities of weeds in the shade and in the sun, or before and after treatment with weed killer. How can you make it more likely that your sample quadrats are truly random samples?

2. Many ideas related to population dynamics are idealized, theoretical constructs that are difficult to see clearly in real populations. Some examples are logistic and exponential growth, and the difference between *r*-selected and *K*-selected populations. Are these ideas useful, even if real-world examples of them are hard to find? Or do they blind us to what is really going on in nature?

Communities and Ecosystems

37

In his classic book *The Closing Circle*, ecologist Barry Commoner explained what he saw as "the four laws of ecology": First, "Everything is connected to everything else." In a biological community, organisms are connected in numerous ways. They compete, they prey on and parasitize one another, and sometimes they even help each other through beneficial symbiotic relationships. Second, "Everything must go somewhere." Matter and energy cannot simply disappear. In an ecosystem, chemicals are recycled and energy from the sun escapes as heat. Third, "Nature knows best." So far, our efforts to alter or create ecosystems (through agriculture, logging, spraying pesticides) have produced unexpected results, and the systems have often functioned less well after our interference. Finally, "There is no such thing as a free lunch." Because all parts of an ecosystem are connected, everything must come from someplace; every gain is balanced by a loss. It will be helpful to keep these simple laws in mind as you study this chapter about the connections in communities and ecosystems.

Organizing Your Knowledge

Exercise 1 (Module 37.1)

This module introduces four major characteristics of biological communities. Each of the italicized phrases in the essay below relates to one of these four characteristics. State whether each phrase relates to biodiversity (B), prevalent form of vegetation (V), response to disturbance (D), or trophic structure (T), by writing the corresponding letter in the parentheses.

The tropical rain forest may hold a *greater variety of species* (^{1}_____), but no place on Earth can rival the great *conifer forests* (^{2}_____) of America's Pacific Northwest for the mass of living things they contain. The major tree species of these forests are *Douglas fir, Sitka spruce, western hemlock, western red cedar, and coastal redwood* (^{3}_____). The trees may be 2 or 3 meters in diameter and more than 60 meters tall. There are groves of redwoods whose tops are 100 meters above the forest floor. A ten-story building would fit comfortably beneath their lowest branches. Along streams and where the tall trees have fallen, a *tangle of bushes and smaller trees* (^{4}_____) reach upward for sunlight, but in the shade under the dense canopy the ground is covered mostly by a thick blanket of needles. Sunlight slants through the canopy and illuminates scattered shrubs and ferns.

The forest may seem quite silent on a warm summer day because much of its activity takes place high in the canopy or in the thick layer of litter on the ground. In fact, many animals spend their entire lives on the trunks and branches of tall Douglas firs. *As many as 1500 species of invertebrates* (^{5}_____)—insects, spiders, mites, and their relatives—have been counted on a single tree. A small rodent called the red tree vole never leaves the canopy, where *it feeds on the needles of Douglas fir* (^{6}_____). The fly-ing squirrel nests in hollow trees but glides to the forest floor *to eat lichens and fungi* (^{7}_____). Some of the fungi combine with tree roots and help trees absorb water and nutrients; the flying squirrel and other small mammals help to spread the spores of these fungi to new areas during their nocturnal feeding forays. Also in the soil and litter under the trees lives a *variety of insects* ^{8}_____ that may rival the insect life of the tropical rain forest canopy.

Many birds inhabit the forest. One of the most common is the Steller's jay, a black-and-dark-blue cousin of the eastern blue jay. Brown creepers and red-breasted nuthatches *pry insects from crevices in bark* (^{9}_____), while the pileated woodpecker, the largest commonly seen woodpecker, chisels ants from tree trunks. In all, this ancient forest is *home to more bird species than any other habitat north of Mexico* (^{10}_____).

Small mammals like tree voles and flying squirrels are *preyed on by carnivores such as the spotted owl* (^{11}_____). In fact, flying squirrels make up the bulk of a spotted owl's diet. Some other predators of this ancient forest are the northern goshawk and the pine marten, a relative of the weasel.

Periodic fires scar the trunks of the big trees. Their thick bark and the large amount of water stored in their trunks usually protect the trees, but perhaps *once every few centuries a stand is destroyed by fire* (^{12}_____). Landslides and floods also can wipe the slate clean. Plants from the surrounding forest gradually invade the clearing, and over a period of hundreds of years a stand of big trees may become reestablished. Natural disasters have given the natural forest a patchy appearance, with groves of an-cient trees adjoining meadows and stands of immature trees.

Now the biggest cause of change in the ancient forest is not fire or flood but the *disturbance caused by logging* (^{13}_____). The knot-free, fine-grained wood of the big trees is valuable, and the trees are clear-cut for lumber and plywood. Only about 4% of the ancient forests of the Pacific Northwest remain undisturbed. The planted forests that replace them may not have *the variety of plant and animal life seen in the natural old stands* (^{14}_____). *Species such as the spotted owl are endangered by the disappearance of the old trees* (^{15}_____) that they depend on for food and nesting sites. Many people are working to preserve undisturbed stands of ancient forest so that the organisms that live there can survive and so that we can continue to marvel at the most massive biological community on Earth.

Exercise 2   (Modules 37.2 – 37.6)

Web/CD Activity 37A   *Interspecific Interactions*

The structure of a community is shaped by interactions among the populations making up the community. The most important kinds of interactions are competition, predation, herbivory, and three kinds of symbiosis—parasitism, commensalism, and mutualism. State which of these five interactions is described in each of the examples below.

_____ 1. Small fish called remoras accompany sharks and dine on scraps left over when sharks feed.

_____ 2. Sheep liver flukes feed on bile and can weaken or kill their hosts. They are passed on to other sheep in the animals' droppings.

_____ 3. Grazing by introduced mountain goats has reduced the numbers of alpine wildflowers in Olympic National Park.

_____ 4. Pest-control specialists have brought in a destructive moth to eat tansy ragwort, a poisonous weed.

_____ 5. Mistletoe obtains nutrients from a tree host.

_____ 6. A small shrimp takes shelter inside a sponge, which is apparently unaffected by its tenant.

_____ 7. Mycorrhizal fungi associated with roots obtain carbohydrates from a tree, while enabling the tree to absorb water and minerals more efficiently.

_____ 8. In many parts of North America, the starling has displaced the bluebird from its nest sites.

_____ 9. A bee pollinates a tropical orchid by being tricked into "mating" with the flower; the bee uses a perfume from the flower to attract a mate.

_____ 10. The influenza virus attacks the lining of the respiratory tract and is passed from person to person by contact or airborne droplets.

_____ 11. Red-winged blackbirds arrive earlier on the breeding grounds but are forced to the edges of a marsh by larger, later-arriving yellow-headed blackbirds.

_____ 12. Lions hunt large herbivorous mammals such as zebras and wildebeest.

Exercise 3   (Modules 37.2 – 37.6)

Web/CD Activity 37A   *Interspecific Interactions*

You can think of an organism's ecological niche as its "role" or "job" in the community. The niche includes the sum of the organism's functions, abilities, and tolerances. It is possible to describe the niche as a sort of "job description" for a species, as you might see in a classified ad: "Applicant will be required to travel in herd, drink through nose, and knock down trees for food" could only describe the job of an elephant!

      Identify the organism whose niche is outlined in each of the following job descriptions. Some are from the text; others are not but will probably (like the elephant) be familiar to you.

_____ 1. "Ability to build trap to catch flying insects. May or may not need to devour mate."

_____ 2. "Ability to cling tightly to rocks, withstand wave action, and outlast competitors at low tide."

_____ 3. "Will be traveling and working outdoors in cold weather. Must have the patience to wait long periods to catch and eat seals through hole in ice. Some swimming ability and camouflage helpful."

_____ 4. "Will be stabilizing tide pool community through predation. Must like mussels."

_____ 5. "Must be able to withstand coastal storms and forest fires while maintaining species' reputation as world's tallest tree."

_____ 6. "Must live in South American rain forest, eat insects, and have poison glands and bright coloration."

_____ 7. "Will work closely with legume. Will be required to fix nitrogen in exchange for daily carbohydrate allowance."

_____ 8. "Important position as keystone predator, consuming sea urchins and evading killer whales."

Exercise 4   (Modules 37.7 – 37.8)

Web/CD Activity 37B   *Primary Succession*

Gradual transition in the species composition of a community that occurs after a disturbance is called ecological succession, described in Module 37.7. Module 37.8 describes how fire—normally considered a disturbance—is a common occurrence in some communities and has an important role in shaping and perpetuating those communities. In this exercise, first state whether each of the following represents a relatively early (E) stage in succession or a relatively late (L) stage. (Hint: Ask yourself if the community were left untouched, whether it would look the same or different in a hundred years. If it would look different, it is in an early stage—there are later stages to come.) Next, for those communities in an early stage, state whether those examples represent primary (P) or secondary (S) succession.

Early or Late?	Primary or Secondary?	
_____	_____	1. Lichen-covered rocks near a melting glacier in Alaska
_____	_____	2. The Pacific Northwest forest described in Exercise 1
_____	_____	3. A vacant lot near your college
_____	_____	4. An oak-maple-beech forest in Ohio
_____	_____	5. A lava flow on the island of Hawaii
_____	_____	6. A lawn in a suburb in New Jersey
_____	_____	7. A cornfield in Virginia

Exercise 5   (Modules 37.9 – 37.10)

Web/CD Activity 37C   *Food Webs*

The trophic structure of an ecosystem is the pattern of feeding relationships by which energy and chemicals flow through the system, from trophic level to trophic level. Name the trophic level of each of the organisms in the following description of a freshwater marsh food web: producer (P), primary consumer (1C), secondary consumer (2C), tertiary consumer (3C), quaternary consumer (4C), or detritivore (D). (Note that a consumer can function on more than one level, depending on what it eats.)

Marshes and other wetlands are among the most endangered of habitats. They are productive "nurseries" for many wildlife species, but many of our wetlands have been drained for agriculture or filled for development.

The freshwater marsh food web starts with plants like cattails, arrowleaf, and various floating or submerged "water weeds" (^{1}_____). They provide food for muskrats (^{2}_____) and mallard ducks (^{3}_____), both of which may in turn be eaten by hawks (^{4}_____) or mink (^{5}_____). Microscopic algae (^{6}_____) make much of the food in the marsh. Small shrimplike crustaceans (^{7}_____) and insect larvae (^{8}_____) graze on the algae. The insects are eaten by ducks (^{9}_____), frogs (^{10}_____), and sunfish (^{11}_____). A frog or sunfish might be eaten by a larger yellow perch (^{12}_____), a great blue heron (^{13}_____), a water snake (^{14}_____), or a mink (^{15}_____). The heron (^{16}_____) also eats perch and snakes, and the hawk (^{17}_____) will also occasionally devour a snake.

This is a highly simplified description of a marsh food web. There might be hundreds of species of large plants and animals making up the community of a marshy roadside pond, not to mention a swarm of microscopic creatures. In addition to this grazing food web, many inconspicuous worms, insect larvae, and snails (^{18}_____) get their food from the dead material produced by the other plants and animals of the marsh.

On a separate sheet of paper, sketch the marsh food web. Write the names of the organisms at their appropriate trophic levels, and connect the names with arrows. It is not necessary to include detritivores. (Where in your diagram should the producers be? The top consumers? Which way do the arrows point? Why?)

Exercise 6   (Module 37.12)

Web/CD Thinking as a Scientist   *How Do Temperature and Light Affect Primary Production?*

Sunlight, moisture, and nutrient availability determine how much food the producers of an ecosystem can make—primary production. After studying the bar graph, rank the following ecosystems in terms of primary production per square meter per year (highest pp = #1, lowest pp = #8).

_____ A. Desert   _____ B. Tropical rain forest   _____ C. Cornfield
_____ D. Beech/maple forest   _____ E. Open ocean   _____ F. Estuary
_____ G. Tundra   _____ H. Kelp bed

Exercise 7 (Modules 37.9 – 37.14)

Web/CD Activity 37C *Food Webs*
Web/CD Activity 37D *Energy Flow and Chemical Cycling*
Web/CD Activity 37E *Energy Pyramids*
Web/CD Thinking as a Scientist   *How Do Temperature and Light Affect Primary Production?*

This flowchart illustrates the movement of energy through an ecosystem. The boxes represent the total mass of organisms at each trophic level. The arrows show the amount of energy passing through each trophic level. Energy enters the producer level as sunlight. Some of this energy is stored in molecules produced in photosynthesis. Energy enters each of the consumer trophic levels when the consumers feed on the level below. Much of the energy in the food entering any level is used to power life processes; the food is used as fuel in cellular respiration, and its energy ends up as heat. Some energy is wasted; it is lost to the detritus food web in the form of dead leaves or droppings. A small portion of the energy is stored up in tissue when organisms grow or reproduce; this production—roughly 5–20% of energy intake at any trophic level—is the only energy available to the next level. Label and color the trophic levels on the diagram: **producers** (green), **primary consumers** (blue), **secondary consumers** (pink), and **tertiary consumers** (orange). Label and color the pattern of energy flow: **sunlight** (yellow), **production energy** (orange), **energy used in cellular respiration** (red), and **energy in wastes** (brown). (Note: Can you see from this diagram why quaternary consumers are quite rare?)

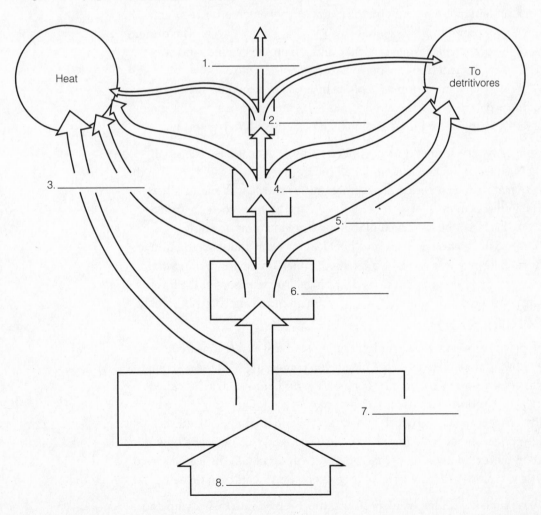

Exercise 8   (Modules 37.15 – 37.21)

Web/CD Activity 37F   *The Carbon Cycle*
Web/CD Activity 37G   *The Nitrogen Cycle*
Web/CD Activity 37H   *Water Pollution from Nitrates*

The biosphere receives a constant supply of energy from the sun, uses this energy for a while, then loses it to space as heat. Unlike energy, the chemicals necessary for life are present on Earth in fixed amounts, and these chemicals are used over and over. These chemicals, such as water, carbon, and nitrogen, occur in various forms and are changed from one form to another by various physical and chemical processes. The story below traces a nitrogen atom as it moves through the various reservoirs and processes of the nitrogen cycle. Fill in the blanks as you follow its journey.

The nitrogen atom, N, had been in the atmosphere for more than 2 years. It was paired with another identical nitrogen atom, forming a molecule of [1]_____ gas, which makes up about 80% of the air. This is by far the largest [2]_____ reservoir of nitrogen. During its time in the atmosphere, N had circled the Earth several times, from the skies over the Philippines, to Africa, to the Antarctic, to South America, and now over a sand dune in North Carolina. There it was captured by [3]_____ in a nodule on the root of a legume called a beach pea. There N was split away from its partner and combined with hydrogen atoms, eventually ending up in an amino group in a protein molecule built by the plant. This N entered (or reentered!) the [4]_____ reservoir of this global [5]_____ cycle. The protein was stored in one of the peas, which ripened, dried, and fell on the ground, and was eaten by a mouse. The amino acid from the pea was incorporated into a [6]_____ molecule in a leg muscle of the mouse. On a moonlit night a month or so later, a great horned owl caught the mouse, and N became part of a protein molecule in one of the owl's feathers. And there it remained until the following spring.

In the spring, as every spring, the owl molted some of its feathers. The small feather containing N ended up in the litter under a pine tree. [7]_____ in the soil broke down the feather over a period of several months, and N was eventually released into the soil, in the form of [8]_____, NH_4^+. Other bacteria, called [9]_____ bacteria, then attached N to three oxygen atoms, forming [10]_____, which was taken up by a huckleberry bush and used to build another protein. The huckleberry was eaten by a cardinal, which broke down the protein and eventually excreted N in the uric acid in its droppings. Decomposers in the forest soil changed this waste product to [11]_____, which nitrifying bacteria quickly converted into [12]_____.

When the patch of forest was cut for lumber, most of the nitrate ions in the soil were washed downhill. A flood deposited the nitrate ion containing N in the soil of a flat river valley. N was soon absorbed by the roots of a buttercup plant—a process called [13]_____. The buttercup used N to make an [14]_____, which was used to build a protein molecule, which ended up in a pollen grain. A beetle collected some of this pollen, and soon N was part of the insect's body. The beetle flew into a bog and landed on the leaf of a Venus flytrap, an insect-eating plant. The leaf snapped shut, and the beetle was slowly digested; the plant used N to make its own proteins.

About a year later, the rotting trunk of a dead tree toppled into the swamp and buried the Venus flytrap. As the plant decomposed, bacteria again incorporated N into ammonium and then nitrate. But this time, in the low-oxygen conditions of the mud, 15 _____ bacteria were able to break down the nitrate ions, releasing N to the air as part of an N_2 molecule. The wind swept the molecule out over the ocean, where . . .

You might find it informative and fun to continue the story from here. You might also find it helpful to make up similar stories for a carbon atom, a phosphorus atom, and a water molecule. This is a good review to do in a study group.

Exercise 9   (Modules 37.14 – 37.21)

Web/CD Activity 37F   *The Carbon Cycle*
Web/CD Activity 37G   *The Nitrogen Cycle*
Web/CD Activity 37H   *Water Pollution from Nitrates*

These modules contrast some processes that occur in natural, "healthy" ecosystems with processes that take place in "unhealthy" systems that have been damaged by human activities. Each of the statements below describes the situation in a healthy ecosystem. Briefly describe the corresponding situation in an unhealthy system affected by human disturbance.

1. The amount of CO_2 released into the atmosphere by respiration equals the CO_2 used in photosynthesis.

2. Forest organisms and soil contain a large store of mineral nutrients such as calcium and phosphorus, which are continuously recycled. A small amount of nutrients are lost each year, but this is balanced by the flow of nutrients into the ecosystem.

3. Over time, nutrients from the surrounding land gradually accumulate in a lake, and the lake becomes more productive, a process called eutrophication.

4. Organic fertilizers release nitrogen and phosphorus gradually, so they are absorbed by crop plants and do not run off to pollute rivers and lakes.

5. Much of the water that falls on tropical forests is returned to the atmosphere by transpiration.

6. Runoff from a cattle feed lot passes through treatment lagoons where microorganisms and plants recycle nutrients.

Testing Your Knowledge

Multiple Choice

1. All of the organisms in a particular area make up
   a. a food chain.
   b. a population.
   c. a community.
   d. a niche.
   e. an ecosystem.

2. When you eat an apple, you are a
   a. primary consumer.
   b. secondary producer.
   c. producer.
   d. secondary consumer.
   e. tertiary consumer.

3. An organism's "trophic level" refers to
   a. the rate at which it uses energy.
   b. where it lives.
   c. what it eats.
   d. whether it is early or late in ecological succession.
   e. the intensity of its competition with other species.

4. The relationship between species A and species B is described as commensalism. This means that
   a. both species suffer.
   b. one species benefits and the other species suffers.
   c. both species benefit.
   d. one species benefits and the other species is unaffected.
   e. any of the above is possible in commensalism.

5. Most plants get nitrogen from
   a. nitrates in the soil.
   b. N$_2$ gas in the air.
   c. proteins.
   d. ammonium in the soil.
   e. rainfall.

6. The energy for nearly every organism in nearly every ecosystem ultimately comes from
   a. minerals in the soil.
   b. the sun.
   c. heat from the earth.
   d. respiration.
   e. decomposition.

7. Why is a diagram of energy flow from trophic level to trophic level shaped like a pyramid?
   a. Organisms at each level store most of the energy and pass little on.
   b. There are more producers than primary consumers, and so on.
   c. Organisms eventually die as they get older.
   d. Most energy at each level is lost, leaving little for the next.
   e. There are always fewer secondary consumers than primary consumers, and so on.

8. The main decomposers in an ecosystem are
   a. plants and animals.
   b. bacteria and viruses.
   c. fungi and bacteria.
   d. bacteria and plants.
   e. plants and fungi.

9. Bacteria are especially important in
   a. the water cycle.
   b. the nitrogen cycle.
   c. ecological succession.
   d. the phosphorus cycle.
   e. recycling of energy.

10. The biggest difference between the flow of energy and the flow of chemical nutrients in an ecosystem is that
    a. the amount of energy is much greater than the amount of nutrients.
    b. energy is recycled, but nutrients are not.
    c. organisms always need nutrients, but they don't always need energy.
    d. nutrients are recycled, but energy is not.
    e. organisms always need energy, but they don't always need nutrients.

11. Which of the following is *not* a property of a community?
    a. prevalent vegetation
    b. biodiversity
    c. response to disturbance
    d. trophic structure
    e. density

12. The niche of an animal is
    a. the number of individuals of the species the environment will support.
    b. the same as its habitat.
    c. the way the animal fits into its environment.
    d. the specific place in the habitat where the animal lives.
    e. its position in the food chain.

13. Which of the following ranks three ecosystems, from highest to lowest, in terms of primary production per square meter?
    a. open ocean—temperate grassland—desert
    b. cultivated land—estuary—deciduous forest
    c. tropical rainforest—cultivated land—coral reef
    d. cultivated land—tropical rain forest—temperate grassland
    e. coral reef—deciduous forest—open ocean

Essay

1. Explain in ecological terms why a given area of farmland can support more people if they eat plants rather than meat.

2. What is decomposition? What are the two major kinds of decomposers in most ecosystems, and how is their role important to the ecosystem?

3. Describe an example of coevolution between predator and prey. Are there ecological relationships other than predator/prey interactions that are shaped by coevolution between two species? Give an example.

4. State whether each of the following is a producer, primary consumer, secondary (or higher) consumer, or detritivore: squirrel, oak tree, mosquito, great white shark, moose, cheetah, mushroom, spider, phytoplankton, grass, and vulture.

5. There is probably a river or stream near where you live. Describe the place in the global water cycle of that river or stream. Where did its water come from? Where is it going?

6. Trace a carbon atom through the carbon cycle. In what chemical form is carbon in the air? How does a carbon atom enter the food chain? In what chemical form might the carbon atom be obtained by a consumer? What chemical process would put the carbon atom back into the atmosphere?

Applying Your Knowledge

Multiple Choice

1. A lichen is actually composed of two organisms—a fungus and an alga. They depend on each other for survival. The most specific term that describes their relationship is
   a. parasitism.
   b. predation.
   c. commensalism.
   d. symbiosis.
   e. mutualism.

2. Which of the following describes mimicry?
   a. An insect's bright colors warn a predator that it tastes bad.
   b. The mottled pattern on a fish looks like dead leaves on the bottom of a pond.
   c. Two species of mice live in the same area and eat the same kinds of seeds.
   d. A harmless frog resembles a poisonous frog.
   e. Both kangaroo rats and jackrabbits hop erratically when escaping from predators.

3. When goats were introduced to an island off the California coast, the goats lived in the same areas and ate the same plants as the native deer. The deer population dwindled, and the deer finally disappeared. This is an example of
   a. commensalism.
   b. succession.
   c. a food chain.
   d. coevolution.
   e. competitive exclusion.

4. Suppose you wanted to establish a self-sustaining ecosystem by sealing some sterilized soil, water, and air in a glass container with a few organisms. You would be most likely to succeed with which of the following?
   a. aphids, bacteria, and spiders
   b. bacteria and ants
   c. clover, bacteria, and grasshoppers
   d. beetles, fungi, and bacteria
   e. spiders, grasshoppers, and grass

5. After clear-cutting, timber companies cannot afford to wait for the long process of _____ to occur naturally; they plant trees right away.
   a. mutualism
   b. succession
   c. coevolution
   d. decomposition
   e. competitive exclusion

6. In an ecosystem the _____ is always greater than the _____ .
   a. number of producers . . . number of primary consumers
   b. biomass of secondary consumers . . . biomass of producers
   c. energy used by primary consumers . . . energy used by secondary consumers
   d. biomass of producers . . . biomass of primary consumers
   e. energy used by primary consumers . . . energy used by producers

7. Under which of the following circumstances would interspecific competition be most obvious?
   a. when resources are most abundant
   b. in a stable, long-established ecosystem
   c. when organisms have quite different ecological niches
   d. among species whose trophic levels are different
   e. when a foreign organism is introduced to a community

8. Milkweed plants produce bad-tasting and poisonous compounds that deter most plant-eaters. But the caterpillars of monarch butterflies are able to eat milkweed leaves without being harmed. In fact, the chemicals obtained from milkweed actually protect the monarch from insect-eating birds. This example illustrates
   a. coevolution.
   b. competitive exclusion.
   c. mutualism.
   d. the effect of a keystone predator.
   e. succession.

9. Two species of birds called cuckoo doves live in a group of islands off the coast of New Guinea. Out of 33 islands, 14 have one species, 6 have the other, 13 have neither, and none has both. What might best explain this? The two species of birds could
   a. be on different trophic levels.
   b. have similar niches.
   c. be detritivores.
   d. have different niches.
   e. be keystone predators.

10. If you wanted to determine whether succession is occurring in a certain area, it might be useful to
    a. determine how many trophic levels are represented.
    b. diagram its food web.
    c. look at its species diversity.
    d. look at old pictures of the area.
    e. measure its productivity.

11. All of the following are found in the soil. Which category includes all the others?
    a. fungi
    b. scavengers
    c. detritivores
    d. bacteria

12. An ecosystem is unlikely to be limited by the supply of ____ because it is obtained from the air.
    a. water
    b. carbon
    c. phosphorus
    d. calcium
    e. nitrogen

13. Seven species of small insect-eating birds called warblers are found to live in and around a maple tree, but they don't compete much because one species tends to stay near the top of the tree, another on lower branches near the trunk, another near the ground under the tree, etc. What is probably going on here?
    a. One species is a keystone species.
    b. The birds occupy different trophic levels.
    c. Resource partitioning is occurring.
    d. Mutualism.
    e. Secondary succession is occurring.

Essay

1. A researcher noted that in many small ponds, fish species A preyed on several smaller species, B, C, D, E, and F. He suspects that A may be a keystone predator in the pond communities. What kind of experiment could you suggest to test whether this hypothesis is valid? If A is a keystone predator, how would you expect the experiment to turn out?

2. A forested area of a new subdivision has been set aside for a park. The developer cuts down most of the trees and plants a lawn. How would you expect the park to compare with the original forest, in terms of prevalent form of vegetation, diversity, trophic structure, and stability?

3. When Gause cultured *Paramecium aurelia* and *Paramecium caudatum* together, *P. caudatum* was invariably driven to extinction (Module 37.2). When *P. aurelia* and *P. bursaria* were cultured together in another experiment, however, both were able to coexist for months. What kinds of similarities and differences must there be among the three species? What do you think would happen if *P. caudatum* and *P. bursaria* were cultured together? Why?

4. Elmer Q. Retchfickle, a civil-defense expert, has developed a design for a fallout shelter to be used during a nuclear war. Elmer proposes to have the inhabitants of the shelter eat mushrooms, which would be grown in the dark on the body wastes of the residents of the shelter.

He thinks that this system would allow shelter residents to remain underground indefinitely. Aside from the questionable nutritional balance of an all-mushroom diet, do you see any problems with this plan from an energy standpoint? Explain.

5. An experimental fish farm consists of algae, crustaceans that eat algae, and fish that eat crustaceans. If the algae grow at a rate of 20 kilograms of algae per square meter of pond per year, how big will the pond have to be to harvest 1,000 kilograms of fish per year? (Assume that energy content is proportional to weight.)

6. In the mountains of California, Stanford ecologist Craig Heller found that in most areas the least chipmunk lived in sagebrush areas and the yellow pine chipmunk lived in higher areas of mixed sagebrush and piñon pines. If the yellow pine chipmunk was absent from an area, the least chipmunk lived in both the sagebrush and sagebrush-piñon pine zones. If the least chipmunk was absent, the yellow pine chipmunk distribution was unchanged. Explain this difference in terms of the concepts discussed in this chapter.

7. The push to develop and settle the Amazon Basin has resulted in the first extensive contact between many groups of indigenous peoples and the outside world. Many of these native people have died from diseases that are seldom fatal in the developed countries, such as measles, chicken pox, and mumps. What kind of coevolution must have occurred between the viruses causing these diseases and their hosts in the developed countries? What will probably occur in the Amazon?

Extending Your Knowledge

1. Imagine that you have been chosen as the biologist for the design team for Biosphere III, a self-contained habitat to be assembled in orbit. It will be stocked with organisms you choose, to form a self-sustaining ecosystem that will support you and three other people for 2 years. What are the main functions that you would expect to be performed by the organisms that you choose? What species would you select, and why? How many of each would you start with, and why?

2. Are you a vegetarian? Have you ever followed a vegetarian diet or considered following such a diet? Do you know anyone who is a vegetarian? What are some advantages and disadvantages of a vegetarian diet? What are some reasons for deciding to be a vegetarian? Are there ecological reasons? What are they?

Conservation Biology

If you need a reminder that biology is relevant to our lives, you only need to look at the daily newspaper headlines relating to the environment: "Study Warns of Rapid Artic Warming," "Tigers Disappearing from the Wild," "Must We Choose Between Dams and Salmon?," "Oil Spill Threatens Coast," "Migratory Bird Numbers Down." While our knowledge of biology grows at an exponential rate, our impact on the biosphere grows even faster. We face a biodiversity crisis of unprecedented proportions, brought about by the growth of the human population and its impact on the living world—destruction of habitat, overexploitation of species, pollution, climate change. We are the most intelligent species on this planet, and therein lies both danger and promise. We are the only species powerful enough to damage the biosphere as much as we have, and also the only species smart enough to fix it. This chapter outlines the dimensions of the biodiversity crisis and the methods we must use to understand, restore, and sustain the biosphere.

Organizing Your Knowledge

Exercise 1 (Introduction, Modules 38.1 – 38.5)

Web/CD Activity 38A   *Connection: Madagascar and the Biodiversity Crisis*
Web/CD Activity 38B   *Connection: Fire Ants as an Exotic Species*
Web/CD Activity 38C   *Connection: DDT and the Environment*
Web/CD Activity 38D   *Connection: The Greenhouse Effect*

These modules outline the extent of the global biodiversity crisis. Review some of the concepts discussed by matching each of the species on the right with its part in the story.

_____ 1. Introduced predator that wiped out most native fish in Lake Victoria

_____ 2. Population drastically reduced by commercial fishing

_____ 3. Exotic species that has replaced native birds in parts of North America

_____ 4. Introduced vine that has become a pest in the southern U.S.

_____ 5. Eggs of this species contaminated by biomagnification of PCBs

_____ 6. A tropical plant valuable in cancer treatment

_____ 7. Predatory mammal now protected in Myanmar

_____ 8. Sought by "bioprospectors" in Yellowstone Park

_____ 9. Starving as global warming shortens its hunting season

_____ 10. May be harmed by UV that gets through ozone "hole"

_____ 11. Home to a third of fish species

_____ 12. Exotic species that has invaded the Great Lakes

_____ 13. Large mammal poached for its horn

_____ 14. Responsible for the current biodiversity crisis

A. Herring gull
B. Rhinoceros
C. Tiger
D. Rosy periwinkle
E. Hot springs prokaryotes
F. House sparrow
G. Phytoplankton
H. Bluefin tuna
I. Kudzu
J. Polar bear
K. *Homo sapiens*
L. Nile perch
M. Zebra mussel
N. Coral

Exercise 2   (Modules 38.1 – 38.5)

Web/CD Activity 38A   *Connection: Madagascar and the Biodiversity Crisis*
Web/CD Activity 38B   *Connection: Fire Ants as an Exotic Species*
Web/CD Activity 38C   *Connection: DDT and the Environment*
Web/CD Activity 38D   *Connection: The Greenhouse Effect*

Review some current threats to biodiversity—and their possible solutions—by filling in the remaining blanks in this chart.

Problem/Cause	*Consequences*	*Solutions*
Sulfur and nitrogen pollutants in the air	1.	2.
3.	Global warming, climate change, flooding, habitat loss	4.
5.	Decline in population of large mammals—tigers, whales, rhinos	6.
7.	Death of top predators—gulls, ospreys, eagles, etc.	8.
Destruction of tropical forests	9.	10.
11.	12.	Reduce use of CFCs
13.	Decline of tuna and other fish populations	14.
15.	16.	Checking the introduction of nonnative species

Exercise 3   (Modules 38.1 – 38.5)

Summarize your knowledge of the biodiversity crisis.

A. List seven general categories of products, benefits, or services that we derive from the biosphere:

B. List four major ways in which humans endanger living things and their habitats:

Exercise 4   (Modules 38.6 and 38.7)

What can we do to defuse the biodiversity crisis and protect the biosphere? The rest of the chapter offers some potential strategies for protecting organisms and ecosystems. Modules 38.6 and 38.7 describe how to nurture threatened populations and species. There are two main strategies at this level—the small-population approach and the declining population approach. Which of the following statements relate to the small-population approach (SP), which relate to the declining population approach (DP), and which relate to both?

_____ 1. Focuses on minimum population size

_____ 2. Is concerned with heading off a small-population vortex

_____ 3. May be necessary to cope with population fragmentation

_____ 4. Focuses on the environmental factors putting pressure on the population

_____ 5. Especially concerned with maintaining genetic diversity of the population

_____ 6. The primary approach used to manage the red-cockaded woodpecker

_____ 7. Must weigh saving the species against conflicting demands

_____ 8. Might be used to protect a keystone species

_____ 9. May involve importing individuals from other areas to increase genetic variation

Exercise 5   (Modules 38.8 – 38.10)

Biologists and conservationists increasingly try to see the bigger picture, and to preserve and restore whole landscapes and ecosystems. They seek to identify biodiversity hot spots, link protected areas (the Yellowstone to Yukon Conservation Initiative) and to rehabilitate damaged ecosystems (the Kissimmee River) when possible. Review managing ecosystems by answering the following questions about wolves and their role in the Yellowstone to Yukon Conservation Initiative.

1. Is the area of the Yellowstone to Yukon Conservation Initiative a global biodiversity hot spot? Why start there?

2. How does this project relate to the subject of landscape ecology?

3. In what way does the project use movement corridors?

4. Why focus on the gray wolf?

5. Why can't species such as wolves thrive in the current system of parks and reserves?

6. How did the decline of the wolf population damage the ecosystems where they used to live?

7. Has reintroduction of wolves presented any problems or possible problems? Describe.

8. Has reintroduction of wolves had an effect on other species in Yellowstone? Explain.

Exercise 6   (Module 38.9)

Some locations harbor much more biodiversity than others. Biologists have identified several trends and patterns associated with the distribution of organisms, and have pinpointed biodiversity hot spots—small areas with large concentrations of species. A biodiversity hot spot would most likely be characterized by which of the descriptions in each pair below?

1. Closer to the equator   Farther from the equator
2. Sensitive to habitat degradation   Resistant to habitat degradation
3. Few endangered species   Many endangered species
4. High plant diversity, but few animals   High plant and animal diversity
5. Likely to be grassland or coniferous forest   Likely to be tropical forest or coral reef
6. Many endemic species   Few endemic species
7. High potential extinction rate   Low potential extinction rate

Exercise 7   (Modules 38.11 – 38.14)

Web/CD Thinking as a Scientist   *Connection: How Are Potential Restoration Sites Analyzed?*
Web/CD Activity 38E   *Conservation Biology Review*

These modules describe methods for restoring, managing, and sustaining ecosystems. Match each description with a method.

_____ 1. Promoting prosperity of both societies and ecosystems

_____ 2. Restoring key nutrients, water, etc., removed from a habitat

_____ 3. The strategy of the Kissammee River Project

_____ 4. Using living organisms to detoxify polluted ecosystems

_____ 5. Surrounding undisturbed areas with areas devoted to low-impact human uses

_____ 6. Pioneered by Costa Rica

_____ 7. The overall strategy of using ecological principles to heal degraded ecosystems

A. Restoration ecology

B. Bioremediation

C. Biological augmentation

D. Establishing zoned reserves

E. Sustainable development

Testing Your Knowledge

Multiple Choice

1. The greatest threat to global biodiversity is
   a. natural disasters such as storms.
   b. pollution.
   c. overexploitation of natural resources.
   d. competition of exotic species with native species.
   e. human alteration of habitats.

2. Which of the following is a greenhouse gas that may contribute to global warming?
   a. methane
   b. carbon dioxide
   c. CFCs
   d. all of the above
   e. a and b

3. Chemical pesticides such as DDT do the greatest harm to top predators such as gulls and pelicans because
   a. these animals are often exposed to chemical spraying.
   b. their metabolism is particularly sensitive to very low concentrations of pesticides.
   c. these animals are usually exposed to more than one pesticide at a time.
   d. pesticide concentrations become magnified as these substances move through food chains.
   e. they tend to seek their prey in areas near human activity.

4. The Kyoto Protocol is an international agreement aimed at
   a. protecting endangered species.
   b. reducing greenhouse gas emissions.
   c. protecting the ozone layer.
   d. reducing acid precipitation.
   e. protecting marine fisheries.

5. Humans have altered _____ of the Earth's land surface.
   a. very little
   b. about 25%
   c. about 50%
   d. about 80%
   e. virtually all

6. Most biodiversity hot spots are in
   a. tropical forests.
   b. mountainous regions.
   c. dry shrublands (chaparral).
   d. wetlands.
   e. coniferous forests.

7. Human activities such as clear-cutting and road building often create edges between areas of different habitat. Edges created by human activities
   a. are usually more abrupt than natural boundaries between habitats.
   b. may have their own species that are not present in adjacent habitats.
   c. may be dominated by relatively few species.
   d. can decrease biodiversity.
   e. are all of the above.

8. Some of the most successful ecosystem restoration projects have been carried out in
   a. desert areas.
   b. wetlands.
   c. old-growth forests.
   d. tropical forests.
   e. coral reefs.

9. The three components of biodiversity are
   a. producer, consumer, and decomposer.
   b. ecosystems, species, and genetic.
   c. community, ecosystem, and biosphere.
   d. plant, animal, and microbe.
   e. polar, temperate, and tropical.

10. Exotic species
    a. are often endangered.
    b. usually increase biodiversity.
    c. often enhance the habitat for native species.
    d. usually reduce biodiversity.
    e. are especially numerous in biodiversity hot spots.

Essay

Imagine that you have been asked to speak to a high school biology class on the topic of "The Global Biodiversity Crisis." At the end of your talk, the students have the following questions. Give short, succinct answers to each question.

1. Why do the industrialized countries have to worry about population growth? Aren't populations growing faster in developing countries?

2. Is destroying tropical rain forests a more serious problem than cutting down forests in the United States? Why?

3. Why should I care what happens in the rain forest? How can it affect me?

4. I heard that the greenhouse effect is a good thing. Is that so? If so, why are people so worried about it?

5. A friend told me that burning fossil fuels is destroying the ozone layer. Is this true?

6. I read that many familiar songbirds are disappearing. If we are more concerned than ever

about protecting them from pollution and habitat destruction, why do many bird populations continue to drop?

7. Why are we so worried that plants and animals are going extinct? Hasn't that happened many times—like when the dinosaurs disappeared?

Applying Your Knowledge

Multiple Choice

1. A conservation group is trying to save Puerto Rico's endemic species. "Endemic" means
   a. they live nowhere else.
   b. predatory.
   c. they are endangered.
   d. little known.
   e. they migrate from place to place.

2. Why does the birth of a German baby have a greater impact on the global environment than the birth of an Indonesian baby?
   a. The population of Germany is smaller, so one baby is a bigger increase.
   b. Europe is closer to environmental disaster than Indonesia.
   c. A German baby will use more resources and produce more pollution.
   d. An Indonesian baby is more likely to die young.
   e. A German is more likely to travel widely.

3. In assessing whether a bird species is endangered, biologists might look at
   a. reproductive rate.
   b. population size.
   c. habitat loss.
   d. population fragmentation.
   e. all of the above.

4. A newspaper headline shouted, "Ecologists Warn that Dead Zone is Growing." This refers to
   a. tropical deforestation.
   b. an increase in ozone depletion.
   c. a biodiversity hot spot.
   d. global warming's effect on vegetation.
   e. an oxygen-depleted area in the ocean.

5. A hybrid car produces less CO_2, which might reduce
   a. global warming.
   b. acid precipitation.
   c. ozone depletion.
   d. bioremediation.
   e. all of the above.

6. Imagine that you have been given the job of setting aside habitat to preserve the diversity of birds in a swath of lowland rainforest. Three plans have been proposed, each of them totaling 200 km^2. Plan A consists of twenty small scattered reserves. Plan B is one large reserve. Plan C is five medium-sized reserves connected by corridors. Which of the following do you think ranks the reserves in order of potential biodiversity, highest to lowest?
   a. A C B
   b. A B C
   c. C A B
   d. B C A
   e. B A C

7. The introduction to Chapter 32 descibed how various plants can be used to "soak up" and remove toxic chemicals from the soil. This is an example of
   a. bioremediation.
   b. biological magnification.
   c. bioprospecting.
   d. biological augmentation.
   e. sustainable development.

8. Which of these countries do you suppose produce the most carbon dioxide per person?
   a. United states
   b. Mexico
   c. Indonesia
   d. India
   e. Tanzania

9. In an effort to save an endangered population, wildlife managers might try to enhance _____ to help reduce _____ .
   a. population fragmentation . . . biodiversity hot spots
   b. biological magnification . . . population fragmentation
   c. movement corridors . . . population fragmentation
   d. biological magnification . . . the number of endemic species
   e. ecosystem processes . . . biodiversity hot spots

10. It is important for us to
    a. protect and preserve diverse species and ecosystems.
    b. appreciate and understand nature.
    c. encourage sustainable development.
    d. improve the human condition.
    e. all of the above.

Essay

1. The top predators in an ecosystem are often regarded as "indicator species"—indicators of the health of the entire system. For this reason, researchers monitor jaguars in the rain forest of Central America and spotted owls in the coniferous forest of the Pacific Northwest. Why do top predators reflect the integrity of their ecosystems?

2. Many factors—both natural and caused by humans—affect the greenhouse effect. State whether you think each of the following would tend to increase or decrease the possibility of global warming, and why.
   a. Conversion from coal-fired to solar and hydroelectric power plants
   b. Large-scale tree-planting programs
   c. Deforestation
   d. Increasing the average efficiency (gas mileage) of automobiles
   e. Slowing world population growth
   f. Warming of climate, speeding up decomposition of organic matter

3. Marine algae need iron in trace amounts and their growth seems to be limited, especially in the Antarctic Ocean, by the lack of iron in sea water. Some biologists have suggested that fertilizing the sea with iron compounds might help reduce global warming. How would this work? Do you think we ought to try it? Why or why not?

4. Explain how the following might be employed in restoring an area of patchy clearcuts into habitat suitable for species that live in larger stands of mature forest, such as spotted owls: bioremediation, the small-population approach, restoration ecology, biological augmentation, the declining-population approach, establishing zoned reserves.

5. Thousands of acres of tropical forest in Central and South America have been cut, burned, and converted to pasture for cattle. Much of the beef raised there is exported, bringing valuable foreign exchange into poor countries. A lot of the beef ends up in the United States, in luncheon meats, hamburger, baby foods, and pet foods. (American fast-food chains used to be major buyers of Latin American beef, but pressure from environmentalists has reduced their reliance on this source.) The land cannot be grazed for very long; heavy rains leach away nutrients, and the rains and overgrazing cause soil erosion. Soon pasture is covered by scrub, and it is uncertain when, if ever, the forest will be able to regrow. But surely a few hamburgers can't affect the forest very much? Or can they?

Using the information that follows, see if you can calculate how much tropical forest is destroyed to produce a four-ounce hamburger: An acre of rain forest consists of about 800,000 pounds of living things. After forest is cut down and grass is planted, there is enough food produced to grow 50 pounds of cattle per acre per year. The pasture land can be used for about 8 years before the soil is exhausted and the land is no longer usable. Only about half a steer can be made into hamburger; the other half is bone, skin, etc. One hamburger contains 4 ounces of beef. One pound is 16 ounces, and one acre is 43,793 square feet. (Based on "Our Steak in the Jungle," by C. Uhl and G. Parker, *BioScience*, Vol. 36, No. 10, p 642.)

Questions: How many pounds of forest life are destroyed for each hamburger? How many square feet of forest are destroyed per hamburger? How many hamburgers do you eat in a year? How much forest could you destroy in a year?

Extending Your Knowledge

1. What aspects of your lifestyle might contribute to global warming? Would you be willing to change the way you live to lessen your contribution to the problem? What kinds of changes could you make that would lessen global warming? Do you think the United States should sign the Kyoto Protocol to reduce greenhouse emissions? Why?

2. What do you think human life will be like if the Earth's population grows to 8 billion in the next two decades and we are still unable to achieve sustainable development? In what ways might your life change? How might life change for a citizen of China or Bangladesh? How might things be different if we are able to achieve sustainability?

Answers

Chapter 1

Organizing Your Knowledge

Exercise 1: 1. All of the organisms of a particular area, plus the nonliving environment that affects them 2. Community 3. A group of interbreeding organisms of one species 4. Organism 5. Organ system 6. A body structure consisting of several tissues, performing particular functions 7. Tissue 8. Cell 9. A functional component of a cell 10. Molecule

Exercise 2:

Exercise 3: The structure and function of a college student's *cells,* and all her other characteristics—from facial features to the functioning of her brain—are determined by the *genetic information in DNA molecules,* which shaped these features during her *growth and development.* The student *utilizes energy* from food to do work—walking, talking, and studying. In performing these actions, she *responds to stimuli* from the environment, and various mechanisms *regulate* her body's activities to adjust to environmental changes. Like her other characteristics, her ability to *reproduce* results from the precise *order* of her molecules, cells, and organ systems. All of these human characteristics have been shaped by natural selection over time—the process of *evolution.*

Exercise 4: 1. Eukarya, Plantae 2. Archaea or Bacteria 3. Archaea or Bacteria 4. Eukarya, Animalia 5. Eukarya, Fungi 6. Eukarya, protists 7. Archaea or Bacteria 8. Eukarya, Animalia 9. Eukarya, Plantae 10. Eukarya, protists

Exercise 5: 1. related 2. Charles Darwin 3. ancestral 4. adaptation 5. population 6. varied 7. heritable 8. predators 9. reproduce 10. offspring 11. natural selection 12. variations 13. environmental 14. reproductive 15. species 16. evolution 17. time 18. theme

Exercise 6: 1. induction 2. discovery 3. hypothesis 4. scientific method 5. deduction 6. observation 7. question 8. hypothesis 9. experiment 10. experimental 11. control 12. variable 13. prediction 14. deduction

Exercise 7: See Module 1.9 for ideas. Recall that science is seeking natural causes for natural phenomena through observation and experiment. Technology is the application of scientific knowledge for a specific purpose.

Testing Your Knowledge

Multiple Choice: 1. e 2. c 3. b 4. d 5. c 6. c 7. e 8. b 9. a 10. c

Essay: 1. Prokaryotes are simple, single-celled organisms whose cells lack a nucleus. The two kinds of prokaryotes are classified in Domain Bacteria and Domain Archaea, two quite distinct groups distinguished mainly on the basis of molecular differences. Organisms in Domain Eukarya—plants, animals, and fungi—all have more complex eukaryotic cells containing a nucleus and other organelles.

2. There are several kingdoms of protists—simple organisms such as algae and protozoa. Like plants, animals, and fungi, protists have eukaryotic cells. But most protists are small and single-celled. Plants, animals, and fungi are all multicellular. Plants, such as pine trees and rosebushes (Kingdom Plantae), differ from the others in that they are photosynthetic and have cells with rigid walls made of cellulose. The fungi of Kingdom Fungi, such as mushrooms

and molds, decompose the remains of dead organisms and absorb the nutrients. Animals (Kingdom Animalia), from worms to human beings, have cells without rigid walls and eat other organisms, and most move actively in search of food.

3. Science deals with questions about natural phenomena that have natural causes. Some examples of questions that potentially could be answered by science are: What particles make up an atom? What makes a cell divide? How do birds find their way when they migrate? Science does not deal with questions that are not subject to experimental test, or that are not concerned with the natural world and its laws: What is the purpose of human existence? Why is there a universe? Is disease caused by evil spirits?

4. Chemical structure—the arrangement of atoms and molecules—is the fundamental level of order that underlies all the properties of life. The specific properties of a cell, organ, or organism arise from the combinations and arrangements of the molecules of which the organism is composed. The structure of DNA—most notably the particular sequence of nucleotides that make up the DNA molecule—provides a blueprint for growth and development and information that cells need to make the molecules involved in their structure, energy use, and response to stimuli. Replication, or copying, of DNA underlies the organism's ability to reproduce. The evolution of the diverse forms of life stems from changes in DNA sequences over time.

Applying Your Knowledge

Multiple Choice: 1. e 2. c 3. d 4. b 5. c 6. c 7. c

Essay: 1. The squirrel depends on an oak tree for shelter, an escape route, and food. It may live in a cavity hollowed out by a woodpecker. It buries nuts, adding to the soil organic matter that might be consumed by insects and worms. Sometimes the buried acorns sprout and grow into trees. Hawks and owls prey on squirrels, and various parasites depend on them. Chipmunks sometimes compete with squirrels for food.

2. The ancestors of beavers may have been ratlike animals that occasionally ventured into the water to escape from predators or obtain food. There may have been heritable variation among the rodents in the population with regard to the shapes of their tails and feet. Those rodents with more flattened tails and more webbed feet were able to swim more efficiently, thus escaping predation and gathering food more effectively, making them more likely to leave offspring than more ratlike individuals. Their offspring inherited their flattened tails and webbed feet, and the proportion of individuals with flattened tails and webbed feet increased over time, until eventually all the individuals in the population looked like beavers. Natural selection occurred as heritable variations were exposed to environmental factors that favored the reproductive success of some individuals over others.

3. Jason's test of MegaGro was not scientifically valid because it was not a controlled experiment. There are any number of variables that might have caused this year's crop of tomatoes to exceed last year's—rainfall, seed type, sunny weather, location of plants, and so on. Jason needs to plant two groups of tomato plants side by side at the same time. Only one variable—the amount of MegaGro—should be allowed to differ between the experimental (MegaGro) and control (no MegaGro) plants. Then if one group of plants produces more tomatoes than the other, Jason can safely conclude that the fertilizer made the difference.

4. Set up an obstacle course for oilbirds to fly through. Divide the birds to be tested into two groups. Plug the ears of the birds in one group so that they cannot hear. Then compare how these experimental birds navigate the obstacle course with the performance of control birds that can hear.

5. The long, narrow, flexible fingers and sensitive nerve endings in the fingertips enable us to grasp and manipulate objects. The flat shape of a leaf and its orientation perpendicular to the rays of the sun allow it to capture sunlight efficiently for the process of photosynthesis. The hawk's strong, curved, sharp beak is used for tearing meat. A frog's long, muscular legs, with their webbed feet, are used in jumping and swimming.

6. You are an organism composed of organ systems such as the circulatory and skeletal systems. Each system is made up of organs, such as the heart or an artery. An organ is made of tissues, and a tissue is made of cells, the smallest living units. Each cell is made up of molecules, which are made of atoms. A cell is alive; it uses energy, responds to stimuli, grows, and reproduces. The molecules that make up a cell are not alive, and they do not have all these properties.

7. Living things are composed of cells. They are complex and highly ordered, and carry out chemical processes, according to the genetic information in their DNA. They take in energy and use it to perform work, and they are able to respond to stimuli and regulate their internal environment. They also grow, develop, reproduce, and evolve. Some nonliving things possess some of these properties, but only living things are characterized by all of them.

8. Human, chimpanzee, and goldfish DNA are all made of the same chemical subunits; they all have the same fundamental double-helix chemical structure; and they all use the same alphabet to encode hereditary information. The four nucleotides in the DNA of chimps and humans are arranged in similar sequences and spell out similar genetic messages. The nucleotides of human DNA and goldfish DNA are less similar in sequence and spell out rather different messages.

Chapter 2

Organizing Your Knowledge

Exercise 1: 1. T 2. F 3. T 4. T 5. T 6. F 7. T 8. T 9. T 10. T

Exercise 2: 1. Mg M 2. O L 3. Zn T 4. H L 5. Cu T 6. I T 7. C L 8. Ca M 9. P M 10. N L 11. Na M 12. Fe T

Exercise 3: 1. C 2. E 3. C 4. C 5. E 6. E 7. C 8. C 9. E 10. C

Exercise 4: 2. N 7 14 7 7 7 3. Cl 17 35 17 18 17 4. O 8 16 8 8 8 5. O 8 17 8 9 8

Exercise 5: 1. The calcium atom gives away 2 electrons, one to each of 2 Cl atoms, producing a Ca^{++} ion and 2 Cl^- ions. These ions form $CaCl_2$.

2. The nitrogen atom shares its electrons with 3 hydrogen atoms, forming ammonia, NH_3.

Exercise 6: 1. chemical 2. cohesion 3. surface tension 4. oxygen 5. hydrogen 6. share 7. covalent 8. nonpolar 9. electronegative 10. negative 11. positive 12. polar 13. neutral 14. negatively 15. positively 16. hydrogen 17. four 18. DNA 19. protein 20. solid 21. liquid 22. gas 23. solvent 24. aqueous 25. solutes 26. heat 27. temperature 28. cup 29. bathtub 30. hydrogen bonds 31. less 32. releases 33. absorbs 34. releases 35. fast 36. evaporate 37. boiling point 38. evaporative 39. less 40. floats

Exercise 7: 1. 4 2. 7 3. 4 4. 7 5. 3 6. 11 7. 2 8. 3 9. 13 10. 8 11. 6 12. 2

Exercise 8: 2 H_2O_2 (reactants) $\longrightarrow$ 2 H_2O + O_2 (products)

Exercise 9:

Testing Your Knowledge

Multiple Choice: 1. b 2. e 3. e 4. d 5. b 6. b 7. e 8. a 9. c 10. c 11. e 12. b

Essay: 1. See Table 2.2 in the text.

2. Iron is an element, and there are atoms of iron. Water is a compound, containing two elements. The smallest particle of water is a molecule, composed of two hydrogen atoms and one oxygen atom. There are no atoms of water; if a water molecule is broken up into hydrogen and oxygen atoms, it is no longer water.

3. Temperature is the average speed of molecules in a body of matter. Heat is the total amount of energy due to movement of molecules. Water molecules in a teakettle move faster, but there are many more of them in a swimming pool, so even though the molecules in a swimming pool are moving slower, their total amount of energy is greater.

4. Acid precipitation forms when air pollutants from burning fossil fuels—sulfur and nitrogen oxides—combine with water vapor in the air to form sulfuric and nitric acids. Acids can break down chemicals in animals, and low pH can interfere with chemical processes, especially in the eggs and young of aquatic organisms. Acid in the soil can lead to mineral imbalances that affect plants.

5. The shared electrons in a water molecule are pulled more strongly toward the oxygen atom than the hydrogen atoms. This makes the oxygen atom partially negative and the hydrogen

atoms partially positive, causing the water molecule to be polar. The + and – charged regions on each water molecule are attracted to oppositely charged regions on adjacent molecules; these attractions are called hydrogen bonds. When water is heated, the heat energy first disrupts the hydrogen bonds and then makes the molecules move faster. Because heat is absorbed as bonds break, water can absorb and store much heat while warming only a small amount.

6. Water evaporates when the fastest molecules move so rapidly that they escape from the liquid and form water vapor. This leaves the molecules still in the liquid with a lower average speed, and as a result the remaining liquid has a lower temperature.

Applying Your Knowledge

Multiple Choice: 1. a 2. e 3. a 4. c 5. e
6. d 7. b 8. a 9. c 10. a

Essay: 1. It will form 2 covalent bonds, because it needs to share 2 electrons with other atoms to have a complete outer shell of 8 electrons. H_2S.

2. Give a plant water containing a radioactive isotope of oxygen H_2O, and look to see whether the sugar made contains the radioactive oxygen $C_6H_{12}O_6$. Carry out a similar test with CO_2.

3. It takes a large amount of heat to break the hydrogen bonds in water before the water warms up much, so water can absorb much heat with a small rise in temperature. As the rock loses heat it cools off a lot, but this same amount of heat, absorbed by the water, does not raise the temperature of the water much.

4. Water molecules can form hydrogen bonds with the positively and negatively charged atoms in a sugar molecule, so the water molecules can surround individual sugar molecules. Water molecules are not attracted to oil molecules; the water molecules stick to each other and do not surround the oil molecules.

5. Each atom is characterized by a certain number of electrons, arranged in energy shells. For most atoms, the outermost shell can hold up to 8 electrons. Carbon has only 4 electrons in its outer shell. It tends to share electrons—form bonds—with four other atoms, filling its outer shell. Thus it tends to form four bonds, and methane is CH_4, not CH_6.

6. Seattle is much closer to the ocean, which stores heat and stabilizes climate. During the summer, the water absorbs heat, cooling Seattle, and during the winter the water releases stored heat to the air, warming the city.

7. Water molecules are more polar than those of most other substances. They are attracted to one another, bonded by hydrogen bonds. These bonds are hard to break, so water must absorb more heat to vaporize than most other substances. The easier-to-vaporize alcohol molecules tend to evaporate first.

Chapter 3

Organizing Your Knowledge
Exercise 1:

1.

Carbon has 4 electrons in an outer shell that holds 8. It completes its outer shell by sharing with 4 other atoms.

2.

C_4H_{10}

3.

or

C_5H_{10}

4.

or

C_5H_{12}. They are isomers.

5.

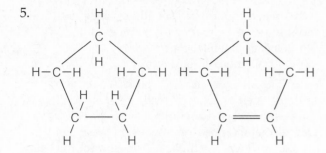

Without double bond: 10 Hs. With double bond: 8 Hs.

Exercise 2: 5 hydroxyls, 4 carbonyls, 3 carboxyls, 5 aminos, 1 phosphate

Exercise 3: 1. D 2. D 3. H 4. H 5. D 6. D 7. H 8. D 9. H 10. H

Exercise 4: 1. monosaccharides 2. polysaccharides 3. proteins 4. fats 5. starch 6. glucose 7. dehydration 8. water 9. hydrolysis 10. disaccharide 11. ring 12. 6 13. cellulose 14. linked 15. digesting 16. fiber 17. disaccharide 18. glucose 19. fructose 20. monosaccharide 21. shapes 22. receptors 23. hydrophilic 24. isomers 25. $C_6H_{12}O_6$ 26. dehydration 27. glycogen 28. starch 29. glucose

Exercise 5:

							A				S							
H	Y	D	R	O	G	E	N	A	T	E	D							
							H				E							
			C	H	O	L	E	S	T	E	R	O	L					
		T					R				O							
	P	H	O	S	P	H	O	L	I	P	I	D						
		G					S				D				S			
	H						C				S				A			
G	L	Y	C	E	R	O	L					F	A	T	T	Y		
	D						E								U			
	T	R	I	G	L	Y	C	E	R	I	D	E			R			
	O						O						W	A	X			
	P			D		S		F		A					T			
	H			O		I		A		N					E			
O	I	L	S		U	N	S	A	T	U	R	A	T	E	D			
	B					B				B								
	I			P	L	A	Q	U	E		O							
	C			E							L							
									L	I	P	I	D	S				
											C							

Exercise 6: 1. transport 2. contractile 3. defensive 4. structural 5. signal 6. storage 7. enzyme

Exercise 7: See Figures 3.12 A, B, and C in the text.

Exercise 8: 1. quaternary DF 2. tertiary AIJ 3. secondary EH 4. primary BCG

Exercise 9: 1. H 2. G 3. J 4. D 5. F 6. B 7. C 8. D 9. J 10. I 11. A 12. E

Exercise 10: 1. fats 2. steroids 3. energy 4. glycerol 5. unsaturated 6. plant oils 7. heart disease 8. butter 9. corn oil 10. cell membranes

Testing Your Knowledge

Multiple Choice: 1. c 2. a 3. c 4. e 5. d 6. c 7. e 8. c 9. d 10. d 11. b 12. e 13. d

Essay: 1. Structural proteins form body parts such as fibers in tendons and ligaments. Contractile proteins in muscles are responsible for movement. Storage proteins such as albumin store amino acids. Defensive proteins called antibodies fight infection. Transport proteins such as hemoglobin carry substances. Some hormones are signal proteins, which convey messages from cell to cell. An enzyme is a protein that catalyzes chemical changes.

2. Animal fats are saturated; their fatty acids contain no double bonds and can pack together tightly and solidify at room temperature. Plant oils are unsaturated; their fatty acids contain double bonds, which cause kinks that prevent them from packing together and solidifying.

3. The secondary and tertiary levels of protein structure are maintained mostly by relatively weak hydrogen bonds and ionic bonds. These weak bonds are easily disrupted by changes in the protein's environment.

4. Glucose: one circle. Maltose: two joined circles. Starch: a chain of linked circles. See Figures 3.4, 3.5, and 3.7 in the text. Maltose is a disaccharide. Starch is a polysaccharide.

5. See Figures 3.14A–3.14D in the text.

Applying Your Knowledge

Multiple Choice: 1. e 2. d 3. d 4. d 5. d 6. b 7. e 8. a 9. a 10. e 11. a 12. a

Essay: 1. Every starch molecule is a polymer of glucose monomers, which are all the same, so starch molecules differ only in length. A protein molecule can be composed of 20 kinds of amino acid monomers. These can be arranged in chains of an almost infinite number of different sequences and lengths, producing many different kinds of proteins.

2. A protein is folded in a specific three-dimensional shape, and a protein's function usually depends on its ability to bind to some other molecule because of its specific shape. The shape of these enzyme molecules allows them to fit a starch molecule. The glucose monomers of

cellulose are linked together in a different orientation, forming a rod instead of a coil. Apparently the enzymes do not fit this structure.

3. The hydrogen and ionic bonds that maintain the three-dimensional shape of a protein are fairly weak. A slight change in the environment can alter these bonds, change the shape of the protein, and impair its function, even if the covalent bonds that hold the amino acids together in the protein are unaffected. The main function of starch is to store glucose; the three-dimensional shape of the molecule is relatively unimportant. A much more drastic change is needed to break the covalent bonds in the starch molecule, separate the glucose subunits, and alter its function.

4. See Figure 3.1 in the text. Make sure every C forms 4 bonds and every H forms 1 bond. Isomers have the same molecular formulas, but their structures differ.

5. The body stores excess glucose in the form of glycogen, a polysaccharide. We can draw on this store of sugar between meals. Since Fred cannot produce glycogen, he cannot store as much sugar and must eat more often to provide his cells with glucose "fuel."

6. The number of possible tripeptides is $20 \times 20 \times 20 = 8000$. The number of possible polypeptides 100 amino acids long is 20^{100}—a very large number indeed!

Chapter 4

Organizing Your Knowledge

Exercise 1: 1. visible light 2. visible light 3. electrons 4. eye lens 5. glass lenses 6. electromagnets 7. none 8. 1,000× 9. 100,000× + 10. dust particle 11. smallest bacterium 12. large atom 13. 0.1 millimeter 14. 0.2 micrometer 15. 2 nanometers 16. no magnification 17. limited magnification and resolving power 18. cannot be used on living specimens

Exercise 2: *Cell 1:* Surface area = 96 μm². Volume = 64 μm³. Surface/volume = 1.5. *Cell 2:* Surface area = 384 μm². Volume = 512 μm³. Surface/volume = 0.75. 1. Cell 2. 2. Cell 2. 3. Cell 1. 4. Cell 1.

Exercise 3: For labels, see Figure 4.4 in the text. Capsule: Sticky coat protects cell
Cell wall: Protects and maintains cell shape
Plasma membrane: Surrounds cell and separates it from environment
Nucleoid: location of bacterial chromosome

Bacterial chromosome: Hereditary material (DNA)
Ribosome: Assembles polypeptides
Prokaryotic flagella: Propel cell
Pili: Attachment to surfaces

Exercise 4: *Prokaryote cell:* prokaryotic, small, plasma membrane, cell wall, cytoplasm, ribosomes, bacterial flagellum. *Plant cell:* eukaryotic, large, membranous organelles, plasma membrane, cell wall, cytoplasm, ribosomes, nucleus, rough ER, smooth ER, Golgi, peroxisome, mitochondrion, chloroplast, central vacuole, cytoskeleton. *Animal cell:* eukaryotic, large, membranous organelles, plasma membrane, cytoplasm, ribosomes, nucleus, rough ER, smooth ER, Golgi, lysosome, peroxisome, mitochondrion, cytoskeleton, flagellum, centriole.

Exercise 5: 1. D 2. I 3. A 4. E 5. B 6. D 7. G 8. A 9. G 10. F 11. C 12. B 13. E 14. G 15. H 16. I 17. F 18. E 19. A 20. B 21. D 22. F 23. F

Exercise 6: See Figure 4.13 in the text. 1. rough ER—vesicle—Golgi—vesicle—plasma membrane—outside 2. rough ER—vesicle—Golgi—lysosome 3. rough ER—vesicle—Golgi—vesicle—plasma membrane 4. smooth ER—vesicle—Golgi—vesicle—plasma membrane—outside

Exercise 7: *Chloroplast:* Found in plants and some protists; carries out photosynthesis; converts energy of sunlight into chemical energy in sugar. *Mitochondrion:* Found in all eukaryotes; carries out cellular respiration; converts chemical energy of foods into chemical energy in ATP.

Exercise 8: *Microfilaments:* solid rods, actin, help change shape, muscle contraction. *Intermediate filaments:* Ropelike structure, fibrous proteins, reinforcing rods, anchor organelles. *Microtubules:* hollow tubes, tubulin, help change shape, move chromosomes, act as tracks, give cell shape, in cilia, in flagella, in centrioles, 9+2 pattern, dynein arms cause bending

Exercise 9: 1. E 2. D 3. A 4. B 5. C 6. F

Exercise 10: See Figures 4.4 and 4.19 in the text.

Testing Your Knowledge

Multiple Choice: 1. e 2. b 3. b 4. e 5. a 6. b 7. e 8. c 9. b 10. b 11. a

Essay: 1. An electron microscope has greater resolving power and is capable of greater magnification than a light microscope, enabling it to reveal finer details than a light microscope. An electron microscope cannot be used to view living specimens, because they must be held in a vacuum and all air and liquids must be removed. A light microscope is preferable for viewing living specimens.

2. Prokaryotic cells are smaller and much simpler than eukaryotic cells. Prokaryotes lack a nucleus and other membranous organelles that compartmentalize eukaryotic cells.

3. There are three major structures present in plant cells but lacking in animal cells. Rigid cell walls containing cellulose surround and support plant cells. Chloroplasts perform photosynthesis. Plant cells usually contain central vacuoles that may help carry out metabolic processes or store water, food, or other chemicals.

4. Membranes separate metabolic processes and allow processes that require different conditions to be carried out simultaneously. They also increase the total surface area of membranes, where many metabolic processes occur.

5. Chloroplasts carry out photosynthesis, capturing the energy of sunlight and storing it in sugar molecules. Mitochondria carry out cellular respiration, converting the chemical energy in sugar and other foods into the chemical energy of ATP, which serves as a cellular fuel.

Applying Your Knowledge

Multiple Choice: 1. d 2. d 3. c 4. a 5. a
6. b 7. b 8. c 9. a 10. e

Essay: 1. The TEM would be best for examining fine details within cell organelles, because it is capable of greater magnification and resolving power and uses a beam of electrons that passes through thinly sliced sections. The SEM would be best for studying bumps on the cell surface, because its electron beam scans the surface and forms an image of the outside of the cell. Electron microscopes are not suitable for observing living specimens because they must be placed in a vacuum and the air and liquids must be removed, so a light microscope would be best for observing changes in the nucleus as the cell prepares to divide, and cell shape as the cell moves.

2. Surface area = 150 μm^2. Volume = 125 μm^3. Surface/volume = 1.2. Surface area = 600 μm^2. Volume = 1000 μm^3. Surface/volume = 0.6. The smaller cell has the greater surface-to-volume ratio, which would enable it to exchange materials more efficiently with its environment via its membrane. The surface-to-volume ratio of the larger cell would equal that of the smaller cell if it were divided up into 8 smaller cubes.

3. The amylase molecules are produced by ribosomes attached to the rough ER and deposited inside the ER. Transport vesicles filled with the protein bud from the ER and fuse with the Golgi apparatus, where the proteins may be modified. Transport vesicles then bud from the Golgi apparatus and fuse with the plasma membrane, dumping amylase outside the cell.

4. The mitochondrion is a sort of cellular "power plant," converting the energy in food molecules into the energy of ATP, which in turn provides energy for many cellular activities. Interfering with the mitochondria could reduce the supply of ATP fuel and thus impair many cellular processes.

5. As a muscle cell grows, ribosomes, ER, and Golgi would work together to produce more protein and cell membrane. Microfilaments, made of the protein actin, are responsible for the contractile activity of muscle and would increase in number in a growing muscle cell. Smooth ER stores calcium that is involved in the contraction process, and it too would grow. Mitochondria provide energy for muscle contraction and other cellular activities and would perhaps increase in number.

Chapter 5

Organizing Your Knowledge

Exercise 1: 1. work 2. kinetic 3. thermodynamics 4. first 5. transformed 6. potential 7. heat 8. second 9. entropy 10. more 11. endergonic 12. exergonic 13. releases 14. ATP 15. ADP 16. phosphate 17. potential 18. hydrolysis 19. phosphate 20. ADP 21. work 22. mechanical 23. transport 24. chemical 25. phosphorylation 26. heat 27. membranes 28. endergonic 29. coupling 30. activation 31. barrier 32. protein 33. catalyst 34. increases 35. changed 36. lower 37. metabolism

Exercise 2: 1. Kinetic energy is energy that is doing work, and potential energy is stored energy. 2. Exergonic reactions release energy, and endergonic reactions consume energy. 3. In a chemical reaction, reactants are changed into products. 4. ADP is phosphorylated (a phosphate group is added), producing ATP. When ATP is hydrolyzed, a phosphate group is removed from ATP, and ADP is produced. 5. Enzymes lower the energy of activation for a reaction, so the reaction with an enzyme requires less energy to get

it started than the reaction without the enzyme.
6. In photosynthesis, the energy of sunlight is used to make glucose from carbon dioxide and water. In cellular respiration, oxygen and glucose combine, and the energy in glucose is used to make ATP. 7. First law: Energy can be transferred or transformed but not created or destroyed. Second law: Energy changes are accompanied by an increase in disorder, or entropy. 8. Mechanical work is movement of the cell or body. Transport work is moving substances through cell membranes. Chemical work is building molecules.

Exercise 3: 1. See Figure 5.6 in the text. Substrate enters enzyme active site and slightly changes site to fit. Products leave and enzyme recycles. 2. Shape of enzyme altered by heat or pH, deforming active site so substrate will not fit. 3. See Figure 5.8 in text. Competitive inhibitor is shaped like substrate and blocks active site. Noncompetitive inhibitor can be any shape; attaches to enzyme and alters shape of active site so substrate will not fit.

Exercise 4: 1. phospholipid 2. glycoprotein 3. carbohydrate I.D. tags 4. extracellular fluid 5. glycolipid 6. transport protein 7. fiber of extracellular matrix 8. cholesterol 9. receptor protein 10. enzyme 11. cytoplasm

Exercise 5: 1. E 2. B 3. A 4. F 5. DGH 6. G 7. C 8. B 9. H 10. E 11. CF 12. D 13. I 14. D 15. C

Exercise 6: 1. Out 2. Out 3. In 4. None 5. In 6. Out 7. In 8. In 9. None 10. Out

Exercise 7: 1. phospholipid 2. nonpolar 3. phospholipid 4. protein 5. fluid 6. glycolipids 7. glycoproteins 8. identity 9. receptor 10. diffusion 11. passive 12. concentration 13. more 14. less 15. isotonic 16. hypertonic 17. out of 18. selectively 19. osmosis 20. transport 21. energy 22. facilitated 23. active 24. ATP 25. endocytosis 26. phagocytosis 27. vesicle 28. lysosome 29. exocytosis

Testing Your Knowledge

Multiple Choice: 1. a 2. c 3. b 4. b 5. d 6. c 7. b 8. a 9. e 10. b 11. d 12. b

Essay: 1. Large and polar molecules cannot easily diffuse through cell membranes. Transport proteins provide pores by which these substances can pass through membranes.

2. The enzyme is shaped in such a way that its active site forms a "pocket" that holds the substrate.

3. Molecules slow down at lower temperatures, slowing the rate of contact between reactant molecules and the enzyme's active site. Higher temperatures denature the enzyme, altering its shape and impairing its specific function.

4. Larger molecules contain more covalent bonds, so they have more potential energy stored in their complex structure. It takes energy to form these bonds, so the reactions by which amino acids join to form proteins are endergonic. The cell obtains energy for endergonic reactions by coupling them to exergonic reactions, the most important of which is the hydrolysis of ATP.

5. A cell will gain water by osmosis if the solution outside the cell is hypotonic, that is, if the outside solution has a lower total concentration of solutes than the cytoplasm. It will lose water if the solution outside is hypertonic, that is, if it has a higher total concentration of solutes than the cytoplasm. Water gain by a plant cell is least serious; the cell wall prevents bursting.

Applying Your Knowledge

Multiple Choice: 1. b 2. e 3. b 4. b 5. a 6. d 7. e 8. b 9. d 10. d 11. e 12. e

Essay: 1. Paper does not burn spontaneously (at room temperature) because energy is required to get this reaction started—energy of activation. This creates an energy barrier that prevents the reaction from happening spontaneously. A flame supplies energy of activation.

2. The hypertonic solution would cause water to move out of the stomach cells by osmosis, causing dehydration.

3. You can't win: The first law of thermodynamics says that energy cannot be created or destroyed, so you can't come out ahead in the energy game. You can't break even: The second law of thermodynamics says that energy changes are accompanied by an increase in disorder, or entropy. Heat, a form of disorder, is created during energy conversions; some energy is always lost to the surroundings as heat, so you can't break even.

4. The insecticide molecule is the same shape as the substrate molecule. It fits into the active site of the enzyme, preventing the enzyme from acting on its normal substrate, and thus interfering with nervous system function.

5. The lecithin molecules coat the droplet, with their nonpolar tails embedded in the fat and the polar heads projecting into the surrounding water.

Chapter 6

Organizing Your Knowledge

Exercise 1: 1. respiration 2. oxygen 3. carbon dioxide 4. cellular respiration 5. glucose 6. oxygen 7. carbon dioxide 8. water 9. ATP 10. 2200 11. sun 12. photosynthesis 13. chemical 14. CO_2 15. H_2O 16. glucose 17. food 18. air 19. aerobically 20. 40 21. anaerobically 22. 2 23. Slow 24. aerobic 25. thin 26. blood 27. mitochondria 28. myoglobin 29. dark 30. fast 31. thicker 32. fewer 33. less 34. glucose 35. lactic acid 36. ache 37. fast 38. slow 39. fast

Exercise 2:

```
E L E C T R O N S                     G
      E     A                         A
      D     D             C           I
      U                   H Y D R O G E N
A D P T                   E           S
      E I                 M     S
P H O S P H O R Y L A T I O N   Y
      Y N                 I     N     P
      D       T R A N S P O R T H     H
      R   C               M     A     O
O X I D A T I O N         O     S     S
      G   R   X           S L O S E S
      E   R   I           I     E     P
      N   I   D           S     S     H
      A   E   I                       A
      S   R   Z           N     T     T
R E D O X     E           A     E
            R E D U C E D
                          H
```

Exercise 3: 1. electrons carried by NADH 2. electrons carried by NADH and $FADH_2$ 3. glycolysis 4. glucose 5. pyruvate 6. citric acid cycle 7. oxidative phosphorylation 8. cytoplasm 9. CO_2 10. CO_2 11. mitochondrion 12. ATP 13. ATP 14. ATP

Exercise 4: 1. I 2. K 3. E 4. C 5. C 6. D 7. B 8. J 9. F 10. A 11. L 12. G 13. H

Exercise 5: 1. NAD^+ 2. NADH 3. pyruvate 4. acetyl CoA 5. CO_2 6. coenzyme A 7. CO_2 8. NADH 9. ATP 10. $FADH_2$

Exercise 6: 1. final 2. ATP 3. oxidative phosphorylation 4. electron 5. inner 6. NADH 7. the citric acid cycle 8. O_2 9. H_2O 10. H^+ ions 11. active transport 12. potential 13. ATP 14. inner compartment of the mitochondrion 15. ATP synthases 16. ADP 17. ATP

Exercise 7: See Figures 6.10 and 6.11 in the text.

Exercise 8: 1. OP 2. G 3. G 4. G 5. OP 6. CA 7. OP 8. OP 9. CA 10. OP 11. OP 12. CA

Exercise 9: 1. facultative 2. yeasts 3. pyruvate 4. muscle 5. oxygen 6. NADH 7. ethanol 8. lactic 9. ATP 10. glycolysis 11. CO_2 12. anaerobes 13. anaerobic 14. less 15. lactate

Exercise 10: 1. carbohydrates 2. fats 3. proteins 4. sugars 5. glycerol 6. fatty acids 7. amino acids 8. glucose

Exercise 11: See Figure 6.15 in the text.

Testing Your Knowledge

Multiple Choice: 1. e 2. b 3. c 4. b 5. d 6. c 7. c 8. a 9. b 10. e 11. e 12. a

Essay: 1. Breathing allows our lungs to supply cells with the oxygen used in cellular respiration and expel the carbon dioxide produced by cellular respiration.

2. Aerobic cellular respiration is more efficient (it produces more ATP per molecule of glucose consumed), but it must take place in an oxygen-containing (aerobic) environment. Fermentation requires no oxygen, but it is much less efficient (it produces much less ATP per glucose).

3. Oxidative phosphorylation occurs in the third (electron transport–chemiosmosis) stage of cellular respiration. Substrate-level phosphorylation occurs in glycolysis and the citric acid cycle. In chemiosmosis, ATP synthase harnesses the flow of H^+ ions down their concentration gradient to make ATP. In substrate-level phosphorylation, an enzyme transfers a phosphate group from an organic molecule to ADP to make ATP. Oxidative phosphorylation makes the most ATP under aerobic conditions. Substrate-level phosphorylation makes the most ATP under anaerobic conditions.

4. Rotenone prevents electrons from passing from an electron carrier molecule to the next. Cyanide and carbon dioxide block the passage of electrons to oxygen. In either case, the energy of moving electrons cannot be captured. Oligomycin blocks the passage of H^+ ions through ATP synthase, so ATP cannot be made. DNP makes the membrane of the mitochondrion leaky to H^+ ions, thus abolishing the H^+ concentration gradient whose energy is harnessed to make ATP.

5. These processes provide small organic molecules that the cell can use in the synthesis of organic molecules not obtained in food.

Applying Your Knowledge

Multiple Choice: 1. b 2. a 3. c 4. c 5. d
6. d 7. c 8. d 9. b 10. a 11. d

Essay: 1. Organisms that rely on aerobic cellular respiration are limited to environments where oxygen is available. Fermenters are able to live in anaerobic environments such as mud, manure piles, and sealed wine vats, free from competition with aerobic organisms.

2. Oxygen picks up electrons at the end of the electron transport chain. When oxygen runs out, there is no place for electrons to go. They "pile up," and electron transport stops. When electron transport stops, NADH and $FADH_2$ pile up, and NAD^+ and FAD are not regenerated. Without NAD^+ and FAD to pick up electrons and hydrogens, the citric acid cycle must stop.

3. Because FAD and NAD^+ are constantly recycled, the cell requires only a small supply of them. Without NAD^+ and FAD to pick up electrons and hydrogens from glycolysis and the citric acid cycle and deliver them to the electron transport chain, cellular respiration would stop.

4. If the victim had suffocated, the lack of oxygen would have caused electrons to pile up, and all the electron carriers would have held electrons. They would have been in the reduced state. Because the carriers were in the oxidized state—lacking electrons—Mickleberry knew that oxygen had been available in the cells to pick them up, so the victim did not suffocate.

5. At first the yeasts in the flask were able to produce sufficient ATP via aerobic cellular respiration. As the yeast population grew, the oxygen in the flask was used up, and they turned to fermentation to produce ATP. Because fermentation is less efficient than aerobic respiration, the yeasts had to consume glucose at a faster rate to maintain their supply of ATP. Ethanol is a waste product of fermentation.

Chapter 7

Organizing Your Knowledge

Exercise 1: pine tree, green bacterium, rosebush, seaweed, moss, alga, grass

Exercise 2:

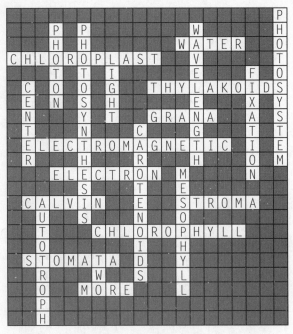

Exercise 3: $6\ CO_2 + 12\ H_2O \longrightarrow C_6H_{12}O_6 + 6\ H_2O + 6\ O_2$. The C and O in $C_6H_{12}O_6$ come from CO_2. The H in $C_6H_{12}O_6$ comes from H_2O. The O in O_2 comes from H_2O. The H in H_2O comes from H_2O and the O comes from CO_2. H_2O is oxidized, and CO_2 is reduced.

Exercise 4: 1. light 2. H_2O 3. outer membrane of chloroplast 4. CO_2 5. light reactions 6. $NADP^+$ 7. Calvin cycle 8. ADP + P 9. stroma 10. granum 11. thylakoids 12. electrons 13. ATP 14. NADPH 15. O_2 16. sugar

Exercise 5: 1. B 2. A 3. H 4. C 5. B 6. A 7. D 8. H 9. G 10. F 11. C 12. E 13. I 14. D 15. A 16. A 17. H 18. G

Exercise 6: a. 5 b. 9 c. 2 d. 6 e. 3 f. 7 g. 8 h. 4 i. 1 Gamma waves have the most energy. Blue and red light are used in photosynthesis.

Exercise 7: 1. photosynthesis 2. chloroplasts 3. mesophyll 4. light 5. chlorophyll 6. thylakoids 7. Calvin 8. stroma 9. carbon 10. photons 11. blue 12. red 13. green 14. carotenoids 15. reaction center 16. electron 17. H^+ 18. stroma 19. thylakoids 20. gradient 21. ATP synthase 22. ATP 23. photophosphorylation 24. water 25. water 26. oxygen 27. oxygen 28. light 29. chlorophyll 30. acceptor 31. NADH 32. Calvin 33. photosynthesis 34. stroma 35. carbon dioxide 36. ATP 37. NADPH 38. G3P 39. glucose 40. respiration 41. ATP

Exercise 8: 1. C_3 2. C_4 3. C_4 4. CAM 5. C_3 6. C_3 7. C_4 or CAM 8. C_4 or CAM 9. CAM 10. C_3

Exercise 9: 1. warm 2. cool 3. cool 4. cool
5. warm 6. cool 7. cool 8. warm 9. cool
10. cool

Exercise 10: For example, "CFCs, used as refrigerants and propellants, are changed by the sun into reactive free radicals, which destroy the ozone that protects life from UV radiation."

Testing Your Knowledge

Multiple Choice: 1. b 2. a 3. a 4. c 5. e
6. b 7. b 8. d 9. b 10. e 11. b

Essay: 1. If a plant is given carbon dioxide containing radioactive ^{18}O, no labeled oxygen atoms appear in the O_2 produced by the plants, showing that the oxygen in O_2 does not come from CO_2. If the plants are given water containing ^{18}O, the O_2 gas they produce does contain labeled O, showing that the oxygen in O_2 comes from H_2O.

2. See Figure 7.5 in the text.

3. Photosynthesis allows producers to manufacture organic food molecules from inorganic reactants. This process produces virtually all the organic matter required as food by plants and all other living things on Earth.

4. Increased burning of fossil fuels (such as oil and gas) releases CO_2 into the atmosphere. Deforestation reduces global photosynthesis, reducing removal of CO_2 from the atmosphere by plants. Both of these activities contribute to an increased greenhouse effect, because more CO_2 in the atmosphere could trap more heat. We could reduce our contribution to the greenhouse effect by using fossil fuels more efficiently, recycling forest products, and replanting trees on deforested land.

Applying Your Knowledge

Multiple Choice: 1. c 2. d 3. e 4. e 5. a 6. c
7. c 8. a 9. c 10. c 11. d

Essay: 1. Plants would grow most slowly under green lights, because green light is not absorbed well by photosynthetic pigments. Green is reflected, which is why plants are green.

2. Carotenoids absorb all colors except yellow and orange. Because carotenoids absorb green light, they are able to capture some light energy not absorbed by chlorophyll and channel this energy to chlorophyll for use in photosynthesis.

3. Energy from the movement of excited electrons along a chain of electron carriers is used to pump H^+ through the membrane into the thylakoid space. This buildup of H^+ lowers the pH of the solution in the thylakoid space.

4. In photosynthesis, high-energy electrons come from chlorophyll whose electrons are excited by light. These electrons are replaced by splitting water. In cellular respiration, glucose is the source of high-energy electrons. In both processes, movement of the electrons through electron carriers is harnessed to move H^+ through a membrane; this energy is ultimately used to make ATP via chemiosmosis. In photosynthesis, the high-energy electrons are also used to make glucose. In cellular respiration, the electrons combine with O_2 and H^+ to form water. Photophosphorylation takes place in the thylakoid membranes of the chloroplast; oxidative phosphorylation takes place along the inner membrane of the mitochondrion. In both processes, ATP synthase harnesses flow of H^+ through the membrane to make ATP.

5. In the light reactions of photosynthesis, sunlight excites chlorophyll in a leaf, causing it to give up excited electrons. High-energy electrons move through electron carriers, which generate an H^+ gradient, which in turn is harnessed to make ATP via chemiosmosis. The electrons and hydrogen are picked up by $NADP^+$, forming NADPH. In the Calvin cycle, energy from the ATP and hydrogen and electrons from NADPH are used to reduce CO_2, forming sugar. When you consume the plant, the sugar is "burned" in cellular respiration. The glucose is oxidized, and its hydrogen and electrons are transferred to NAD^+ and FAD, forming NADH and $FADH_2$. These molecules deliver the high-energy electrons to the electron transport system in the mitochondrion. The electrons move through electron carriers, which generate an H^+ gradient, which in turn is harnessed to make ATP via chemiosmosis. The ATP is then hydrolyzed to ADP and P to power muscle movement.

6. Van Helmont was correct in concluding that the soil did not contribute much material to the growth of the tree, but he was wrong in thinking that the weight gain must have come from the water. What Van Helmont couldn't know (but we do) is that water supplies only the hydrogen atoms used in photosynthesis to build the carobohydrates that make up most of a plant. The bulk of the atoms in carbohydrate molecules—all the carbon and oxygen—come from CO_2 that the plant gets from the air!

Chapter 8

Organizing Your Knowledge

Exercise 1: 1. like 2. asexual 3. DNA 4. DNA 5. chromosomes 6. identical 7. mother 8. binary fission 9. circular 10. smaller 11. simpler 12. plasma membrane 13. wall 14. genomes 15. egg 16. sperm 17. fertilization 18. variation

Exercise 2:

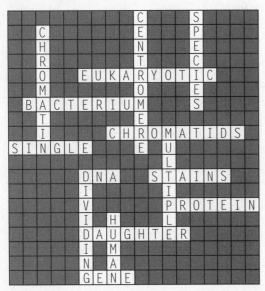

Exercise 3: 1. interphase, activity between divisions 2. G_1, cell growth following division 3. S, growth and DNA synthesis 4. G_2, growth and activity between DNA replication and division 5. mitotic phase, mitosis plus cytokinesis 6. mitosis, division of nucleus and chromosomes 7. cytokinesis, division of cytoplasm

Exercise 4: *Nucleus:* I—membrane-bounded, P—envelope breaks down, M—none, A—none, T—daughter nuclei and envelopes form. *Spindle:* I—none, P—spindle forms from microtubules, M—fully formed; some fibers attached to chromosomes, A—fibers move chromosomes to poles, T—breaks down. *Chromosomes:* I—duplicated but dispersed as chromatin, P—chromatin coils to form shorter, thicker pairs of chromatids joined at centromeres, M—line up on metaphase plate, A—sister chromatids separate, T—chromosomes uncoil to form chromatin. *Cell size and shape:* I—rounded, doubles in volume, P—rounded, M—rounded, A—spindle elongates cell, T—elongation continues; cytokinesis splits cell into two smaller rounded cells. *Sketch:* See Module 8.6 in the text.

Exercise 5: 1. interphase 2. telophase 3. metaphase 4. prophase 5. anaphase

Exercise 6: 1. E 2. A 3. C 4. B 5. D

Exercise 7: For example: In animals, microfilaments produce a cleavage furrow that pinches the cell apart, while in plants, vesicles align in a cell plate, where a cell wall grows to split the cell.

Exercise 8: 1. asexual 2. growth 3. liver 4. repairs 5. skin 6. digestive tract 7. replaced 8. numbers 9. types 10. mitosis 11. cultures 12. surface 13. anchorage 14. density-dependent inhibition 15. growth factors 16. control 17. stops 18. G_1 19. S 20. receptor 21. nerve or muscle 22. cancer 23. inhibiting 24. 20–50 25. tumor 26. benign 27. malignant 28. metastasis 29. Sarcomas 30. Leukemias 31. radiation 32. chemotherapy 33. cell division 34. spindle

Exercise 9: 1. Homologous pairs of chromosomes called autosomes are found in the somatic cells of both males and females. Sex chromosomes are an additional pair of chromosomes—homologous in females (who have 2 X chromosomes) but mostly not homologous in males (who have X and Y chromosomes). 2. The two chromosomes of a homologous pair carry genes for the same inherited traits at the same place, or locus. One is inherited from the father and one from the mother. 3. The two chromatids of a single chromosome are exactly identical, having been produced via the process of DNA replication during the S phase of interphase. 4. A diploid cell contains two homologous sets of chromosomes. A haploid cell contains one chromosome set. Diploid cells divide to form haploid cells in meiosis. 5. Gametes are sperm and eggs. Somatic cells are body cells other than gametes. In humans, somatic cells are diploid. 6. An egg is haploid. A zygote, formed by fusion of a sperm and egg, is diploid. 7. In meiosis, a diploid cell divides to form haploid cells. In fertilization, haploid cells (egg and sperm) fuse to form a diploid cell (zygote). 8. Mitosis produces two cells that are genetically identical to the mother cell. Meiosis is the division of a diploid cell to produce four haploid cells that are not genetically like the mother cell or each other. 9. These are the sex chromosomes. They are mostly not homologous. A female mammal has two X chromosomes; a male has an X and a Y. 10. A gene that determines a particular characteristic is located at a particular place (called its locus) on a chromosome.

Exercise 10: See Module 8.14 in the text for appearance of chromosomes, homologous chromosomes, sister chromatids, and crossing over.
1. interphase I   2. prophase I   3. metaphase I
4. anaphase I   5. telophase I and cytokinesis
6. prophase II   7. metaphase II   8. anaphase II
9. telophase II and cytokinesis   10. meiosis I
11. meiosis II

Exercise 11: 1. produces daughter cells identical to parent cell   2. involves two cell divisions
3. produces four daughter cells   4. sister chromatids of each chromosome separate   5. homologous pairs (tetrads) line up at metaphase plate
6. crossing over occurs between homologous chromosomes   7. provides for asexual reproduction, growth, replacement, repair

Exercise 12:

1.

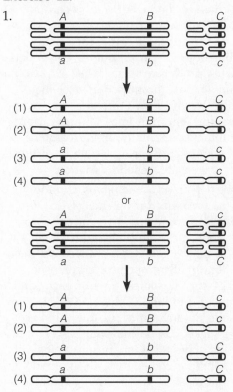

2.

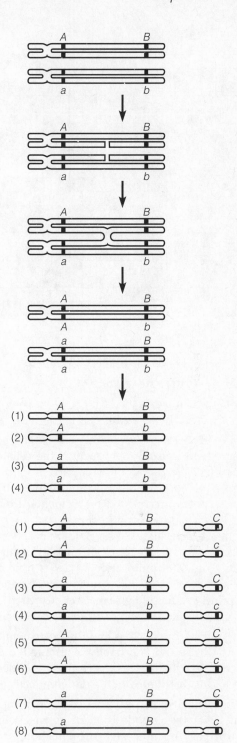

8 combinations are possible

Exercise 13:

```
S T E R I L I T Y                       F
          H                   T U R N E R
    F E W E R         H       A G E     M
          E           O           S     A
        L E U K E M I A           T     L
          E     M A L E N E S S   E     E
        Z       O               S       S
        Y       G   G A M E T E S
R       G     N O N D I S J U N C T I O N
E   B   O     H   S           O
T   A   T     O               S
A B O R T E D R           B I R T H
R R R   O     R                       A
D N I   W             F E T A L       L
E S S             R           E       F
D O M                 S I S T E R
  K A R Y O T Y P E           S
```

Exercise 14:

1. C, X 2. D, W 3. B, Z 4. A, Y

Testing Your Knowledge

Multiple Choice: 1. d 2. a 3. d 4. b 5. a 6. c 7. d 8. b 9. a 10. c 11. c 12. b 13. e 14. d 15. b

Essay: 1. In asexual reproduction, the cells or buds that give rise to offspring are produced by mitosis. Because in this process genetic information is copied and passed on to identical daughter cells, the offspring produced by asexual reproduction are exactly like the single parent and each other. In sexual reproduction, each offspring inherits a unique combination of genes from both parents. Because of this genetic recombination, the offspring produced by sexual reproduction are not exactly like either parent or each other.

2. In mitosis, the nucleus and duplicated chromosomes divide and are equally distributed to daughter cells. In cytokinesis, the cytoplasm divides in two and is distributed to the daughter cells.

3. Cancer cells escape from the control mechanisms that regulate division in normal cells. Cancer cells do not need to be anchored to a solid surface to grow. They seem unaffected by density-dependent inhibition and will pile up on one another in cell culture. They either don't require growth factors or manufacture their own growth factors. Unlike normal cells, cancer cells can divide indefinitely. If they stop dividing, they stop at random points in the cell cycle, not at the restriction point.

4. Mitosis produces two cells that are genetically identical to the mother cell. Meiosis is the division of a diploid cell to produce four haploid cells that are not genetically like the mother cell or each other. In humans, mitosis is responsible for producing diploid somatic cells for growth, replacement of worn-out cells, or body repair. Meiosis produces haploid gametes (eggs and sperm) for sexual reproduction. Many (but not all) somatic cells undergo mitosis; special cells in ovaries and testes undergo meiosis.

5. Because sister chromatids are produced when DNA is duplicated, the genes of sister chromatids are identical. Homologous chromatids carry genes for the same traits (eye color, for example) at corresponding places, or loci, but they might carry different versions of the gene (blue or brown) at these loci.

6. The two chromosomes—maternal and paternal—of a homologous pair carry different genetic information. Because the two chromosomes of each pair line up independently of other pairs at metaphase I of meiosis, many different combinations of chromosomes are possible in the gametes produced by meiosis. Also in meiosis, homologous chromosomes exchange corresponding segments in the process of crossing over, creating chromosomes with a variety of gene combinations. Finally, many combinations of eggs and sperm are possible at fertilization, further increasing genetic variation.

7. Sometimes a pair of homologous chromosomes fails to separate during meiosis I. One of the cells resulting from this division ends up with 24 chromosomes, the other with 22 chromosomes. All of the gametes produced by meiosis II have abnormal numbers of chromosomes. Some have 24 chromosomes—one of each kind plus an extra of one kind. Others have 22 chromosomes—one of each kind but one missing. Sometimes meiosis I is normal, but a pair of sister chromatids stays together during meiosis II. In this case, some gametes have the normal number of chromosomes, but some have one extra and some are one short.

8. An extra copy of chromosome 21 is usually not fatal, but it does produce Down syndrome. In most other situations in which an abnormal number of chromosomes occurs, the offspring is spontaneously aborted, so these abnormalities must be so serious as to be fatal.

Applying Your Knowledge

Multiple Choice: 1. d 2. b 3. c 4. c 5. a 6. b 7. a 8. d 9. d 10. d 11. d 12. b 13. b 14. d

Essay: 1. Cancer is uncontrolled cell division. Chemotherapy (and other cancer therapies like radiation) slows tumor growth by interfering with cell division. Cell division is most rapid in the skin and digestive tract lining, where it replaces lost cells, so the side effects of chemotherapy are greatest on skin and intestinal lining.

2. Cells with between 100 and 200 units were in S phase, in the process of replicating their DNA.

3. These data suggest that prophase is the longest stage of mitosis, making up about half the process of mitosis. Telophase is also fairly long, lasting about a third of mitosis. Metaphase and anaphase take a relatively short time to occur.

4. There would be 39 different kinds of chromosome—39 homologous pairs. (In female mammals, the two sex chromosomes, as well as each pair of autosomes, look alike.)

5. Possible alignments: *ABC* and *abc*, *ABc* and *abC*, *Abc* and *aBC*, and *AbC* and *aBc*. These same eight combinations are possible in eggs or sperm. There are $8 \times 8 = 64$ combinations possible in the zygotes.

6. Parental-type gametes: *QR* and *qr*. Recombinant-type gametes: *Qr* and *qR*.

7. *XXY*, *XYYY*, and *XXYY* would produce male phenotypes; A single *Y* chromosome produces "maleness"; the absence of a *Y* chromosome produces "femaleness."

Chapter 9

Organizing Your Knowledge

Exercise 1: 1. S 2. G 3. C 4. Q 5. U 6. K 7. W 8. L 9. A 10. T 11. H 12. X 13. D 14. E 15. V 16. B 17. Y 18. M 19. N 20. F 21. I 22. Z 23. J 24. R 25. P 26. O

Exercise 2: 1. a. green, yellow b. *GG*, *gg* c. all *Gg* 2. a. 1 *GG*:2 *Gg*:1 *gg* b. 3 green:1 yellow 3. a. both *Bb* b. white *bb*

Exercise 3: 1. *BBSS* 2. *bbss* 3. *BS* 4. *bs* 5. *BbSs* 6. *BS* 7. *bS* 8. *Bs* 9. *bs* 10. *BS* 11. *bS* 12. *Bs* 13. *bs* 14. *BBSS* 15. *BbSs* 16. *BBSs* 17. *BbSs* 18. *BbSS* 19. *bbSS* 20. *BbSs* 21. *bbSs* 22. *BBSs* 23. *BbSs* 24. *BBss* 25. *Bbss* 26. *Bbss* 27. *bbSs* 28. *Bbss* 29. *bbss* 30. $\frac{9}{16}$ 31. brown, short-haired 32. $\frac{3}{16}$ 33. brown, long-haired 34. $\frac{3}{16}$ 35. white, short-haired 36. $\frac{1}{16}$ 37. white, long-haired

Exercise 4: 1. a. white b. brown 2. a. *bb* b. *Bb* or *BB* 3. white 4. *Bb* 5. all brown, 0 white 6. heterozygous 7. homozygous

Exercise 5: 1. 0 2. 1 3. 1 4. $\frac{1}{6}$ 5. $\frac{5}{6}$ 6. independent 7. product 8. $\frac{1}{6}$ 9. $\frac{1}{6}$ 10. $\frac{1}{36}$ 11. multiplication 12. $\frac{1}{2}$ 13. $\frac{1}{2}$ 14. homozygous 15. multiplication 16. $\frac{1}{2}$ 17. $\frac{1}{2}$ 18. $\frac{1}{4}$ 19. addition 20. sum 21. $\frac{1}{36}$ 22. $\frac{1}{36}$ 23. $\frac{2}{36}$ ($\frac{1}{18}$) 24. $\frac{1}{2}$ 25. $\frac{1}{2}$ 26. $\frac{1}{2}$ 27. $\frac{1}{2}$ 28. $\frac{1}{4}$ 29. $\frac{1}{4}$ 30. $\frac{1}{4}$ 31. $\frac{1}{4}$ 32. $\frac{1}{2}$

Exercise 6: 1. none 2. *Ww* and *Ww* 3. $\frac{1}{4}$

Exercise 7: 1. *ss* 2. ? 3. ? 4. ? 5. *Ss* 6. *Ss* 7. *Ss* 8. *ss* 9. ? 10. *ss* 11. *Ss* 12. ? 13. *Ss* 14. *SS* 15. ? 16. ? 17. half colored 18. half colored 19. ? 20. half colored 21. fully colored 22. ? 23. ? 24. half colored 25. half colored 26. blank 27. fully colored 28. fully colored

Exercise 8: 1. T 2. F (one-fourth) 3. F (thousands) 4. F (normal parents who are heterozygous) 5. T 6. F (Caucasians) 7. F (not evenly distributed) 8. T 9. T 10. F (not necessarily more common) 11. T 12. F (less common) 13. T 14. F (There is debate among geneticists.) 15. T 16. F (cystic fibrosis) 17. T 18. T

Exercise 9: 1. counseling 2. carrier 3. fetal 4. AFP 5. Tay-Sachs 6. one-fourth 7. phenylketonuria 8. amino acid 9. diet 10. amniocentesis 11. amniotic 12. cells 13. karyotype 14. Down 15. chorionic villus 16. ultrasound 17. 35

Exercise 10: 1. B 2. A 3. D 4. B 5. C 6. C 7. E 8. D 9. E 10. A

Exercise 11: 1. Many genetic tests detect conditions that are not yet treatable. 2. Tests should always be accompanied by counseling, but this is not always the case. 3. A "bad" result might stigmatize an individual. Might she lose her job, or be denied health insurance? 4. Will genetic testing only be available to those who can afford it?

Exercise 12: 1. *B* 2. *b* 3. *S* 4. *s* 5. *B* 6. *b* 7. *S* 8. *s* 9. *B* 10. *S* 11. *b* 12. *s* 13. *S* 14. *B* 15. *S* 16. *B* 17. *s* 18. *b* 19. *s* 20. *b* 21. *b* 22. *B* 23. *S* 24. *s* 25. *b* 26. *B* 27. *S* 28. *s* 29. *b* 30. *S* 31. *B* 32. *s* 33. *S* 34. *b* 35. *S* 36. *b* 37. *s* 38. *B* 39. *s* 40. *B*

Exercise 13: 1. B 2. C 3. E 4. A 5. D

Exercise 14: 1. E 2. D 3. D 4. A 5. C 6. B 7. A

Exercise 15: 1. *X* 2. dominant 3. X^CY 4. X^cY 5. daughters 6. sons 7. *X* 8. *Y* 9. carriers 10. X^CX^c 11. half 12. half 13. half 14. all 15. father 16. carriers 17. mother 18. *X* 19. *c*

Testing Your Knowledge

Multiple Choice: 1. c 2. e 3. d 4. b 5. e
6. b 7. b 8. a 9. d 10. e 11. c 12. a 13. b
14. b 15. a 16. b

Essay: 1. Mendel studied peas, which are easy to grow and come in many readily distinguishable varieties. He studied seven characteristics that each occur in two distinct forms. He first made sure that he had true-breeding varieties. Mendel carefully controlled which plant mated with which and kept scrupulous records of the matings and their results. Finally, he analyzed and interpreted his results mathematically, devising hypotheses that he tested and retested experimentally.

2. Each flip is an independent event, with a probability of $\frac{1}{2}$. The probability of getting two heads is the product of the individual probabilities, $\frac{1}{2} \times \frac{1}{2} = \frac{1}{4}$. The probability of coin 1 coming up heads and coin 2 coming up tails is also the product of the individual probabilities, $\frac{1}{2} \times \frac{1}{2} = \frac{1}{4}$. But there is another way that you can get one head and one tail—coin 1 could come up tails and coin 2 could come up heads. The combined probability of this occurring is also $\frac{1}{2} \times \frac{1}{2} = \frac{1}{4}$. There are two different ways to get one head and one tail. The probability of an event that can occur two different ways is the sum of the separate probabilities of the different ways, in this case $\frac{1}{4} + \frac{1}{4} = \frac{1}{2}$.

3. Peas and fruit flies are easy to raise in a small space at little expense. They have many easy-to-spot characteristics, and it is easy to control matings between different varieties. They both produce many offspring in a relatively short time.

4. Humans who inherit *X* chromosomes from both mother and father *(XX)* are female. Humans who inherit an *X* chromosome from the mother and a *Y* chromosome from the father *(XY)* are male. A gene on the *Y* chromosome appears to trigger testis development. In the absence of this gene, an individual develops ovaries and is female. In grasshoppers and crickets, an individual with one sex chromosome *(XO)* is male, and one with two sex chromosomes *(XX)* is female. Sperm cells with or without sex chromosomes determine the sex of the offspring. In certain fishes, butterflies, and birds, a *ZZ* individual is male, a *ZW* individual is female, and eggs determine the sex of offspring. In ants and bees, diploid individuals are female and haploid individuals are male.

Applying Your Knowledge

Multiple Choice: 1. c 2. d 3. c 4. b 5. a
6. d 7. c 8. c 9. c 10. c 11. d 12. b

Essay: 1. The parents are both heterozygous, *Aa*. The daughter is homozygous recessive, *aa*. The probability is $\frac{1}{4}$.

2. $\frac{9}{16}$ purple green, $\frac{3}{16}$ purple yellow, $\frac{3}{16}$ white green, $\frac{1}{16}$ white yellow

3. The man is heterozygous, *Ff*. His wife is homozygous recessive, *ff*. The children with freckles are *Ff;* those without freckles are *ff*.

4. Pink × pink yields $\frac{1}{4}$ red, $\frac{1}{2}$ pink, $\frac{1}{4}$ white. Pink × white yields $\frac{1}{2}$ pink, $\frac{1}{2}$ white.

5. The man is $I^A i$, the woman $I^B i$. The probability of a type B child is $\frac{1}{4}$.

6. All four individuals have about the same skin color. Couple 2 could have children with the widest range of skin colors, ranging from *AABBCC* to *aabbcc*. (Couple 1 could have children ranging only from *aaBBCC* to *aabbCC*.)

7. Prediction: $\frac{1}{4}$ red eyes and straight wings, *RrSs;* $\frac{1}{4}$ red eyes and curled wings, *Rrss;* $\frac{1}{4}$ pink eyes and straight wings, *rrSs;* and $\frac{1}{4}$ pink eyes and curled wings, *rrss*. The actual results of the cross indicate that the genes for eye color and wing shape are linked. In the heterozygous parent, *R* and *S* are on one chromosome, and *r* and *s* on the homologous chromosome. Most gametes were *RS* and *rs*, but crossing over resulted in a small percentage of *Rs* and *rS* gametes being produced.

8. Genes *s* and *b* are farther apart than *h* and *b*. The order of the genes on the chromosome is *shb* (or *bhs*).

9. The genes for sex-linked disorders are on the *X* chromosome. If both parents are normal but have a child with a sex-linked disorder, the mother must carry the allele for the disorder on one of her *X* chromosomes, and she will pass it on to half of her children. Half of her daughters will carry the allele for coloboma iridis, but they will be of normal phenotype because they will inherit the normal allele from their father. Half of her sons will also inherit the coloboma iridis allele. Because they inherit *Y* chromosomes from their father (and no normal allele that could mask the coloboma iridis allele), half the sons will exhibit coloboma iridis. Thus, one-fourth of their children—all boys—will have coloboma iridis.

10. If Diane's brother has hemophilia (and presumably her father does not), her mother must be a carrier of the hemophilia allele, and she would pass this allele on to half her daughters. The probability that Diane is a carrier is ½. If Diane is a carrier, there is a probability of ¼ that she and Craig will have a child—a son—with the disease. The probability both that Diane is a carrier and that Diane and Craig will have a child with hemophilia is ½ × ¼ = ⅛.

Chapter 10

Organizing Your Knowledge

Exercise 1: 1. N 2. D 3. DR 4. SPB 5. B 6. I 7. D 8. SP 9. ACGT 10. A 11. K 12. L 13. S 14. D 15. W 16. C 17. E 18. B (or N) 19. L 20. F 21. J 22. O 23. H 24. ACGU 25. U 26. Q (or D or R) 27. V

Exercise 2: 1. sugar-phosphate backbone 2. phosphate group 3. sugar (deoxyribose) 4. double helix 5. complementary base pair 6. adenine (A) 7. guanine (G) 8. thymine (T) 9. hydrogen bond 10. cytosine (C) 11. nucleotide 12. polynucleotide 13. pyrimidine bases 14. purine bases

Exercise 3: See Figure 10.4A in the text.

Exercise 4: 1. D 2. D 3. A 4. F 5. E 6. B 7. C

Exercise 5:

									C			D		
						T	H	O	U	S	A	N	D	S
						A			M			A		
						M			P			P		
	P	R	O	T	E	I	N	S	L			O		
				R		N			E			L		
	M	E	T	A	B	O	L	I	S	M		Y		
				N		S			E			P		
C	O	D	O	N	S		P	H	E	N	O	T	Y	P	E
				C					T			P		
	G			R	N	A		B	A	S	E	S		
	E			I					R			I		
G	E	N	O	T	Y	P	E	E	N	Z	Y	M	E	S
	E			I					R			E		
				O			M							
				N			O							
	T	R	A	N	S	L	A	T	I	O	N			
				D			D							

Exercise 6: 1. nucleus 2. RNA polymerase 3. promoter 4. nucleotides 5. terminator 6. messenger RNA 7. processed 8. introns 9. exons 10. cytoplasm 11. amino acid 12. transfer RNA 13. amino acid 14. codon 15. enzyme 16. anticodon 17. Ribosomes

18. ribosomal RNA 19. polypeptide 20. initiator 21. start codon 22. polypeptide 23. amino acid 24. amino acids 25. stop codon 26. ribosome 27. ribosome 28. protein

Exercise 7: 1. transcription 2. translation 3. DNA 4. mRNA 5. RNA polymerase 6. amino acid 7. tRNA 8. anticodon 9. large ribosomal subunit 10. initiator tRNA 11. initiation 12. small ribosomal subunit 13. mRNA 14. start codon 15. polypeptide 16. peptide bond 17. elongation 18. codons 19. termination 20. polypeptide 21. stop codon

Exercise 8: 1. Met-Pro-Asp-Asn-Ile-Lys 2. Met-Pro-Asp-His-Ile-Lys 3. one base changed 4. substitution (10th base changed from A to C) 5. one amino acid changed (Asn to His) 6. Met-Pro-Asp-Glu-Tyr 7. one base changed 8. Insertion (10th base, G, inserted between C and A) 9. two amino acids changed and chain shortened by one amino acid, due to change to stop codon 10. effect of mutation in mRNA 3 greatest, due to reading frame shift and alteration of all codons following insertion

Exercise 9: 1. E 2. G 3. B 4. O 5. R 6. Y 7. U 8. A 9. J 10. D 11. V 12. U 13. T 14. I 15. F 16. L 17. H 18. Z 19. U 20. P 21. N 22. L 23. C 24. Q 25. X 26. K 27. W 28. M 29. S

Exercise 10: 1. B 2. D 3. B 4. D 5. C 6. A 7. B 8. A 9. A 10. D 11. C 12. A 13. A 14. A

Testing Your Knowledge

Multiple Choice: 1. a 2. c 3. c 4. b 5. c 6. e 7. a 8. b 9. a 10. d 11. c 12. c 13. b 14. e 15. d 16. a

Essay: 1. See Figure 10.3D in the text.

2. A single-ringed pyrimidine, such as T, must pair with a double-ringed purine, such as A. Two pyrimidines—A and C—would not be large enough to reach across the double helix. T specifically pairs with A and not G because T and A have complementary chemical side groups that form hydrogen bonds.

3. Twenty amino acids are used in building proteins. There are only 4 nucleotides in DNA, so a one-base code could specify only 4 amino acids. A two-base code could specify only 4^2, or 16 amino acids. In a triplet code, 4^3, or 64, combinations of bases are possible, more than enough to code for 20 amino acids.

4. A mutation is a change in the nucleotide sequence of DNA. Some mutations are spontaneous,

but many are caused by X rays, ultra-violet light, and chemicals. Most mutations are harmful because they alter protein amino acid sequences and impair the function of proteins. Some mutations do not alter amino acid sequence; others lead to an improved protein or one with new capabilities that enhance an organism's success. Mutations create genetic diversity that makes evolution by natural selection possible.

5. A base substitution can result in no change to a protein or, most often, a change in one amino acid. Effects on the organism may be minimal. A base deletion alters the sequence of triplet groupings in a gene, causing a drastic change in the amino acid sequence downstream from the deletion. This can have a profound effect on the protein and the organism.

6. Messenger RNA (mRNA) is transcribed from a gene and carries the instructions for making a particular polypeptide. A ribosome is the site of translation, the process of making a polypeptide according to the mRNA message. Transfer RNAs act as translators, each kind matching a particular amino acid with a particular codon in the mRNA. The ribosome "reads" the mRNA one codon at a time, and tRNAs deliver their amino acids, which are added to the polypeptide chain one at a time.

Applying Your Knowledge

Multiple Choice: 1. a 2. a 3. d 4. e 5. c
6. d 7. b 8. a 9. a 10. c 11. c 12. d

Essay: 1. Extract DNA from dead type A *E. coli* and mix it with live type V *E. coli*. If the type V *E. coli* absorb the DNA and genes are made of DNA, the type V bacteria will be able to grow on the simple medium without the vitamin. For comparison, extract protein from dead type A *E. coli* and mix it with live type V *E. coli*. If the type V *E. coli* absorb the protein and genes are made of protein, the type V bacteria will be able to grow on the simple medium without the vitamin. Prediction: DNA will transform the type V bacteria, but protein will not.

2. In the DNA double helix, T and A bases pair up; wherever there is a T nucleotide in one strand, there is a complementary A nucleotide in the other. Thus the amount of T is equal to the amount of A. The specific nucleotide sequences of goldfish and human DNA are different, so the number of A-T pairs is different in the two species.

3. They are both right. Eric is talking about phenotype and Renee about genotype. The nucleotide

sequences of your genes—genotype—determine the amino acid sequences of proteins that cause your hair to be curly or straight—your phenotype.

4. Met-Glu-Leu-Ser-Ile-Asp. (Remember that translation always starts at the AUG start codon.)

5. Because a 3-base codon specifies each amino acid, the message coding for the polypeptide would consist of 233×3, or 699 nucleotides (actually 702, if the stop codon is included). The rest of the nucleotides are extras. Translation of protein does not have to start and stop at the ends of the mRNA. It starts at a start codon (AUG) and stops at a stop codon (UAA, UAG, or UGA). The nucleotides from the start codon to the stop codon are the only ones that spell out the amino acid sequence of the polypeptide.

6. The virus may be able to insert its genes into a nerve cell's DNA and remain latent within the cell as a provirus. From time to time the provirus may begin reproducing complete viruses, causing disease symptoms.

7. In a eukaryotic cell, the RNA transcript is processed before translation. Noncoding introns are removed and exons are spliced together to produce the mRNA. A prokaryote does not process its RNA before translation, so the introns from the eukaryotic gene are not removed. When this unedited RNA is translated, the wrong polypeptide is produced.

8. Some bacteria are combining their DNA with the DNA from the other strain, producing bacteria with the characteristics of both strains. It could be that some bacteria are dying and their DNA is being taken up from the medium by living bacteria—a process called transformation. Bacterial genes could be transferred from bacterium to bacterium by a bacteriophage—transduction. Or the bacteria could be undergoing conjugation—bacterial mating—in which DNA is transferred from one bacterium to another.

Chapter 11

Organizing Your Knowledge

Exercise 1: *lac* operon: 1. E 2. G 3. A 4. B 5. F 6. C 7. H 8. D *trp* operon: 1. E 2. A 3. G 4. B 5. F 6. C 7. H 8. D

Exercise 2: *Stomach gland cell:* citric acid cycle enzyme gene, digestive enzyme gene. *Hair follicle cell:* keratin gene, citric acid cycle enzyme gene. *Stem cell:* hemoglobin gene, citric acid cycle enzyme gene.

Exercise 3:

T O R T O I S E S H E L L
B
A
D
S
A C T I V A T O R S
H
I
S
E   F   I N A C T I V E
N   A   T   O
H   A   R   O   H E L I C A L   P
A   T   A   C   S   O
N   O   N   L   L
C   N   S   E   Y
E X P R E S S I O N   M
R   C   S   E
S   C H R O M O S O M E   R
        I   M   P   A
        P   R E G U L A T O R Y O   S
        T   E   M   E
        I   N   O
        O   E   S
P A C K I N G   S C A T T E R E D
                                E
        S U P E R C O I L

Exercise 4: 1. E 2. F 3. D 4. B 5. C 6. A
7. G

Exercise 5: 1. DNA unpacking and changes
2. transcription 3. addition of cap and tail
4. splicing 5. flow through nuclear envelope
6. mRNA breakdown 7. translation 8. cleavage/modification/activation 9. protein
breakdown

Exercise 6: 1. F 2. A 3. D 4. C 5. E 6. G
7. B

Exercise 7: 1. head 2. tail 3. genes 4. receptor
5. transcription 6. translated 7. activates
8. mRNA 9. head 10. tail 11. mitoses
12. protein 13. segments 14. expression
15. homeotic 16. homeobox 17. amino acids
18. DNA 19. Mutations 20. development
21. order 22. head 23. tail 24. early

Exercise 8: 1. cell cycle 2. division 3. growth
4. carcinogens 5. oncogenes 6. proto-oncogenes
7. growth 8. division 9. mutation 10. tumor-
suppressor 11. signal 12. protein 13. growth
14. cell division 15. absence 16. transcription
17. increased 18. mutation 19. descendants
20. repair

Exercise 9: 1. C 2. F 3. D 4. F 5. C 6. C
7. C 8. A 9. E

Testing Your Knowledge

Multiple Choice: 1. d 2. a 3. c 4. a 5. e
6. e 7. d 8. a 9. b 10. e 11. e 12. b 13. e
14. e

Essay: 1. Differentiated cells can be made to
dedifferentiate and regrow the entire organism.
Differentiation apparently does not irreversibly
change the DNA.

2. Most genes in plants and animals consist of
regions that code for polypeptides, called exons,
interrupted by long noncoding segments, called
introns. Both introns and exons are transcribed
from DNA into RNA, then the introns are removed and the remaining exons linked together—a process called RNA splicing. Introns
may contain nucleotide sequences that regulate
gene activity. The splicing process may help
control flow of mRNA from nucleus to cytoplasm. In some cases splicing can occur in more
than one way, producing different mRNA molecules from the same transcript.

3. A homeotic gene is a master control gene
that functions during development. It produces
a polypeptide that acts as an activator or repressor for a battery of genes, switching them on or
off together and thus shaping large-scale aspects
of body plan.

4. Flies, frogs, and humans have evolved separately for many millions of years. Since they
share homeoboxes, this suggests that homeoboxes arose early in the history of life and have
remained remarkably unchanged for a long time.

5. A gene in the egg cell codes for a protein
that signals surrounding follicle cells. The protein binds to the membrane of a follicle cell and
through a series of relay proteins activates transcription factors in the target cell. This triggers
transcription and translation of specific genes
into proteins. These follicle cell proteins in turn
act on the egg cell, causing it to localize a kind
of mRNA at the end of the egg that will later
become the head. After fertilization, the "head"
mRNA is translated into a regulatory protein
that acts on other genes, which trigger the pattern of gene expression that divides the embryo
into segments from head to tail.

6. Cancer cells escape from normal controls that
regulate division and growth, and multiply ex-
cessively. A cell becomes cancerous when a mutation occurs in a proto-oncogene, a gene that
normally makes a protein that helps stimulate
cell division. The mutation changes the gene
into an oncogene that codes for a hyperactive
protein—one that stimulates cell division more
than normal. A second kind of mutation occurs
in a tumor-suppressor gene. This mutation
keeps a protein that normally blocks cell division from being made. The combined effect of

these mutations is the uncontrolled cell division characteristic of cancer.

7. Animal cloning is achieved by replacing the nucleus of an egg or zygote with the nucleus of a somatic cell. The cell formed then divides repeatedly, forming a ball of cells called a blastocyst. In the right environment (i.e., the uterus of a surrogate mother), this blastocyst may grow into a new individual, genetically identical to the nuclear donor. This is called reproductive cloning. Alternatively, embryonic stem cells can be harvested from the blastula. These cells can be grown indefinitely in the laboratory, and have the potential for differentiating into virtually any kind of somatic cell. Such embryonic stem cells may be used to repair or replace injured or diseased organs—a procedure called therapeutic cloning.

8. To avoid cancer a person should avoid carcinogens—UV radiation, unnecessary X rays, tobacco, alcohol, and so on. A high-fiber, low-fat diet, with foods high in vitamin C, vitamin E, and substances related to vitamin A, also reduces the risk of cancer.

Applying Your Knowledge

Multiple Choice: 1. c 2. b 3. c 4. a 5. e 6. c 7. a 8. c 9. c

Essay: 1. Normally, when lactose is absent, the repressor binds to the operator site, blocking gene transcription, and no enzymes for using lactose are made. When lactose is present, it binds to the repressor and changes the shape of the repressor in such a way that it can no longer bind to the operator. Genes are then transcribed and enzymes for using lactose are made. If the mutation altered the repressor in such a way that it could no longer bind to the operator, the genes would be transcribed and enzymes for using lactose would be made all the time, whether or not lactose was present.

2. Many answers are possible, but the following is an example: Liver cells are small and metabolically active, making and breaking down many substances. Muscle cells are long and thin and have the ability to contract or shorten. Salivary gland cells form saclike clusters and secrete saliva. Liver cells make fibrinogen, a blood-clotting protein. Genes for making fibrinogen would be active in liver cells but not in the other cells. The genes for building contractile proteins would be active in muscle cells, and the gene for making amylase, a digestive enzyme, would be active in salivary gland cells.

All the cells perform glycolysis, the first process in breaking down sugar, so genes that code for glycolysis enzymes would be active in all the cells. The gene that codes for the blood protein hemoglobin would not be active in any of the cells.

3. The minimum number of nucleotides needed to code for the protein is $258 \times 3 = 774$ (or 777, if you wish to include the 3 nucleotides of the stop codon). There are numerous nucleotides upstream from the AUG start codon and downstream from the stop codon. Only the 777 nucleotides from start to stop actually code for the protein. The mRNA is much shorter than the gene because noncoding introns in the gene are cut out and the remaining exons are joined together to form the final mRNA, in the RNA splicing process that occurs in the nucleus.

4. A gene might code for a repressor or activator protein that binds to the DNA of other chromosomes and turns off or enhances transcription of genes at those sites. A gene may code for production of a hormone that travels through the bloodstream to a distant site, enters a cell, and similarly affects gene transcription.

5. Different carcinogens can cause the same kind of cancer because they all might cause mutations in bone marrow cells that change protooncogenes into oncogenes. The oncogenes, no matter how they are activated, have the same effect—to stimulate the uncontrolled cell division characteristic of this kind of cancer.

6. Cloning a woolly mammoth is not as far-fetched as it sounds (and a lot easier than cloning a dinosaur!). You would have to replace the nucleus of an elephant egg or zygote with a nucleus obtained from a frozen mammoth somatic cell, for example a skin cell. The resulting "zygote" would then be injected into the uterus of a surrogate mother elephant. It is possible that the two species are closely enough related that the mammoth nucleus and the elephant egg cytoplasm would be compatible, and if the mammoth nucleus was undamaged, the zygote might be able to develop into a baby mammoth. Wow!

Chapter 12

Organizing Your Knowledge

Exercise 1: 1. F 2. H 3. B 4. L 5. I 6. G 7. J 8. D 9. M 10. A 11. O 12. N 13. C 14. K 15. E

Exercise 2: 1. vectors 2. gene 3. restriction
4. sequences 5. recognition 6. sticky 7. plasmids 8. complementary 9. ligase 10. replication
11. recombinant 12. transformation 13. cloning
14. library 15. shotgun 16. viruses 17. introns
18. reverse 19. exons 20. retrovirus 21. bacteria
22. genomic 23. expressed 24. protein 25. probe
26. UUCACAUC 27. radioactive 28. microarray
29. complementary 30. expression 31. secrete
32. hormones 33. cancer 34. vaccines 35. viruses
36. Yeast 37. integrate 38. sugars 39. milk

Exercise 3: 1. isolating plasmid from *E. coli*
2. obtaining copies of gene and protein from
cloned bacteria 3. cloning recombinant DNA
4. using a nucleic-acid probe to find a gene
5. cutting DNA with restriction enzyme
6. extracting DNA from a eukaryotic cell
7. joining plasmid and DNA fragment using
DNA ligase 8. inserting a plasmid into a bacterium via transformation 9. using reverse transcriptase to make an artificial gene 10. separating DNA fragments via gel electrophoresis
11. mixing plasmids and DNA fragments with
sticky ends 12. using a DNA microarray to find
a base sequence

Exercise 4: 1. Indicate on diagram, within each
CCGG sequence 2. Perpetrator—two places;
Sam—one place; Joe—two places 3. Perpetrator—
three fragments; Sam—two fragments; Joe—three
fragments 4. Perpetrator—large, medium, and
small fragments; Sam—two fairly large fragments; Joe—large, medium, and small fragments
5. Perpetrator and Joe: A large fragment near the
top, a medium fragment in the middle, and a
small fragment at the bottom; Sam: Two fairly
large fragments near the top. Like this:

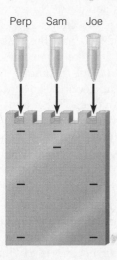

6. Joe's fingerprint matches the evidence from the
crime scene; he should be tried for the crime.
Sam should be set free. (Looking at the DNA

base sequences, your first impulse might have
been to accuse Sam. The trick here is that the
perpetrator's DNA and Joe's match *exactly*, but
Joe's DNA sequence is shown *upside down* relative
to the perpetrator's. There is no upside down to
real DNA or the enzyme, which cuts both into 3
pieces, implicating Joe as the murderer.)

Exercise 5: 1. Technical issues: Using the right
virus vector to get the corrected gene into the
cell, inserting the gene into the correct part of
the genome, building control mechanisms so the
gene will produce the proper amount of gene
product at the right time in the right places,
and inserting genes so they do not harm other
life functions 2. Ethical issues: Who will have
access to such an expensive and difficult procedure? Should gene therapy be reserved for treating serious diseases, or could we use it to
"enhance" a healthy person? How much risk
(the possibility of causing cancer, etc.) is acceptable? Should we "tamper with evolution" by
eliminating genetic defects in eggs, sperm, and
zygotes, thus altering the genetic diversity of
future generations?

Exercise 6: 1. Fingerprints, hair, or blood from
crime scenes 2. Tissue samples from fossils or
mummies 3. Single cells from embryos to be
tested for genetic defects 4. Samples from cells
infected with small amounts of hard-to-detect
viruses

Exercise 7: 1. E 2. F 3. O 4. I 5. H 6. N
7. M 8. K 9. L 10. D 11. G 12. B 13. A
14. C 15. J 16. P

Exercise 8:

Testing Your Knowledge

Multiple Choice: 1. b 2. e 3. d 4. a 5. c
6. a 7. d 8. e 9. c 10. d 11. a

Essay: 1. Transposons are segments of DNA that are able to move from one site in an organism's genome to another. They often insert in the middle of other genes, disrupting their normal functions. This could alter genes in such a way as to increase genetic diversity, contributing to evolution.

2. A restriction enzyme is used to cut DNA at a particular nucleotide sequence, called a recognition sequence. It cuts the two strands of DNA unevenly, producing DNA fragments with single-stranded ends, called "sticky ends." The sticky ends produced by a restriction enzyme are complementary, and complementary fragments produced by the enzyme can hydrogen bond to one another by their single-stranded ends. An enzyme called DNA ligase can join these loosely linked pieces of DNA by catalyzing the formation of covalent bonds between adjacent nucleotides. A fragment of DNA can be spliced into a plasmid, a small, circular bacterial DNA molecule. A bacterium can take up the plasmid from solution by transformation. When the bacterium replicates its larger chromosome, it also replicates the plasmid. It divides repeatedly, producing a clone of cells containing the recombinant plasmid.

3. Bacteria and yeasts have plasmids that can be used as gene vectors, and they can rapidly and cheaply be grown in large quantities. They will secrete protein products into their growth medium, where they can be collected and then purified.

4. Recombinant DNA technology can be used to produce genetically modified (GM) crop plants that resist herbicides, insects, and disease and can grow in poor soil. DNA technology also produces drugs, vaccines, and hormones and has been used to produce transgenic animals.

5. Several human proteins, such as insulin and growth hormone, are produced by recombinant bacteria and yeasts and used to treat diseases. Genetically engineered viruses and bacteria are also used in the production of vaccines and drugs use against various diseases. Diagnostic tests look for genetic diseases, cancer genes, or DNA of infectious organisms. Recombinant DNA technology is beginning to be used in gene therapy—alteration of genes in human cells to remedy genetic diseases.

6. Genetic engineering could create hazardous pathogens that could infect humans, animals, or plants. New organisms could also transfer genes to other species, harming them or making them harmful. Recombinant organisms could also compete with and displace wild species, with unpredictable consequences. Plants with herbicide genes might spread out of control, or those with genes that kill insects kill useful or harmless species. Strict laboratory procedures prevent accidental infection and escape of genetically engineered organisms. The organisms themselves are genetically crippled, so they cannot survive outside the laboratory. Some dangerous experiments are banned. Genetically modified crop plants and animals are tested under controlled conditions before being released.

7. The potential dangers of recombinant DNA technology present some ethical questions: Are possible benefits worth the risks? How do we weigh the possible benefits against the risks? How do we feel about altering species? About adding new species to an already precarious environment? Do we have a right to forgo recombinant DNA technology if it can solve environmental and medical problems? Should we alter our own genes? What uses will be made of genetic information? One of the biggest ethical questions concerning human gene therapy is whether we should alter evolution by eliminating genetic "defects" in our descendants.

8. Restriction enzymes cut DNA into fragments at specific sequences called restriction sites. The cut-up DNA is placed on a gel and pulled through the gel by the attraction of the negatively charged phosphate groups on the DNA to a positive electrode. This method is called electrophoresis. Smaller fragments move through the gel faster, and this separates fragments of different sizes. Fragments of different sizes accumulate in bands in the gel, and the resulting DNA fingerprint looks a bit like a supermarket bar code. Each individual has a unique DNA fingerprint, because the nucleotide sequence of each person's DNA is unique. This means that restriction sites differ among different individuals, and restriction enzymes cut the DNA into different-sized fragments, which produce a different restriction fragment band pattern.

9. DNA data might be used to deny health insurance or specific jobs or positions to individuals or certain ethnic groups. It could be used to coerce people into medical tests or treatment. It could also lead to ill-conceived efforts to "improve" the human species.

Applying Your Knowledge

Multiple Choice: 1. b 2. b 3. c 4. c 5. b 6. b
7. a 8. c 9. e 10. d 11. a 12. e

1. Knowing the amino acid sequence of EGF, you could use the genetic code chart to figure out the corresponding DNA nucleotide sequence, and then program the machine to synthesize a gene coding for the protein. Use a restriction enzyme to cut the gene and plasmid, producing complementary sticky ends. Combine the gene and plasmid, and use ligase to bond them. Introduce the plasmid into *E. coli* via transformation. The bacteria will replicate their DNA and divide in the growth medium, producing a clone of bacteria with the EGF gene. When you have grown a sufficient quantity of the bacteria, extract the EGF.

2. Shotgun cloning might produce clones of bacteria with altered genes and new combinations of genes, giving them unpredictable and potentially dangerous properties. For example, bacteria could be produced that are capable of infecting human cells, or producing a dangerous toxin, or containing cancer genes. Careful!

3. She could use a nucleic-acid probe—an RNA or single-stranded DNA with the complementary base sequence. Such a probe might be an RNA fragment with the sequence UACCGAUAG, for example. This probe contains nucleotides labeled with radioactive isotope or dye. She could mix it with each of the clones. It bonds to the desired complementary base sequence, and the radioactivity or color tags the clone with the sequence she is trying to find.

4. This bacterium could grow only under laboratory conditions. If it accidentally escaped from the lab, it would not be able to survive and thus would not cause harm in the outside environment.

5. The remains are probably those of Manuel, because the DNA fingerprint of the remains contains bands matching some of those in the father's DNA fingerprint and some of those in the mother's DNA fingerprint, but no bands that are found in niether of the parents' DNA fingerprints.

6. The lack of blood enzyme is a better candidate for gene therapy because bone marrow cells can be removed from the patient; their genes can be altered; and the cells can be reintroduced to the patient. The reintroduced cells can continue to proliferate through life, curing the disease. At this time it is not possible to genetically engineer specialized cells such as nerve cells that do not continue to proliferate, because it is not possible to get the corrected gene into the cells and have them function properly.

Chapter 13

Organizing Your Knowledge

Exercise 1: 1. D 2. E 3. G 4. F 5. B 6. H
7. A 8. C

Exercise 2:

Exercise 3: 1. comparative embryology 2. molecular biology 3. fossil record 4. comparative anatomy 5. molecular biology 6. biogeography 7. fossil record

Exercise 4: 1. Charles Darwin 2. evolution 3. vary 4. heritable 5. overproduction 6. resources 7. fit 8. survive 9. reproduce 10. more 11. natural selection 12. pesticide 13. differential 14. population 15. population 16. isolated 17. gene pool 18. adapted 19. genes 20. microevolution 21. population

Exercise 5: B. *RR, Rr, rr* C. 36, 48, 16 D. 36%, 48%, 16% E. 72, 48, 0 F. 0, 48, 32 G. $p = 0.6$, $q = 0.4$ H. 0.6, 0.4 J. *RR, Rr, rr* K. 36%, 48%, 16% L. 72, 48, 0 M. 0, 48, 32 N. $p = 0.6$, $q = 0.4$ O. The genotype and allele frequencies stayed the same in the second generation and will continue to do so in succeeding generations. This illustrates that the frequency of each allele in the gene pool will remain constant unless acted on by other agents. Sexual reproduction alone does not lead to microevolution—the so-called Hardy-Weinberg principle.

Exercise 6: 1. The population is not isolated; there is gene flow—migration of individuals with yellow alleles into the population. 2. A mutation alters the frequency of red and yellow alleles in the gene pool. 3. Because the population is small, a chance event such as a landslide can significantly alter the gene pool. This is called genetic drift. 4. Individuals differ in reproductive success, changing the relative frequencies of red and yellow alleles in the gene pool. This is natural selection, and it results in adaptation to the environment. 5. Mating is not random because a yellow plant does not stand an equal chance of mating with every other plant in the population. It is more likely to mate with a yellow plant.

Exercise 7:

Exercise 8: 1. struggle 2. fittest 3. fitness
4. phenotype 5. phenotype 6. genotype
7. gene pool 8. population 9. context
10. dimorphism 11. secondary 12. mate
13. bell 14. stabilizing 15. directional
16. change 17. diversifying 18. intermediate
19. Natural selection 20. compromises

Testing Your Knowledge

Multiple Choice: 1. e 2. b 3. d 4. d 5. d
6. a 7. c 8. e 9. b 10. b 11. e 12. d

Essay: 1. Because organisms produce more offspring than the resources in their environment can support, only a fraction of these offspring survive and reproduce themselves. Individuals in a population vary in their characteristics, and many of these varying traits are inherited. Those individuals with traits that fit them best to their environment are most likely to survive and reproduce, and thus these individuals leave more offspring than less fit individuals. Over time, the fitter types make up a larger fraction of the population, and through this process of natural selection the population adapts to its environment.

2. Biogeography, the geographical distribution of species, suggests that species change when they move into new habitats. Animals on islands resemble those on the nearest land mass more than those in more distant localities. The fossil record shows sequential changes in living things over time. Comparative anatomy and comparative embryology show the relationships among organisms descended from a common ancestor. Molecular biology shows that organisms thought to be more closely related on the basis of other criteria also have a greater proportion of their DNA and proteins in common.

3. A population is a group of individuals of a particular species living in a particular place at a particular time. A species is all the populations whose individuals can potentially interbreed. The white-tailed deer on an island constitute a population; all white-tailed deer are members of a particular species.

4. Darwin would have said that both horses and zebras are descended from a common ancestral species that lived in the past. As the descendants of this species spread into different environments, they adapted in different ways to different conditions. Horses and zebras eventually became distinct species, but their similarities show that they are related—descended from the same ancestor.

5. In a small population, a chance event might alter the frequency of alleles in the gene pool so as to increase the frequency of an allele that would otherwise be disadvantageous. In a diploid population, a recessive allele affects the phenotype only when two copies of it are present in a homozygous individual. Many harmful recessive alleles in a population might be "hidden" in heterozygous individuals, not exposed to the effects of natural selection. Sometimes natural selection actually favors heterozygotes—a phenomenon known as heterozygote advantage. In this case, heterozygotes might have greater reproductive success than either homozygote, further increasing the frequency of an allele that is harmful in the homozygous recessive state.

6. All these phenomena alter allele and genotype frequencies in the gene pool, but in different ways. In a small population, a chance event could kill some individuals and significantly alter the gene pool. If individuals prefer some types of potential mates to others—that is, if mating is nonrandom—certain alleles are more likely to be passed on than others, and the gene pool can change. Migration of individuals in or out of the population can add new alleles or remove alleles from the gene pool. Mutation creates a new allele, altering allele frequencies. If a certain allele enables individuals with that allele to better survive and reproduce, that allele is more likely to be passed on to the next generation than others,

and its frequency in the gene pool will increase. This is natural selection, and it is the only circumstance that leads to adaptations.

Applying Your Knowledge

Multiple Choice: 1. e 2. a 3. a 4. c 5. b 6. d 7. c 8. b 9. e 10. d 11. b

Essay: 1. Lamarck would have said that the ancestral butterfly had a shorter proboscis. It stretched the proboscis to reach the nectar deep in the flower, and this change was passed on to its offspring. In this way, the proboscis got longer and longer. Darwin would have said that the ancestral butterflies varied—some had longer proboscises than others—and that proboscis length is inherited. Those butterflies with the longest proboscises were able to get the most food from the flowers and were therefore better able to survive and reproduce. A greater proportion of butterflies in the next generation inherited longer proboscises, and over time, this survival of the fittest resulted in an increase in the average length of the proboscis.

2. Darwin saw that by selecting individuals with desired characteristics as breeding stock, humans play the role of the environment and bring about differential reproduction. If this artificial selection can bring about great change in pigeons in a few hundred years, natural selection must be able to cause even greater changes over millions of years.

3. An individual cannot adapt in an evolutionary sense. Each plant has particular characteristics; it can either stand heat or it cannot. Trina recognized that individual plants probably varied in their ability to withstand heat. Perhaps plants with thinner leaves or shallower roots are more likely to wilt and die. Those with thicker leaves or deeper roots might withstand heat and drought better, and are better able to survive and reproduce, passing their genes for the characteristics that enable them to withstand the heat on to the next generation. In the next generation of wildflowers, a greater proportion of individuals will be of the more fit type. Thus individual plants do not adapt, but the population does.

4. Use the formulas $p + q = 1$ and $p^2 + 2pq + q^2 = 1$ to answer these questions. Nine percent of the individuals in the population are homozygous recessive (white). Thus, $q^2 = .09$, and $q = 0.3$. Since $p = 1 - q$, $p = 0.7$. Frequencies of homozygous dominant, heterozygous, and homozygous recessive individuals are 49%, 42%, and 9%, respectively. Frequency of the green allele is 0.7. Frequency of the white allele is 0.3. The genotype and gene frequencies will stay the same in the next generation if the population is at Hardy-Weinberg equilibrium.

5. The population could be made to deviate from Hardy-Weinberg equilibrium in several ways. One could randomly scoop out some fish. In such a small population, this could probably alter allele and genotype frequencies by chance—genetic drift. One could cause gene flow by adding white fish to the aquarium. Subjecting the fish to X rays might cause mutations that would alter the makeup of the gene pool. One could segregate the fish so that mating is nonrandom—only allowing fish of the same color to mate with each other, for example. Putting in a predator that can see and catch the white fish more easily than the green ones would result in natural selection in favor of the green fish and adaptation of the fish population to their environment.

6. The data indicate that the owls are catching a larger proportion of the dark mice than of the light mice. Perhaps the dark mice are easier to see against the white sand. If coloration is inherited, the light mice are more likely to pass their genes on to future generations, so a larger proportion of the mice will be light colored. Directional selection is occurring, and the mouse population is adapting to its environment.

Chapter 14

Organizing Your Knowledge

Exercise 1: 1. Because both are extinct, we must rely on appearance rather than the biological species concept. 2. Do tigers and lions interbreed under natural conditions? Is a tiglon fertile? 3. Appearance can be deceptive; they can interbreed. But would a Chihuahua really be able to mate with a Saint Bernard? 4. The criterion of interbreeding is useless for asexual organisms; we have to classify them by appearance and biochemical features. 5. A and C are becoming separate species, but there is still some gene flow between them via B. 6. If it is difficult to distinguish species on the basis of appearance, it might be difficult to know whether isolated populations are of the same or different species. 7. Are the birds the same species if they look alike? If the populations are separated, it might not be possible to determine whether they could interbreed in nature.

Exercise 2: 1. prezygotic, temporal isolation 2. postzygotic, hybrid breakdown 3. postzygotic, hybrid sterility 4. prezygotic, mechanical isolation 5. prezygotic, behavioral isolation 6. prezygotic, gametic isolation 7. prezygotic, habitat isolation 8. postzygotic, hybrid inviability

Exercise 3: There are many possible answers for each of these questions, for example: 1. ocean, mountain range 2. mountain range, river 3. separate river basins, dry land between lakes 4. mountain range, ocean 5. ocean, mountain range 6. land bridge, deep water between coral reefs

Exercise 4: 1. yes, yes, 36 2. no, no, 33 3. no, no, 35 4. yes, yes, 36 5. yes, yes, 28 6. no, no, 21

Exercise 5: There are many possible examples, such as the following: 1. As a large lake dries up, fish or frogs could be isolated in smaller ponds or springs. 2. Wildflowers or beetles could evolve on isolated mountains surrounded by lowlands. 3. Insects could evolve in isolated caves. 4. Unique plants could evolve on rock outcrops that have different minerals than the surrounding soil.

Exercise 6: 1. smooth, jumpy 2. Darwin, recent 3. gradual transitions, sudden appearance and disappearance 4. slow, fast 5. continuous, quick change and stability

Exercise 7: 1. C 2. G 3. E 4. F 5. A 6. B 7. D

Exercise 8: 1. population 2. species 3. geographical barrier 4. genetic drift 5. natural selection 6. interbreed 7. pre 8. behavioral isolation 9. post 10. hybrid inviability 11. species 12. isolating 13. allopatric 14. punctuated equilibrium 15. polyploid 16. tetraploid 17. Nondisjunction 18. isolated 19. species 20. sympatric

Exercise 9: 1. B 2. D 3. E 4. A 5. C

Testing Your Knowledge

Multiple Choice: 1. b 2. b 3. d 4. d 5. c 6. a 7. d 8. a 9. e 10. b

Essay: 1. It is not possible to see whether fossil species could have interbred. Isolated populations may not naturally have the opportunity to interbreed. The interbreeding criterion does not apply to asexual organisms.

2. A splinter population is cut off from the parent population by a geographical barrier such as a mountain range, a canyon, or a stretch of ocean. Changes in allele frequencies occur in the splinter population, caused by mutation, selection, and genetic drift. These changes are undiluted by gene flow from the parent population. Finally, the splinter population has diverged so much from the parent population that individuals from the two populations can no longer interbreed. The presence of prezygotic or postzygotic barriers tells us that they have become separate species.

3. Imagine two species of flowers: species A, with a diploid number of 14 ($2n = 14$), and species B, with a diploid number of 12 ($2n = 12$). Two things have to happen to produce a fertile hybrid—hybridization and then nondisjunction in meiosis. Pollen from species A ($n = 7$) fertilizes an egg from species B ($n = 6$). This produces a zygote with 13 chromosomes. This zygote can develop into a hybrid plant, but the hybrid would be sterile because its chromosomes cannot pair during meiosis. However, a failure during meiosis—nondisjunction—could produce gametes with 13 chromosomes. Self-fertilization among the gametes would produce a zygote with 26 chromosomes ($2n = 26$). This hybrid would be fertile, because all the chromosomes would be able to pair in meiosis.

4. A small, isolated population is more likely to undergo speciation, because it is more likely to be affected by natural selection and genetic drift in a local area and less affected by gene flow that would make it more like other populations.

5. According to the gradualist model, isolated populations gradually evolve differences as they adapt to local environments. There is a smooth, gradual process of change from one species to the next, through an incremental accumulation of microevolutionary differences. The punctuated equilibrium model suggests that evolution occurs in spurts. New species rapidly diverge from the parent stock and then remain unchanged for long periods. The fossil record more often shows the sudden appearance, persistence, and sudden disappearance of species predicted by the punctuated equilibrium model, rather than the gradual transitions predicted by the gradualist model (although not all changes show up in fossils).

Applying Your Knowledge

Multiple Choice: 1. a 2. e 3. e 4. d 5. a 6. e 7. b 8. c 9. b 10. c

Essay: 1. The western United States has much more rugged terrain than the east. There are many mountain ranges and canyons that would be formidable geographical barriers, isolating populations and promoting allopatric speciation.

2. A cell from a mule would contain 63 chromosomes, 32 from the horse gamete and 31 from the donkey gamete. A mule is sterile because, in meiosis, its chromosomes are unable to pair up and then separate to form haploid cells.

3. This is a gray area, because dogs and coyotes probably would not interbreed under natural conditions in the wild. If dogs and coyotes are different species, this shows us that they are closely related, having recently evolved from a common ancestral species.

4. Dry stretches of desert between streams constitute geographical barriers that isolate fish populations in desert streams, much as stretches of ocean separate land animals on islands. Isolated fish populations might undergo different mutations and genetic drift and be subjected to different kinds of natural selection. They might evolve into different species, like animals isolated on oceanic islands.

5. Flies in newer populations might be less picky because they have not been separated from other populations as long, so they still share some characteristics (courtship signals, and so on) and can be "fooled" by flies of other species. Perhaps females are pickier than males because females only mate once and males can mate repeatedly. Females who make mistakes are not as successful at reproducing, so there is strong selection in favor of females who can tell the difference between their own species and other species.

6. According to the punctuated equilibrium model, species evolve rapidly and then remain unchanged for long periods. If the banana-eating moths are native to Hawaii, and bananas have only been there for 1000 years, this suggests that a new species of moth evolved there in 1000 years or less. This is a short time by evolutionary standards and is a bit of evidence in favor of the punctuated equilibrium model.

7. This criticism is not valid. Formation of a new species by allopatric speciation might take thousands of years. We have not been able to study species for long enough to see the whole process happen, but there is a lot of circumstantial evidence—Darwin's finches, desert pupfish, and so on. More important, we have seen many species of plants appear virtually overnight, via sympatric speciation—hybridization and polyploidy. Hugo de Vries saw this happen in primroses, for example.

8. Evolutionary biologists believe that humans and chimpanzees descended from a common apelike ancestor. One of the changes that sets humans apart from chimps is the way our skulls and brains develop. The skull of an adult human is more rounded and larger in proportion to the body than the skull of an adult chimp. In fact, the shape and proportions of the human skull are more like those of a baby chimp. This retention of human characteristics in an adult is called paedomorphosis, and it was important in the development of the much larger adult human brain. Our hands and binocular vision are exaptations—adaptations that evolved in one context that proved useful for another function. Grasping hands and binocular vision developed as adaptations to life in the trees by the ancestors of chimps and humans. Our hands and eye–hand coordination subsequently made possible the fashioning and use of tools. A glance at the human (hominid) family tree shows that it has many branches—humans of different proportions, with different brain capacities and capabilities. At times, several species of hominids coexisted. All but one, our own species, are now extinct. This may be a result of species selection—competition and unequal survival among different species.

Chapter 15

Organizing Your Knowledge

Exercise 1: A. 8, Phanerozoic, Mesozoic
B. 10, Phanerozoic, Cenozoic C. 3, Proterozoic
D. 5, Phanerozoic, Paleozoic E. 2, Proterozoic
F. 9, Phanerozoic, Cenozoic G. 4, Phanerozoic, Paleozoic H. 6, Phanerozoic, Paleozoic
I. 1, Archaean J. 7, Phanerozoic, Paleozoic

Exercise 2: 1. The mammoth is approximately 17,190 years old. (After 5730 years, the ^{14}C-to-^{12}C ratio would be half as great. After 11,460 years it would be ¼ as great, and after 17,190 years it would be ⅛ as great.)The minimum and maximum ages of the fossil are 15,471 to 18,909 years.

2. Half the potassium-40 would disappear in the first 1.3 billion years, leaving 0.5 grams. Half of this would disappear by 2.6 billion years, leaving 0.25 grams.

Exercise 3: 1. L 2. S 3. G 4. J 5. N 6. O
7. A 8. P 9. R 10. U 11. E 12. F 13. C
14. K 15. I 16. B 17. M 18. D 19. Q
20. H 21. T

Exercise 4: 1. Systematics 2. phylogeny
3. homologies 4. convergent 5. analogy
6. Molecular 7. blood 8. base 9. closely
10. mitochondria 11. ribosomal 12. 550

13. prokaryotes 14. clock 15. homoplasies
16. genomes 17. taxonomists 18. binomial
19. species 20. domain 21. phylogenetic
22. cladistics 23. clades 24. birds

Exercise 5: 1. species 2. genus 3. family
4. order 5. class 6. phylum 7. kingdom
8. domain

Exercise 6: 1. mosses, ferns, conifers, flowering
plants 2. green algae 3. homologous features
4. shared derived characters; shared derived
characters 5. chlorophyll *b* 6. protected embryos 7. chlorophyll *b*, or protected embryos, or
vascular tissues 8. seeds 9. shared derived
characters 10. outgroup: green algae, mosses;
ingroup: ferns, conifers, flowering plants
11. clade or monophyletic taxon 12. conifers
and flowering plants 13. green algae, mosses,
ferns, conifers, flowering plants, plants with
seeds, plants with vascular tissues, plants, organisms with chlorophyll *b*

Exercise 7: 1. B or F 2. F or B 3. D 4. A or C
5. C or A 6. E

Exercise 8: 1. 5-kingdom 2. 3-domain
3. 5-kingdom 4. 3-domain 5. 5-kingdom
6. 3-domain 7. 5-kingdom 8. 3-domain
9. both 10. both

Testing Your Knowledge

Multiple Choice: 1. e 2. e 3. a 4. b 5. d
6. a 7. a 8. b 9. b 10. b

Essay: 1. When the land masses fused, the total
amount of shoreline and shallow coastal areas
was reduced, and many marine species that inhabit shallow waters may have died as their
habitats shrank. Ocean current would have
shifted, altering climate. Less land would have
been near the moderate coastal climate, so the
climate in the interior may have become more
harsh, leading to extinction of land species.

2. These events occurred in the following order:
first prokaryotes, origin of eukaryotes, origin of
animals, appearance of first vertebrates, movement of plants and animals onto land, dominance of dinosaurs and cone-bearing plants,
origin of flowering plants, diversification of
mammals, appearance of humans.

3. Boundaries between geological eras and periods are marked by major transitions in the forms
of life present in the fossil record. For example,
many forms of marine life died out at the end of
the Paleozoic era, and dinosaurs first appeared in
the Mesozoic. These major changes in the history
of life have apparently occurred at irregular intervals, so the geological time scale is uneven.

4. When plates collide, their edges may crumple
and form mountain ranges. The Himalayas
formed this way when India collided with
Eurasia. Volcanoes, such as Mount St. Helens,
often arise where plates grind past each other.

5. Eukarya—Domain, Animalia—Kingdom,
Chordata—Phylum, Vertebrata—Subphylum,
Mammalia—Class, Primates—Order,
Hominidae—Family, *Homo*—Genus, *sapiens*—
species.

Applying Your Knowledge

Multiple Choice: 1. d 2. c 3. a 4. b 5. c
6. a 7. c 8. e 9. d 10. c

Essay: 1. There would be 200 mg after 5,730
years, and 100 mg after 11,460 years. The tree
was buried 11,460 years ago.

2. Figure 15.3B in the text shows that Australia
(with Antarctica) broke away from Gondwana
while the other continents were still joined. This
might have happened after the marsupials got
to Australia but before placental mammals
evolved, so the placentals were never able to
reach Australia and displace the marsupials.

3. Late in the Cretaceous, the climate was
changing. Many food plants used by the dinosaurs died out before the dinosaurs did.
Continental drift may have caused climatic
changes, but were these changes alone enough
to cause the extinction of the dinosaurs? Rocks
in India indicate that at the same time, massive
volcanic activity released particles into the
atmosphere, perhaps contributing to the cooling.
Sediments from the end of the Cretaceous
contain a layer of iridium, an element rare on
Earth but common in meteorites. The iridium
layer may be fallout from a huge cloud of dust
produced when a large meteorite or asteroid hit
the Earth, probably near Mexico, where a large
crater has been discovered. The dust could have
blocked light and disrupted the climate for an
extended period, killing off many plants and the
dinosaurs that depended on them. Finally, it is
possible that a combination of factors is responsible for the demise of the dinosaurs. Many
forms of marine life also perished; any of these
hypotheses could account for the extinction of
both dinosaurs and marine species.

4. One way to determine how closely related
the birds are would be to compare the amino
acid sequences of proteins from the birds.
Samples of hemoglobin, for example, could be
sequenced; the more the sequences are alike,
the closer the relationship between the birds.

Similarly, samples of DNA could be sequenced. The greater the similarity in nucleotide sequences, the closer the relationship.

5. Apparently, the similarities between the two groups of vultures are analogous, not homologous. The birds probably have similar lifestyles and live in similar environments. They have apparently been similarly shaped by natural selection, even though they are not closely related. This is called convergent evolution.

Chapter 16

Organizing Your Knowledge

Exercise 1: A. 6 B. 4 C. 2 D. 3 E. 8 F. 1 G. 9 H. 10 I. 5 J. 7 K. 11

Exercise 2: 1. simple molecules 2. energy from lightning, UV radiation 3. small organic molecules 4. polymerization 5. polypeptides 6. simple RNA replication 7. self-replicating RNA molecules 8. lipid/polypeptide spheres 9. RNA-polypeptide "co-ops" 10. enclosure of co-ops 11. natural selection 12. the first prokaryotic cells

Exercise 3: 1. B 2. A 3. B 4. A 5. A 6. A 7. A 8. B 9. B 10. B 11. A 12. B 13. A

Exercise 4: 1. photoheterotroph 2. chemoautotroph 3. chemoheterotroph 4. photoautotroph

Exercise 5:

Exercise 6: 1. prokaryotes 2. eukaryotes 3. O_2 4. pathogens 5. 50 6. poisons (toxins) 7. exotoxins 8. diarrhea 9. anthrax 10. endotoxins 11. *Salmonella* 12. spores 13. water-treatment 14. sewage 15. antibiotics 16. nutrients 17. legumes 18. nitrogen 19. cyanobacteria 20. organic 21. wastes 22. dead organisms

23. treatment 24. anaerobic 25. sludge 26. aerobic 27. copper 28. bioremediation 29. tick

Exercise 7: 1. Small prokaryotes may have lived inside larger prokaryotes or may have been ingested as food; the engulfed cells and host cell may have come to depend on each other.
2. mitochondrion and chloroplast 3. Alpha proterobacteria are the closest relatives of mitochondria, and cyanobacteria are the closest relatives of chloroplasts. 4. Mitochondria and chloroplasts are like bacteria in several ways: They have their own DNA, RNA, and ribosomes, which are similar to those of bacteria. They transcribe their own RNA and make their own polypeptides. They replicate their own DNA and reproduce by a process like binary fission. They have two membranes, and the inner one resembles a bacterial membrane. 5. Some organelles may have first developed as infoldings—pockets or pouches projecting inward from an ancestral prokaryote's plasma membrane. 6. nuclear envelope and ER 7. nucleus

Exercise 8: 1. LMP 2. B 3. FGH 4. FJK 5. O 6. FJK 7. A 8. F 9. E 10. H 11. IJK 12. N 13. E 14. BGHNO 15. O 16. CDE 17. G 18. HNO 19. B 20. H 21. C 22. JK 23. A 24. D 25. F

Exercise 9: 1. H 2. F 3. E 4. A 5. C 6. D 7. G 8. B

Testing Your Knowledge

Multiple Choice: 1. d 2. a 3. b 4. c 5. a 6. c 7. e 8. e 9. c 10. c 11. a 12. b 13. d 14. d 15. c

Essay: 1. Present-day conditions are different from conditions on Earth when life first appeared. Earth's present atmosphere is rich in oxygen, which tends to disrupt chemical bonds. Early Earth probably had a reducing atmosphere, instead of an oxidizing one. It would have added electrons to molecules, causing chemical bonds to form and causing simple molecules to combine to form more complex ones.

2. Miller's experiment showed how simple organic compounds could have formed from inorganic chemicals on the primitive Earth. A flask containing warm water simulated the sea, while CH_4, H_2, and NH_3 simulated the early atmosphere. Electrical sparks introduced into the flask simulated lightning. A condenser cooled water vapor and caused "rain" that washed dissolved compounds into the "sea." As material circulated through the apparatus, the sparks triggered

chemical reactions that formed a variety of organic compounds, including a mixture of amino acids.

3. Small nucleic acid molecules—probably RNA at first—may have developed the ability to replicate. Polypeptides—simple proteins—might have acted as enzymes to catalyze this process, while the RNA "genes" may have come to code for the simple proteins. Polypeptides tend to self-assemble into fluid-filled spheres, which could have acted as a protective, semipermeable membrane, enclosing these gene-enzyme "co-ops."

4. Bacteria cause tuberculosis, cholera, bubonic plague, and typhoid fever. Most bacteria cause illness via poisons—exotoxins secreted from cells or endotoxins that are components of cell walls.

5. Prokaryotes are important in the cycling of chemical elements between living things and the nonliving environment. Cyanobacteria restore oxygen to the atmosphere. Some bacteria take nitrogen from the air and convert it to forms in the soil that can be used by plants to make proteins and nucleic acids. Prokaryotes also act as decomposers, breaking down organic matter in dead organisms and body wastes and converting them to inorganic forms that are available to organisms for reuse. Some prokaryotes might be useful in similarly breaking down spilled petroleum and toxic wastes.

6. Primitive cyanobacteria, such as those that formed fossil stromatolites, dominated the Earth from about 3 billion to about 1 billion years ago. Their photosynthesis released enough oxygen (O_2) gas to make the atmosphere aerobic, setting the stage for the evolution of aerobic organisms such as animals.

7. The endosymbiosis hypothesis suggests that eukaryotic cells may have started out as cooperative communities of prokaryotic cells. An ancestral prokaryote may have ingested small heterotrophic bacteria that were able to use O_2 to release energy from organic molecules via cellular respiration. These cells remained alive inside the host cell and continued to perform respiration. They eventually became mitochondria. Similarly, small photosynthetic prokaryotes may have taken up residence inside host cells and become chloroplasts.

8. Multicellularity probably arose several times, with today's seaweeds, plants, animals, and fungi evolving from different kinds of protists. Unicellular protists may have formed colonies of interconnected cells. Such a colony might form when a cell divided and its offspring remained connected. The cells gradually became more specialized and interdependent, with specialized cells

coming to perform specific tasks, such as locomotion, feeding, gas exchange, and reproduction.

Applying Your Knowledge
Multiple Choice: 1. c 2. c 3. b 4. c 5. a 6. e 7. a 8. c 9. b 10. c

Essay: 1. This may be a problem for theories about the origin of life, because many different organic compounds make up a living thing. If a given set of starting conditions produce only certain amino acids, but not ATP or sugars, it is hard to see how this could lead to the origin of life. On the other hand, it is possible that different chemicals could have been formed under different conditions at different times or in different locations. Given enough time, this mixture of chemicals might have collected in one spot, giving rise to life.

2. There is no reason to believe that the chemical reactions that gave rise to organic molecules on the primitive Earth are restricted to our own neighborhood. It is possible that similar reactions could happen elsewhere, even in clouds of gas and dust in space. Organic matter in meteorites leads scientists to believe that the reactions that form organic molecules might be quite likely to happen in space or on other planets. Organic matter from space may have had a role in the origin of life on Earth, and it may indicate that life has arisen on other planets. (Of course, it is possible that organic compounds originating on the Earth could have contaminated the meteorites after they got here.)

3. The fact that UV light breaks chemical bonds poses some problems for the hypothesized role of UV in the origin of organic compounds. If UV light provided energy for the reactions that caused inorganic compounds to combine and form simple organic compounds, why did it not then cause these same organic compounds to break apart?

4. If the diphtheria bacterium is like most bacteria, it causes illness by producing a toxin (in this case probably an exotoxin secreted by the bacteria) that can cause damage or interfere with body functions at places removed from the site of infection.

5. The number of bacteria would double every 20 minutes, so they would double 18 times in 6 hours. After 20 minutes, there would be 200, after 40 minutes 400, after an hour 800, and so on. There would be more than 26 million bacteria at the end of 6 hours!

Chapter 17

Organizing Your Knowledge

Exercise 1: Differences: 1. Green algae are supported by water; leaves and stems of plants usually contain rigid supporting elements. 2. Whole body of green alga exposed to water, light, minerals; only above-ground parts of plant exposed to light and roots in contact with soil nutrients and water. 3. Green alga does not have true organs with specialized functions; plant has discrete organs—roots, stems, and leaves. 4. Plants have growth-producing regions at the tips of roots and stems. Algae lack these specialized regions. 5. Green algae lack mycorrhizae; most plants have mycorrhizae. 6. Green algae have no cuticle; plants have a waxy cuticle that prevents drying. 7. Green algae lack stomata; gas exchange occurs via stomata in plants. 8. Green algae lack vascular tissues; plants have vascular tissues that transport nutrients and water. 9. In green algae, the embryo develops in water; in plants the embryo develops in the female gametangium. 10. Many plants have resistant spores and/or seeds that aid in dispersal; green algae do not. Similarities: 1. Have similar mechanism for making cell walls of cellulose. 2. Have similar mechanism for forming cell plate that divides cytoplasm in cell division. 3. Their peroxisomes contain similar enzymes. 4. Have similar flagellated sperm. 5. Nuclear and chloroplast genes are similar.

Exercise 2: A. 3, gymnosperms, angiosperms B. 1, bryophytes C. 4, angiosperms D. 2, bryophytes, ferns

Exercise 3: 1. sporophytes 2. haploid 3. diploid 4. gametophyte 5. gametes 6. gametangia 7. sperm 8. flagella 9. fertilization 10. zygote 11. sporangium 12. spores 13. meiosis

Exercise 4: 1. meiosis 2. spores 3. mitosis and development 4. gametophyte 5. sperm 6. egg 7. fertilization 8. zygote 9. mitosis and development 10. sporophyte 11. sporangia 12. haploid phase 13. diploid phase

Exercise 5: 1. pine tree 2. male cones, female cones 3. male cones, pollen grains 4. female cones, ovules 5. ovule 6. ovule 7. eggs, sperm 8. pollen grain, sperm, egg 9. zygote, embryo, seed 10. seed, embryo 11. pine tree

Exercise 6: 1. D 2. E 3. I 4. A 5. F 6. C 7. G 8. H 9. B

Exercise 7: 1. meiosis 2. pollen grain (male gametophyte) 3. female gametophyte 4. egg

5. pollination 6. stigma 7. pollen tube 8. sperm 9. fertilization 10. zygote 11. seed 12. embryo 13. seed coat 14. food supply 15. fruit 16. sporophyte 17. ovary 18. ovule 19. anther 20. diploid phase 21. haploid phase

Exercise 8: 1. cones 2. flowers 3. wind 4. insects, and so on 5. a year or more 6. 12 hours or so 7. more than a year 8. days or weeks 9. seeds unprotected 10. seeds protected and dispersed by fruit

Exercise 9: 1. A 2. G, A 3. F 4. F 5. G 6. M 7. A 8. M 9. M, F 10. G 11. A 12. F 13. A 14. G, A 15. F 16. G 17. M 18. G 19. A 20. M, F 21. A 22. A

Exercise 10: 1. 50 2. 50 3. 20 4. 40 5. 25 6. 5,000, 295,000 7. 25

Exercise 11: 1. J 2. E 3. P 4. Q 5. M 6. C 7. T 8. W 9. H 10. U 11. S 12. F 13. D 14. L 15. V 16. K 17. O 18. G 19. I 20. B 21. R 22. A 23. N

Exercise 12: 1. three 2. spores 3. hyphae 4. mycelium 5. mating types 6. nuclei 7. heterokaryotic 8. fruiting body 9. diploid 10. meiosis

Exercise 13:

Testing Your Knowledge

Multiple Choice: 1. a 2. d 3. b 4. d 5. a 6. b 7. c 8. b 9. b 10. e 11. d 12. e 13. d 14. c 15. c

Essay: 1. Algae are supported by water, and they lack supporting tissues. The leaves and stems of plants contain rigid supporting elements. The whole body of an alga is exposed to water, light, and minerals, while only aboveground parts of the plant are exposed to light and only the roots are in contact with a reliable supply of nutrients and water. Plants have discrete organs—roots, stems, and leaves, and specialized growing regions at the tips of stems and roots—while algae do not have true organs with specialized functions. Plants have mycorrhizae, but algae do not. Plants are covered by a cuticle that prevents drying, pierced by stomata for gas exchange, but algae have no need for such a covering or stomata. Plants have vascular tissues that transport nutrients and water, but algae do not. In plants, zygotes and embryos develop within the female gametangium.

2. Many people think all are plants because they are rigid, "rooted" in one spot, and cannot move. Seaweeds lack the specialized interdependent tissues and organs seen in plants. Fungi are heterotrophic, while plants and seaweeds are autotrophic—capable of making their own food via photosynthesis.

3. Gymnosperms and angiosperms—the cone-bearing and flowering plants—both produce seeds. Seed plants do not require water for fertilization. Instead of sperm that swim to the eggs, seed plants produce pollen grains that protect sperm and carry them through the air to the female parts of the plant. The seed is a survival packet for the embryo—a protective coat and a food supply that enable the embryo to get a good start.

4. Gymnosperms—conifers such as pines and firs—supply lumber for construction, wood pulp for making paper, and most other forest products. Forests are also important for wildlife habitat, watersheds, and recreation. We can reduce the cutting of forests by recycling lumber and paper, and by cutting trees no faster than they can be replanted.

5. See Figure 17.9B in the text.

6. Animals assist angiosperms by carrying pollen from anthers to carpels and by dispersing seeds. Flowers produce nectar and pollen, which are attractive to animals, and advertise their presence with showy shapes, colors, and scents. Fruits sometimes cling to the fur of animals or attract animals to eat them and thereby disperse seeds in their droppings.

7. The threadlike hyphae of fungi grow through their food and secrete enzymes that digest the food outside their bodies. The fungal cells then absorb the digested nutrient molecules through their cell membranes.

8. A lichen consists of a fungus growing in a mutualistic relationship with green algae or cyanobacteria. Their exact roles are not completely understood, but it is known that the fungus receives food from the algae or cyanobacteria, and the photosynthetic cells receive protection, water, and certain minerals from the fungus.

Applying Your Knowledge

Multiple Choice: 1. e 2. d 3. c 4. a 5. b 6. e 7. b 8. b 9. c 10. d

Essay: 1. Like seed plants, ferns have true roots, stems, and leaves; vascular tissues that transport water and nutrients; and a life cycle in which the sporophyte stage predominates. Like mosses, ferns have flagellated sperm that swim to the egg.

2. Mosses and ferns are dispersed in the form of spores. Gymnosperms and angiosperms are dispersed in the form of seeds. Spores are very small and can be carried by the wind for long distances, but they are relatively vulnerable and short-lived. Seeds are bigger and cannot travel as far as fast (at least without help from animals), but they are protected by a seed coat and carry their own food supply that the embryo can draw on for a good start in life.

3. What probably happened is that the farmer killed beneficial mycorrhizal fungi along with the harmful parasites. Without the help of the mycorrhizae, the trees are not able to absorb water and soil minerals effectively.

4. Fungi can actually grow into other cells, where they are protected. They grow very rapidly, which enables them to come back quickly if not completely killed off. (Also, since they are eukaryotes, their metabolism is more like our own cells than is the metabolism of pathogenic bacteria. This makes them harder to kill than bacteria, because chemicals that injure fungi might injure human cells as well.)

5. Naming a lichen is tricky because the lichen consists of two organisms—a fungus and an alga or cyanobacterium. Does the lichen get one name or two?

Chapter 18

Organizing Your Knowledge

Exercise 1: The black line goes between prokaryotes and protists; prokaryotes have simple cells lacking nuclei and other organelles, while the other organisms have more complex eukaryotic cells. The blue line goes above protists; they are primarily unicellular, and those above them multicellular. The green line goes between plants and fungi; plants are autotrophs, and fungi and animals are heterotrophs. The red line goes between fungi and animals; fungi digest their food and then absorb the nutrients; animals ingest food and then digest it.

Exercise 2: See Figure 18.2 in the text. A colony of cells may have become a hollow sphere. Cells of the colony then specialized and folded inward, forming two layers of cells performing various functions.

Exercise 3: 1. flatworms, crustaceans, insects, humans, and so on 2. parts arranged in a circle around a central axis 3. no dorsal and ventral surfaces 4. anterior and posterior ends 5. distinct head 6. sedentary or passively drifting 7. head with sense organs contacts environment first

Exercise 4: See Figure 18.3D in the text.

Exercise 5: 1 Porifera 2. bilateral 3. Cnidaria 4. Platyhelminthes 5. pseudocoelom 6. no notochord or backbone 7. Chordata 8. jointed legs 9. no jointed legs 10. Arthropoda

Exercise 5: 1. Porifera 2. Cnidaria 3. sea 4. radial 5. pores 6. two 7. flagella 8. enters 9. exits 10. bacteria 11. skeletal 12. nerves 13. tissues 14. protists 15. jellies 16. digestive 17. nervous 18. tissues 19. two 20. organ systems 21. polyp 22. medusa 23. jellies 24. hydra 25. tentacles 26. cnidocytes 27. underneath 28. gastrovascular 29. wastes

Exercise 6: 1. Cnidaria; Platyhelminthes; Nematoda 2. jelly, hydra; planarian, tapeworm; *Caenorhabditis, Trichinella* 3. radial; bilateral; bilateral 4. 2; 3; 3 5. medusa or polyp; flattened ribbon or leaf shape; cylinder 6. none; none; pseudocoelom 7. incomplete; incomplete or none; complete 8. fresh water and salt water; fresh water, salt water, soil, and parasites in animals; water, soil, and parasites in animals and plants 9. build coral reefs; human and animal parasites; human, animal, and plant parasites

Exercise 7: 1. E 2. G 3. C 4. G 5. D 6. C 7. B 8. F 9. G 10. E 11. D 12. A 13. C 14. D 15. D 16. E 17. E 18. G 19. C 20. E

Exercise 8: 1. Annelida 2. segments 3. sea 4. fresh water 5. walls 6. nervous 7. excretory 8. digestive 9. closed 10. movement 11. flexibility 12. mobility 13. circular 14. feces (castings) 15. polychaetes 16. ocean 17. tubes 18. appendages 19. feeding 20. leeches

21. blood 22. carnivores 23. fresh water 24. teeth 25. blood clots 26. Arthropods

Exercise 9:

Exercise 10: Circle the following: 1, 4, 6, 8, 9, 10, 12, 13, 14, 15, 16.

Exercise 11: 1. dorsal hollow nerve cord 2. notochord 3. pharyngeal slits 4. muscular postanal tail See Figure 18.14B in the text.

Exercise 12: 1. H 2. I, J 3. I 4. H 5. A, B 6. J 7. C, D 8. F, G 9. I 10. J 11. F 12. G 13. J 14. J 15. I 16. H, I, J 17. A 18. F 19. J 20. I 21. E, F, G 22. B 23. J 24. G 25. J 26. H 27. I 28. D 29. E 30. H 31. H 32. I 33. A 34. E

Exercise 13: 1. G 2. B 3. F 4. H 5. E 6. C 7. A 8. I 9. D 10. K 11. M 12. N 13. O 14. J 15. L

Exercise 14: 1. G 2. B 3. D 4. E 5. A 6. A 7. C 8. F

Testing Your Knowledge

Multiple Choice: 1. b 2. b 3. c 4. a 5. a 6. b 7. b 8. a 9. c 10. e 11. c 12. e 13. d 14. c 15. d

Essay: 1. Animals are eukaryotic, multicellular heterotrophs that lack cell walls. They ingest their food and usually digest it in a digestive tract. Animals are diploid, except for their haploid eggs and sperm. The animal zygote usually goes through a blastula stage and a gastrula stage, and there may be a larval stage that metamorphoses into the adult.

2. Some characteristics important in determining an animal's phylum are number of cell layers, body symmetry, presence of tissues and organs, presence and type of body cavity, embryonic development of the body cavity, and presence or absence of segmentation.

3. All animal phyla discussed in this chapter have some kind of body cavity, except for sponges, cnidarians, and flatworms. The body cavity aids in movement, cushions internal organs, and may stiffen the body and help in circulation.

4. The mantle secretes the shell, but it can have other functions. In some bivalves, such as scallops, sensory structures such as eyes are arranged along the edge of the mantle. It functions in waste disposal in some mollusks. The mantle cavity houses the gills, and it is modified as a kind of lung in land snails. In squids, the mantle cavity can expel water for jet propulsion. Snails have a one-piece coiled shell. The two-part hinged shell of a clam can be closed for protection or opened for feeding. Squids have a small internal shell that stiffens the body.

5. Insects are arthropods, so they are covered by a tough exoskeleton that protects them from injury and drying out. They have jointed appendages that are modified for various functions—locomotion, sensing the environment, and feeding. Many insects have highly specialized mouthparts. Insects are small and can reproduce rapidly. They are the only invertebrates with wings, and they use their flying ability to find food, escape predators, and disperse to new habitats.

6. Reptile skin is covered by scales made of keratin, a waterproof protein that keeps the body from drying out. Reptiles are also able to lay their amniotic eggs on land. The reptile embryo develops within a hard shell and a protective, fluid-filled sac called the amnion. The young reptile does not go through an aquatic larval stage but rather emerges from the egg as a miniature adult, ready to cope with life on dry land. Amphibians have thin, moist skin that dries out easily. They must lay their jelly-covered eggs in water, and their tadpole larvae must develop in water.

7. Birds have feathers, which form a lightweight covering and shape the wings. They have many modifications to reduce their weight: no teeth, a reduced tail, hollow bones and feathers, and air sacs in the body. They have large flight muscles attached to a keeled breastbone. They are endothermic, maintaining a high constant temperature and metabolic rate. The bird's circulatory system and lungs are highly efficient in delivering food and oxygen for the rapid metabolism needed for flight.

Applying Your Knowledge

Multiple Choice: 1. d 2. c 3. a 4. e 5. c
6. a 7. c 8. b 9. a 10. a 11. c

Essay: 1. Like all other animals, sponges are eukaryotic, multicellular heterotrophs that lack cell walls, ingest their food, and are mostly diploid.

2. The temporary digestive cavity of *Trichoplax* suggests a hypothesized stage in animal evolution when a relatively unspecialized colony of cells may have folded inward to form a digestive cavity, much as a blastula becomes a gastrula during animal development.

3. Pigs and people are infected when they eat undercooked meat. If meat is thoroughly cooked, the worms will be killed. Pigs can also become infected by ingesting the feces of infected humans or animals. A way to prevent the spread of the worms might be to use good sanitation and keep the pigs away from sewage.

4. Your criteria might include number of species or individuals or how widespread the animals are. By these criteria, arthropods and chordates are very successful. You might discuss particular structural, functional, or behavioral features or adaptations that contribute to their success.

5. The larvae suggest that mollusks and annelids are related. These two phyla are on adjacent branches of the animal phylogenetic trees. They are both bilaterally symmetrical animals with a coelom that forms within masses of cells in the middle tissue layer.

6. A dog is a chordate because it has a dorsal hollow nerve cord (the brain and spinal cord), a flexible supporting rod called a notochord, pharyngeal slits behind the mouth, and a muscular post-anal tail (a tail posterior to the anus). It is a vertebrate because it has an internal skeleton made of bone, consisting of a skull, a backbone of vertebrae, and bones supporting its appendages—the legs. In the adult dog, the notochord is replaced by the vertebral column, and the pharyngeal slits disappear, but these structures are present in a dog embryo.

7. Birds and mammals are endothermic. Unlike reptiles and amphibians, they maintain a constant high rate of metabolism and body temperature, enabling them to function in a cold environment.

Chapter 19

Organizing Your Knowledge

Exercise 1: 1. sensitive grasping hands with opposable thumbs 2. eyes positioned for depth perception, limber shoulder joints, and opposable thumbs 3. limber shoulder joints and opposable thumbs 4. eyes positioned for depth perception and sensitive hands with opposable thumbs 5. limber shoulder and hip joints, depth perception 6. eyes positioned for depth perception and sensitive hands with opposable thumbs 7. eyes positioned for depth perception, limber shoulder joints, sensitive hands

Exercise 2: 1. I 2. EFGHI 3. EFGHI 4. CDEFGHI 5. C 6. D 7. AB 8. B 9. A 10. H 11.A 12. C 13. D 14. E 15. F 16. G 17. C 18. D 19. EFGHI 20. C 21. D 22. E 23. A 24. CDEFGHI 25. C 26. D 27. H 28. K 29. P 30. R 31. O 32. J 33. Q 34. M

Exercise 3: 1. Bipedal walking 2. Shorter jaw and flatter face 3. Larger skull and brain 4. Language 5. Symbolic thought 6. Manufacture and use of complex tools 7. Long-term pair bonding between mates 8. Reduced sexual dimorphism 9. Longer period of parental care

Exercise 4: 1. C 2. B 3. A 4. B 5. E 6. D 7. B 8. F 9. C 10. G 11. F 12. G 13. G 14. D 15. G 16. E 17. B 18. E 19. G

Exercise 5: 1. Researchers have found and studied fossils of hundreds of ancient hominids, dating back seven million years. They have identified dozens of species of ancient hominids. In addition, they can now use DNA comparisons to track our origins. The "missing link" is no longer missing. 2. Humans and chimps are closely related; their genomes are 99% identical. But humans are not the descendants of modern-day chimps. An ape that lived 5–7 million years ago appears to have been the common ancestor of both chimps and humans. 3. The hips and skulls of *Australopithecus* fossils show that hominids walked upright when their brains were still chimp-sized. 4. The human evolutionary tree is a many branched bush. At times in the past, several species of hominids coexisted. All but one of the branches died out—the one leading to us—so most fossil hominids were not our ancestors. 5. DNA analysis of fossil Neanderthal bones proves pretty convincingly that there was little or no interbreeding between Neanderthals and our ancestors, and thus they made little or no genetic contribution to present-day humans. 6. It is true that hominids first appeared in Africa and *Homo erectus* spread around the globe. But then *Homo sapiens* appeared in Africa, and in a second wave of migration, displaced earlier hominids all over the world.

Exercise 6: 1. Hunting/gathering/scavenging 2. Machines 3. Hunting/gathering/scavenging 4. Hunting/gathering/scavenging 5. Agriculture 6. Machines 7. Agriculture 8. Machines 9. Hunting/gathering/scavenging 10. Agriculture 11. Hunting/gathering/scavenging

Testing Your Knowledge

Multiple Choice: 1. c 2. d 3. b 4. d 5. e 6. e 7. c 8. d 9. b 10. b

Essay: 1. Primates have limber hip and shoulder joints and grasping hands and feet, useful for climbing and manipulating objects. They have an acute sense of touch in their hands and feet. Their snouts are short, and their eyes are set close together on the front of their face, which enhances depth perception.

2. Chimpanzees (and closely related bonobos) are the living primates most closely related to humans. They are apes and share numerous anatomical similarities with humans. Chimpanzees and humans share behavioral similarities as well. Chimps are very intelligent. They can make and use simple tools; they raid other social groups of their own species; they can learn sign language; and their behavior in front of mirrors indicates that they are, to some degree, self-aware. Biochemical evidence also shows the close relationship between chimps and humans; their genomes are 99% identical.

3. a. Anthropoids have relatively larger brains and depend more on sight and less on sense of smell than prosimians. b. The nostrils of New World monkeys face forward and those of Old World monkeys are narrow and point downward, and their tails are not prehensile. c. Monkeys have tails; apes do not. Monkeys have front limbs that are equal in length to their hind limbs; the front limbs of apes are longer than their hind limbs. d. Most apes walk with their knuckles on the ground; humans are bipedal, walking upright. Humans have shorter jaws and larger brains than other apes. (There are other differences, such as language and use of tools, but chimps share these traits to some degree.)

4. The three major milestones in our evolution were: (1) evolution of our bipedal stance; (2) enlargement of the brain; (3) evolution of a prolonged period of parental care.

5. Early African hominids may have driven several species of saber-toothed cats to extinction. *Homo sapiens* may have been responsible for wiping out many large animals in Europe, Australia, and the New World. Intensive farming and overgrazing in the Middle East depleted the soil and exposed it to erosion, leaving much of the area a desert.

6. Humans have not changed much physically in the last 100,000 years. But our culture—accumulated knowledge, customs, beliefs, arts, and especially technology—has grown and changed enormously. We alter our environment to suit our needs and desires, but changing the Earth much faster than we or other living things can adapt biologically. The human population continues to grow, and all life on Earth is threatened by the changes we have wrought.

7. DNA analysis of modern humans shows that African lineages are the oldest branches in the human family tree. Studies of mitochondrial DNA (which is passed on maternally) and Y chromosomes (passed on from father to son) in modern human populations show that all living humans inherited their mitochondrial DNA from a woman who lived 150,000 years ago. The Y chromosome points to an African ancestor 100,000 years ago. The oldest fossils of *Homo sapiens*—160,000 years old—have been found in Africa (Ethiopia), so fossils support genetic data.

Applying Your Knowledge

Multiple Choice: 1. b 2. a 3. d 4. b 5. c 6. b 7. e 8. d 9. e 10. e

Essay: 1. Humans must contend with a tradeoff between a pelvis adapted for bipedal walking, which limits the width of the hips and birth canal, and our enlarged brain. If fetal development were longer, the baby's head would not pass through the birth canal. This means babies are born helpless and need a long period of parental care, but this long childhood contributes to learning and transmission of culture.

2. Chimpanzees display all of these "human" characteristics. They sometimes strip a blade of grass and use it as a tool to fish for termites. Chimps can learn human sign language, although it is not known how much they use symbolic language in the wild. A chimpanzee looking in a mirror displays behaviors—examining its body, making faces—that suggest it has a concept of self.

3. Hominids are thought to have diverged from other apes 6 to 8 million years ago and started walking upright. A large brain evolved later. A small-brained bipedal hominid 5.0 million years old fits current theory. A large-brained hominid that retained its quadrapedal ways 2.5 million years ago would be harder to explain.

4. When did humans diverge from the apes? What was the common ancestor of apes and humans like? Are the australopiths our ancestors, or simply another branch of the hominid family tree? Can all modern humans be traced back to a single origin? When? Where? Which hominid was our immediate ancestor? What is the role of the Neanderthals in human evolution, and what happened to them? Might there have been other so-far unknown species of hominids that once co-existed with *Homo sapiens*? What triggered the rapid cultural evolution of the last 100,000 years? What next?

Chapter 20

Organizing Your Knowledge

Exercise 1: 1. The tail is flattened into broad flukes, which propel the whale through the water. 2. The hummingbird's long, thin beak enables it to sip nectar from tubular flowers. 3. The sensitive nerve endings and finely controlled muscles of the hand enable you to grasp and manipulate objects. 4. The frog's long legs enable it to leap to safety, and its webbed feet make it a good swimmer. 5. The mosquito's mouthparts allow it to pierce the skin and suck blood. 6. The stomach chambers carry out various steps in breaking down hard-to-digest grass.

Exercise 2: 1. cells 2. tissue 3. epithelial 4. connective 5. nervous 6. organ 7. system 8. organism 9. organism 10. systems 11. organs 12. tissues 13. cells

Exercise 3: 1. stratified squamous epithelium 2. multiple layers of flattened cells 3. lining of esophagus, epidermis of skin 4. covers and protects surfaces subject to abrasion 5. simple cuboidal epithelium 6. single layer of cube-shaped cells 7. tubular passageways where urine forms in kidneys 8. simple squamous epithelium 9. linings of lungs and blood vessels 10. exchange of materials by diffusion 11. simple columnar epithelium 12. single layer of elongated, cylindrical cells 13. secretes and absorbs in walls of digestive tract

Exercise 4:

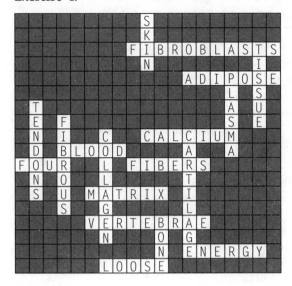

Exercise 5: 1. B 2. D 3. C 4. B 5. D 6. C
7. D 8. C 9. A 10. A 11. A 12. D 13. C
14. B 15. A

Exercise 6: 1. nervous 2. muscular 3. skeletal
4. digestive 5. respiratory 6. circulatory
7. lymphatic 8. integumentary 9. immune
10. excretory 11. endocrine 12. reproductive

Exercise 7: 1. X-rays 2. examination of hard tissues—bones, cartilage, dense tumors 3. X-ray cross sections combined by computer 4. cross sections and three-dimensional views of normal and abnormal hard and soft tissues and changes in brain and elsewhere 5. magnets, radio pulses, and computer 6. visualizes soft tissue such as brain, but does not visualize bone 7. magnets, radio pulses, computer 8. 3-D microscopic images of small structures 9. radiation from radioactive isotope picked up by detector 10. measuring metabolic activity, especially of brain 11. magnets, radio pulses, computer 12. tracks changes in blood flow

Exercise 8: See Figure 20.12B in the text.

Exercise 9: 1. increase in temperature 2. eating a meal 3. hypothalamus (brain) 4. pancreas 5. increase in blood temperature 6. increase in blood sugar level 7. nerve impulses 8. hormone—insulin 9. sweat glands and blood vessels in skin 10. cells 11. sweat glands secrete sweat, and blood vessels dilate and fill with warm blood, cooling the body 12. cells take up sugar and lower blood sugar level 13. around 37°C 14. 70 to 110 mg of sugar per 100 mL of blood

Testing Your Knowledge

Multiple Choice: 1. b 2. d 3. c 4. d 5. b
6. e 7. b 8. a 9. d 10. e 11. b

Essay: 1. A wing is light—mostly feathers—but strong. Powerful breast muscles move the wing. Its broad surface catches the air and generates lift that keeps the bird aloft.

2. Epithelial tissue consists of sheets of closely packed cells that cover and line body surfaces. It protects, secretes, and absorbs. Connective tissue consists of sparse cells scattered in a matrix, which usually is a web of fibers embedded in a liquid, jelly, or solid. Most connective tissues form a framework that supports and protects the organs of the body. Muscle tissue consists of bundles of elongated cells called fibers. Muscle cells can contract and are responsible for body movement. Nervous tissue consists of cells called neurons and other supporting cells that form a communication and coordination system within the body. Neurons have elongated extensions that transmit signals to other cells.

3. The digestive system ingests food, breaks it down, and absorbs it into the bloodstream. The respiratory system supplies the blood with oxygen and disposes of carbon dioxide. The circulatory system transports nutrients and oxygen to cells and carries away wastes. The lymphatic system returns fluid to the blood and plays a role in body defense. The immune system protects the body from disease and foreign substances. The excretory system disposes of metabolic wastes. The endocrine system and nervous system control and coordinate body activities via hormones and nerve signals. The male and female reproductive systems produce gametes and support the growth of the developing embryo. The muscular system moves the body and its parts; the skeletal system supports and protects the body; and the integumentary system covers the body.

4. Computerized tomography uses less powerful (and therefore less potentially damaging) X-rays than conventional X-rays. A CT scan reveals more of the fine detail of soft tissues. It yields cross-sectional views, which can reveal features that might block one another on conventional X-rays, and these cross sections can be combined by computer to give three-dimensional views.

5. In a house, the stimulus is a change in room temperature, and in the body it is a change in the temperature of the blood. The control center

is the thermostat in a house, and in the body it is a "thermostat" in the brain. The home thermostat sends a signal to the furnace—the effector—which responds to an increase in room temperature by turning off and allowing the house to cool a bit. When blood temperature goes up, the brain sends a signal to body effectors—sweat glands and blood vessels in the skin. The sweat glands respond by secreting sweat, which increases evaporative cooling. The blood vessels dilate to bring warm blood to the surface, where it loses its heat and cools the body.

6. Small animals have a larger surface-to-volume ratio than larger animals, and most of their cells may be in direct contact with their surroundings, so their body surfaces may be sufficient to exchange materials with the environment. Larger animals have smaller surfaces in relation to overall volume, so they require specialized surfaces for exchange of materials with the environment.

Applying Your Knowledge

Multiple Choice: 1. b 2. c 3. c 4. e 5. c 6. c 7. e 8. c

Essay: 1. A woodpecker displays many features that illustrate the correlation between structure and function. Its keen hearing enables it to detect insects inside a tree. It uses its sharp, chisel-like bill to cut a hole in the tree and its long tongue to probe for insects. Two toes on each foot face forward and two backward, enabling the woodpecker to cling to a vertical tree trunk, and its stout tail is used as a brace as it drills for insects.

2. An anatomist studies the structure of the body and so might be interested in the shape of the fish's fins and body, the structure and arrangement of the penguin's feathers, and how the insect's exoskeleton protects it from its enemies. A physiologist studies body function and so might be interested in how the fish's muscles propel it through the water, how the penguin generates body heat, and how the insect's senses warn it about potential predators.

3. Bone cells are embedded in a matrix of rope-like collagen fibers and hard calcium salts. This combination makes bone hard without being brittle. Simple squamous epithelium is a single layer of thin cells, forming the lining of the lungs and blood vessels. Gases and liquids can easily diffuse through this thin layer, facilitating exchange. Blood cells are small and round and

suspended in a liquid matrix, enabling them to be pumped around the body to transport oxygen and defend against infection.

4. The digestive system takes in, breaks down, and absorbs food into the blood. The respiratory system draws in air, and oxygen enters the blood in the lungs. The food and oxygen are transported to the brain by the circulatory system—the heart, blood, and blood vessels. The nervous and endocrine systems control these activities.

5. When blood calcium gets too high, the parathyroid glands slow absorption of calcium by the intestine and speed up excretion of calcium by the kidneys. When blood calcium gets too low, the parathyroids increase absorption by the intestine and slow excretion by the kidneys. This illustrates negative feedback because a change in the concentration of calcium triggers the control mechanisms to counteract further change in the same direction.

Chapter 21
Organizing Your Knowledge

Exercise 1: 1. herbivore, bulk feeder 2. omnivore, substrate feeder 3. herbivore, fluid feeder 4. carnivore, suspension feeder 5. omnivore, bulk feeder 6. carnivore, absorptive feeder 7. carnivore, bulk feeder 8. carnivore, fluid feeder 9. omnivore, suspension feeder 10. herbivore, bulk feeder

Exercise 2: Label as in Figure 21.3B in the text. Color as follows: *Earthworm:* mouth, pharynx, esophagus, and crop yellow; gizzard red; intestine red and green; anus blue. *Grasshopper:* mouth, esophagus, and crop yellow; gizzard red; stomach red and green; gastric pouches green; intestine green and blue; anus blue. *Bird:* mouth, esophagus, and crop yellow; stomach and gizzard red; intestine red and green; anus blue.

Exercise 3: See Figure 21.4 in the text.

Exercise 4: 1. incisor 2. liver 3. circular 4. anus 5. gallbladder 6. pyloric sphincter 7. pharynx 8. liver 9. molars 10. longitudinal 11. duodenum (small intestine) 12. pancreas 13. tongue 14. large intestine 15. canines 16. oral cavity 17. stomach 18. bolus 19. salivary glands 20. small intestine 21. large intestine 22. epiglottis 23. stomach 24. esophagus 25. rectum 26. oral cavity 27. pharynx (trachea is respiratory)

Exercise 5: 1. esophagus 2. peristalsis 3. acid chyme 4. gastric glands 5. gastric juice 6. gastrin 7. mitosis 8. hydrochloric acid 9. pepsin

10. proteins 11. small intestine 12. protects
13. mucous 14. mucus 15. heartburn 16. gastric ulcer 17. mucus 18. infection 19. pyloric sphincter 20. small intestine 21. ulcer
22. bile 23. pancreas 24. neutralizes 25. liver
26. fats 27. emulsification 28. lipase 29. proteins 30. amino acids 31. starch 32. monosaccharides 33. absorption 34. villi
35. microvilli 36. lymph 37. capillaries
38. diffuse 39. glycogen 40. large intestine
41. water 42. feces

Exercise 6: 1. B 2. E 3. A 4. C or F 5. D
6. C or F

Exercise 7: Here's a possible answer, in exactly 25 words: "Every animal must obtain from its diet fuel for metabolism, raw materials for making the animal's own molecules, and substances the animal cannot make itself."

Exercise 8: Answers for this exercise will depend on your body weight.

Exercise 9:

```
                  L  A D I P O S E
            C     E     B
            O   P O T A S S I U M          C
    A   P   M       S     O               H
    C   H   P   W A T E R                  H
    I   O   L         P R O T E I N        L        N
    D O S A G E       I     T     D     L E P T I N
    S   P   X   A S C O R B I C   R
        H       S     O     N          C
    C O E N Z Y M E S   N I N E   E   F A T S
    R       N     M         N     L     A
    U       I     I I O D I N E   O
    S     V I T A M I N         E       R
    O F I L     E     A     R I
    B E A N S     R   I N G R E D I E N T S
    E   M I       A     I     A S
    S   I O   C A L C I U M     S
    I   N N     S     A
  F A T E       M A L N O U R I S H E D
    Y
```

Exercise 10: 1. ↓ 2. ↑ 3. ↓ 4. → 5. ↑ 6. → 7. ↓
8. ↓ 9. ↓ 10. → 11. ↑ 12. ↑ 13. ↑ 14. ↓ 15. ↑
16. ↓ 17. ↓ 18. ↓ 19. →

Testing Your Knowledge

Multiple Choice: 1. d 2. a 3. b 4. c 5. c
6. e 7. d 8. c 9. a 10. e 11. b 12. b 13. c
14. c 15. c

Essay: 1. The large polymer molecules in food—proteins, carbohydrates, and fats—are too large to pass through cell membranes. They cannot be taken up by cells unless they are digested first. Also, an animal needs to break down these large molecules to obtain monomers to make its own polymers.

2. Both a grasshopper and a human have alimentary canals with a mouth and an anus. In both, food is cut and chewed by the mouth before it passes through the esophagus. In a grasshopper, chewed food is stored in the crop, ground by the gizzard, digested largely by the stomach, and absorbed in the stomach and gastric pouches. The intestine mainly compacts wastes. Humans have no crop or gizzard. The human stomach begins chemical digestion, but most digestion and absorption occur in the small intestine. The human large intestine has much the same functions as the grasshopper intestine.

3. Saliva contains a slippery glycoprotein that lubricates food. Buffers neutralize food acids and help prevent tooth decay. Antibacterial agents kill food bacteria, and an enzyme in saliva begins the digestion of starch.

4. Herbivores generally have longer, more complex alimentary canals than carnivores, because plants are harder to digest. Horses and elephants house cellulose-digesting microbes in the colon and a large cecum. Some of the nutrients are absorbed by the cecum and colon, but some are lost. Rabbits compensate for this loss by reingesting their fecal pellets, extracting additional nutrients as the already-digested material passes through the alimentary canal a second time. Ruminant mammals such as cattle have four-chambered stomachs. Food is digested by bacteria in two of the compartments, then the softened material (the "cud") is regurgitated and rechewed, then swallowed and passed to another compartment where digestion is completed. The nutrients absorbed by the cow's intestine come from both plant material and the rapidly reproducing bacteria in the stomach.

5. Fad diets—low-carbohydrate diets, low-fat diets, and formula diets—may be inadequate in essential nutrients, may result in loss of body protein, may cause health problems, and may ultimately prove unsatisfying and result in bingeing and regaining lost weight. A balanced

diet of 1200 calories or more, combined with moderate exercise, is best for healthy, gradual weight loss.

6. Food labels provide two kinds of information—a list of ingredients and nutrients present in the food. Such information is useful in planning a healthy, balanced diet—for example, obtaining the proper balance of vitamins, minerals, and other nutrients, or avoiding foods high in calories, sugar, fat, and sodium.

Applying Your Knowledge

Multiple Choice: 1. c 2. b 3. b 4. a 5. e 6. a 7. b 8. b 9. a 10. b

Essay: 1. Muscle is meat, so meat protein will supply James with the right mix of amino acids needed by the body for building muscle. Meat is also high in saturated fats, however, which can contribute to weight gain and heart disease. Plant foods are low in saturated fats, and contain fiber and vitamins that reduce cancer risk. But because no single plant contains all the amino acids needed to build muscle, Michael must be careful to consume a variety of plants to get all the amino acids his body needs.

2. It would be most difficult to live without the small intestine, because it is the portion of the digestive system where most food molecules are digested and where nearly all absorption of nutrients takes place.

3. If trypsin and chymotrypsin were manufactured in their active form, they would digest proteins in the cells where they are made and stored. Like pepsin in the stomach, they are produced in an inactive form that does not digest protein, and they are converted to active, protein-digesting enzymes once they enter the intestine.

4. If 4.83 kcal of food energy are liberated for every liter of oxygen consumed, an astronaut would consume 2500 kcal/4.83 kcal/L = 518 liters of oxygen per day.

5. He could try eliminating all scandium from the diet and looking for ill effects. This might be a problem because this might harm human experimental subjects. He could experiment with animals, but they might not require scandium as he thinks humans do. If he eliminates scandium from the diet, he might accidentally eliminate other elements as well, and then it would be difficult to determine what effects are due to scandium alone. Finally, if scandium is required, it must be in minute quantities, or this requirement probably would have been discovered already. If this mineral is required in tiny concentrations, how do we make sure none is present?

6. Calculate by using the conversion factors in Module 21.15 in the text.

Chapter 22

Organizing Your Knowledge

Exercise 1: 1. transport 2. breathing 3. breathing 4. transport 5. exchange 6. breathing 7. exchange 8. exchange

Exercise 2: 1. C, P 2. A, R 3. D, Q 4. B, S 5. A, R 6. A, R 7. B, S

Exercise 3: 1. skin 2. oxygen 3. carbon dioxide 4. respiratory 5. water 6. moist 7. large 8. thin 9. gills 10. moist 11. less 12. lamellae 13. blood vessels 14. ventilate 15. opposite 16. blood 17. countercurrent exchange 18. oxygen 19. more 20. more 21. diffusion 22. air 23. higher 24. easier 25. dry 26. tracheal 27. tracheal 28. circulatory 29. cells 30. oxygen 31. carbon dioxide 32. lungs 33. diaphragm 34. nose 35. filtered 36. nasal 37. pharynx 38. larynx 39. vocal cords 40. tense 41. less tense 42. trachea 43. bronchi 44. bronchioles 45. mucus 46. cilia 47. alveoli 48. millions 49. blood capillaries 50. carbon dioxide

Exercise 4: See Figure 22.5A in the text.

Exercise 5:

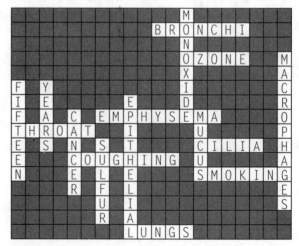

Exercise 6: 1. I 2. I 3. E 4. I 5. I 6. I 7. E 8. E 9. I 10. E 11. E 12. E

Exercise 7: 1. ↑ 2. ↓ 3. ↑ 4. ↓ 5. ↑ 6. ↑ 7. ↓ 8. ↑ 9. ↑

Exercise 8: 1. circulatory 2. O_2 3. CO_2 4. low 5. high 6. alveoli 7. brain 8. drop 9. CO_2 10. diaphragm 11. ribs 12. CO_2 13. O_2 14. O_2 15. CO_2 16. CO_2 17. higher 18. lower 19. O_2

20. soluble 21. hemoglobin 22. red 23. iron
24. O_2 25. muscles 26. higher 27. higher
28. out of 29. into 30. red 31. hemoglobin
32. water 33. bicarbonate 34. plasma
35. buffering 36. rises 37. CO_2

Exercise 9: A. 5 B. 6 C. 1 D. 11 E. 2 F. 4
G. 9 H. 7 I. 3 J. 8 K. 10

Testing Your Knowledge

Multiple Choice: 1. b 2. d 3. a 4. e 5. a
6. c 7. b 8. d 9. a 10. a 11. b

Essay: 1. Air contains a higher concentration of oxygen than water, and it is easier to move across the respiratory surface, but it tends to dry the respiratory surface. There is no danger of the respiratory surface drying out in water, but water contains less oxygen than air and is harder to move.

2. Gills and lungs are both divided into small units that greatly enlarge their surface area for gas exchange. They both have a thin, moist surface through which oxygen passes into numerous blood vessels. They usually have a mechanism for ventilating the respiratory surface. The primary difference between gills and lungs is that gills project outward from the body and lungs are folded inward.

3. There are millions of alveoli in each lung, forming a large respiratory surface. Oxygen dissolves easily in the moisture inside an alveolus and passes easily through the thin wall of the alveolus and into the network of capillaries that cover its surface.

4. The two vocal cords are located in the larynx, or voice box, which is between the pharynx and the trachea. When you exhale, air rushes past the vocal cords. If they are tensed, they vibrate and produce sounds. The more the muscles in the larynx tense the vocal cords, the faster they vibrate, and higher-pitched sounds are produced. When the cords are less tense, they produce lower-pitched sounds.

5. When you inhale, the diaphragm contracts and moves downward and muscles between the ribs contract and expand the rib cage. This enlarges the chest cavity, and the lungs expand with it. This reduces the pressure in the alveoli to less than that of the atmosphere. Air rushes in from the area of higher pressure outside to the area of lower pressure in the alveoli. Thus, the lungs expand first, causing air to rush in. (This is called "negative pressure" breathing.)

6. Water passes through the gill in a direction opposite to the flow of blood within each of the many thin lamellae that make up the gill's respiratory surface. As blood picks up more and more O_2, it comes into contact with water that has more and more O_2 available. Thus, a concentration gradient is maintained that favors diffusion of O_2 along the entire length of the lamella. This allows the gill to extract a large percentage of O_2 from the water flowing through it.

Applying Your Knowledge

Multiple Choice: 1. b 2. c 3. b 4. a 5. b
6. d 7. c 8. d 9. b 10. d

Essay: 1. When you inhale, air passes through your nose; into the nasal cavity; through the pharynx, larynx, and trachea; and into the bronchi that lead into each lung. The air continues through smaller and smaller bronchioles, which end in tiny alveoli. An O_2 molecule dissolves in the moisture on the surface of an alveolus and diffuses through the thin epithelium of the alveolus and into a capillary. The O_2 molecule dissolves in blood plasma, diffuses through the membrane of a red blood cell, and attaches to one of the iron atoms in a hemoglobin molecule. The blood returns to the heart, which pumps it through a blood vessel to a capillary in the brain. There the O_2 molecule detaches from the hemoglobin molecule and diffuses out of the red blood cell, into the plasma, through the wall of the capillary, and into a brain cell.

2. The respiratory surface should have a large area through which gas exchange can occur. This could be accomplished in a relatively small space by folding or dividing the surface into many smaller tubes, plates, or sacs—like the alveoli of the lungs or lamellae of fish gills. The surface must be moist so that gas molecules will dissolve readily, and it must be thin so that gases can diffuse through it easily and rapidly. The respiratory surface should be protected in some way—perhaps folded inside the robot's body—to keep it from drying out or being damaged. Finally, there should be some mechanism for ventilating the respiratory surface—similar to breathing or movement of a fish's gill covers—to move fresh air over the surface to maximize gas absorption.

3. The carbon monoxide molecules would attach to hemoglobin and reduce its ability to carry oxygen. Cells cannot use carbon monoxide as they use oxygen, so they might be starved for oxygen.

4. 20 cigarettes/day × 5 minutes/cigarette × 365 days/year × 40 years = 1,460,000 minutes. There

are 365 × 24 × 60 = 525,600 minutes in one year, so this will shorten the smoker's life by 1,460,000/525,600 = 2.8 years.

5. The respiratory centers in the brain normally regulate breathing rate by monitoring blood pH. CO_2 combines with water in the blood to form carbonic acid, so if the concentration of CO_2 in the blood increases, the blood becomes more acidic, and pH drops. The brain detects this change and speeds up breathing, which removes carbon dioxide and returns blood pH to the acceptable range. In the submarine, CO_2 removal was functioning normally, so there was no buildup of CO_2 (and no drop in pH) in the blood of the sailors—and no response on the part of their respiratory centers—even though oxygen was getting dangerously low.

Chapter 23

Organizing Your Knowledge

Exercise 1: 1. G 2. F 3. H 4. J 5. I 6. A 7. K 8. B 9. D 10. C 11. L 12. E

Exercise 2: A: a. 1 b. 3 c. 4 d. 2 B: a. 1 b. 3 c. 2 d. 6 e. 5 f. 4

Exercise 3: *Labels:* a. superior vena cava b. capillaries of upper body c. pulmonary artery d. pulmonary artery e. capillaries of right lung f. aorta g. capillaries of left lung h. pulmonary vein i. pulmonary vein j. right atrium k. left atrium l. inferior vena cava m. aorta n. right ventricle o. left ventricle p. capillaries of lower body. *Colors and blood flow:* See Figure 23.4B in the text.

Exercise 4: 1. superior vena cava 2. right atrium 3. right ventricle 4. pulmonary arteries 5. pulmonary veins 6. left atrium 7. left ventricle 8. aorta 9. capillaries 10. inferior vena cava 11. right 12. pulmonary 13. left 14. systemic 15. artery

Exercise 5: 1. arteries 2. heart 3. capillaries 4. veins 5. capillaries 6. heart 7. thin 8. thick 9. in-between 10. epithelium (and basement membrane) 11. epithelium, smooth muscle, connective tissue 12. epithelium, smooth muscle, connective tissue 13. no 14. no 15. yes

Exercise 6: 1. AV 2. SL 3. SL 4. AV 5. SL 6. AV 7. AV 8. SL

Exercise 7: Answers will depend on your heart rate—typically on the order of 5,000 ml/minute.

Exercise 8: 1. cholesterol and saturated fat 2. pacemaker (SA node) 3. heart attack 4. AV node 5. angioplasty 6. atherosclerosis

7. bypass 8. brain 9. epinephrine 10. defibrillator (AED) 11. CT and MRI 12. coronary artery 13. stent 14. artificial pacemaker

Exercise 9: 1. G 2. C 3. B 4. A 5. A or B 6. D 7. D 8. A or B 9. D or E 10. A 11. A or B 12. F 13. B 14. F or G 15. F and G

Exercise 10: Some possible changes that might cause increased blood flow through muscle capillaries are increased temperature, increased CO_2 concentration, increased concentration of other waste products, decreased O_2 concentration, and nerve impulses from the brain.

Exercise 11: The red arrow is oxygen-rich blood flowing into the capillary; the blue arrow is oxygen-poor blood leaving. At the arterial end of the capillary, the blood pressure arrow is bigger than the osmotic pressure arrow, so the net flow is out. At the venous end, the osmotic pressure arrow is bigger than the blood pressure arrow, so the net flow is in.

Exercise 12: 1. in-between 2. blood clotting 3. erythrocytes 4. medium-sized 5. most numerous 6. largest 7. defense and immunity

Exercise 13: 1. high 2. pulse 3. arterioles 4. erythrocytes 5. oxygen 6. leukocytes 7. phagocytes 8. plasma proteins 9. platelets 10. capillary 11. epithelium 12. interstitial 13. connective tissue 14. platelets 15. fibrinogen 16. fibrin 17. coronary arteries 18. heart attack 19. venule 20. vein 21. valve

Exercise 14: 1. E 2. G 3. H 4. C 5. I 6. J 7. A 8. B 9. F 10. D

Exercise 15: 1. atrium (The others are all blood vessels.) 2. fibrin (The others are blood cells.) 3. fibrinogen (The others relate to blood pressure.) 4. aorta (The others are parts of the pulmonary circuit.) 5. leukocyte (The others are involved in blood clotting.) 6. semilunar valve (The others are parts of the heart's electrical system.) 7. leukocyte (The others relate to heart contraction and relaxation.) 8. epithelium (The others are concerned with delivery of oxygen to cells.) 9. pulmonary artery (The others relate to the systemic circuit.)

Testing Your Knowledge

Multiple Choice: 1. d 2. e 3. e 4. a 5. d 6. b 7. b 8. e 9. a 10. a 11. a 12. b 13. b

Essay: 1. A fish has a single circuit of blood flow. The fish's two-chambered heart receives and pumps only oxygen-poor blood from body tissues. After leaving the heart, the blood passes through capillaries in the gills, where it picks up oxygen. In the gills, the blood slows down considerably, then it continues on to capillaries in various body tissues. A mammal, a terrestrial vertebrate, has two blood circuits instead of one. The right side of the heart receives oxygen-poor blood returning from the tissues and pumps it through capillaries in the lungs. The left side of the heart receives this oxygen-rich blood and pumps it to capillaries in body tissues. Blood passing through the lung capillaries loses speed and pressure, but the second trip through the heart gives it a boost, so the oxygen-rich blood is rapidly and efficiently delivered to the tissues. This supports the higher metabolic rate characteristic of many land vertebrates.

2. When you are exercising, cardiac output can be increased by increasing the heart rate and the amount of blood the heart pumps with each beat.

3. High blood pressure causes the heart to work harder, which may result in an enlarged and weakened heart. It can also cause small tears in the lining of blood vessels and increase the tendency for vessel-blocking plaques and clots to form there. High blood pressure increases the risk of heart attacks, heart disease, stroke and kidney failure.

4. Blood pressure is highest in the aorta and only slightly lower in the arteries, because the heart is pushing on the blood with full force in these vessels. As blood passes through the narrower arterioles, pressure drops sharply, mainly due to increased resistance to blood flow caused by friction between the blood and the extensive inner surface of the many arterioles. Pressure continues to drop as blood flows through the capillaries, due to the resistance and huge surface area of the capillary walls. By the time blood gets to the veins, its pressure is near zero. Reduced pressure in the chest due to breathing, squeezing action of skeletal muscles, and venous valves help keep blood moving back to the heart.

5. The large number of capillaries and their minute size give them a tremendous combined surface area through which materials can be exchanged between blood and interstitial fluid. Capillary walls consist of a single thin layer of epithelial cells, and there are gaps between these cells, also facilitating exchange.

Applying Your Knowledge

Multiple Choice: 1. c 2. a 3. c 4. b 5. b 6. a 7. d 8. e 9. d 10. a

Essay: 1. Cardiac output is equal to the amount of blood pumped by each heartbeat times the number of beats per minute. The runner's cardiac output is 90 mL per beat × 160 beats per minute = 14,400 mL, or 14.4 L, per minute.

2. In an artery, blood pressure increases during systole, when the heart is contracting and pushing blood into the artery. Pressure drops during diastole, when the heart relaxes. An arterial pressure of 130/80 tells you that the systolic pressure is 130 mm of mercury, and the diastolic pressure is 80 mm of mercury. The numerous arterioles and capillaries offer much resistance to blood flow. By the time blood gets to a vein, the force of the beating heart no longer propels it, so there are no pulses of pressure, just a slow, steady flow at a low, constant pressure.

3. At the capillaries, blood pressure tends to force fluid out of the blood and into the tissues. The osmotic pressure of blood causes fluid to leave the tissues and enter the blood. Blood pressure tends to win out at the arterial end of a capillary, and osmotic pressure is greater at the venous end, but overall these forces are just about in balance. In a person with high blood pressure, blood pressure would overcome osmotic pressure, and there would be a net movement of fluid into the tissues, causing swelling.

4. If the blood vessels were reversed as described, the right ventricle would pump blood to the tissues, and this blood would soon return to the right atrium and then to the right ventricle, where this oxygen-poor blood would again be pumped to the tissues, without being sent to the lungs to pick up oxygen. Similarly, the left atrium and left ventricle would keep sending the same blood to the lungs, without ever pumping this oxygen-rich blood to the tissues that need it. Once the baby is born and the lungs are functioning, there would be no way for oxygen-rich blood to get to the tissues, which would soon die from lack of oxygen.

Chapter 24

Organizing Your Knowledge

Exercise 1: 1. H 2. M 3. E 4. I 5. F 6. A 7. C 8. L 9. G 10. B 11. J 12. D 13. K

Exercise 2: A. bacteria B. histamine C. swelling D. fluid E. leakiness F. pathogens G. cell debris

Exercise 3: 1. An antigen is a foreign molecule that elicits an immune response. An antibody is a protein that attaches to and counters a particular antigen.   2. T cells mature in the thymus and carry out cell-mediated immunity. B cells mature in bone marrow and carry out humoral immunity.   3. Antibodies secreted by B cells carry out humoral immunity, sending out antibodies to attack bacteria and viruses in body fluids. In cell-mediated immunity, T cells attack body cells invaded by bacteria and viruses. 4. The lymphatic system consists of lymph vessels, lymph nodes, and other organs such as the thymus and spleen. It returns a fluid called lymph to the blood vessels and helps defend the body. The circulatory system consists of the heart, blood vessels, and blood.   5. Active immunity occurs when the body is stimulated by antigens and secretes antibodies in its own defense. In passive immunity, the body does not secrete its own antibodies but rather "borrows" antibodies temporarily.   6. Both blood capillaries and lymphatic capillaries are narrow tubes with thin walls, but blood capillaries are open at both ends, whereas lymphatic capillaries end in dead ends.   7. Innate immunity—defenses such as the skin, digestive juices, and antimicrobial proteins—is nonspecific, always ready to act against any threat. Acquired immunity—carried out by the immune system—fully develops only after exposure to a specific pathogen.

Exercise 4: 1. innate 2. immune 3. antigens 4. antibodies 5. antigen-binding sites 6. lymph 7. selection 8. clone 9. effector 10. antibodies 11. primary 12. days 13. immune 14. secondary 15. memory 16. decades

Exercise 5: 1. antigens 2. B cells 3. primary immune response 4. antibodies 5. effector cells 6. memory cells 7. antigens 8. secondary immune response 9. antibodies

Exercise 6: 1. activation of complement 2. precipitation of dissolved antigens 3. neutralization 4. agglutination of microbes 5. cell lysis 6. phagocytosis

Exercise 7: For example, monoclonal antibodies might be used to identify human blood proteins in criminal investigations. Wildlife managers might use them to differentiate between wild and hatchery fish. Monoclonal antibodies might also be used to identify particular materials in archaeological digs—skins, animal remains, or food, for example.

Exercise 8: 1. H, V 2. D, P 3. F, T 4. A, S 5. G, R 6. J, Y 7. C, W 8. I, U 9. E, Q 10. B, X

Exercise 9: 1. HIV 2. condoms 3. blood 4. African 5. AZT 6. helper T cells 7. ten years 8. cancer 9. a vaccine 10. education

Exercise 10: 1. In immunodeficiency diseases, components of the immune system are defective and the body is susceptible to infection.   2. An allergy is an overreaction by the immune system to normally harmless antigens, such as those carried by cat hair. This triggers the inflammatory response, which produces allergy symptoms. 3. Cancer can cause cells to produce abnormal antigens. The immune system usually recognizes such cells as foreign and destroys them. Cancer cells may sometimes fool or evade the immune system, enabling a tumor to grow to fatal proportions.   4. The immune system of a heart transplant recipient recognizes the MHC markers of the transplanted heart as foreign and attacks the organ.   5. In rheumatoid arthritis, some of the patient's own antibodies attack the joints, where they cause inflammation and damage.   6. HIV, the AIDS virus, attacks helper T cells, disabling the immune system and leaving the victim defenseless against infections and certain cancers. 7. Adrenal stress hormones suppress immune cells. Thus psychological stress can affect immune function. It has been shown that students produce less interferon during exams.

Testing Your Knowledge

Multiple Choice: 1. d 2. e 3. d 4. e 5. a 6. d 7. e 8. c 9. c 10. b 11. a 12. c 13. e 14. d 15. a 16. e

Essay: 1. A vaccine contains a harmless form of a disease-causing microbe. It cannot make you sick, but it stimulates lymphocytes to become effector cells and mount a primary immune response. At the same time, it stimulates some lymphocytes to become memory cells, which are responsible for active immunity and are able to mount a rapid and massive secondary immune response if you are ever exposed to the "real" harmful microbe, even years later.

2. B cells carry out humoral immunity against microbes in body fluids. T cells carry out cell-mediated immunity against body cells infected by microbes or altered by cancer. When B cells are activated, they produce antibodies that circulate in body fluids and attach to microbe antigens. This clumps and inactivates the microbes and tags them for destruction by complement and phagocytes. Receptors on activated cytotoxic T cells enable them to bind to antigens on infected body cells. The T cells discharge a substance called perforin that destroys infected cells.

3. Immune system cells do not analyze antigens and produce antibodies to fit. Instead, the body contains small numbers of many types of

immune cells, each type capable of making antibodies that fit a certain antigen. When an antigen appears, only the clone of cells capable of responding to that antigen are selected—a mechanism known as clonal selection. The cells that respond have the ability to make an antibody that fits the antigen before the antigen appears. Antibodies are not made to order.

4. They are lymph nodes. Lymph carries the microbes responsible for your sore throat to the lymph nodes, where T and B lymphocytes are activated. The activated B cells proliferate and produce antibodies that are carried by body fluids to sites of infection. T cells also travel to sites of infection. The rapid production of lymphocytes makes your lymph nodes swollen and tender.

5. When tissues are damaged, cells release chemical alarm signals such as histamine, which trigger various defensive mechanisms. Histamine causes blood vessels to dilate and become leakier. Blood flow increases, and plasma passes out of the blood vessels and into the damaged tissues. Other chemicals attract white blood cells, which engulf bacteria and debris. Blood clotting proteins and platelets wall off the damaged area and allow repair to begin. Local increases in blood flow, fluid, and cells produce the redness, heat, and swelling characteristic of inflammation. Inflammation disinfects and cleans injured tissues and allows healing to begin.

Applying Your Knowledge

Multiple Choice: 1. a 2. b 3. d 4. c 5. e 6. b 7. b 8. a 9. a 10. d

Essay: 1. The first time a person is exposed to an allergen such as bee venom, B cells make special antibodies that attach to mast cells—cells that make histamine and other chemicals that trigger an immune response. By the time this slow sensitization process occurs, the venom is gone. The next time the individual is stung, the antibodies are already in place, and the mast cells suddenly release large amounts of inflammatory chemicals. This triggers an abrupt dilation of blood vessels, causing a sudden drop in blood pressure—anaphylatic shock—that can be fatal.

2. It is possible that each time you catch a cold it is a different virus. You get sick for a while, but eventually the primary immune response fights off the cold virus. Memory cells make you immune to that virus but not to other cold viruses, which probably have different antigens. A 2-year-old gets lots of colds because a 2-year-old has not yet acquired immunity to many of the cold viruses. By the age of 80, an individual has been exposed to and developed immunity to most cold viruses, so colds are less frequent.

3. The injection contained antibodies against hepatitis, so it conferred passive immunity to the disease. The antibodies soon disappeared from John's blood. Because he was not exposed to hepatitis antigens, his body never developed memory cells and active immunity against the disease.

4. A virus may have several different antigens on its surface. In addition, antigens are large molecules, and antibodies usually recognize localized regions, called antigenic determinants, on the surface of antigen molecules. It is possible that the same virus could activate lymphocyte clones that respond to different antigens or antigenic determinants in different people.

5. An antibody can attach to two different antigen molecules or microbes at once, and more than one antibody could attach to a given antigen or microbe. In this way, antibodies stick antigens and microbes together—processes called precipitation (antigens) and agglutination (microbes). The clumps of antigens or microbes formed in this way are easily engulfed by phagocytes.

6. The body directs the cell-mediated immune response, carried out by T cells, at transplanted organs, whose antigens the body senses as foreign. Transplant recipients are given drugs that suppress immunity, particularly cell-mediated immunity. This makes recipients more prone to infections, especially viral infections, because cell-mediated immunity is primarily directed against virus-infected cells.

Chapter 25

Organizing Your Knowledge

Exercise 1: 1. ectotherms 2. endotherm 3. ectotherms 4. ectotherm 5. endotherms 6. endotherms 7. endotherm 8. ectotherm

Exercise 2: 1. D, S 2. B, Q, T 3. B, Q 4. C, T 5. D, S, T 6. A, P 7. D, R 8. A, T 9. D, T 10. B, R 11. A, P 12. B, R 13. D, S, T 14. C, T 15. E, T 16. B, P

Exercise 3: 1. osmoconformer 2. osmoregulator 3. osmoregulator 4. osmoregulator 5. neither 6. gain 7. lose 8. lose 9. lose 10. lose 11. gain 12. lose 13. actively transports ions into cells 14. kidneys excrete dilute urine; digestive system and gills take up ions 15. drinks seawater; expels salts through gills and concentrated urine 16. drinks water and consumes salts; behavior and waterproof skin and eggs save water

Exercise 4: 1. most aquatic animals, including many fishes 2. mammals, amphibians, some fishes 3. birds, insects, many reptiles, some amphibians, land snails 4. readily diffuses into water; no energy expended to make it 5. highly soluble in water, less toxic than ammonia 6. not toxic, requires little water for disposal; can be stored in shelled egg 7. highly toxic, cannot be stored in body, requires much water for disposal 8. energy expended to make it; requires some water for disposal 9. energy expended to make it

Exercise 5: 1. C 2. H 3. J 4. F 5. K 6. G 7. L 8. B 9. A 10. I 11. D 12. E

Exercise 6: See Figure 25.9A in the text.

Exercise 7: See Figure 25.9D in the text.

Exercise 8: 1. A 2. F 3. B 4. E 5. D 6. C

Exercise 9: 1. B 2. B 3. C 4. D 5. A 6. B 7. A 8. D 9. B

Exercise 10: 1. Dialysis is better if kidney damage is expected to be temporary. There are no problems with rejection or infection. Over the long term, dialysis is expensive, time-consuming, and not always available. It is no substitute for a real kidney, because dialysis restricts diet and lifestyle, and wastes build up between treatments.

2. A transplanted kidney works continuously, with no buildup of wastes as between dialysis treatments. The donor can live with one kidney, and over the long run a transplant is less expensive than dialysis. There may be a long wait for a transplant, and there is a chance of transplant rejection and infection due to drugs that suppress rejection.

Exercise 10: 1. hepatic portal vessel 2. glycogen 3. glucose 4. broken down 5. ammonia 6. urea 7. inferior vena cava 8. renal 9. nephrons 10. glomerulus 11. Bowman's capsule 12. filtrate 13. reabsorption 14. secretion 15. excreted 16. active transport 17. water 18. loop 19. renal vein 20. secretion 21. urine 22. ureters 23. urinary bladder

Testing Your Knowledge

Multiple Choice: 1. a 2. d 3. b 4. c 5. b 6. a 7. c 8. e 9. c 10. e 11. b 12. e 13. c 14. c 15. a

Essay: 1. When an animal is too warm, nerves signal blood vessels in the skin surface to dilate, so more blood flows from the warm body core to the surface, and its heat can escape to the surroundings. Heat loss can be further increased by evaporative cooling, such as sweating or panting. When the body is too cold, sweating and panting are reduced, and nerves signal blood vessels in the skin to constrict, decreasing movement of heat from the warm body core to the surface and reducing heat loss.

2. A goose's legs must contain countercurrent heat exchangers. In such a heat exchanger, warm blood from the core of the body and cold blood from the foot flow in opposite directions in adjacent blood vessels. Heat from the body warms the blood from the foot. Cold blood is thus warmed as it moves up the leg, so it does not chill the body, and warm blood is cooled as it flows to the foot, so there is little heat left to lose when it gets there.

3. The internal fluids of a freshwater fish have a higher solute concentration than the surrounding water. This causes water to enter the fish by osmosis, mainly through the gills. The fish does not drink water (except with its food). To get rid of the excess water, the fish's kidneys produce large amounts of dilute urine. Some solutes are lost with the urine. To replace them, the digestive system absorbs ions from food, and the gills take up salt from the surrounding water. A saltwater fish has the opposite problem. The solute concentration of its body fluids is less than that of the surrounding water, so it tends to lose water by osmosis. To compensate, it drinks salt water and disposes of the excess salts via its gills and its kidneys, which excrete small amounts of concentrated urine.

4. An animal can bask in the sun or huddle together with other animals to warm up. It can turn in a direction that exposes less surface to the sun, move into the shade, hide in a damp burrow, or bathe to cool off.

5. In filtration, water and other small molecules are forced by blood pressure from the porous glomerulus into Bowman's capsule, the first part of the nephron tubule. The process of reabsorption is reclaiming valuable solutes and water from the filtrate. Solutes are reabsorbed by active transport mainly in the proximal and distal tubules, and water is reabsorbed by osmosis in the loop of Henle and the collecting duct. Some substances, such as excess H^+ ions, are secreted by active transport from the blood into the filtrate in the proximal and distal tubules. After the filtrate is refined by the processes of absorption and secretion, it is called urine. It leaves the kidney and is expelled from the body—a process called excretion.

Applying Your Knowledge

Multiple Choice: 1. a 2. c 3. c 4. e 5. d 6. c
7. a 8. b 9. e 10. d

Essay: 1. The windchill factor takes into account not only the temperature but also the movement of air, which accelerates heat loss. The body loses heat by conduction, convection, radiation, and evaporation. In comparison with still air, moving air increases heat loss by convection and evaporation. Thus, on a windy day, the body loses heat faster. The windchill factor reflects the fact that a windy day seems colder than a day without wind.

2. A horned lizard is an ectotherm. It warms itself by absorbing heat from its surroundings, so it expends a lot of effort moving in and out of the sun to maintain the proper body temperature. It does not generate heat from metabolism, so it does not eat much. A mouse is an endotherm. It gets most of its heat from its own metabolism, so it does not need to expend much effort changing its position or location relative to the sun or outside energy sources. But because it maintains a constant high temperature, it needs to expend more effort than the horned lizard hunting for food to "burn."

3. The liver occupies a critical location in the body—between the intestine and the heart. Substances that are ingested (or inhaled—the blood from the lungs eventually circulates through the liver) pass first through the liver before being pumped by the heart throughout the body. The liver has the metabolic machinery for altering and inactivating a large number of poisonous substances. Unfortunately, the liver also tends to take in substances that are poisonous to the liver itself. Thus, the liver is the first organ to be damaged by many harmful substances, such as mushroom poison or solvents in glue.

4. Mammals and birds are endotherms, which generate body heat via their own metabolism. They are able to stay warm enough to function even when there is little heat available from the outside environment. Crocodiles and frogs are ectotherms, which obtain heat from their surroundings. There is little heat available in the polar regions; an Antarctic crocodile would not be able to obtain enough heat to function in this environment.

5. After the subject eats a candy bar, blood entering the liver would be high in glucose. The liver would store up some of this glucose as glycogen, so blood leaving the liver would have a lower glucose content. The liver removes and alters toxic substances such as alcohol. After a drink, blood leaving the liver would be lower in alcohol than blood entering the liver. Between meals, the liver breaks down its stored glycogen and adds glucose to the blood, so the blood leaving the liver would contain more glucose than the blood entering the organ.

Chapter 26

Organizing Your Knowledge

Exercise 1: 1. neurosecretory cell 2. blood vessel 3. target cell 4. endocrine cell 5. secretory vesicles 6. blood vessel 7. target cell 8. nerve cell 9. nerve signals 10. nerve cell. For hormones and neurotransmitter molecules, see Figure 26.1 in the text.

Exercise 2: 1. both 2. N 3. both 4. N 5. both 6. E 7. N 8. E 9. N 10. E 11. E 12. N

Exercise 3: 1. steroid hormone 2. plasma membrane 3. target cell 4. receptor protein 5. nucleus 6. hormone-receptor complex 7. DNA 8. new protein 9. cellular response 10. epinephrine 11. receptor protein 12. target cell 13. plasma membrane 14. relay molecules 15. signal-transduction pathway 16. enzyme 17. cellular response

Exercise 4: See Figure 26.3A and Table 26.3 in the text.

Exercise 5: 1. thyroid gland 2. T_3 3. T_4 4. lowers blood calcium 5. estrogens 6. stimulate uterus; female sex characteristics 7. promotes uterine lining growth 8. adrenal medulla 9. epinephrine 10. norepinephrine 11. adrenal cortex 12. increase blood glucose 13. mineralocorticoids 14. raises blood glucose 15. insulin 16. lowers blood glucose 17. parathyroid glands 18. parathyroid hormone 19. pineal gland 20. involved in body rhythms 21. anterior lobe of pituitary 22. growth hormone 23. stimulates thyroid gland 24. stimulates milk production 25. ACTH 26. stimulates ovaries and testes 27. follicle-stimulating hormone 28. stimulates contraction of uterus and mammary gland 29. antidiuretic hormone 30. thymosin 31. stimulates T-cell development 32. testes 33. support sperm formation; development of male characteristics 34. hypothalamus

Exercise 6: 1. P 2. H 3. A 4. H 5. P 6. A 7. A 8. P 9. A (or H) 10. H 11. P 12. A 13. H 14. H 15. A 16. P

Exercise 7: 1. E 2. C 3. B 4. D 5. A 6. E 7. A 8. D 9. B

Exercise 8: 1. blood clotting 2. parathyroid glands 3. parathyroid hormone 4. increased 5. increased 6. release of calcium from 7. calcitonin 8. thyroid gland 9. decreased 10. deposition of calcium in

Exercise 9: 1. increase 2. increases 3. increase 4. increase 5. decrease 6. decrease 7. decreases 8. increase 9. increases 10. increase 11. decreases 12. increases 13. decrease 14. decrease

Exercise 10: 1. E 2. E 3. M or G 4. M or G 5. M 6. E 7. E 8. G 9. M or G 10. M or G 11. G 12. M 13. E 14. E 15. G

Exercise 11: There is no single correct answer for a concept map; you learn by constructing it; many arrangements are possible.

Exercise 12: 1. All regulate reproduction 2. Both are produced by the posterior lobe of the pituitary. 3. All act to raise blood glucose. 4. ACTH stimulates the adrenal cortex to secrete glucocorticoids. 5. All target the kidney. 6. All are produced by the thyroid gland. 7. Both are secreted by the pancreas and have opposing effects on blood glucose. 8. All are secreted in response to stress. 9. All are secreted by the anterior lobe of the pituitary. 10. They have opposing effects on blood calcium. 11. Both affect milk secretion.

Testing Your Knowledge

Multiple Choice: 1. d 2. e 3. e 4. d 5. b 6. c 7. a 8. b 9. c 10. e 11. e 12. c 13. b 14. c 15. d

Essay: 1. The pituitary gland controls many other glands, but not all of them, and the pituitary itself is controlled by the hypothalamus, which is the real "master gland."

2. The adrenal medulla secretes epinephrine and norepinephrine, which trigger the fight-or-flight response to short-term stress. The adrenal cortex secretes mineralocorticoids and glucocorticoids in response to prolonged stress.

3. Hormones and neurotransmitters are both chemical signals between cells. Hormones are carried by the blood to distant cells, while neurotransmitters affect nearby cells directly.

4. For example: Gigantism—abnormally large size—is caused by oversecretion of growth hormone during development. Type 1 diabetes results from undersecretion of insulin. In type 2 diabetes, insulin levels are normal, but cells are unable to respond to it. In either type of diabetes, cells are unable to take up and use glucose

and so burn fats and proteins instead. Glucose builds up in the blood, and numerous complications result.

5. Glucocorticoids are used in treating injuries because they reduce swelling and pain. But because they decrease pain, their use can result in further injury. They also suppress the inflammatory response and immunity to infection, so they may slow healing.

6. Insulin lowers blood sugar, and glucagon raises it. Parathyroid hormone raises blood calcium, and calcitonin lowers it.

Applying Your Knowledge

Multiple Choice: 1. c 2. c 3. d 4. b 5. c 6. d 7. b 8. e 9. a 10. e

Essay: 1. When a person with hypoglycemia eats sugar, hyperactive beta cells in the pancreas secrete too much insulin, which causes blood sugar to drop too low. Eating sugar would just make this problem worse.

2. In diabetes, cells break down fat for energy because they have not received the insulin signal to take in and use glucose.

3. The enlarged thyroid gland might secrete more T_3 and T_4 than normal, boosting metabolic rate and body temperature. A pituitary tumor might cause the pituitary to secrete too much TSH, causing the thyroid gland to secrete too much T_3 and T_4.

4. An organ is affected by a hormone only if it has the proper receptor. Many cells have insulin receptors, but only kidney cells have receptors for ADH.

5. Receptors could vary in various tissues in different animals. An animal that lacks receptors that respond to dioxin might not be affected at all. In another animal with a large number of receptors in the liver, cells might sustain serious liver damage, while a different animal whose immune cells have some receptors might suffer slight immune system impairment.

Chapter 27

Organizing Your Knowledge

Exercise 1: 1. hermaphroditism 2. sperm 3. asexual reproduction 4. zygote 5. gamete 6. regeneration 7. external fertilization 8. ovum 9. copulation 10. budding 11. internal fertilization 12. fragmentation

Exercise 2: 1. oviduct 2. ovary 3. uterus 4. vagina 5. follicles 6. corpus luteum 7. wall of uterus 8. endometrium 9. cervix 10. seminal vesicle 11. ejaculatory duct 12. prostate gland 13. bulbourethral gland 14. epididymis 15. testis 16. scrotum 17. vas deferens 18. erectile tissue 19. urethra 20. glans 21. prepuce

Exercise 3: 1. seminiferous tubules of testes 2. follicles in ovary 3. daily and continuously 4. before birth 5. two secondary spermatocytes 6. one secondary oocyte and a first polar body 7. four immature sperm cells 8. one ovum and a second polar body 9. four 10. one 11. millions per day 12. one per month

Exercise 4: 1. hypothalamus 2. follicle-stimulating hormone (FSH) 3. luteinizing hormone (LH) 4. follicle 5. ovary 6. estrogen 7. endometrium 8. uterus 9. ovum 10. follicle 11. negative 12. FSH 13. LH 14. preovulatory 15. pituitary 16. FSH 17. LH 18. ovulation 19. fourteenth 20. oviduct 21. postovulatory 22. corpus luteum 23. progesterone 24. follicles 25. progesterone 26. fertilized 27. endometrium 28. menstruation

Exercise 5: A. 3, R B. 4, P C. 1, S D. 2, Q

Exercise 6:

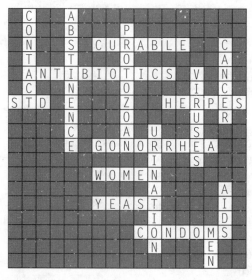

Exercise 7: 1. F 2. I or J 3. I or J 4. L 5. B 6. G 7. A 8. K 9. B 10. A 11. E 12. K 13. D 14. A 15. M 16. E 17. C 18. H

Exercise 8: 1. A sperm swims to the egg, propelled by a flagellum powered by ATP. 2. The acrosome in the sperm head bursts on contact, releasing enzymes that digest an opening in the egg's coat. 3. Species-specific protein molecules on the surface of the sperm head bind with proteins on the egg. 4. If an egg were fertilized by more than one sperm, the resulting zygote nucleus would contain too many chromosomes and could not develop normally. 5. Fusion of sperm and egg membranes causes the egg membrane to quickly become impenetrable to other sperm, and then another barrier, the fertilization membrane, forms, which also keeps other sperm out. 6. Cellular respiration and protein synthesis speed up, and the egg and sperm nuclei fuse, forming the diploid zygote nucleus.

Exercise 9: A. 8 B. 5 C. 6 D. 4 E. 11 F. 1 G. 2 H. 9 I. 7 J. 10 K. 3 L. blastocoel M. digestive cavity (archenteron) N. blastocoel O. blastula P. neural tube Q. coelom R. somite S. zygote T. digestive cavity (archenteron) U. gastrula V. blastopore W. notochord X. ectoderm Y. mesoderm Z. endoderm

Exercise 10: 1. endoderm 2. mesoderm 3. endoderm 4. ectoderm 5. mesoderm 6. mesoderm 7. endoderm 8. ectoderm

Exercise 11: 1. induction 2. pattern formation 3. cell migration 4. programmed cell death 5. changes in cell shape 6. differentiation

Exercise 12: 1. I 2. M 3. J 4. G 5. A 6. K 7. P 8. T 9. R 10. S 11. B 12. H 13. N 14. Q 15. D 16. O 17. E 18. F 19. C 20. L

Exercise 13: There are many possible combinations; for example, a man's sperm could fertilize a donated egg in vitro. This egg could then be planted in his partner's uterus. A woman's eggs could be fertilized with her partner's sperm and the embryo implanted in the uterus of a surrogate mother.

Testing Your Knowledge

Multiple Choice: 1. e 2. c 3. d 4. c 5. b 6. c 7. b 8. a 9. b 10. d 11. d

Essay: 1. When food is abundant and the environment is unchanging, asexual reproduction would enable a single aphid to produce many offspring quickly. The populations could expand rapidly and exploit available resources. Sexual reproduction increases genetic variability among offspring. This might make the species more adaptable when the environment is changing and food supplies are less predictable.

2. The outermost extraembryonic membrane, the chorion, forms the embryo's part of the placenta. Extensions of the chorion, called chorionic villi, absorb oxygen and nutrients from the mother's blood. The amnion forms a fluid-filled sac around the embryo. The fluid absorbs shock

and keeps the embryo from drying out. The yolk sac produces the embryo's first blood cells and germ cells, which later give rise to gamete-producing cells in the gonads. The allantois forms part of the umbilical cord and part of the embryo's urinary bladder.

3. Primary oocytes form in a female embryo and begin dividing to form eggs before birth. At maturity, FSH and LH cause a single follicle containing an egg to mature and rupture each month, and fertilization triggers the final meiotic cell division that completes the egg's development. Sperm production does not begin until a male reaches maturity, but after that point primary spermatocytes divide and mature at a rate of millions per day, stimulated by FSH. A primary oocyte divides to form a single ovum and some polar bodies, which are discarded. A primary spermatocyte divides to form four sperm.

4. See Figures 27.10 and 27.11 in the text.

Applying Your Knowledge

Multiple Choice: 1. c 2. d 3. b 4. b 5. e 6. a 7. a 8. d 9. e 10. e

Essay: 1. The fish that reproduce sexually would probably adapt most successfully to a changing environment, because sexual reproduction produces more variation among offspring than does asexual reproduction. Among a variety of offspring there is more likelihood of some individuals being suited to changed environmental conditions.

2. Sperm cannot develop at body temperature. The testes are located in the scrotum, a sac outside the abdominal cavity, where the sperm-forming cells can be kept cool enough to function normally. Extended periods of time soaking in a hot tub might interfere with sperm development.

3. The eggs of aquatic animals are usually fertilized externally, in the water. The male might court the female for her to lay her eggs, but it is not necessary for a male and a female to be close to one another for fertilization to occur. Land animals usually use internal fertilization. The male and the female must copulate so that sperm are deposited inside the body of the female. This requires closer cooperation between the male and the female, which is accomplished by more elaborate courtship behaviors.

Chapter 28

Organizing Your Knowledge

Exercise 1: See Figure 28.2 in the text.

Exercise 2:

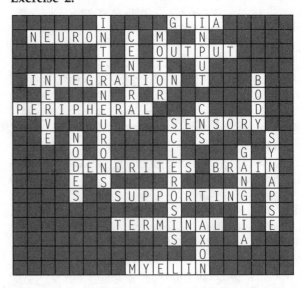

Exercise 3: A. 3, T B. 1, P C. 2, Q D. 4, R E. 5, S

Exercise 4: 1. neurons 2. electrical 3. neurotransmitters 4. cleft 5. axon 6. plasma membrane 7. receptor 8. receiving cell 9. channels 10. action potentials 11. enzyme

Exercise 5: 1. Y 2. N 3. Y 4. Y 5. N 6. N 7. N

Exercise 6: 1. Reduced level of serotonin would result in lower level of stimulation and depressed mood. 2. Increased inhibition would result in lower level of stimulation and depressed mood. 3. This would simulate stimulation by serotonin, elevating mood. 4. Increased inhibition would result in lower level of stimulation and depressed mood. 5. Decreased inhibition by GABA would elevate mood. 6. Increased stimulation by serotonin would elevate mood. 7. This would increase serotonin, elevating mood. 8. This would increase serotonin, elevating mood.

Exercise 7: 1. a. E b. D c. C d. A e. F f. B 2. A, B, C, E   3. D, F

Exercise 8: 1. brain 2. somatic nervous system 3. autonomic nervous system 4. ventricles 5. meninges 6. cranial nerves 7. white matter 8. cerebrospinal fluid 9. peripheral nervous system 10. gray matter 11. central nervous system 12. spinal cord 13. spinal nerves 14. blood-brain barrier

Exercise 9: 1. stimulates activity 2. inhibits activity 3. constricts 4. relaxes 5. stimulates erection 6. promotes ejaculation 7. inhibits 8. accelerates 9. stimulates saliva production 10. inhibits saliva production 11. constricts 12. dilates

Exercise 10: 1. cerebral cortex, E 2. thalamus, F 3. hypothalamus, G 4. pons, B 5. midbrain, C 6. cerebellum, A 7. medulla, D

Exercise 11: 1. rapid eye movement (REM) 2. midbrain 3. motor 4. frontal 5. motor 6. cerebellum 7. limbic 8. right 9. sensory 10. spinal cord 11. thalamus 12. somatosensory 13. parietal 14. left 15. corpus callosum 16. short 17. long 18. amygdala 19. hippocampus 20. left 21. association 22. left 23. medulla oblongata 24. pons

Exercise 12: 1. E 2. B and C 3. B 4. A and D 5. C 6. E 7. A and E 8. D 9. A 10. C 11. E 12. A 13. E 14. A 15. C 16. D

Testing Your Knowledge

Multiple Choice: 1. b 2. d 3. d 4. d 5. a 6. c 7. c 8. d 9. a 10. c 11. a

Essay: 1. See Figure 28.6 in the text. The transmitting and receiving neurons are separated by a narrow synaptic cleft. A chemical neurotransmitter is contained in small vesicles in the synaptic terminals at the end of the axon of the transmitting cell. When an action potential arrives at the end of the transmitting cell's axon, the vesicles fuse with the plasma membrane. Neurotransmitter molecules are released into the synaptic cleft, rapidly diffuse across, and bind to receptor molecules in the membrane of the receiving cell. This opens ion channels, allowing ions to diffuse through the membrane and initiating new action potentials in the receiving cell. Then the neurotransmitter is broken down by an enzyme, and the ion channels close.

2. The parasympathetic and sympathetic divisions are both parts of the autonomic nervous system, which exerts control over many involuntary (mostly internal) body functions. In general, the two systems oppose each other. The sympathetic system prepares the body for intense activities by speeding up the heart, dilating breathing passages, and stimulating glucose release by the liver, but slowing down activities of the digestive tract. The parasympathetic system primes the body for rest and digestion by slowing the heart, constricting the breathing passages, and stimulating the digestive system.

3. The relative size of the vertebrate brain increased in birds and mammals. The forebrain and hindbrain became subdivided into subregions with specific functions, such as the medulla, cerebellum, and cerebrum. The third trend was an increase in the power of the forebrain, specifically the cerebrum, which is much larger in birds and mammals than in other vertebrates.

4. The three functions of the nervous system are sensory input, integration, and motor output. As you read the question, sensory receptors in your eyes send signals to processing centers in the brain—sensory input. The brain interprets the signals, comparing them with memory and determining their symbolic and conceptual meaning—integration. Finally, the brain sends signals to the muscle cells in your arm and hand, directing their movements as you write an answer—motor output.

5. Some transmitting neurons deliver excitatory neurotransmitters, which open Na^+ channels and thus trigger action potentials in the receiving cell. Other transmitting cells deliver inhibitory neurotransmitters, which open channels for Cl^- or K^+ and decrease the receiving cell's tendency to transmit action potentials. Only if excitatory signals outweigh inhibitory signals will the receiving cell transmit action potentials.

Applying Your Knowledge

Multiple Choice: 1. e 2. a 3. e 4. c 5. a 6. d 7. b 8. a 9. c 10. d

Essay: 1. a. Acetylcholine stimulates receiving neurons to transmit action potentials. If the receptors for acetylcholine are blocked, acetylcholine will not be able to stimulate receiving neurons, and the action potentials will be stopped at the synapse. b. If acetylcholine release is inhibited, transmitting cells will not be able to signal receiving cells, and action potentials will be stopped at the synapse. c. If DF prevents acetylcholine from being broken down, acetylcholine will remain in receptors in the receiving cell, and the receiving cell will continue to transmit signals, without being stimulated by the sending cell.

2. When dominoes topple, each domino affects the next one in the row. The dominoes do not travel along the row, but the topple is relayed along the row, one domino to the next. Similarly, when sodium channels open in a neuron, movement of sodium ions into the neuron triggers channels in the next portion of the membrane to open, letting sodium in at that point, which

triggers channels to open farther down, and so on. Sodium does not travel along the neuron, just the change in the neuron membrane. Thus, the action potential travels along the neuron, much the way the topple travels along the row of dominoes. The biggest difference between the dominoes and the neuron is that the dominoes can only topple once. The neuron membrane can repolarize after each action potential and thus quickly transmit many action potentials in succession.

3. Action potentials from the brain pass through the spinal cord on their way to the nerves leading to the arms and legs. Apparently, Andrew's injury damaged the spinal cord between the brain and the place where nerves to the legs originate, so that signals to the legs were interrupted. His injury must have been below the connection between the spinal cord and nerves to the arms, so action potentials to and from the arms were not affected. The brain is not involved in the knee-jerk reflex. Signals travel to the spinal cord via a sensory neuron, across a synapse, and back to leg muscles via a motor neuron. His spinal cord is injured above these neurons, so signals from sensory neurons do not reach the brain, and Andrew cannot feel his legs. But he doesn't have to feel his legs for the reflex to function.

4. The cerebrum is responsible for higher mental functions such as interpreting sensory information, language, and control of voluntary movements. Centers in the brainstem and midbrain, such as the medulla and hypothalamus, control involuntary vital functions such as heart rate, breathing, and body temperature. These centers may continue to do their jobs even if the cerebrum is damaged.

Chapter 29

Organizing Your Knowledge

Exercise 1: A. 2, M B. 4, P C. 3, N D. 5, L E. 1, Q F. 6, O

Exercise 2: 1. electromagnetic receptor 2. near eyes of rattlesnake 3. mechanoreceptor 4. skin and hairs 5. electroreceptor 6. electromagnetic receptor 7. skin of electric fish 8. photoreceptor 9. electromagnetic receptor 10. light 11. pain receptor 12. hair cell 13. sound 14. thermoreceptor 15. heat 16. skin 17. stretch receptor 18. mechanoreceptor 19. skeletal muscle 20. chemoreceptor

21. chemicals in external environment 22. electromagnetic receptor 23. Earth's magnetic field

Exercise 3: 1. sclera, K 2. ciliary body, G 3. suspensory ligament, I 4. cornea, B 5. iris, E 6. pupil, L 7. aqueous humor, C 8. lens, F 9. vitreous humor, J 10. choroid 11. retina, D 12. fovea, H 13. optic nerve, A 14. artery and vein 15. optic disk (blind spot)

Exercise 4: See Figure 29.6 in the text.

Exercise 5: 1. farsightedness (hyperopia) 2. nearsightedness (myopia) 3. astigmatism 4. distant 5. nearby 6. varies 7. nearby 8. distant 9. varies 10. eyeball too short or lens less elastic 11. eyeball too long 12. misshapen lens or cornea 13. corrective lens thicker in the middle than at the edges 14. corrective lens thinner in the middle than at the edges 15. lens that compensates for asymmetry in eye

Exercise 6: *Rods:* 3, 5, 6, 8, 9, 10, 12, 13, 15, 16, 17, 19 *Cones:* 1, 2, 4, 5, 7, 8, 11, 13, 14, 15, 18

Exercise 7: 1. stirrup, G 2. anvil, G 3. hammer, G 4. auditory canal, F 5. semicircular canals, H 6. auditory nerve, D 7. eardrum, E 8. pinna, C 9. oval window 10. cochlea, A 11. Eustachian tube, B 12. outer ear 13. middle ear 14. inner ear

Exercise 8: 1. B 2. F 3. E 4. G 5. H 6. C 7. D 8. A 9. D 10. E 11. E

Testing Your Knowledge

Multiple Choice: 1. c 2. b 3. a 4. e 5. c 6. d 7. b 8. b 9. a 10. a

Essay: 1. Rattlesnakes have infrared receptors near their eyes that enable them to detect the heat given off by warm-blooded prey. Some animals may be able to detect the magnetic field of the Earth and use this information to find their way when migrating. Certain fishes find their prey by sensing the electrical fields produced by the muscles of their prey.

2. Sensory receptors send action potentials to the brain, and the brain is made aware of sensory stimuli. This is called sensation. Perception is meaningful interpretation of these sensory data.

3. Farsightedness is a vision defect that occurs when the eyeball is too short, and the lens cannot bend light rays sharply enough to bring an image into focus on the retina. It can also occur when the lens loses its elasticity and cannot curve enough to bring an image into focus. It can be corrected with a lens that is thicker in

the middle than at the edges, making light rays converge slightly before entering the eye.

4. A stimulus usually interacts with a receptor molecule on or in a sensory receptor cell. This activates a signal transduction pathway that opens ion channels in the cell membrane. Positively charged ions flow into the cell, altering the membrane potential, increasing it to a level called the receptor potential. If the receptor potential is big enough, it triggers action potentials that are sent to the central nervous system.

5. The senses of smell and taste both rely on chemoreceptors. There are different kinds of smell and taste receptors, each stimulated by molecules of particular types. The particular combination of receptors stimulated gives rise to our perception of different tastes and smells. Smells are carried by the air, and their molecules attach to proteins on olfactory receptors in the nasal cavity. Taste receptors are usually stimulated by direct contact between chemicals and proteins on receptor cells in taste buds. There are five kinds of taste sensations, but thousands of smells.

Applying Your Knowledge

Multiple Choice: 1. d 2. c 3. e 4. c 5. a 6. c 7. a 8. c 9. c 10. d

Essay: 1. Receptors for both the sense of hearing and balance are located in the inner ear. Sound receptors are in the cochlea, and balance receptors are in the semicircular canals, saccule, and utricle. Ménière's disease must impair the inner ear in some way.

2. Blue cones are stimulated when we see blue; yellow cones are stimulated when we see yellow. White stimulates all three types of cones, and black stimulates none of them.

3. One method that has been tried is to blindfold people, drive them around an unfamiliar area, spin them around to make sure they are completely mixed up, then ask them to point toward their starting point. Two groups of people were tested: One group wore helmets containing powerful magnets (which would presumably interfere with a magnetic sense); the other group wore nonmagnetic helmets. (The results were tantalizing but inconclusive.)

4. Changes in the environment tend to be more important than ongoing "background" stimuli. Natural selection has enabled the brain to sense a sudden change more readily than a continuing stimulus. Pain is an important message about danger, injury, or disease. Natural selection would not favor an animal that gets used to pain. It is important for the animal to be reminded of pain and deal with its cause.

5. Hair cells are the receptors for hearing; if they were damaged, they would not be able to respond to vibrations in the cochlea. A brain injury that affects the auditory portion of the cerebral cortex would impair the brain's ability to sense or perceive sounds. Arthritis of the middle-ear bones might interfere with their movement and prevent them from transmitting vibrations to the cochlea. A torn eardrum might not be able to pick up vibrations from the air and transmit them to the ear bones. Earwax in the auditory canal would block sound waves from setting the eardrum in motion.

Chapter 30

Organizing Your Knowledge

Exercise 1: 1. friction 2. gravity 3. tripod 4. momentum 5. peristalsis 6. friction 7. gravity 8. streamlined 9. up and down 10. airfoils 11. greater 12. lift 13. gravity

Exercise 2: 1. earthworm, *Hydra*, jellies 2. insect, crab, clam 3. sponge, sea star, human 4. fluid under pressure in a body compartment 5. armorlike covering or shell over outside 6. hard supporting elements embedded inside soft tissues 7. fluid (water) 8. insect: chitin; clam: calcium carbonate 9. sponge: calcium carbonate or silica spicules and protein fibers; human: bone and cartilage 10. cushions, gives body shape, peristaltic movement 11. protection, support, movement 12. support, movement, protection 13. cannot support locomotion in which body is held off ground; offers little protection 14. insect exoskeleton does not grow with animal; must be molted, leaving animal vulnerable 15. does not offer strong protection of exoskeleton

Exercise 3: See Figure 30.3A in the text.

Exercise 4: 1. B, E 2. C, F 3. A, D

Exercise 5: 1. G 2. L 3. H 4. B 5. O 6. K 7. B 8. E 9. D 10. A 11. B 12. H 13. N 14. F 15. M 16. I 17. J 18. C

Exercise 6: See Figures 30.7 and 30.8 in the text.

Exercise 7: 1. capillaries 2. tendon 3. nerves 4. motor unit 5. neuromuscular junction 6. fiber 7. acetylcholine 8. action potential 9. tubules 10. myofibrils 11. Z 12. sarcomeres

13. thin 14. actin 15. thick 16. myosin
17. slide 18. calcium 19. heads 20. ATP
21. shortening 22. calcium 23. endoplasmic
reticulum

Exercise 8: 1. Aerobic 2. Anaerobic
3. Anaerobic 4. Aerobic 5. Anaerobic
6. Anaerobic 7. Aeriobic 8. Aerobic
9. Aerobic 10. Aerobic

Testing Your Knowledge

Multiple Choice: 1. c 2. b 3. e 4. d 5. c
6. d 7. b 8. d 9. e 10. b

Essay: 1. A wing is an airfoil, a shape that is thicker at the leading edge and more curved on top than underneath. When a bird flaps its wings, the air passing over the wings has to travel a longer distance than the air moving under the wings. Air molecules are spaced farther apart above a wing than below, resulting in greater air pressure below than above, which lifts the wing.

2. A skeleton supports the body against gravity; this function is illustrated by the vertebral column. The skeleton protects organs, as the ribs protect the lungs and heart. The skeleton works with muscles in movement; this is illustrated by the leg bones and their opposing sets of muscles.

3. Arthropods—animals such as insects, crabs, and spiders—have jointed exoskeletons. An exoskeleton is strong and effective at protecting the organs inside. A drawback of an exoskeleton is that it does not grow with the animal and must be molted periodically. This is wasteful and leaves the animals temporarily vulnerable.

4. The two main forces that must be overcome by a moving animal are friction and gravity. Aquatic animals are buoyed up by the water and are not affected much by gravity, but the frictional resistance of water is a problem for them. Land animals must expend energy supporting themselves against the pull of gravity, but air offers much less resistance than water, so friction is not a serious problem for them.

5. Check your drawings against Figure 30.9A in the text.

Applying Your Knowledge

Multiple Choice: 1. c 2. e 3. d 4. a 5. a
6. c 7. b 8. d 9. b 10. b

Essay: 1. The frog's long fingers and toes support webs that enable the frog to swim. Its hind legs, much longer than its front legs, are used for hopping. The frog's short, rigid backbone stiffens its body for hopping. Flexible cartilage in its joints and sternum absorb shock on landing.

2. An earthworm is segmented, and each of its segments is filled with fluid. Muscles in different segments can perform different movements simultaneously. So, for example, circular muscles can contract in anterior segments, lengthening them and pushing the front of the worm forward. At the same time, longitudinal muscles can shorten posterior segments, drawing the rear of the worm up behind. This enables the worm to make varied crawling and turning movements. The whole body of the roundworm is one big fluid-filled compartment, so the whole worm has to move the same way at once.

3. The short wings of a songbird are easy to flap and enable the bird to fly fast and make tight turns. An eagle uses its broad wings to lift its prey from the ground. The elongated wings of a gull are more suited to gliding long distances with minimal effort.

4. A motor unit is a neuron and the muscle fibers it controls. In the hand, the motor units are small; one neuron might control only one muscle fiber. In the leg, the motor units are bigger; one neuron might control hundreds of muscle fibers. Thus, even though the hand is smaller than the leg, it takes more neurons to control the hand, because of its one-to-one arrangement of neurons and muscle fibers. Therefore, more area in the brain is devoted to controlling the hands than the legs.

5. Walking with the legs splayed to the sides is hard work. A lot of energy is devoted simply to fighting gravity and keeping the body off the ground. Animals that walk this way are slower than those whose legs are directly under the torso. If dinosaurs' legs were under the body, like those of horses and elephants, they probably were a lot faster than alligators or lizards.

Chapter 31

Organizing Your Knowledge

Exercise 1: There are many possible answers to this question. As I write this, I can see landscape plants such as Douglas firs, rhododendrons, and roses that shade the house, produce oxygen, hold soil in place, and beautify my surroundings. There are lots of papers, made from trees, on my oak desk. There is a pack of chewing gum on the desk; chicle for making gum comes from a tropical tree. I am wearing a cotton shirt and jeans and drinking a cup of tea.

As mentioned in Module 31.1, plant researchers might study resistance to cold or resistance to pathogens to improve crop plants. Other possibilities might be to look into growth to make trees grow faster so we can get more wood from less land, or to study flowering so farmers can grow more fruit.

Exercise 2: *Monocots:* D—complex array of vascular bundles, E—floral parts usually in multiples of three, F—veins usually parallel, H—one cotyledon, I—fibrous root system. *Dicots:* C—vascular bundles in ring, G—flower parts usually in multiples of four or five, B—veins usually netlike, J—two cotyledons, A—taproot usually present

Exercise 3:

```
              G                                    C
    V   T A P R O O T                        C H
H A I R     H       T              B U L B   B
  C U       I Z     I              O R       C
C O M P A N I O N   S H O O T      E   P E T I O L E
  L E       M       S U E              P L A
    C O R T E X     L A M E L L A      L   A
      A P I C A L       N O D E   X I
        E           A       M   I S T
        M E S O P H Y L L       L E
      S E     I     E   W A L L S
      S C   P A R E N C H Y M A   A S
      L E   I V   O     F I B E R   R
    V E I N D   E   L   U   T   O
    R O   E R   T U B E R V A   D E L O
    E O   D M       N S   L I G N I N
S C L E R E I D   C U T I C L E   D
    H R   R S   P I T H   U S   R
    Y M   I     Y   L       S P I N E
    M I   D R   M   A   L
    A   S T O M A   G U A R D   R O O T
```

Exercise 4: See Figure 31.6 in the text.
a. monocot stem　b. root　c. leaf　d. dicot stem

Exercise 5: 1. Q　2. I　3. R　4. S　5. M　6. L 7. N　8. A　9. D　10. F　11. O　12. E　13. G 14. K　15. B　16. H　17. C　18. J　19. P

Exercise 6: See Figure 31.7B in the text.

Exercise 7: 1. E A D C F I　2. C　3. J G H (A D) C B F I　4. G C　5. J G H　6. B　7. E (A D)

Exercise 8: 1. bark　2. secondary phloem 3. cork　4. cork cambium　5. vascular cambium 6. sapwood　7. heartwood　8. secondary xylem (colors from inside out: red, pink, green, light blue, dark blue)

Exercise 9: See Figure 31.9A in the text. The carpel (stigma and ovary) should be pink and the stamens blue.

Exercise 10: 1. water　2. germination　3. seed coat　4. embryo　5. cotyledons　6. endosperm 7. sporophyte　8. meiosis　9. spores　10. gametophytes　11. sepals　12. petals　13. anthers 14. stamens　15. stigmas　16. carpels　17. male 18. sperm　19. pollination　20. ovary　21. sperm 22. an ovule　23. embryo sac　24. gametophyte 25. egg　26. zygote　27. endosperm　28. fertilization　29. seed　30. embryo　31. cotyledons 32. ovary　33. an aggregate fruit　34. sugar 35. seed coats　36. dormancy

Exercise 11: 1. E　2. A　3. D　4. B　5. C

Testing Your Knowledge

Multiple Choice: 1. e　2. c　3. a　4. d　5. e 6. b　7. b　8. a　9. b　10. c　11. c

Essay: 1. The root anchors the plant, absorbs water and minerals, transports water and minerals to other parts of the plant, and stores food. The epidermis, especially root hairs, functions mainly in absorption; the vascular cylinder carries out transport; and the cortex stores food.

2. See Figure 31.6 in the text. The epidermis makes up the dermal tissue system, xylem and phloem the vascular tissue system, and pith the ground tissue system. The vascular cylinder is in the center of the root, surrounded by a thick cortex and the epidermis. The center of a stem is filled by pith, with the vascular tissues forming a ring, and a thin cortex on the outside, covered by epidermis.

3. The diploid sporophyte stage is the full-grown flowering plant, with roots, stems, and leaves. Its flowers contain anthers and ovules, in which cells undergo meiosis to produce haploid spores. These spores divide by mitosis to develop into pollen grains (the male gametophytes) and embryo sacs (female gametophytes) —both haploid. The gametophytes produce haploid gametes—sperm and eggs—by mitosis.

4. A meristem consists of unspecialized parenchyma cells that divide and generate new cells and tissues. Apical meristems occur in the tips of roots and the apical and axillary buds of shoots. These are responsible for primary— lengthwise—growth. An apple tree also has lateral meristems—cylinders of vascular cambium and cork cambium just inside its trunk—which enable the tree to grow in diameter. This is secondary growth.

5. Sclerenchyma cells have thick, rigid secondary walls hardened with lignin, which enables them to form a sturdy framework for the stem. The water-conducting xylem cells in a root die when mature and form long, hollow pipes that carry water. Epidermal cells are thin and transparent, letting light into the leaf. They secrete a waxy cuticle, which keeps the leaf from drying out.

6. The vascular cambium forms a cylinder just underneath the bark of a tree, starting out between the primary xylem and the primary phloem. The vascular cambium is a meristem; its cells divide and give rise to xylem cells and phloem cells. As cambium cells divide and differentiate, new xylem cells are added to the inside of the cambium and new phloem cells are added to the outside of the cambium. This causes the tree to grow in diameter.

7. Pollination is transfer of a pollen grain from a stamen to the stigma of a carpel. The pollen grain germinates, and its tube cell gives rise to a pollen tube that grows down into the ovary to an ovule. Its generative cell divides to form two sperm cells, which enter the embryo sac. One of the sperm cells fertilizes the egg, forming the plant zygote. The other sperm fertilizes a large diploid cell, which divides to form the triploid endosperm, food stored in the seed for the embryo. Fertilization of the haploid egg by one sperm and a diploid cell by another sperm is called double fertilization, and it occurs only in flowering plants. Fertilization in animals involves only a single haploid egg and a single haploid sperm.

8. Some plants, such as turnips, have large roots that store food. Stems are modified in many ways. Horizontal runners and stolons enable grasses and strawberries to reproduce and spread vegetatively. Enlarged horizontal stems called tubers, such as potatoes, store food. A bulb is also a stem that functions in storage and reproduction by fragmentation. The leaves of some plants, such as celery, have enlarged petioles that store food and water. Leaves may be modified to form tendrils, which help plants climb. Cactus spines are modified leaves that protect the plant from grazing animals.

Applying Your Knowledge

Multiple Choice: 1. a 2. c 3. c 4. d 5. c 6. d 7. a 8. d 9. a 10. d

Essay: 1. Most of a tree trunk is xylem, and xylem cells are dead at functional maturity, so the cells that make up most of a tree are not alive.

2. The spike would pass through layers of cork, cork cambium, and secondary phloem, which make up the tree's bark. Then it would penetrate the vascular cambium, the functional secondary xylem of the sapwood, and the nonfunctional secondary xylem of the heartwood.

3. Garlic—bulb (modified stem); walnut—seed; cabbage—leaves; carrot—root; cauliflower—flower buds; artichoke—leaves and flower bud; asparagus—stem; cucumber—fruit and seeds; broccoli—flower buds; rice—fruit (mostly seed); potato—tuber (underground stem).

4. Apical meristems at the tips of roots and shoots contain dividing, elongating, differentiating cells that are responsible for plant growth. After Grandpa nailed the horseshoe to the tree, the tips of its branches continued to grow, and other meristems caused the tree to expand in girth, but the part of the trunk where he nailed the horseshoe, 5 feet from the ground, no longer contained embryonic cells capable of elongation.

5. It would be difficult to kill a palm by girdling it. An oak tree is a dicot. Its vascular tissues are arranged in a ring, with the phloem just under the cork layer of the bark, vulnerable to damage. The palm is a monocot, with its vascular bundles scattered throughout the diameter of the trunk, and therefore less vulnerable to damage.

6. The vascular cylinder is in the middle of a root. Lateral roots must start growing from the inside of the main root so that the vascular cylinders of the main root and lateral roots will connect.

7. Seeds contain all the nutrients needed to sustain the plant embryo until its leaves and roots are functional. Plant cells and human cells need much the same nutrients, so seeds are a good source of nutrition.

Chapter 32

Organizing Your Knowledge

Exercise 1: 1. carbon dioxide (CO_2) 2. water (H_2O) 3. oxygen (O_2) 4. inorganic ions (minerals) 5. CO_2 6. air 7. H_2O 8. soil 9. nitrogen 10. nitrogen 11. magnesium 12. nitrogen 13. phosphorus 14. minerals 15. O_2

Exercise 2: See Figure 32.2B in the text.

Exercise 3: A. 4, diffuse (evaporate) B. 1, xylem, osmosis C. 5, cohesion, adhesion D. 2, root pressure E. 6, hydrogen F. 3, guard, osmosis, stomata, transpiration G. 7, transpiration-cohesion-tension

Exercise 4: 1. source 2. sink 3. source
4. source 5. source 6. source 7. source
8. source 9. sink 10. source 11. sink 12. sink
13. sink 14. sink

Exercise 5: 1. C, H, O, N, S, P, Ca, P, Mg 2. C,
H, O, N, S, P 3. Fe, Cl, Cu, Mn, Zn, Mo, B, Ni
4. N 5. Ca 6. K 7. Fe 8. P 9. Mg

Exercise 6: 1. B 2. H 3. K 4. J 5. D 6. I
7. C 8. A 9. E 10. F 11. G

Exercise 7: 1. Humus content of the soil can become depleted, reducing its fertility, and fertilizer runoff can pollute water. 2. Crop rotation maintains fertility. Different crops have different needs and contribute different qualities (nutrients, texture) to soil. 3. This could channel runoff, reduce penetration of water into the soil, and increase soil erosion. 4. Salts that are left behind when irrigation water evaporates can make soil too salty for crops. 5. This adds humus, restoring the natural texture and fertility of soil and reducing dependence on inorganic fertilizers and pesticides. 6. Drip irrigation uses less water, allows plants to absorb most of the water, reduces salt buildup, and reduces water and nutrient loss from evaporation and drainage. 7. These plants are rich in protein, but are expensive and demand a great deal of fertilizer.

Exercise 8: 1. Nitrogen 2. proteins 3. nucleic acids 4. air 5. ammonium 6. nitrate 7. nitrogen-fixing 8. ammonifying 9. nitrifying
10. amino acids 11. mycorrhizae 12. mutually
13. growth factors 14. antibiotics 15. sugar
16. dodder 17. mistletoe 18. Carnivorous
19. Venus flytrap 20. nitrogen 21. insects
22. legume 23. nodules 24. nitrogen-fixing
25. organic compounds 26. N_2 27. rotate
28. corn or wheat

Testing Your Knowledge

Multiple Choice: 1. b 2. a 3. e 4. d 5. d
6. a 7. c 8. e 9. a 10. b 11. c 12. c

Essay: 1. Root cells actively pump mineral ions into the xylem. Water follows by osmosis and may exert a slight push, called root pressure, that may move water a few meters up a tree. The primary force is a pull exerted by the transpiration of water molecules from the tree's leaves. The water molecules cling to each other and cell walls by hydrogen bonds—phenomena known as cohesion and adhesion. As each water molecule evaporates from a leaf, it pulls on the next, which pulls on the next. This pull, exerted through a column of water molecules in xylem extending all the way to the roots, is what moves water from roots to leaves.

2. If you grew plants in hydroponic culture, bathing them in solutions of known composition, you could compare plants that received a complete nutrient solution with plants receiving the same solution minus molybdenum.

3. Phloem transports sugars, via a pressure-flow mechanism, from sugar sources to sugar sinks. In the summer, a leaf is a sugar source and a potato is a sugar sink. At the leaf, where sugar is being made, it is pumped into the phloem by active transport. Water follows by osmosis, raising the water pressure at this end of the phloem "pipeline" and pushing phloem sap toward the underground potato tubers. In a potato, sugar is transported out of the phloem and stored, and water follows. This lowers the water pressure at this end of the phloem tube, maintaining the pressure gradient that keeps phloem sap moving from leaves to potatoes. Early in the spring, when the potato is full of carbohydrates and the leaves are developing but not yet performing photosynthesis, their roles are reversed. The potato is a sugar source and the leaf a sugar sink, and phloem sap flows upward from potato to leaf.

4. Clay particles are negatively charged, and because opposite charges attract, positively charged nutrient ions tend to cling to them. Root hairs obtain these ions by cation exchange. The root hairs release hydrogen (H^+) ions into the soil solution. The H^+ ions take the place of the nutrient ions on the clay particles, and the root hair can then absorb the free ions.

5. Humus holds water but keeps the topsoil porous enough to allow aeration of roots. The nutrients in humus are relatively insoluble and released gradually, so they are not washed from the soil and do not pollute water.

Applying Your Knowledge

Multiple Choice: 1. c 2. e 3. d 4. e 5. d
6. e 7. b 8. d 9. c 10. d

Essay: 1. Spraying with antibiotics probably killed beneficial nitrogen-fixing bacteria growing in nodules on the roots. Without their help, the bean plants became nitrogen deficient. The plants and root fungi probably formed a mutually beneficial mycorrhizal association. The fungi helped the plants absorb water and nutrients and may have protected them from disease. Without the fungi, the plants suffered.

2. Bubbles in the xylem would interfere with transport of water to the leaves. Transpiration of water from the leaves pulls water through the xylem from the roots. Bubbles in the xylem tubes would disrupt the cohesion of water molecules and reduce this transpiration-adhesion-cohesion pull.

3. Most plants open their stomata during the day. This allows them to take in the CO_2 used in photosynthesis and transpire water, which pulls water and minerals up from the roots. In its hot, dry environment, the jade plant might lose too much water if its stomata remained open during the day. By opening its stomata at night, it reduces water loss.

4. Potassium is needed as an activator of several enzymes, and it is also the main solute that regulates water balance in a plant—the opening and closing of guard cells, for example. The plant does not build molecules containing potassium, but it needs potassium as a regulator of its activities.

5. Acid and heavy rain would be most likely to deplete the nitrogen in the soil. Organic matter decays slowly in acid soil, making little nitrogen available. Unlike phosphate or magnesium ions, nitrate is negatively charged. The rain would be most likely to wash these negative ions from the soil, because they would not cling to negatively charged clay particles. Denise and Roger could enrich the soil by planting and plowing under legumes, such as clover.

Chapter 33

Organizing Your Knowledge

Exercise 1: 1. left 2. left 3. straight 4. right (then left to light) 5. left 6. straight 7. left 8. straight 9. no growth 10. no growth 11. straight 12. left 13. straight 14. right 15. no growth

Exercise 2: 1. auxin 2. cytokinins 3. IAA 4. apical meristem 5. roots 6. embryos 7. fruits 8. xylem sap 9. Auxin 10. cell walls 11. inhibits 12. stimulates 13. inhibits 14. axillary 15. bushy 16. cytokinins 17. auxin 18. cytokinins 19. inhibit 20. stimulate 21. lower 22. cytokinins 23. root 24. shoot

Exercise 3: 1. ABA 2. G 3. ABA 4. ABA 5. G 6. G 7. ABA 8. ABA 9. G 10. ABA 11. G 12. G 13. ABA 14. G

Exercise 4: 1. ethylene 2. aging 3. softening 4. ripening 5. tomatoes 6. CO_2 7. ethylene 8. deciduous 9. water 10. pigments 11. chlorophyll 12. shorter 13. cooler 14. auxin 15. abscission

Exercise 5: 1. gibberellins 2. ethylene 3. auxin 4. cytokinins

Exercise 6: 1. C 2. A 3. D 4. E 5. B

Exercise 7: 1. Growth of a shoot toward light. 2. Auxin migrates from light to dark side of shoot tip, causing cells on dark side to elongate faster. 3. gravitropism 4. Gravity may pull organelles downward, affecting auxin distribution and slowing the rate of growth on lower side of root. 5. thigmotropism 6. Growth movement in response to touch

Exercise 8: For example: Plants may exhibit circadian rhythms, daily cycles of stomata movement or opening. A rhythm may be controlled by a biological clock, perhaps a gene producing protein on a daily cycle.

Exercise 9: 1. yes 2. no 3. no 4. no 5. yes 6. no 7. yes 8. yes 9. yes 10. no

Exercise 10: 1. B 2. D 3. H 4. E 5. A 6. G 7. C 8. F

Testing Your Knowledge

Multiple Choice: 1. e 2. a 3. e 4. b 5. c 6. b 7. c 8. b 9. b 10. e 11. c

Essay: 1. Responses called tropisms enable a plant to alter its pattern of growth to adjust to environmental changes. Phototropism is bending toward light. A hormone called auxin migrates from the light side to the dark side of a shoot tip, causing cells on the dark side to elongate faster. Gravitropism, response to gravity, also appears to be controlled by auxin. Gravity may pull organelles downward, redistributing auxin in the process. This causes the upward side of a root to grow faster, even in the dark.

2. Circadian rhythms continue even when a plant is sheltered from environmental changes—kept in constant darkness, for example. A biological clock does not keep perfect time, so circadian rhythms tend to drift somewhat in relation to the external environment. Environmental cues such as the daily cycle of light and dark enable plants to reset the clock and keep in sync with the environment.

3. As fruit ripens and ages, it produces ethylene gas, which can diffuse through the air from fruit to fruit. Ethylene accelerates ripening, so the ethylene from damaged or overripe fruit can speed up the softening of crisp apples.

4. The signals for leaf drop are the shorter, cooler days of autumn. Auxin prevents leaf drop, but as a leaf ages, it produces less auxin. Cells in the abscission layer, at the base of the

leaf stalk, begin to produce ethylene. The ethylene causes enzymes to digest cell walls in the abscission layer. Eventually, the stalk breaks away, and the leaf falls.

5. Long-day plants flower when the night is shorter than a critical length. Short-day plants flower when the night is longer than a critical length. Phytochrome, a colored protein, absorbs light and helps the plant set its biological clock so that it can measure the length of the night. Light changes a form of phytochrome called P_r to P_{fr}. P_{fr} may be converted back into P_r during the night. The biological clock may measure the timing of P_{fr}-to-P_r changes to measure the length of the night.

Applying Your Knowledge

Multiple Choice: 1. b 2. d 3. b 4. a 5. e 6. d 7. c 8. b 9. e 10. d

Essay: 1. The gas probably contained ethylene, which causes changes in the abscission layer at the base of a leaf that cause the leaf to drop from the stem.

2. Different concentrations of plant hormones can have quite different effects on the same plant. Relatively low auxin concentrations stimulate elongation of stems, but higher concentrations inhibit stem growth.

3. Normally, production of auxin by stems would be balanced by production of cytokinins by roots. Auxin tends to inhibit axillary buds, while cytokinins stimulate them. Grafting a shoot to a large root system would alter this balance. The excess of cytokinins from the roots would stimulate development of axillary buds, which would grow into lateral stems. This would make the plant more bushy, producing a shoot system proportional to the root system.

4. Auxin, produced by the leaf, normally offsets the effects of ethylene, produced by the stalk. Ethylene causes the stalk to drop off. Smearing auxin on the stalk should offset the effects of ethylene and cause the stalk to remain attached to the stem.

5. Apparently, the age or size of soybean plants has little influence on whether or not they flower. Soybeans must be short-day (long-night) plants, which flower in the fall, when the night exceeds a critical length. Once the plants are mature, they wait until the nights are a certain length to flower.

6. Abscisic acid (ABA) acts as a growth inhibitor. It causes seed dormancy by inhibiting germination. ABA might be a good place to start.

7. You could grow a shoot straight up, then turn the plant so that the shoot is horizontal. Place a light under the horizontal shoot. Will it grow toward the light or away from gravity?

Chapter 34

Organizing Your Knowledge

Exercise 1: 1. biosphere 2. ecosystems 3. abiotic components 4. biotic components 5. nutrients 6. populations 7. habitats 8. ecology 9. interactions 10. groups belonging to same species 11. organisms

Exercise 2: 1. scarcity of water 2. penguin 3. high body temperature 4. hydrothermal vent organism 5. heat 6. fire 7. heat and dryness 8. cold and wind 9. light 10. pronghorn 11. cold

Exercise 3: 1. directly 2. North Pole 3. tilt 4. summer 5. winter 6. warms 7. moisture 8. rises 9. rain 10. descends 11. 30 12. desert 13. moisture 14. rotation 15. prevailing 16. trade winds 17. northeast 18. currents 19. winds 20. rotation 21. continents 22. Gulf Stream 23. oceans 24. rainfall 25. rain shadows

Exercise 4:

Exercise 5: See Modules 34.9–34.18 in the text.

Exercise 6: 1. desert 2. temperate grassland 3. temperate forest 4. tropical forest 5. coniferous forest 6. tundra

Exercise 7: 1. E 2. C 3. G 4. F 5. B 6. A 7. B 8. D 9. G 10. H 11. A 12. B 13. E 14. G 15. C 16. H 17. D 18. E 19. F 20. C

Exercise 8: See Figure 34.9 in the text.

Testing Your Knowledge

Multiple Choice: 1. d 2. b 3. a 4. a 5. c
6. c 7. d 8. b 9. d 10. c 11. b 12. a

Essay: 1. There are many possible examples. Beavers dam streams and create ponds. Coral animals build coral reefs, which create habitats for many other species. Overgrazing might create a desert. Trees create shade that might alter the temperature and soil moisture.

2. During the postwar era, new technologies such as chemical fertilizers and pesticides came into widespread use. These increased food production and promised to wipe out insect-borne diseases such as malaria. In the 1960s it became clear that technology had a downside. Pesticide residues threatened wildlife, and pests became resistant to chemicals. Now many people are concerned about such negative effects of human activities on the biosphere.

3. The sun's rays strike the equator directly, warming the air and evaporating moisture. Warm air rises, and as it rises, it cools. Its moisture condenses and falls as rain, watering tropical forests. Cool air is heavier than warm air, so the dry air descends, 30° north and south of the equator, creating deserts.

4. In North America, the temperate forest and temperate grassland have been most extensively altered, mostly by agriculture and urban development.

5. The intense cold, relatively light precipitation, and the short growing season of the far north and high mountains prevent the growth of trees. The ground is permanently frozen—permafrost. This kind of environment is characterized by the tundra biome—shrubs, grasses, mosses, and lichens—species that can live under these extreme conditions. Tropical rain forests are characteristic of warm, humid equatorial areas where there is abundant rainfall and little seasonal variation in climate.

Applying Your Knowledge

Multiple Choice: 1. e 2. d 3. a 4. d 5. d
6. b 7. d 8. b 9. b 10. c

Essay: 1. Temperature, rainfall, soil nutrients, wind, and fire are all abiotic factors that would affect a tree. Predation by herbivorous insects and competition with other trees are biotic factors that might affect the tree. Day to day, the tree might adapt to changes in these factors by adjusting its rate and pattern of growth. For example, it might grow more slowly in a colder location, or it might drop its leaves in the winter. Over many generations, the tree has become adapted to a particular range of environmental factors via natural selection.

2. Water absorbs and stores heat, so the range of temperatures in the ocean is much narrower than on land. Since most of the ocean is within the same narrow temperature range, temperature is not an important factor shaping the distribution of life in the sea. Temperature is more variable on land, so it has a bigger impact on the distribution of land life. Most of the ocean is without light, but most places on land are light for about half of each day. Thus, light is not a factor in where living things can exist on land, but it is an important factor in the sea.

3. The routes of sailing ships depend on the prevailing winds. On their way to North America, ships from Europe first sailed south to catch the northeast trade winds. They first sailed north to catch the westerlies on the return to Europe.

4. The biomes characteristic of warm environments, such as most deserts and grasslands, will probably expand toward the poles. Those characteristic of colder climates, such as coniferous forest and tundra, will probably shrink.

5. The photic zone, where light is available for photosynthesis, is only a thin layer at the surface of the sea. Here there is plenty of light but few nutrients needed for the growth of phytoplankton. Sediments at the bottom of the sea, the benthic zone, are rich in nutrients, but here there is no light for photosynthesis. Life is abundant in the shallow waters of the continental shelves, where sunlight reaches the bottom, but the continental shelves make up only a small portion of the oceans.

Chapter 35

Organizing Your Knowledge

Exercise 1: 1. *Proximate:* Movement of a small object. *Ultimate:* Fitness of frogs has been enhanced by their ability to detect and capture insects. The fact that it is "automatic" suggests it is genetic. 2. *Proximate:* Shape and pattern of egg. *Ultimate:* Reproductive success of gulls has been enhanced by an ability to recognize and retrieve their own eggs. The pattern of spots must vary, so the pattern on a particular egg must be learned by experience, but looking for it could be genetic. 3. *Proximate:* The appearance and location of the hive. *Ultimate:* Bees that could find their way back to the hive were more successful than those who couldn't. The location of the hive can vary, so finding it each day results from experience. 4. *Proximate:* Red wing patches. *Ultimate:* Those males that identify themselves and defend their territories have greater reproductive success than those who do not. Based on the information given, this behavior could be genetic or learned. 5. *Proximate:* The locked cage. *Ultimate:* Those monkeys able to use their intelligence to solve these kinds of problems have the most offspring. The monkey has to figure out the lock, so it has not seen it before. The ability to open the lock must be a result of experience. 6. *Proximate:* The appearance or sound of a bee. *Ultimate:* Toads that could not learn how to avoid harmful animals soon died out. The behavior of the toad is modified by experience. 7. *Proximate:* Colder weather or shorter days of autumn. *Ultimate:* Monarchs that flew south for the winter had greater reproductive success. Because the butterflies have not made the trip before, the behavior must be genetically programmed. 8. *Proximate:* Moving "hawk" silhouette. *Ultimate:* Chicks that run for cover live to reproduce. Because newly hatched chicks do it, the behavior must be genetic.

Exercise 2: 1. edge of cliff 2. turns away from edge 3. chicken 4. runs for cover 5. bird 6. stuffs food into mouth 7. red-winged blackbird 8. threatens and attacks intruder 9. frog 10. moving insect 11. human infant 12. adult face 13. gull or graylag goose 14. rolls egg into nest 15. male prairie vole 16. mating with female

Exercise 3: 1. association 2. imprinting 3. imprinting 4. habituation 5. cognitive map 6. problem-solving 7. imprinting 8. problem-solving 9. cognitive map 10. association 11. social learning 12. habituation (probably association too) 13. imprinting 14. spatial learning 15. social learning

Exercise 4: See Chapter 3, Exercise 10, for a discussion of constructing concept maps.

Exercise 5: 1. Catch some turtles off Brazil and tag them in some way. Later catch some turtles laying eggs on Ascension Island and look for the tags. 2. See which way worms move when placed on the boundary between dark and light areas. 3. First see if the crabs retain their rhythmic behavior if moved into a laboratory away from the ocean. If so, move some crabs to a lab in a different time zone, where tides would be different. See if they stay on the schedule of their original habitat or that of the new location. 4. Blindfold a homing pigeon and see if it can still get home. 5. See whether birds in the lab, at a constant temperature, still become restless

as fall approaches. 6. See whether the rat retains a rhythm of activity under constant light. 7. Keep newly hatched flies isolated until ready to mate, then see if they mate successfully. 8. Add or remove flowers or nectar and see if aggressiveness (time spent fighting) changes accordingly.

Exercise 6: 1. specialists 2. dominant 3. search image 4. efficient 5. optimal foraging 6. territory 7. scent 8. food 9. offspring 10. dominance 11. control 12. fighting 13. communication 14. scent 15. visual 16. sounds 17. monogamous 18. pair bond 19. paternity 20. agonistic 21. ritual 22. courtship 23. agonistic 24. altruism 25. kin 26. genes 27. reciprocal

Exercise 7: Honeybee—near the center, but closer to movements. Ring-tailed lemur—in the center. Wolf—in the center. Cricket—near sounds. Human—between sounds and movements, but closer to sounds. Domestic dog—in the center, but closer to chemicals. Common loon—between movements and sounds, but closer to movements. White-crowned sparrow—between movements and sounds, but closer to sounds.

Exercise 8: 1. C 2. M 3. Q 4. O 5. D 6. L 7. B 8. A 9. E 10. F 11. H 12. G 13. R 14. N 15. J 16. P 17. I 18. K

Testing Your Knowledge

Multiple Choice: 1. c 2. c 3. c 4. a 5. d 6. d 7. b 8. b 9. b 10. e 11. b 12. d

Essay: 1. Newly hatched geese are genetically programmed to learn, during a short critical period after hatching, to follow a moving object (usually their mother). Similarly, a songbird is programmed to learn its species' song during a critical period, then sing that song later. Other birds imprint on their own eggs or offspring. Imprinting provides potential mates, parents, and offspring with a quick way to identify each other, saving time and energy. The energy saved can be used in finding food and rearing young, increasing the chances of reproductive success.

2. A dominance hierarchy assures access to food, water, mates, and nesting or roosting sites for higher-ranking individuals. It establishes order within the group, so individuals can devote energy to finding food, defending against predators, locating mates, or caring for young, instead of fighting each other. This contributes to fitness.

3. Courtship rituals signal that potential mates are not threats to each other. They also confirm that individuals are of the correct species and sex and are in condition to breed. Also, courtship rituals allow individuals to assess each others' health, and thus the contribution of a partner to healthy offspring and potential reproductive success.

4. Social behavior provides organization within animal populations, but for animals to coordinate their activities, they must be able to signal each other about their intentions and actions. Bees communicate the location of nectar by "dancing." Wolves and cheetahs mark their territories with scents. Loon courtship involves a complex series of movements. Lemurs wave their scented tails to communicate aggression. A frog attracts a mate by croaking.

5. Some birds use the position of the sun as a reference point, adjusting for its motion across the sky. Other birds that migrate at night use the fixed north star as a directional reference.

6. Fixed action patterns enable animals to perform behaviors correctly the first time, with no time required for learning. FAPs seem to be most important in animals with short life spans, such as insects. In birds and mammals, FAPs are important in mating and parental care, where there is little chance for practice and little margin for behavioral error.

7. Wolves, chimpanzees, and bees cooperate in social groups. Social cooperation aids in hunting and foraging for food, mating, care of the young, and defense.

Applying Your Knowledge

Multiple Choice: 1. d 2. b 3. c 4. c 5. b 6. d 7. a 8. e 9. e 10. b

Essay: 1. Group hunting allows animals to capture prey that are too fast, too wary, or too big for one predator to bring down alone. The biggest disadvantage of group hunting is that each predator must share food with others in the group.

2. The proximate cause of the diving behavior is an environmental cue—the sound of a bat's sonar. The ultimate cause of the diving behavior has to do with natural selection. The fitness (reproductive success) of moths is enhanced by diving to the ground when they hear bat sonar.

3. This is a question of optimal foraging. Alone, a wagtail does not catch as many insects, so it

does not obtain as much energy, but it does not go as far, so it does not expend as much energy either. In a flock, a wagtail gets more insects and energy, but it must expend more energy covering more territory to get them. Foraging alone or in a flock depends on which provides the most energy gain for energy expended.

4. This is an example of association. The patients learn to associate particular foods eaten just before treatment with the "punishment" of chemotherapy, and subsequently they want to avoid those foods. This kind of learning would help an animal quickly learn to avoid harmful foods.

5. Newly hatched songbirds, such as sparrows, cannot feed themselves and need a constant food supply. In this situation, the male might achieve more reproductive success by helping a single mate raise her young than by seeking more mates. As soon as they hatch, baby chicks can feed on their own. The rooster does not need to help; he achieves more reproductive success by mating with a number of hens.

6. Altruism might be fatal for an individual bird, but if it saves the lives of its relatives, its genes will live on through them. An individual might enhance its fitness by helping its relatives. This idea is called kin selection. Alternatively, reciprocal altruism might be involved here. A favor extended today might be repaid by the recipient tomorrow. In this way, unrelated birds might enhance each other's fitness.

7. A sociobiologist might say that this behavior is at least partially a consequence of our evolution. Adults are genetically programmed to be attracted to and take good care of babies. An individual is likely to be around his or her own babies the most. Individuals who like their babies enhance their reproductive success, and liking babies is passed on via their offspring.

Chapter 36

Organizing Your Knowledge

Exercise 1: 1. 2 hectares 2. This figure will vary but will probably be about 30 shrubs. 3. This figure will vary but will probably be about 15 shrubs per hectare. 4. This figure will vary but will probably be about 250 shrubs. 5. One could take more samples, or count every shrub, but this would take more time and cost more money.

6. The shrubs are clumped. This could be due to variations in moisture or soil nutrients, or perhaps because offspring sprout close to parent plants. 7. Mice are mobile; they do not hold still to be sampled. 8. Using the method in the Web/CD Activity, $N = (40 \times 40)/10 = 160$ mice. 9. $160/16.8 =$ about 10 mice per hectare. 10. You have to assume that a marked mouse has the same chance of being caught as an unmarked individual. You could be wrong. If the mice are territorial and you put the traps in the same locations both times, you are more likely to recapture the same mice. If the captured mice learn to avoid the traps, you are then less likely to recapture them.

Exercise 2: 1. B 2. A 3. E 4. A 5. E 6. D 7. D 8. C 9. C

Exercise 3: 1. 6 aphids 2. 20 aphids 3. 8, 28 4. 11, 39 5. 55, 77, 108, 151, 211, 295 6. The graph should look a bit like Figure 36.4A in the text. 7. It is J-shaped. 8. This kind of growth could not continue. Eventually, some environmental factor, such as limited food, would slow the growth of the aphid population. 9. The graph is S-shaped. 10. It looks as though the population briefly exceeded carrying capacity, then dropped a bit. 11. The population leveled off at carrying capacity—the number of individuals the environment could support over the long term as food or space became limited. 12. around 700 deer 13. logistic growth

Exercise 4: 1. biotic 2. biotic 3. abiotic 4. biotic 5. abiotic 6. biotic 7. biotic 8. biotic 9. abiotic 10. abiotic 11. biotic 12. abiotic 13. abiotic 14. biotic 15. abiotic 16. biotic 17. biotic 18. biotic 19. abiotic 20. biotic 21. biotic 22. biotic

Exercise 5: 1. rapid population growth when conditions are favorable 2. small 3. large 4. many 5. few 6. older 7. quantity 8. many insects and weeds 9. many large land vertebrates, such as polar bears

Exercise 6: 1. 6.4 2. 500 million 3. plague 4. 8 5. birth 6. death 7. zero population growth 8. developed 9. large 10. transition 11. developing 12. births 13. deaths 14. 98% 15. age 16. developed 17. two 18. social security 19. health care 20. bottom-heavy 21. food 22. schools 23. unemployed 24. double 25. water 26. predators 27. carrying capacity 28. fisheries 29. faster 30. ecological footprint 31. exceeds 32. 20 33. renewable 34. biological 35. meat 36. education 37. birth 38. carrying capacity 39. birth 40. death

Testing Your Knowledge

Multiple Choice: 1. e 2. c 3. d 4. d 5. d 6. a 7. a 8. c 9. c 10. d 11. e 12. b

Essay: 1. The exponential growth curve is J-shaped and describes a population growing at an ever-accelerating rate under conditions that do not restrain population growth. The rate of growth depends only on how big the population is, and as the population increases, it grows faster. The logistic growth curve is S-shaped and describes the growth of a population that is limited by its environment. When the population is small, it grows slowly. Growth accelerates as the population increases, but growth slows down and eventually ceases as the population approaches and reaches carrying capacity, the number of organisms that the environment can support.

2. Carrying capacity depends on species and habitat. The carrying capacity of a desert for foxes might be quite different from the carrying capacity of a forest for mice. The animals are different sizes, they require different kinds and amounts of food, and desert and forest habitats produce different amounts of food.

3. Abiotic factors include temperature, rainfall, fire, and floods. Biotic factors include food supply and predation. Abiotic factors are not affected by the density of the population. Biotic factors are density dependent; they affect a greater fraction of the population as the population increases.

4. Human culture, not biology, is responsible for our rapid and sustained population growth. Nutrition, health care, and sanitation have enabled us to maintain a birth rate that exceeds our death rate. Agricultural and industrial technology have enabled us to more effectively exploit the Earth's resources and inhibit the mechanisms that would normally slow population growth. In effect, our population has continued to grow because we have increased the Earth's carrying capacity.

5. Some populations, such as the lynx and the snowshoe hare, undergo regular ups and downs in population density, called population cycles. Predator population cycles probably depend on the ups and downs of their prey. Prey population cycles are probably caused by changes in the availability and quality of their plant food or overexploitation of prey by predators. It is also possible that stress from overcrowding may reduce fertility, leading to population fluctuations.

6. Organisms with life histories shaped by *r*-selection are usually small in size, mature

quickly, produce many offspring, and do not invest much parental care in their offspring. Examples are dandelions and aphids. They usually live in unpredictable environments and multiply exponentially when conditions are favorable. They emphasize producing large numbers of offspring quickly. Organisms whose life histories are shaped by *K*-selection are large in size, mature later, and produce fewer offspring, but give their offspring more care. Their populations may be quite stable, held near carrying capacity by biotic factors. They emphasize production of better-endowed offspring that can compete in a well-established population. Polar bears and chimpanzees are examples of such organisms.

Applying Your Knowledge

Multiple Choice: 1. e 2. b 3. e 4. a 5. d
6. a 7. c 8. b 9. b 10. c

Essay: 1. Using the mark-recapture method described in Web/CD Activity 36A, there are 2,000 smallmouth bass in the lake.

2. The total size of the population is different from population density. Density is the number of individuals of a species per unit of area. The population of the United States is bigger than the population of Japan, but the people of Japan are packed into a much smaller area, so Japan has a higher population density.

3. A population that grows logistically grows slowly when the population is either large (limited by food supply) or small (limited by "breeding stock"), and grows fastest when the population is at an intermediate level. (The S-shaped growth curve is steepest at around half of carrying capacity.) The salmon population will grow fastest when it is kept at 50,000.

4. Temperature is an abiotic limiting factor. Its effect alone on the robin population is not influenced by the size of the population. One would expect the same percentage of robins to be killed whether the population is large or small. Competition for shelter is a biotic factor that has density-dependent effects on birth and death rates. It affects a bigger percentage of the robins when the population is larger.

5. These populations were introduced to environments where they are not subjected to their normal limiting factors, such as predators and parasites. Without these limits, the populations grew larger than would otherwise be the case.

6. Populations on islands or in lakes are easier to define than in other areas. The organisms are limited by natural geographical boundaries, and there is no mixing, interbreeding, or confusion with organisms from elsewhere. They might also be easier to find and sample than populations with less well-defined boundaries.

Chapter 37

Organizing Your Knowledge

Exercise 1: 1. B 2. V 3. V 4. V 5. B 6. T 7. T 8. B 9. T 10. B 11. T 12. D 13. D 14. B 15. D

Exercise 2: 1. commensalism 2. parasitism 3. herbivory 4. herbivory 5. parasitism 6. commensalism 7. mutualism 8. competition 9. mutualism 10. parasitism 11. competition 12. predation

Exercise 3: 1. black widow spider 2. barnacle 3. polar bear 4. sea star 5. redwood 6. poison-arrow frog 7. nitrogen-fixing bacterium 8. sea otter

Exercise 4: 1. EP 2. L 3. ES 4. L 5. EP 6. ES 7. ES

Exercise 5: 1. P 2. 1C 3. 1C 4. 2C 5. 2C 6. P 7. 1C 8. 1C 9. 2C 10. 2C 11. 2C 12. 3C 13. 3C 14. 3C 15. 3C 16. 4C 17. 4C 18. D. Your food web should look something like Figure 37.10 in the text, with producers on the bottom, quaternary consumers on the top, and arrows pointing up to show direction of energy flow.

Exercise 6: A. 8 B. 2 C. 5 D. 4 E. 7 F. 3 G. 6 H. 1

Exercise 7: 1. production energy 2. tertiary consumers 3. energy used in cellular respiration 4. secondary consumers 5. energy in wastes 6. primary consumers 7. producers 8. sunlight

Exercise 8: 1. N_2 2. abiotic 3. nitrogen-fixing bacteria 4. biotic 5. biogeochemical 6. protein 7. bacteria 8. ammonium 9. nitrifying 10. nitrate 11. ammonium 12. nitrate 13. assimilation 14. amino acid 15. denitrifying

Exercise 9: 1. Burning of wood and fossil fuels adds CO_2 to the air faster than it is used and defrestation slows removal of CO_2 from the air. This is causing a buildup of CO_2 in the atmosphere and global warming. 2. When a forest is clear-cut, vegetation holds less moisture, and the increased runoff of water washes mineral nutrients from the soil. 3. Runoff from deforested land, fertilizer from farmland, animal wastes from pastures and stockyards, or sewage from cities can add nutrients to a lake, causing rapid eutrophication. The lake can become choked with vegetation and algae, and their decomposition can deplete oxygen, kill fish, and reduce species diversity. 4. Using large amounts of inorganic fertilizers increases crop yields, but excess fertilizer runs off, over fertilizing lakes and rivers and causing "blooms" of algae and cyanobacteria. 5. Destruction of tropical forests reduces transpiration and may alter the amount of water vapor in the air, changing weather patterns. 6. Untreated runoff from feedlots enters fresh water, causing blooms of algae and cyanobacteria, and potential human disease from intestinal microbes in the waste.

Testing Your Knowledge

Multiple Choice: 1. c 2. a 3. c 4. d 5. a 6. b 7. d 8. c 9. b 10. d 11. e 12. c 13. e

Essay: 1. At each trophic level in an ecosystem, organisms fail to ingest some food, and some of the food they do eat passes through their systems undigested. Much of the food they do assimilate is burned in cellular respiration, releasing energy that is used to carry out body activities and is eventually given off as heat. This leaves little energy—an average of 10%—available to the next trophic level. Ninety percent of the energy is lost to the next level. If humans eat grain, they get 10 to 100 times the energy that would be available if they fed the grain to cattle and then ate beef, because then most of the energy in the grain is lost in passing through the additional trophic level.

2. Decomposition is the breakdown of organic materials—plant litter, animal wastes, dead organisms, and so on—into inorganic compounds. Bacteria and fungi are an ecosystem's most important decomposers. They recycle organic material and release inorganic nutrients such as ammonia, nitrate, and phosphate, which are then available for plants to use for growth.

3. The speed of antelope is an adaptation to pursuit by fast predators. The cooperative hunting of a lion pride is an adaptation to their speedy prey. Predator and prey have shaped each other's evolution. Symbiotic relationships are also shaped by coevolution. For example, plants and their pollinators often coevolve. The shape and color of a flower might adapt to the behavior of a particular insect, and the life cycle and mouthparts of the insect might adapt to the flower.

4. Squirrel—primary consumer; oak tree—producer; mosquito—secondary or higher consumer;

great white shark—secondary or higher consumer; moose—primary consumer; cheetah—secondary or higher consumer; mushroom—detritivore (decomposer); spider—secondary or higher consumer; phytoplankton—producer; grass—producer; vulture—detritivore

5. Water evaporates from oceans and other bodies of water and becomes water vapor. Transpiration by plants also adds water vapor to the air. Winds blow water vapor over the land, and the water condenses and forms rain. Precipitation runs off the land, forming streams and rivers, which return water to the sea, and the cycle is repeated.

6. A fern captures a CO_2 molecule in photosynthesis and uses the carbon atom it contains to build a glucose molecule, which ends up in a leaf. An insect eats the leaf and burns the glucose molecule in cellular respiration, releasing CO_2 to the air. A tree captures the CO_2 and uses it to build a protein molecule in its bark. The carbon atom is released in CO_2 when the tree burns in a forest fire. A tree a hundred miles away again captures the CO_2 in photosynthesis and uses it to build a cellulose molecule. When the tree topples and dies, decomposers such as bacteria could break down the cellulose in cellular respiration and release the carbon in CO_2, but this tree does not decompose. It is buried and forms coal. Millions of years later, the coal is burned in a power plant, and the carbon atom again ends up in CO_2.

Applying Your Knowledge

Multiple Choice: 1. e 2. d 3. e 4. c 5. b 6. c 7. e 8. a 9. b 10. d 11. c 12. b 13. c

Essay: 1. You could remove A—the predator species—from some of the ponds and see if this had any effect on the abundance of other species. If A is a keystone predator, you would expect its removal to alter the relative sizes of its prey populations. For example, species B might crowd out the other species.

2. Various species of trees are the prevalent form of vegetation in the forest, but grass is the prevalent form of vegetation in the park. The number of species would probably be reduced in the park, so its species diversity would be lower. The trophic structure of the park would be altered, with fewer large producers (trees) and more small producers (grass); fewer primary consumers that feed on trees and more primary consumers that feed on grass; and a different mix of higher consumers that feed on

them. Even the detritivores would probably be different. The forest community is probably more stable; left alone it would probably persist unchanged for many years. The park is less stable; left alone, ecological succession would turn it into a forest. Also, the park is less stable in terms of disruption by disease or catastrophe. A disease that affected one of the few species in the park would probably have more impact than a disease that affected one of the many species in the forest.

3. *P. aurelia* and *P. caudatum* have similar niches; their needs and abilities are similar, so they compete with each other. *P. bursaria* apparently has a different niche from *P. aurelia*; they do not compete as much. If *P. aurelia* and *P. caudatum* are similar, and *P. aurelia* can coexist with *P. bursaria*, then *P. caudatum* could probably coexist with *P. bursaria* if they were cultured together.

4. Much of the energy in the food consumed by the inhabitants of the shelter is used in cellular respiration to power body activities; this energy is lost as heat. Thus, the body wastes of the inhabitants contain only a fraction of the energy that was originally in their food. The mushrooms, like the shelter inhabitants, use much of the energy in these wastes in their own cellular respiration, storing little of it in their tissues. As energy is lost from the system, the mushroom crops would rapidly dwindle, and the inhabitants would quickly starve. Energy flows through an ecosystem; it is degraded to heat and dissipated, not recycled. Without energy from outside—light (from the sun or electric lights) to grow crops, or canned food, the system quickly runs out.

5. If 10% of the energy on one trophic level is transferred to the next level, 20 kilograms of algae will grow 2 kilograms of crustaceans, and this will grow 0.2 kilograms of fish. If 1 square meter of pond grows 0.2 kilograms of fish and you want 1000 kilograms of fish, the pond will have to have an area of 1000/0.2 = 5000 square meters.

6. The least chipmunk can live in a broader range of habitats—both the sagebrush and the sagebrush-piñon pine zones. The yellow pine chipmunk is narrower in its requirements and is confined to the sagebrush-piñon pine zone. The two species have somewhat similar niches because they compete. If both species are present, the yellow pine chipmunk is apparently a better competitor, and it excludes the least chipmunk from the sagebrush-piñon pine zone—an example of competitive exclusion. (This is very similar to the barnacles described in Module 37.2.)

7. Coevolution shapes host-parasite relationships. The deadliest strains of viruses do not spread as readily because they die with their hosts. Less deadly strains are more successful and displace the more deadly strains. The parasites thus become less deadly over time. Among the hosts, those individuals with the greatest natural resistance to the viruses survive the epidemic and pass their resistance on to their children, increasing the resistance of the host population over time. This will probably happen among the indigenous peoples, but at great human cost.

Chapter 38

Organizing Your Knowledge

Exercise 1: 1. L 2. H 3. F 4. I 5. A 6. D 7. C 8. E 9. J 10. G 11. N 12. M 13. B 14. K

Exercise 2: 1. Acid precipitation 2. Increase efficiency in use of fossil fuels; burn cleaner fuels 3. CO_2 buildup from burning of fossil fuels to power industry, vehicles, home heating; combined with deforestation 4. Reducing fossil-fuel use; renewable energy sources; replanting forests 5. Habitat destruction and hunting 6. Ban hunting, set aside well-protected reserves. 7. Pollution and biological magnification of chemical pesticides such as DDT, PCBs 8. Reduce use of pesticides and other non-biodegradable chemicals 9. Reduction of biodiversity; environmental changes; loss of potential useful products 10. Conservation of forests; managing forest without destroying it 11. Damage to the atmosphere's protective ozone layer 12. CFC pollutants released from air conditioners, refrigerators, etc. 13. Overfishing and damage to wetlands and reefs where fish reproduce. 14. Limit fishing; protect fish "nurseries" 15. Introduction of exotic species 16. Predation on or competition with native species

Exercise 3: A. 1. Food, medicines, wood products 2. Recycling of nutrients and wastes 3. Stabilization of climate 4. Purification of air and water 5. Flood control 6. Pollination of crops 7. Control of pests 8. Protection from UV rays 9. Moral and esthetic values
B. 1. Destruction of habitat (deforestation, damaging coral reefs) 2. Overexploitation (tigers, tuna) 3. Introduction of exotic species (kudzu, starlings) 4. Pollution (DDT, acid precipitation) 5. Climate change (CO_2 and global warming)

Exercise 4: 1. SP 2. SP 3. both 4. DP 5. SP 6. DP 7. both 8. both 9. SP

Exercise 5: 1. On a global scale, the Rocky Mountain area does not rank as a biodiversity hot spot. (Most are in the tropics.) The area is of interest because it is large, relatively undeveloped, and already contains many parks and reserves. 2. Landscape ecology is the application of ecological principles to the study of the structure and dynamics of a collection of ecosystems, such as the Yukon-Yellowstone region. One of the priorities of landscape ecology is to study human land use and to make biodiversity conservation a priority, as is being done in the initiative area. 3. One of the goals of the initiative is to connect existing parks and reserves in the region, so that wide-ranging species such as wolves and grizzlies can move about safely. 4. The gray wolf is a keystone species, with many effects on other species in its community. 5. Radio-tracking studies have shown that wolves range over a huge area. They do not stay within the boundaries of existing parks and reserves, and are exposed to dangers such as traffic and hunting when they leave protected areas. 6. When wolves were removed, populations of large herbivores such as elk and deer grew unchecked, damaging vegetation and habitat for smaller animals. 7. Wolves don't stay inside reserves and are hit by cars. Some recreationists and ranchers fear wolf attacks. Some wolves have been shot. 8. Elk, moose, and deer populations dropped, and grasses and trees returned. Stream banks recovered, and birds and water-dwellers such as beavers made a comeback.

Exercise 6: 1. closer to equator 2. sensitive to habitat degradation 3. Many endangered species 4. high plant and animal diversity 5. likely to be tropical forest or coral reef 6. many endemic species 7. high potential extinction rate

Exercise 7: 1. E 2. C 3. A 4. B 5. D 6. D 7. A

Testing Your Knowledge

Multiple Choice: 1. e 2. e 3. d 4. b 5. c 6. a 7. e 8. b 9. b 10. d

Essay: 1. Each person in an industrialized country uses a much greater share of the resources and produces more pollution than an individual in a developing country.

2. Tropical forests harbor a disproportionately large number of the world's species of plants and animals. Most biodiversity hot spots are in tropical forests. Besides, most of the forest land in the United States has already been cut.

3. The rain forest produces oxygen, uses CO_2, and affects the water cycle. Destroying the forest could alter global climate. We may be able to obtain useful products from tropical species.

4. The greenhouse effect warms the Earth enough to make life possible. But the CO_2 added by human activities might increase the greenhouse effect and make it too warm.

5. It is pollution of the atmosphere by CFCs from air conditioners and refrigerators that destroys the ozone layer. Combustion of fossil fuels intensifies the greenhouse effect, but has little effect on ozone.

6. Many songbirds migrate between the U.S. and areas outside the U.S., where they may be threatened.

7. Never before has a mass extinction of so many species been caused by a single species— in this case, humans—and we're supposed to be "intelligent"!

Applying Your Knowledge

Multiple Choice: 1. a 2. c 3. e 4. e 5. a 6. d 7. a 8. a 9. c 10. e

Essay: 1. Top predators depend on all the species below them in the food web. If a disturbance alters the vegetation or affects a primary or secondary consumer, the disturbance also affects the top predators, who depend on the lower trophic levels for food. A good example is biological magnification of pesticides. Additionally, many top predators such as the wolf, are keystone species, with disproportionate influence on ecosystem dynamics

2. a. Conversion to solar and hydroelectric power would decrease global warming, because this would reduce CO_2 emissions. b. Growing trees use CO_2, so this would reduce global warming. c. Burning and decomposition of logs add CO_2 to the air, and trees remove it, so deforestation would increase global warming. d. More efficient cars would burn less gasoline and produce less CO_2 per mile, so this would reduce global warming. e. Fewer people would use less fossil fuels and cut forests at a slower rate, slowing CO_2 buildup and global warming. f. Decomposition produces CO_2, so this would increase global warming.

3. Given enough iron, the algae might grow faster, speed up photosynthesis, and take CO_2 from the atmosphere at a faster rate. Since CO_2 traps heat in the atmosphere, the drop in CO_2 would reduce the rate of global warming. But

trying this might have unintended effects, such as disrupting ocean habitats and food chains.

4. Restoration ecology is our overall approach—using ecology to repair a damaged ecosystem. The small-population approach might reveal that the spotted owl population is too small to sustain itself, and through the lens of the declining population approach, we would study how changes in habitat (removal of large trees) impact the owls. Mature forest species such as owls might be helped by retaining or establishing coridors to link areas of big trees and reduce habitat edges. Via bioaugmentation, we might introduce organisms to restore depleted soil, or use organisms to remove toxics (bioremediation). Ultimately we could protect the trees and the owls on a zoned reserve—undisturbed forest surrounded by forest that could be harvested in a controlled manner.

5. 1000 pounds of forest life are destroyed for each hamburger; this amounts to about 55 square feet of forest. If you eat 100 hamburgers a year, you could destroy 55 x 100 = 5500 square feet of rain forest, roughly the area of three suburban houses. (Explanation: 50 pounds of beef per year for 8 years is 400 pounds. Half is usable meat, so that totals 200 pounds. There are 4 4-ounce burgers in a pound, so 200 pounds of meat yields 800 burgers per acre. 800,000 pounds of forest/800 = 1000 pounds. 43,793 square feet of forest/800 = 55 square feet.)